Symmetry through the Eyes of a Chemist

Second Edition

István Hargittai

Magdolna Hargittai

Symmetry through the Eyes of a Chemist

Second Edition

Symmetry through the Eyes of a Chemist

Second Edition

István Hargittai

Budapest Technical University
and Hungarian Academy of Sciences
Budapest, Hungary

Magdolna Hargittai

Hungarian Academy of Sciences
Budapest, Hungary

Plenum Press • New York and London

Library of Congress Cataloging-in-Publication Data

Hargittai, István.
Symmetry through the eyes of a chemist / István Hargittai, Magdolna Hargittai. -- 2nd ed.
p. cm.
Includes bibliographical references and indexes.
ISBN 0-306-44851-3 (hc). -- ISBN 0-306-44852-1 (pbk.)
1. Molecular theory. 2. Symmetry (Physics) I. Hargittai, Magdolna. II. Title.
QD461.H268 1995
541.2'2--dc20 95-30533
CIP

ISBN 0-306-44851-3 (Hardbound)
ISBN 0-306-44852-1 (Paperback)

A Division of Plenum Publishing Corporation
233 Spring Street, New York, N. Y. 10013

10 9 8 7 6 5 4 3 2 1

The first edition of this book was published by VCH, Weinheim, Germany, 1986

Printed in the United States of America

Preface to the Second Edition

We have been gratified by the warm reception of our book, by reviewers, colleagues, and students alike. Our interest in the subject matter of this book has not decreased since its first appearance; on the contrary. The first and second editions envelop eight other symmetry-related books in the creation of which we have participated:

I. Hargittai (ed.), *Symmetry: Unifying Human Understanding*, Pergamon Press, New York, 1986.

I. Hargittai and B. K. Vainshtein (eds.), *Crystal Symmetries. Shubnikov Centennial Papers*, Pergamon Press, Oxford, 1988.

M. Hargittai and I. Hargittai, *Fedezzük föl a szimmetriát!* (Discover Symmetry, in Hungarian), Tankönyvkiadó, Budapest, 1989.

I. Hargittai (ed.), *Symmetry 2: Unifying Human Understanding*, Pergamon Press, Oxford, 1989.

I. Hargittai (ed.), *Quasicrystals, Networks, and Molecules of Fivefold Symmetry*, VCH, New York, 1990.

I. Hargittai (ed.), *Fivefold Symmetry*, World Scientific, Singapore, 1992.

I. Hargittai and C. A. Pickover (eds.), *Spiral Symmetry*, World Scientific, Singapore, 1992.

I. Hargittai and M. Hargittai, *Symmetry: A Unifying Concept*, Shelter Publications, Bolinas, California, 1994.

We have also pursued our molecular structure research, and some books have appeared related to these activities:

I. Hargittai and M. Hargittai (eds.), *Stereochemical Applications of Gas-Phase Electron Diffraction*, Parts A and B, VCH, New York, 1988.
R. J. Gillespie and I. Hargittai, *The VSEPR Model of Molecular Geometry*, Allyn and Bacon, Boston, 1991.
A. Domenicano and I. Hargittai (eds.), *Accurate Molecular Structures*, Oxford University Press, Oxford, 1992.
M. Hargittai and I. Hargittai (eds.), *Advances in Molecular Structure Research*, Vol. 1, JAI Press, Greenwich, Connecticut, 1995.

For this second edition, we have revised both text and illustrative material. It gives us pleasure to acknowledge the kind assistance from several colleagues, including Lawrence F. Dahl (University of Wisconsin, Madison), Avitam Halevi (Technion, Haifa), Lionel Salem (University of Paris, Orsay), Péter Surján (Eötvös University, Budapest), and Richard Wiegandt (Mathematical Research Institute, Budapest).

We are grateful to István Fábri and Judit Szűcs for their dedicated technical assistance.

For over a quarter of a century, our research work in structural chemistry has been supported by the Hungarian Academy of Sciences. Scientific meetings and lecture invitations have taken us to many places, and these travels have helped us build up the illustrative material of this book. We have enjoyed the friendship and enthusiastic interest of our colleagues all over the world.

István and Magdolna Hargittai

Budapest, Hungary

From the Preface to the First Edition

This book surveys chemistry from the point of view of symmetry. We present many examples from chemistry as well as from other fields, in order to emphasize the unifying nature of the concepts of symmetry.

We hope that all those chemists, both academic and industrial, who take broader perspectives will benefit from our work.

We hope that readers will share some of the excitement, aesthetic pleasure, and learning that we have experienced during its preparation. In the course of our work we have become ever more conscious of the diverse manifestations of symmetry in chemistry, and in the world at large. We believe that consciousness will also develop in the reader.

Despite its breadth, our book was not intended to be comprehensive or to be a specialized treatise in any specific area.

We would like especially to note here two classics in the literature of symmetry which have strongly influenced us: Weyl's *Symmetry* and Shubnikov and Koptsik's *Symmetry in Science and Art*.

Our book has a simple structure. After the introduction (Chapter 1), the simplest symmetries are presented using chemical and nonchemical examples (Chapter 2). Molecular geometry is then discussed in qualitative terms (Chapter 3). Group-theoretical methods (Chapter 4) are applied in an introductory manner to the symmetries of molecular vibrations (Chapter 5), electronic structure (Chapter 6), and chemical reactions (Chapter 7). These chapters are followed by a descriptive discussion of space-group symmetries (Chapter 8), including the symmetry of crystals (Chapter 9).

The general perception of symmetry that most people have is sufficient for reading Chapters 1, 2, 3, 8, and 9. However, in order to appreciate Chapters 5, 6, and 7, the introduction to group theory given in Chapter 4 is necessary. Chapter 4 also deals with antisymmetry.

We express our thanks to those distinguished colleagues who have read one or more chapters and helped us with their criticism and suggestions. They include James M. Bobbit (University of Connecticut), Russel A. Bonham (Indiana University), Arthur Greenberg (New Jersey Institute of Technology), Joel F. Liebman (University of Maryland), Alan L. Mackay (University of London), Alan P. Marchand (North Texas State University), Kurt Mislow (Princeton University), Ian C. Paul (University of Illinois), Péter Pulay (University of Arkansas), Robert Schor (University of Connecticut), and György Varsányi (Budapest Technical University).

We thank those authors and copyright owners who gave us permission to use their illustrations in our book. We made all efforts to identify the sources of all illustrative materials and regret if, inadvertently, we missed anything in doing so.

Most of the final version was compiled during our stay at the University of Connecticut, 1983–85, and we greatly benefited from the school's creative and inspiring atmosphere. We express our gratitude to Dean Julius A. Elias, to IMS Director Leonid V. Azaroff, and to our colleagues in the departments of Chemistry and Physics.

We dedicate this book to the memory of József Pollák (1901–1973), who was the stepfather of one of us (IH). He was an early and decisive influence in stimulating the interests which eventually led to the creation of this book.

István and Magdolna Hargittai

Storrs, Connecticut, and *Budapest, Hungary*

Contents

Symmetry through the Eyes of a Chemist

Second Edition

Chapter 1

Introduction

Fundamental phenomena and laws of nature are related to symmetry, and, accordingly, symmetry is one of science's basic concepts. Perhaps it is so important in human creations because it is omnipresent in the natural world. Symmetry is beautiful, although alone it may not be enough for beauty, and absolute perfection may even be irritating. Usefulness and function and aesthetic appeal are the origins of symmetry in the worlds of technology and the arts.

Much has been written, for example, about symmetry in Béla Bartók's music [1-1]. It is not known, however, whether Bartók consciously applied symmetry or was simply led intuitively to the golden ratio so often present in his music. Another unanswerable question is how these symmetries contribute to the appeal of Bartók's music, and how much of this appeal originates from our innate sensitivity to symmetry. Bartók himself always refused to discuss the technicalities of his composing and liked merely to state, "We create after Nature."

Nature abounds in symmetries, and they are present not only in the inanimate world but in the living world as well. Curiously, only recently has related research intensified, probing into the significance of symmetry in, for example, mate selection and other biological actions. A long article titled "Why Birds and Bees, Too, Like Good Looks" in the *New York Times* in 1994 [1-2] is a sign of growing public interest in symmetry matters.

The above examples illustrate how we like to consider symmetry in a broader sense than how it appears just in geometry. The symmetry concept provides a good opportunity to widen our horizons and to bring chemistry closer to other fields of human activities. An interesting aspect of the relationship of chemistry with other fields was expressed by Vladimir Prelog in his Nobel lecture [1-3]:

> Chemistry takes a unique position among the natural sciences for it deals not only with material from natural sources but creates the major parts of its objects by synthesis. In this respect, as stated many years ago by Marcelin Berthelot, chemistry resembles the arts; the potential of creativity is terrifying.

Of course, even the arts are not just for the arts' sake, and chemistry is certainly not done just for chemistry's sake. However, in addition to creating new healing medicines, heat-resistant materials, pesticides, and explosives, chemistry is also a playground for the organic chemist to synthesize exotica including propellane and cubane, for the inorganic chemist to prepare compounds with multiple metal–metal bonds, for the stereochemist to model chemical reactions after a French parlor trick (cf. Section 2.7.3), and for the computational chemist to create undreamed-of molecules and to write exquisitely detailed scenarios of as yet unknown reactions, using the computer. Symmetry considerations play no small role in all these activities. The importance of blending fact and fantasy was succinctly expressed by Arthur Koestler [1-4]: "Artists treat facts as stimuli for the imagination, while scientists use their imagination to coordinate facts." An early illustration of an imaginative use of the concept of shape is furnished by C. A. Coulson [1-5], citing Lucretius from the first century B.C., who wrote that "atoms with smooth surfaces would correspond to pleasant tastes, such as honey; but those with rough surfaces would be unpleasant."

Chemical symmetry has been noted and investigated for centuries in crystallography, which is at the border between chemistry and physics. It was probably more physics when crystal morphology and other properties of the crystal were described, and more chemistry when the inner structure of the crystal and the interactions between the building units were considered. Later, discussion of molecular vibrations, the selection rules, and other basic principles in all kinds of spectroscopy also led to a uniquely important place for the concept of symmetry in chemistry with equally important practical implications.

The discovery of the handedness, or chirality, of crystals and then of molecules led the symmetry concept nearer to the real chemical laboratory. It was still, however, not the chemist, in the classical sense of the profession, who was most concerned with symmetry, but the stereochemist, the structural chemist, the crystallographer, and the spectroscopist. Symmetry used to be considered to lose its significance as soon as molecules entered the most usual chemical change, the chemical reaction. Orbital theory and the discovery of the conservation of orbital symmetry have removed this last blindfold. The awarding of the 1981 Nobel Prize in chemistry to Fukui [1-6] and Hoffmann [1-7] signifies these achievements (Figure 1-1).

During the past dozen or so years, two important discoveries in molecular science and solid-state science have been intimately connected with symmetry. One is the C_{60} buckminsterfullerene molecule [1-8] and the whole emerging fullerene chemistry. The other is quasicrystals [1-9]. Buckminsterfullerene will be mentioned again in Section 3.7, and the quasicrystals in Section 9.8. Some general considerations [1-10], however, are presented here.

Geometry, and especially physical geometry, was central to Buckminster Fuller's (1895–1983) natural philosophy. He was not a chemist, but had a high esteem for chemistry, and quoted Avogadro's law (Figure 1-2) to illustrate that chemists consider volumes as material domains and not merely as abstractions. Fuller (Figure 1-3) recognized the importance of synergy for chemistry and gave this explanation for it [1-11]:

> Chemists discovered that they had to recognize synergy because they found that every time they tried to isolate one element out of a complex or to separate atoms out, or molecules out, of compounds, the isolated parts and their separate behaviors never explained the

Figure 1-1. Kenichi Fukui (left). Photograph (1992) by Tsuneo Ide. Courtesy of Professor Fukui. Roald Hoffmann (right). Photograph (1982) by the authors.

Figure 1-2. Avogadro and Avogadro's law on Italian stamp.

associated behaviors at all. It always failed to do so. They had to deal with the wholes in order to be able to discover the group proclivities as well as integral characteristics of parts. The chemists found the Universe already in complex association and working very well. Every time they tried to take it apart or separate it out, the separate parts were physically divested of their associative potentials, so the chemists had to recognize that there were associated behaviors of wholes unpredicted by parts; they found there was an old word for it—synergy.

In a different, though not entirely unrelated, context, Avogadro has been proposed [1-12] to be the ultimate godfather of buckminsterfullerene for he was

Figure 1-3. R. Buckminster Fuller and his Geodesic Dome, the U.S. Exhibition Hall at the 1967 Montreal Expo. Photographs (1973) courtesy of Lloyd Kahn, Bolinas, California. The arrow indicates a pentagon among the hexagons.

the inventor of the whole concept of monoelemental compounds. This proposal was made by D. E. H. Jones, who originally brought up the idea of the hollow-shell graphite molecule almost 20 years prior to the discovery of buckminsterfullerene [1-13]. In a synergistic move, Jones also referred to a biological analogy of what would be considered today a model of a giant fullerene molecule. A few pentagons are seen interspersed in the generally hexagonal pattern of *Aulonia hexagona*, shown in Figure 1-4 [1-14]. The similarity to Fuller's Geodesic Dome at the 1967 Montreal Expo (Figure 1-3) is striking. This Geodesic Dome did indeed play an important role in leading the discoverers of buckminsterfullerene, Kroto, Smalley (Figure 1-5), and associates [1-8], to the right hypothesis about its molecular structure. Kroto has eloquently described [1-15] how remembering his visit to the Dome, almost two decades before, assisted him and his colleagues in arriving at the highly symmetrical truncated icosahedral geometry; (see, e.g., Ref. [1-16] and Section 2.8 on polyhedra) during the exciting days following the crucial experiment.

Mathematicians have, of course, known for a long time (see, e.g., Ref. [1-17]) that one can close a cage having an even number of vertices with any number of hexagons (except 1), provided that 12 pentagons are included in the network. The truncated icosahedron has 12 pentagons and 20 hexagons, and it is one of the semiregular solids of Archimedes (see Section 2.8). Leonardo da Vinci (1452–1519) drew a hollow framework of this structure to illustrate the book *De Divina Proportione* by Luca Pacioli (Figure 1-6). All such carbon substances whose cage molecules contain 12 pentagons and various numbers of

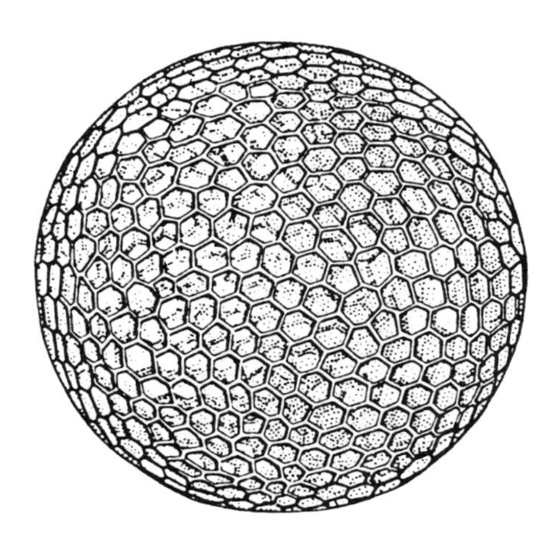

Figure 1-4. Häckel's *Aulonia hexagona* in D'Arcy W. Thompson, *On Growth and Form* [1-14].

Figure 1-5. H. W. Kroto. Photograph courtesy of Professor Kroto (left). R. E. Smalley and models of buckminsterfullerene. Photograph courtesy of Professor Smalley (right).

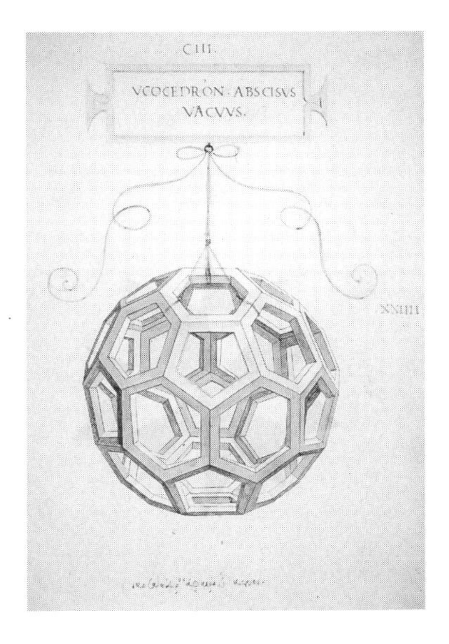

Figure 1-6. Leonardo da Vinci's truncated icosahedron, drawn for Luca Pacioli's *De Divina Proportione*.

hexagons are called fullerenes, of which C_{60} has the special name buckminsterfullerene.

There is another early and beautiful example of the fullerene-type structures. Lion sculptures are common in China as guards in front of important buildings [1-18]. They appear in pairs. The female has a baby lion under the left paw, and the male has a sphere under the right paw. This sphere is said to represent a ball made of strips of silk which was a favorite toy in ancient China. The surface of the ball is usually decorated by a regular hexagonal pattern. We know, however, that it is not possible to cover the surface of the sphere by a regular hexagonal pattern. There are, indeed, considerable chunks of the sphere hidden by the lion's paw and the stand itself on which the lion and the sphere stand. There is at least one lion sculpture (Figure 1-7a) under whose paw the sphere is decorated by a hexagonal pattern interspersed by pentagonal shapes (Figure 1-7b), not unlike Fuller's Geodesic Dome structure. This sculpture stands in front of the Gate of Heavenly Purity in the Forbidden City and dates back to the reign of Qian Long (1736–1796) of the Qing dynasty.

Incidentally, balls made of strips of silk are popular decorations for display in Japan. They are called *temari*, and Figure 1-8 shows one with the pattern of the buckminsterfullerene structure [1-19].

In conclusion, two theoretical studies are quoted, in which Osawa [1-20]

a

b

Figure 1-7. (a) Gold-plated lion sculpture in front of the Gate of Heavenly Purity (Qianqingmen) in the Forbidden City, Beijing. (b) Close-up of the sphere under the lion's paw. Several pentagonal shapes are seen interspersed among the hexagonal pattern decorating the surface of the sphere. Photographs by the authors.

Figure 1-8. Japanese *temari* displaying a pattern of truncated icosahedron [1-19]. Temari courtesy of Kiyoko Urata.

and Bochvar and Gal'pern [1-21] described the I_h-symmetric C_{60} molecule in the early 1970s. These predictions, however, were not followed up by experimental work. The papers, published originally in Japanese and in Russian, had gone into oblivion long before the appearance of buckminsterfullerene, though they were graciously rediscovered afterward.

The other important symmetry-related discovery was the quasicrystals. Both the truncated icosahedral structure of buckminsterfullerene and the regular but nonperiodic network of the quasicrystals are related to fivefold symmetry. In spite of this intimate connection between them at an intellectual level, their stories did not really cross. The conceptual linkage between them is provided by Fuller's physical geometry, and this is also what relates them to the icosahedral structure of viruses (see Section 9.5.2).

The actual experimental discovery of quasicrystals was serendipitous [1-9], notwithstanding some previous predictions (see, e.g., Ref. [1-22]). It has been a rock-solid fundamental dogma of crystallography that fivefold symmetry is a noncrystallographic symmetry. We shall return to this question in Section 9.3. Suffice it to quote here an illustration of the mosaic coverage of a surface by the first four regular polygons (Figure 1-9) from a beautiful early paper on fivefold symmetry [1-23]. There have been many attempts to cover the surface with regular pentagons without gaps and overlaps, and some examples [1-24–1-26] are shown in Figure 1-10. Then, Penrose [1-27] found two elements which, by appropriate matching, could tile the surface with long-range pentagonal symmetry though only in a nonperiodic way (Figure 1-11).

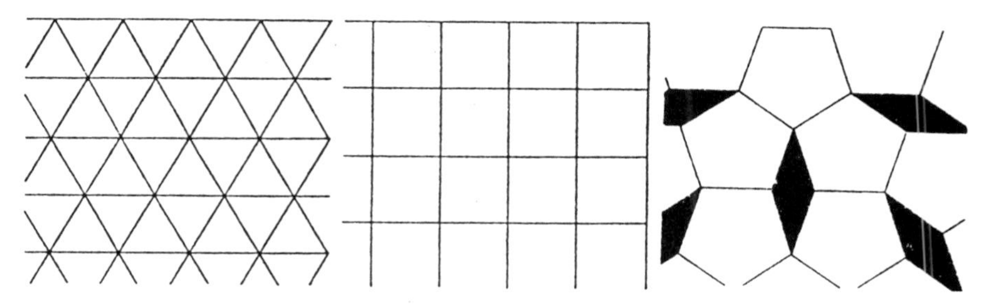

Figure 1-9. Mosaic coverage of the surface by the equilateral triangle and the regular hexagon, the square, and the regular pentagon, after Breder [1-23].

This pattern was extended by Alan Mackay (Figure 1-12a) into the third dimension, and even a simulated diffraction pattern was produced [1-22] which showed tenfoldedness (Figure 1-12b). It was about the same time that Dan Shechtman (Figure 1-13a) was experimenting with metallic phases of various alloys cooled with different speeds and observed tenfoldedness in an actual electron diffraction experiment (Figure 1-13b) for the first time. The discovery of quasicrystals has given an added perspective to crystallography and the utilization of symmetry considerations.

The question may also be asked as to whether "chemical symmetry" differs from any other kind of symmetry? Symmetries in the various branches of the sciences are perhaps characteristically different, and one may ask whether they could be hierarchically related. The symmetry in the great conservation laws of physics (see, e.g., Ref. [1-28]) is, of course, present in any chemical system. The symmetry of molecules and their reactions is part of the fabric of biological structure. Left-and-right symmetry is so important for living matter that it may be matched only by the importance of "left-and-right" symmetry in the world of the elementary particles, including the violation of parity, as if a circle is closed, but that is, of course, a gross oversimplification.

When we stress the importance of symmetry considerations, it is not equivalent to declaring that everything must be symmetrical. In particular, when the importance of left-and-right symmetry is stressed, it is the relationship of left and right, rather than their equivalence, that has outstanding significance.

We have already mentioned that symmetry considerations have continued their fruitful influence on the progress of contemporary chemistry. This is so for contemporary physics as well. It is almost surprising that fundamental conclusions with respect to symmetry could be made even in this century. It was related by C. N. Yang [1-29] that Dirac considered Einstein's most

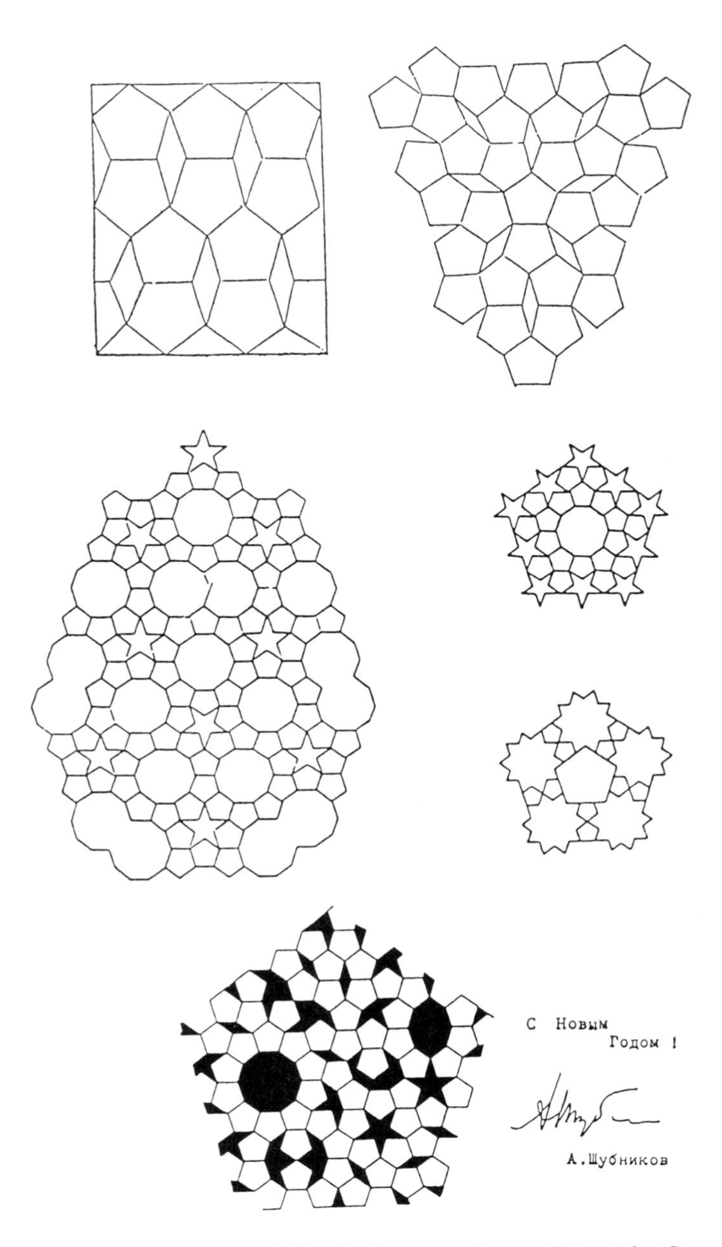

Figure 1-10. Attempts of pentagonal tiling by, from top to bottom, Dürer (after Crowe [1-24]), Kepler (after Danzer *et al*. [1-25]) and Shubnikov (after Mackay [1-26]).

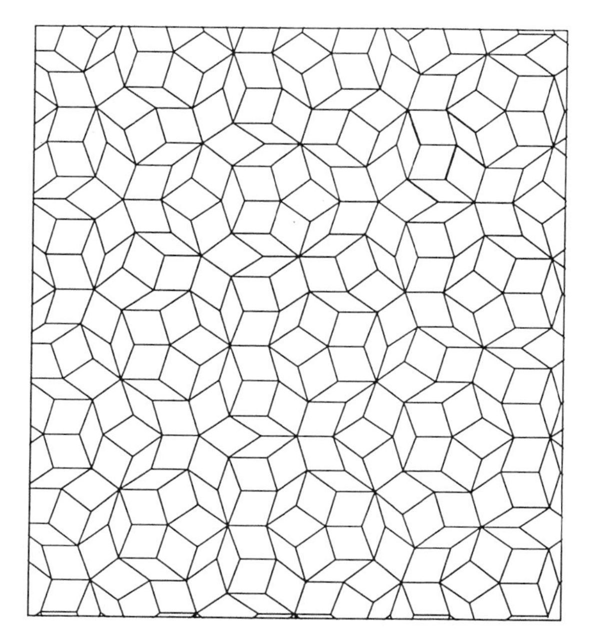

Figure 1-11. Penrose tiling.

important contributions to physics to be "his introduction of the concept that space and time are symmetrical." Dirac also had the prescience to write already in 1949 [1-30] that "I do not believe that there is any need for physical laws to be invariant under reflections." Yet most physicists were surprised by the discovery of the nonconservation of parity in 1957 (cf. Ref. [1-31]). Since then, *broken symmetries* have received increasing attention. The term relates to situations in which symmetries that are expected to hold are valid only approximately or fail completely [1-32]. The three basic possibilities are incomplete symmetry, symmetry broken by circumstances, and spontaneously broken symmetry.

But what is symmetry? We may not be able to answer this question satisfactorily, at least not in all its possible aspects. According to the crystallographer (and symmetrologist) E. S. Fedorov, "Symmetry is the property of geometrical figures to repeat their parts, or more precisely, their property of coinciding with their original position when in different positions" [1-33]. According to the geometer H. S. M. Coxeter [1-34], "When we say that a figure is 'symmetrical' we mean that there is a congruent transformation which leaves a 

Figure 1-12. A. L. Mackay. Photograph (1982) by the authors. Simulated "electron diffraction" pattern of two-dimensional Penrose tiling, after Mackay [1-22]. Photograph courtesy of Professor Mackay.

it unchanged as a whole, merely permuting its component elements." Fedorov's definition is cited here after another symmetrologist (and crystallographer), A. V. Shubnikov [1-33], who added that while symmetry is a property of geometrical figures, obviously, "material figures" may also have symmetry. He further said that only parts which are in some sense equal among themselves can be repeated and noted the existence of two kinds of equality, to wit congruent equality and mirror equality. These two equalities are the subsets of the metric equality concept of Möbius, according to whom "figures are

a 

Figure 1-13. D. Shechtman. Photograph (1991) by the authors. Electron diffraction pattern with tenfold symmetry. Photograph courtesy of Professor Shechtman.

equal if the distances between any given points on one figure are equal to the distances between the corresponding points on another figure" [1-34].

Symmetry also connotes harmony of proportions, however—a rather vague notion according to Weyl [1-35]. This very vagueness, at the same time, often comes in handy when relating symmetry and chemistry or, generally speaking, whenever the symmetry concept is applied to real systems. Mislow and Bickart [1-36] published an epistemological note on chirality in which much of what they have to say about chirality, as this concept is applied to geometrical figures versus real molecules, solvents, and crystals, is true about the symmetry concept as well. Mislow and Bickart argue that "it is unreasonable to draw a sharp line between chiral and achiral molecular ensembles: in contrast to the crisp classification of geometric figures, one is dealing here with a fuzzy borderline distinction, and the qualifying 'operationally' should be implicitly or explicitly attached to 'achiral' or 'racemic' whenever one uses these terms with reference to observable properties of a macroscopic sample." Further, Mislow and Bickart [1-36] state that "when one deals with natural phenomena, one enters 'a stage in logic in which we recognize the utility of imprecision' [1-37]." The human ability to geometrize nongeometrical phenomena greatly helps to recognize symmetry even in its "vague" and "fuzzy" variations. In accordance with this, Weyl [1-35] referred to Dürer, who "considered his canon of the human figure more as a standard from which to deviate than as a standard toward which to strive."

Symmetry in its rigorous sense helps us to decide problems quickly and qualitatively. The answers lack detail, however [1-38]. On the other hand, the vagueness and fuzziness of the broader interpretation of the symmetry concept allow us to talk about degrees of symmetry, to say that something is more symmetrical than something else. An absolutist geometrical approach would allow us to distinguish only between symmetrical and asymmetrical, possibly with dissymmetrical thrown in for good measure. So there must be a range of criteria according to which one can decide whether something is symmetrical, and to what degree. These criteria may very well change with time. A case in point is the question as to whether or not molecules preserve their symmetry upon entering a crystal structure or upon the crystal undergoing a phase transition. Our notion about structures and symmetries may evolve as more accurate data become available (though the structures and symmetries are unchanged, of course, by our notions). A whole new approach is developing to analyze symmetry properties in terms of a continuous scale rather than of a discrete "yes/no" [1-39] and to quantify chirality [1-40].

The remarkable phenomenon of statistical symmetry was noted by Loeb [1-41]. There are some apparently totally asymmetrical structures in which characteristic parameters are, however, subject to certain well-defined constrained patterns when averaged according to some system.

Recognizing structural and other kinds of regularities has always been important in chemistry. It has been argued, for example, that at the time of the first edition of *The Nature of the Chemical Bond* [1-42], Linus Pauling had access to less than 0.01% of the structural information of 50 years later, yet his ideas on structure and bonding have stood the test of time [1-43].

The history of periodic tables, following Mendeleev's seminal discovery, also demonstrates chemists' never-ending quest for beauty and harmony. Dmitri I. Mendeleev was looking for a simple system for presenting the elements as he was writing a general chemistry text for his students. The Soviet stamp block, issued for the centennial of the periodic table, depicts its earliest version (Figure 1-14). Approximately 700 periodic tables were published during the first one hundred years after the original discovery in 1869. E. G.

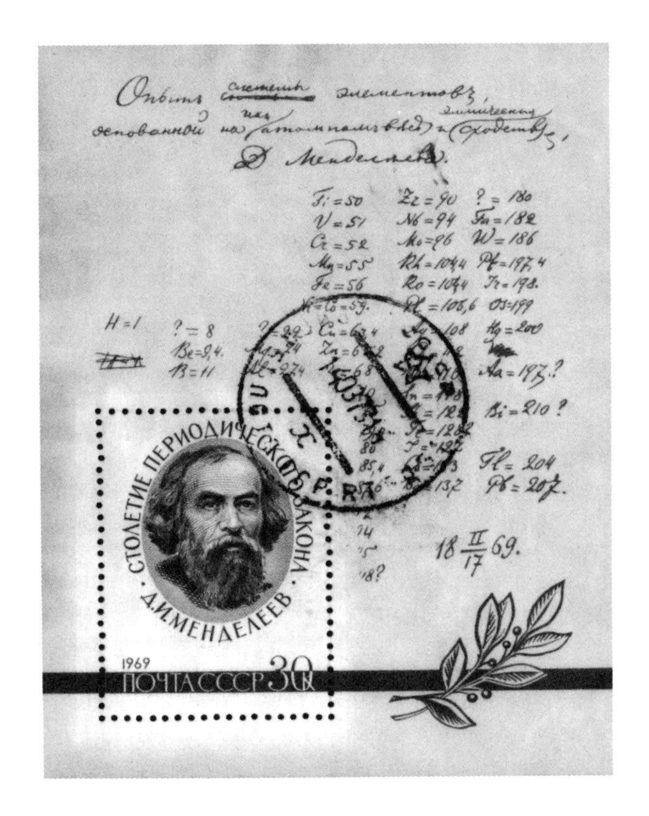

Figure 1-14. Soviet stamp block issued to commemorate the centenary of Mendeleev's periodic system.

Mazurs [1-44] has collected, systematized, and analyzed them in a unique study. Classification of all the tables reduced their number to 146 different types and subtypes which are described by such terms as "helices, space lemniscates, space concentric circles, space squares, spirals, series tables, zigzags, parallel lines, step tables, tables symmetrical about a vertical line, mirror image tables, tables of one revolution and of one row, tables of planes, revolutions, cycles, right side as well as left side electronic configuration tables, tables of concentric circles and parallel lines, right side as well as left side shell and subshell tables." Figure 1-15 shows the traditional, rectangular-shaped table in the form of a wall decoration displayed on the facade of the St. Petersburg college building where Mendeleev used to work. The characteristic symmetry of this arrangement is periodicity itself. The two tables of Figure 1-16 were drawn after Mazurs [1-44], one with concentric circles in space,

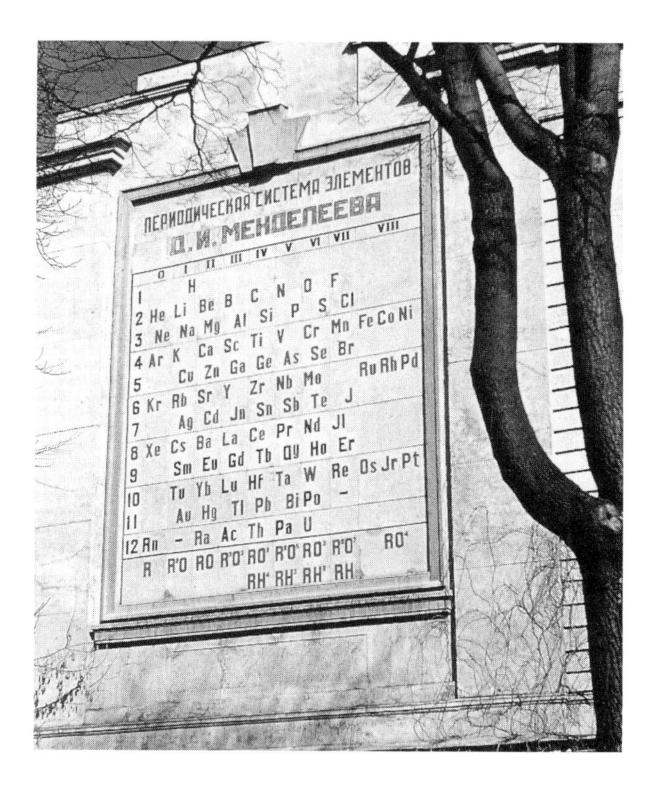

Figure 1-15. Mendeleev's periodic system on the facade of the college building where Mendeleev used to work. Photograph courtesy of Dr. A. Belyakov, St. Petersburg.

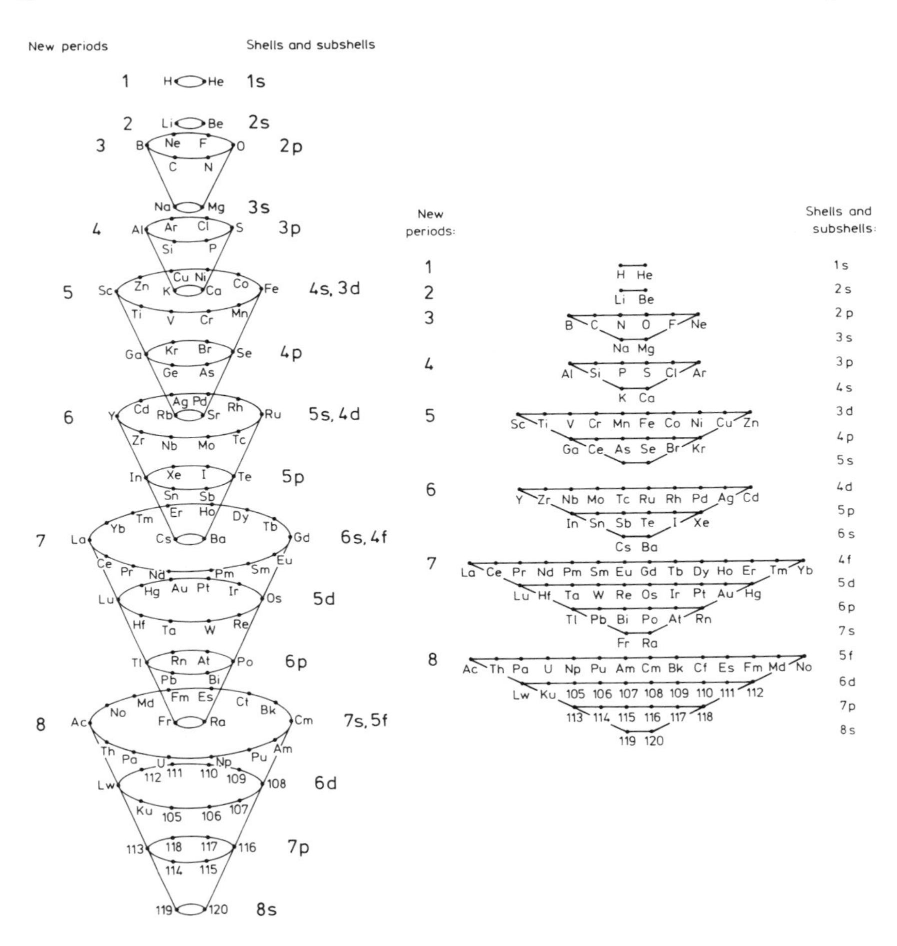

Figure 1-16. Two periodic tables drawn after Mazurs [1-44].

representing subshells and period cones stretched vertically; and the other with parallel lines in the plane with bilateral symmetry. Figure 1-17 [1-45] is a spiral representation of the periodic system (drawn proportionally to the increasing mass of the elements, prior to the understanding of the foundation of the system in the electronic structure of the elements), and Figure 1-18 is an artistic representation of the spiral nature of the system by a chemist sculptor [1-46].

The quest for symmetry and harmony has, of course, contributed more than mere aesthetics in establishing the periodic table of the elements. Beauty and reason blend in it in a natural fashion. C. A. Coulson, theoretical chemist

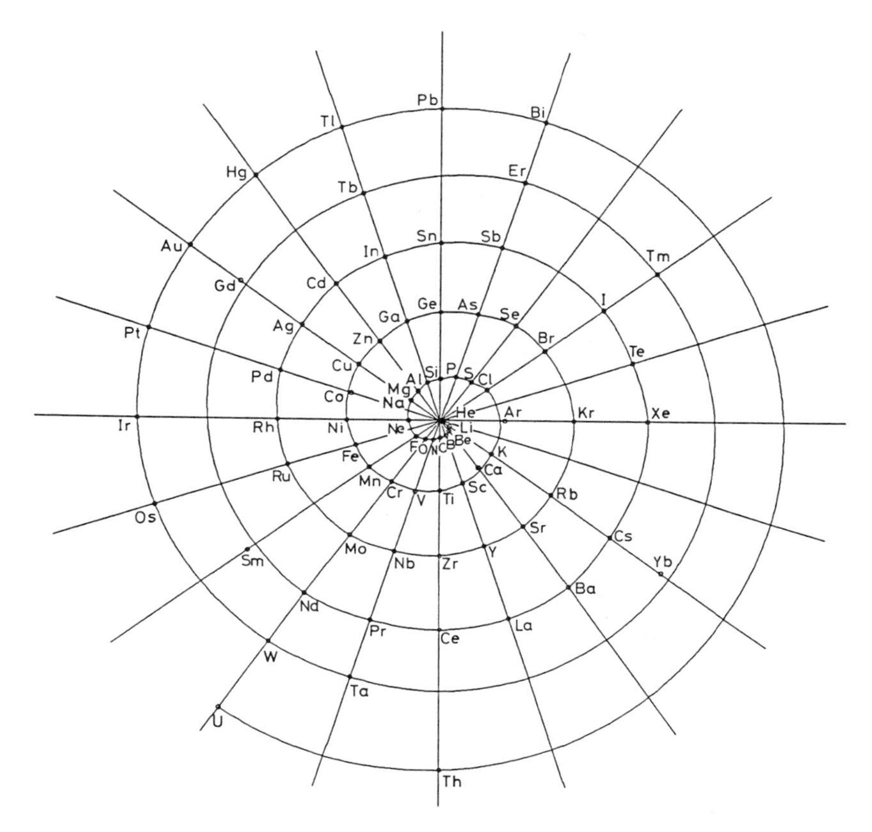

Figure 1-17. Spiral periodic system, after Erdmann [1-45]. Computer graphics constructed by Judit Molnar, Budapest.

and professor of mathematics, concluded his Faraday lecture on symmetry [1-5] with the words:

> Man's sense of shape—his feeling for form—the fact that he exists in three dimensions—these must have conditioned his mind to thinking of structure, and sometimes encouraged him to dream dreams about it. I recall that it was Kekulé himself who said: "Let us learn to dream, gentlemen, and then we shall learn the truth." Yet we must not carry this policy too far. Symmetry is important, but it is not everything. To quote Michael Faraday writing of his childhood: "Do not suppose that I was a very deep thinker and was marked as a precocious person. I was a lively imaginative person,

Figure 1-18. Artistic spiral representation of the periodic system by B. Vizi [1-46]. Photograph courtesy of Dr. Vizi, Veszprém.

and could believe in the *Arabian Nights* as easily as in the Encyclopedia. But facts were important to me, and saved me." It is when symmetry interprets facts that it serves its purpose: and then it delights us because it links our study of chemistry with another world of the human spirit—the world of order, pattern, beauty, satisfaction. But facts come first. Symmetry encompasses much—but not quite all!

REFERENCES

[1-1] E. Lendvai, in *Module, Proportion, Symmetry, Rhythm* (G. Kepes, ed.), George Braziller, New York (1966).
[1-2] N. Angier, *The New York Times*, February 8, 1994, p. C1.
[1-3] V. Prelog, *Science* **193**, 17 (1976).
[1-4] A. Koestler, *Insight and Outlook*, Macmillan, London (1949).
[1-5] C. A. Coulson, *Chem. Br.* **4**, 113 (1968).
[1-6] K. Fukui, *Science* **218**, 747 (1982).

[1-7] R. Hoffmann, *Science* **211**, 995 (1981).
[1-8] H. W. Kroto, J. R. Heath, S. C. O'Brien, R. F. Curl, and R. E. Smalley, *Nature* **318**, 162 (1985).
[1-9] D. Shechtman, I. Blech, D. Gratias, and J. W. Cahn, *Phys. Rev. Lett.* **53**, 1951 (1984).
[1-10] I. Hargittai, *Per. Mineral.* **61**, 9 (1992).
[1-11] R. B. Fuller, *Synergetics: Explorations in the Geometry of Thinking*, Macmillan, New York (1975), p. 4.
[1-12] D. E. H. Jones, *Philos. Trans. R. Soc. London* **343**, 9 (1993).
[1-13] D. E. H. Jones, *New Scientist* **35**, 245 (1966).
[1-14] D. W. Thompson, *On Growth and Form*, Cambridge University Press (1961).
[1-15] H. W. Kroto, *Angew. Chem. Int. Ed. Engl.* **31**, 111 (1992).
[1-16] H. W. Kroto, in *Symmetry 2, Unifying Human Understanding* (I. Hargittai, ed.) Pergamon Press, Oxford, (1989), p. 417.
[1-17] P. C. Gasson, *Geometry of Spacial Forms*, Ellis Horwood, Chichester, (1983), pp. ix–x.
[1-18] I. Hargittai, *Math. Intell.* **17**, 34 (1995).
[1-19] I. Hargittai, *Forma* **8**, 327 (1993).
[1-20] E. Osawa, *Kagaku (Kyoto)* **25**, 854 (1970).
[1-21] D. A. Bochvar and E. G. Gal'pern, *Dokl. Akad. Nauk SSSR* **209**, 610 (1973).
[1-22] A. L. Mackay, *Physica* **114A**, 609 (1982).
[1-23] C. M. Breder, *Bulletin of the American Museum of Natural History* **106**, 173 (1955).
[1-24] D. W. Crowe, in *Fivefold Symmetry* (I. Hargittai, ed.), World Scientific, Singapore (1992), p. 465.
[1-25] L. Danzer, B. Grünbaum, and G. C. Shephard, *Am. Math. Monthly* **89**, 568 (1982).
[1-26] A. L. Mackay, *Sov. Phys. Crystallogr.* **26**, 517 (1981).
[1-27] R. Penrose, *Math. Intell.* **2**, 32 (1979/80).
[1-28] R. Feynman, *The Character of the Physical Law*, The MIT Press, Cambridge, Massachusetts (1967).
[1-29] C. N. Yang, in *The Oscar Klein Memorial Lectures, Vol. 1*, (G. Ekspong, ed.), World Scientific, Singapore (1991), p. 11.
[1-30] P. A. M. Dirac, *Rev. Mod. Phys.* **21**, 392 (1949).
[1-31] R. H. Dalitz and R. Peierls, *Biogr. Mem. Fellows Roy. Soc.* **32**, 159 (1989). We thank Professor J. D. Dunitz, Zürich, for this reference.
[1-32] R. Peierls, *Contemp. Phys.* **33**, 221 (1992).
[1-33] A. V. Shubnikov, *Simmetriya i antisimmetriya konechnykh figur*, Izd. Akad. Nauk SSSR, Moscow (1951).
[1-34] H. S. M. Coxeter, *Regular Polytopes*, 3rd ed., Dover Publications, New York (1973).
[1-35] H. Weyl, *Symmetry*, Princeton University Press, Princeton, New Jersey (1952).
[1-36] K. Mislow and P. Bickart, *Israel J. Chem.* **15**, 1 (1976/77).
[1-37] M. Scriven, *J. Philos.* **56**, 857 (1959).
[1-38] R. G. Pearson, *Symmetry Rules for Chemical Reactions, Orbital Topology and Elementary Processes*, Wiley-Interscience, New York (1976).
[1-39] H. Zabrodsky, S. Peleg, and D. Avnir, *J. Am. Chem. Soc.* **114**, 7843 (1992); **115**, 8278 (1993); H. Zabrodsky and D. Avnir, in *Advances in Molecular Structure Research, Vol. 1* (M. Hargittai and I. Hargittai, eds.), JAI Press, Greenwich, Connecticut (1995).
[1-40] A. B. Buda, T. Auf der Heyde, and K. Mislow, *Angew. Chem. Int. Ed. Engl.* **31**, 989 (1992).
[1-41] A. L. Loeb, *Space Structures. Their Harmony and Counterpoint*, Addison-Wesley Publishing Co., Reading, Massachusetts (1976).
[1-42] L. Pauling, *The Nature of the Chemical Bond*, Cornell University Press, Ithaca, New York (1939) [2nd ed., 1940; 3rd ed., 1960].

[1-43] P. Murray-Rust, in *Computer Modelling of Biomolecular Processes* (J. M. Goodfellow and D. S. Moss., eds.), Ellis Horwood, New York (1992).
[1-44] E. G. Mazurs, *Graphic Representations of the Periodic System during One Hundred Years*, The University of Alabama Press, University, Alabama (1974).
[1-45] H. Erdmann, *Lehrbuch der anorganischen Chemie*, Dritte Auflage, F. Vieweg und Sohn, Braunschweig (1902).
[1-46] B. Vizi, *Chemistry in Sculptures*, Hungarian Chemical Society, Budapest, 1990.

Chapter 2

Simple and Combined Symmetries

2.1 BILATERAL SYMMETRY

The simplest and most common of all symmetries is bilateral symmetry. Yet at first sight it does not appear so overwhelmingly important in chemistry as in everyday life. The human body has bilateral symmetry, except for the asymmetric location of some internal organs. A unique description of the symmetry of the human body is given by Thomas Mann in *The Magic Mountain* [2-1] as Hans Castorp is telling about his love to Clawdia Chauchat:

> How bewitching the beauty of a human body, composed not of paint or stone, but of living, corruptible matter charged with the secret fevers of life and decay! Consider the wonderful symmetry of this structure: shoulders and hips and nipples swelling on either side of the breast, and ribs arranged in pairs, and the navel centered in the belly's softness, and the dark sex between the thighs. Consider the shoulder blades moving beneath the silky skin of the back, and the backbone in its descent to the paired richness of the cool buttocks, and the great branching of vessels and nerves that passes from the torso to the arms by way of the arm pits, and how the structure of the arms corresponds to that of the legs!

The bilateral symmetry of the human body is emphasized by the static character of many Egyptian sculptures (Figure 2-1). Mobility and dynamism, however, do not diminish the impression of bilateralness of the human body (Figure 2-2).

Figure 2-1. Egyptian sculpture from 2700 B.C. Photo of Lehnert & Landrock Art Publishers, Cairo. Used by permission.

Already Kepler [2-2] noted in connection with the shape of the animals that the

> upper and lower depends on their habitat, which is the surface of the earth . . . The second distinction of front and back is conferred on animals to put in practice motions that tend from one place to another in a straight line over the surface of the earth . . . bodily existence entailed the third diameter, of right and left, should be added, whereby an animal becomes so to speak doubled.

The "three diameters" of Kepler suggest a Cartesian coordinate system [2-3].

Bilateral symmetry is indeed very common in the animal kingdom. It always appears when *up* and *down* as well as *forward* and *backward* are different, whereas leftbound and rightbound motion have the same probability. As translational motion along a straight line is the most characteristic for the vast majority of animals on Earth, their bilateral symmetry is trivial. This symmetry is characterized by a *reflection plane*, or *mirror plane*, hence its usual label m.

Figure 2-2. Mobility does not diminish the perception of bilaterality of the human body: swimmer (MTI-Foto Archive, Budapest) and gymnast (photograph by T. Szigeti, Budapest). Used by permission.

Bilateral symmetry is widespread in the animal world (Figure 2-3) and in some flowers (Figure 2-4). It may be only accidental for a tree (Figure 2-5a). Generally, however, trees as well as other plants have radial, or cylindrical, or conical symmetries with respect to the trunk or stem. Although these symmetries are very approximate, they can be recognized without any ambiguity (Figure 2-5b).

Bilateral symmetry, and symmetry in general, often appears in expressions of religion (see, e.g., Figure 2-6).

The symmetry plane of the human face is sometimes emphasized by artists. Some examples are cited in Figure 2-7. Of course, there are minute variations, or even considerable ones, between the left and right sides of the human face (see, e.g., Figure 2-8).

The origin and meaning of the deviations from bilateral symmetry of the human face have generated considerable research interest (see, e.g., Ref. [2-4]). It may even be that the two sides of our face differ in expressing emotions, and one may be more "private" and the other may be more "public."

Figure 2-3. Animals: Dog, monkey (photographs by the authors), insect (on stamp), and butterfly (on stamp).

Differences between the left and right hemispheres of the brain have been the subject of intensive studies, and several monographs have appeared (see, e.g., Refs. [2-5] and [2-6]). Hemispheric asymmetry has received so much attention that recently it was suggested that "the time has come to put the brain back together again" [2-6].

Bilateral symmetry has outstanding importance in man-made objects. It has a functional purpose. The bilateral symmetry of various vehicles, for example, is determined by their translational motion. On the other hand, the cylindrical symmetry of the Lunar Module is consistent with its function of vertical motion with respect to the moon's surface. It has been noted [2-7] that

Figure 2-4. Flowers: Orchids from Hawaii. Photographs by the authors.

the motorcycle with a sidecar may be disappearing because its shape suggests circular rather than translational motion.

Examples of cylindrical symmetry related to the preferential importance of the vertical direction are the salt columns in the Dead Sea (Figure 2-9a) as well as the stalactites and the stalagmites in caves (Figure 2-9b), both formed of calcium carbonate.

The occurrence of radial type symmetries rather than more restricted ones necessitates a spatial freedom in all relevant directions. Thus, for example, the copper formation in Figure 2-10a has a tendency to form cylindrically symmetric structures. On the other hand, the solidified iron dendrites obtained from iron–copper alloys, after dissolving away the copper, display bilateral symmetry in Figure 2-10b.

Both folk music and music by master composers are rich in various symmetries. Figure 2-11 shows two examples with bilateral symmetry. The first example (Figure 2-11a) is from a Hungarian folk song entitled *Crunchy Cherries Are Ripening*. The sequence is A, A^5/A^5_v, A, where the upper index indicates a 5-note shift to higher frequencies and the lower index v indicates some minute variation. Another example (Figure 2-11b) is *Unisono No. 6* from Bartók's *Microcosmos* series, written specifically for children. Figure 2-11b illustrates a mirror plane which includes a note.

The introductory piece of the *Microcosmos* is depicted in Figure 2-12a. It has only approximate bilateral symmetry though the two halves are markedly present. When some schoolchildren in their early teens were asked to express their impressions in drawing while listening to this piece of music for the first

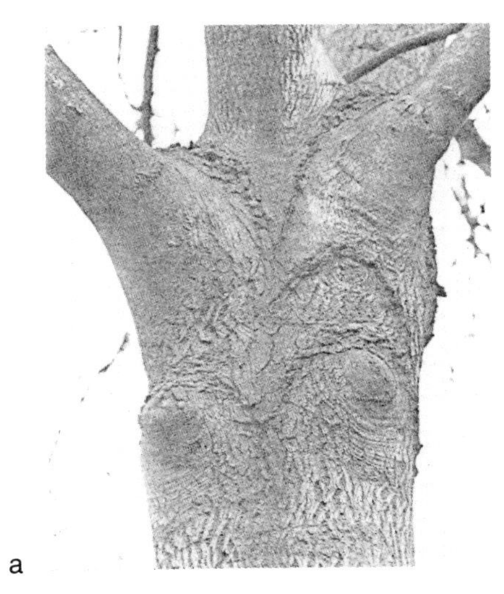

a

b

Figure 2-5. (a) Accidental bilateral symmetry, *Tree near Aveley*, Essex. Photograph used by permission of C. T. Ballard. (b) Conical and radial symmetries of trees. Photographs by the authors.

Figure 2-6. Artistic and religious expressions of bilaterality: Photographs from Venice, Italy, and Zagorsk, Russia, by the authors.

a

b

c

Figure 2-7. Human face in artistic expression: (a) Henri Matisse, *Woman's Portrait*. Reproduced by permission from The Hermitage, St. Petersburg; (b) George Buday, *Miklós Radnóti*, woodcut, 1969. Reproduced by permission of George Buday, R. E; (c) Jenő Barcsay, *Woman's Head*, 1961. Reproduced by permission of Ms. Barcsay.

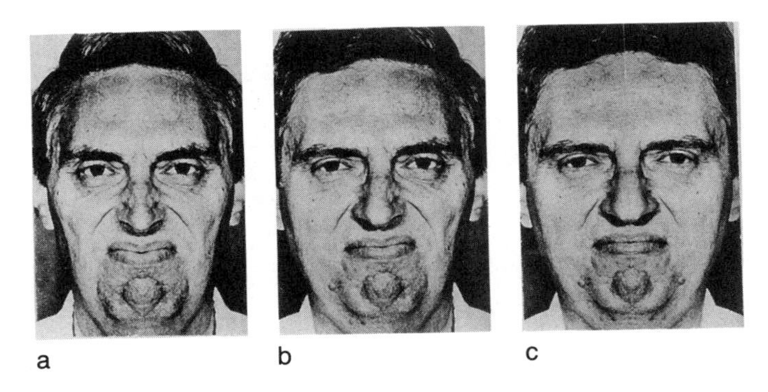

Figure 2-8. Face expressing distaste [2-4]. Reproduced by permission, copyright (1978) by the American Association for the Advancement of Science. (a) Left-side composite; (b) original; (c) right-side composite.

Figure 2-9. (a) Salt columns in the Dead Sea. Drawing by Ferenc Lantos after a color slide of Palphot, Ltd. Herzlia, Israel. (b) Calcium carbonate stalactites and stalagmites in a cave in southern Germany. Photographs by the authors.

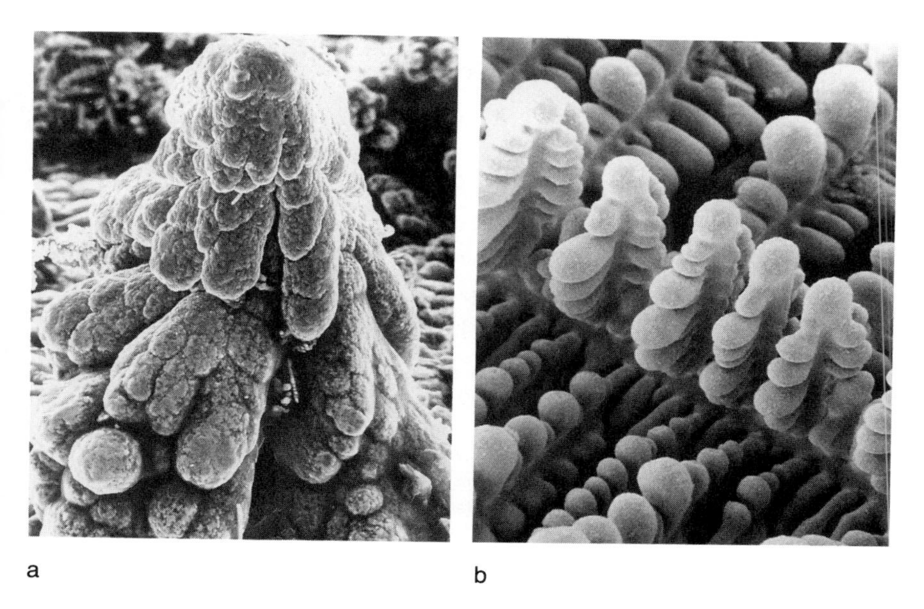

Figure 2-10. (a) Electrolytically deposited copper, magnification × 1000. Courtesy of Dr. Maria Kazinets, Ben Gurion University, Beer Sheva. (b) Directionally solidified iron dendrites from an iron–copper alloy after the copper has been dissolved away, magnification × 2600. Courtesy of Dr. J. Morral, The University of Connecticut.

Figure 2-11. (a) Hungarian folk song *Crunchy Cherries Are Ripening*. (b) Bartók's *Microcosmos, Unisono No. 6*.

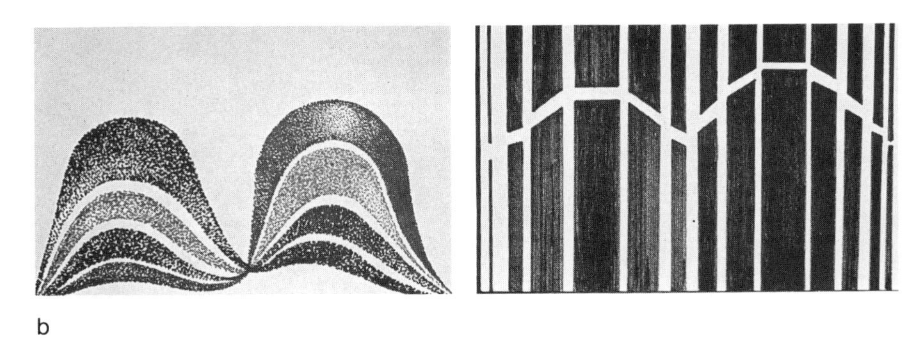

Figure 2-12. (a) Bartók's *Microcosmos, Unisono No. 1*. (b) Drawings inspired by the *Unisono No. 1* by students in their early teens, Komló Music School. Courtesy of Mária Apagyi.

time, they invariably produced patterns with bilateral symmetry. Two of the drawings are reproduced in Figure 2-12b.

Changing from *Microcosmos* to the "macro"cosmos, a typical galaxy of the universe would display bilateral symmetry if viewed edge-on as shown in Figure 2-13.

Weyl [2-9] calls bilateral symmetry also heraldic symmetry as it is so common in coats of arms. Characteristically, the Hapsburg and the Romanov eagles were double-headed (Figure 2-14), and there are occurrences of double heads elsewhere as well (Figure 2-15).

Figure 2-13. Edge-on view of a typical galaxy [2-8]. Reproduced by permission of R. Jastrow.

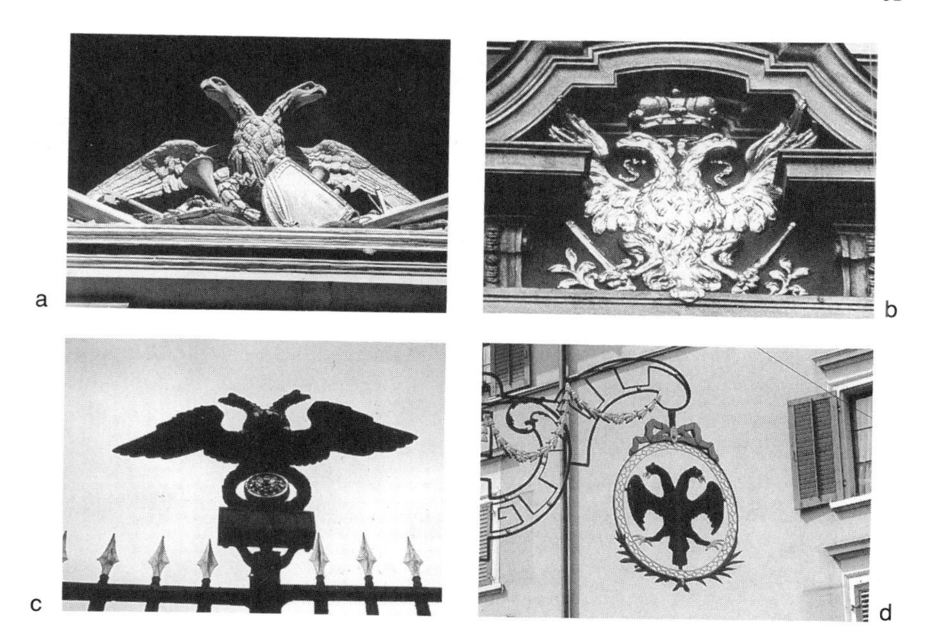

Figure 2-14. Double-headed eagles. Photographs by the authors. (a) Vienna; (b) Prague; (c) St. Petersburg; (d) Zurich.

Figure 2-15. Double-headed animals: (a) Belgian ad in Brussels; (b) Chinese decoration in Beijing. Photographs by the authors.

2.2 ROTATIONAL SYMMETRY

Staying with heraldry, the contour of the simple and powerful Oriental symbol yin yang of the South Korean coat of arms is shown in Figure 2-16a. It has twofold rotational symmetry in that a half rotation about the axis perpendicular to the midpoint of the drawing brings back the original figure. This rotation axis is a symmetry axis. The Taiwanese stamp with two fish, reminiscent of yin yang, in Figure 2-16c and the logo in the recycling ad in Figure 2-16b both have twofold rotational symmetry.

The *order* of a rotation symmetry axis tells us how many times the original figure reoccurs during a complete rotation. The *elemental angle* is the smallest angle of rotation by which the original figure can be reproduced. Thus, for twofold rotational symmetry, the order of the rotation axis is obviously two, and the elemental angle is 180°. The corresponding numbers for threefold and fourfold rotational symmetries are three and 120° and four and 90°, respectively. In Figure 2-17, rotational symmetries are illustrated by sculptures of interweaving fish and dolphins.

Figure 2-18 further illustrates threefold rotational symmetry. Fourfold

Figure 2-16. (a) Contour of yin yang. (b) Logo of Reynolds Recycling painted on a track in Honolulu, Hawaii. Photograph by the authors. (c) Two fish on a Taiwanese stamp.

Figure 2-17. Interweaving fish and dolphins; (a) Twofold in Washington, D.C.; (b) Threefold in Prague; (c) Fourfold in Linz, Austria. Photographs by the authors.

rotational symmetries are illustrated in Figure 2-19. This is the symmetry of the swastika, an ornament since prehistoric times but also associated with Nazism. It is illustrated by John Heartfield's anti-Nazi poster from 1934 (Figure 2-19a). American quilts provide a wealth of symmetries. Exclusively rotational symmetries are generally rare but they can be found, for example, in the so-called friendship quilts (Figure 2-19b) which were made by exchanging patterns among a circle of friends and were believed to have strength and dignity as well as simplicity [2-12].

Machinery parts, performing rotational motion only, such as propellers, have rotational symmetry only. An example is the four-blade propeller in Figure 2-19c.

It is very rare in the living world to find creatures which have *only* rotational symmetry. An example from the animal world is the jellyfish (Figure 2-19d [2-7]). Such exclusively rotational symmetry may be a consequence of preferential rotational motion in capturing food.

Fivefold rotational symmetry is displayed by the flowers in Figure 2-20a. Other examples of fivefold rotational symmetry in Figure 2-20 are the NASA

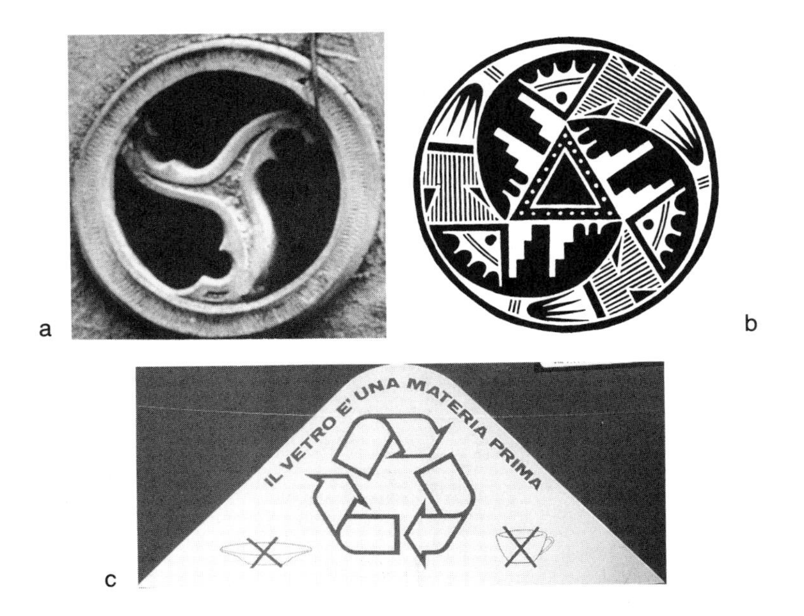

Figure 2-18. Examples of threefold rotational symmetry: (a) Italian decoration; photograph by the authors; (b) American Indian pottery decoration [2-10]; (c) Italian logo for recycling; photograph by the authors.

logo and the hubcap, while sixfold rotational symmetry characterizes the Star of David and the six-blade windmill in Figure 2-21.

The order of rotation axes (n) may be 1, 2, 3, . . . , up to infinity, ∞; thus, it may be any integer. The order 1 means that a complete rotation is needed to bring back the original figure; thus, there is a total absence of symmetry, which means asymmetry. A onefold rotation axis is an identity operator. The other extreme is the infinite order. This means that any, even infinitesimally small, rotation leads to congruency. Some examples of rotational symmetry with increasing order n are shown in Figure 2-22.

Decorations displaying exclusively rotational symmetry occur often. The otherwise widespread symmetry plane in decorations is easily eliminated by interweaving the motifs (see, e.g., the Star of David in Figure 2-21).

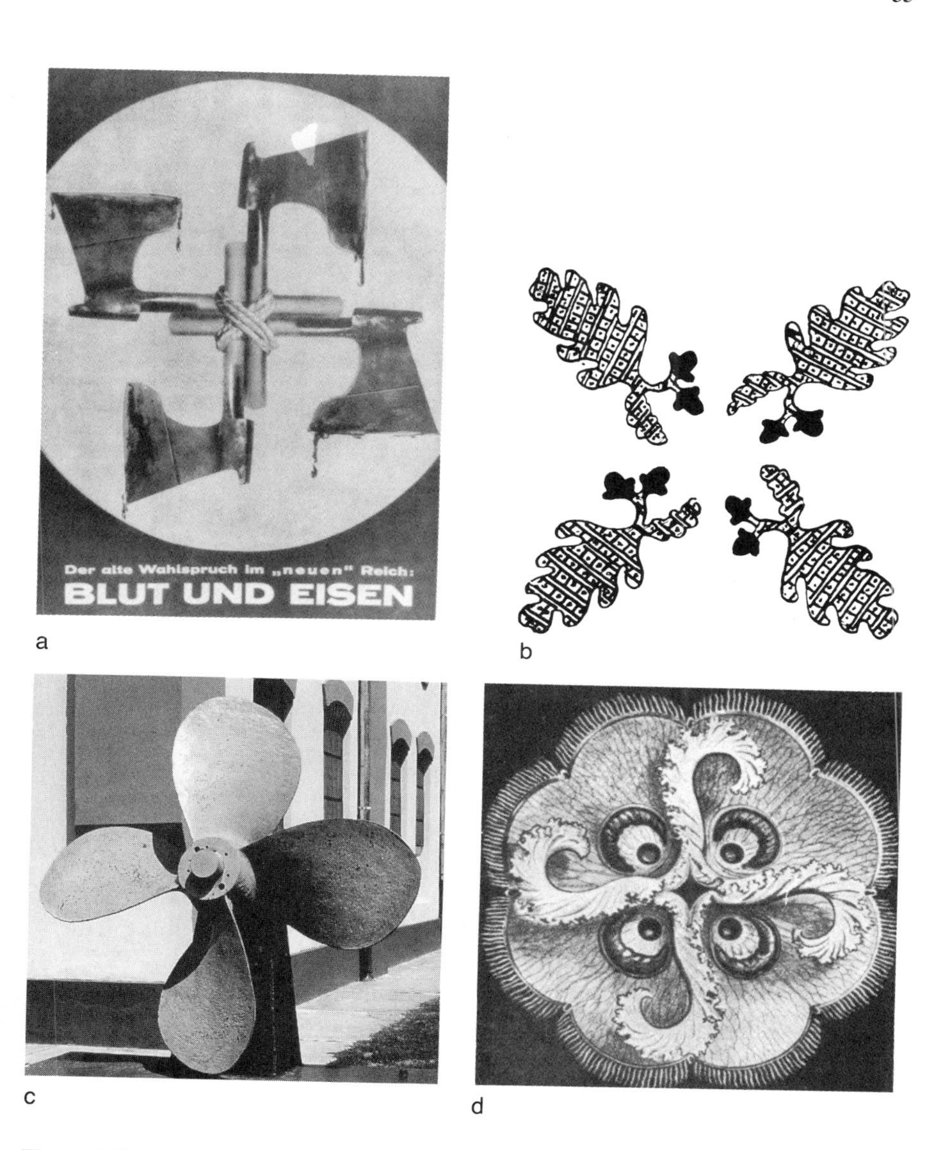

Figure 2-19. Examples of fourfold rotational symmetry: (a) John Heartfield's *Blood and Iron*, anti-Nazi poster from 1934 (cf. Ref. [2-11]), used by permission; (b) decoration of friendship quilt [2-12]; (c) four-blade propeller in Budapest; photograph by the authors; (d) jellyfish *Aurelia insulinda* from Ref. [2-7], used by permission.

Figure 2-20. Examples of fivefold rotational symmetry: (a) Hawaiian flowers; (b) NASA logo, Florida Space Center; (c) hub-cap. Photographs by the authors.

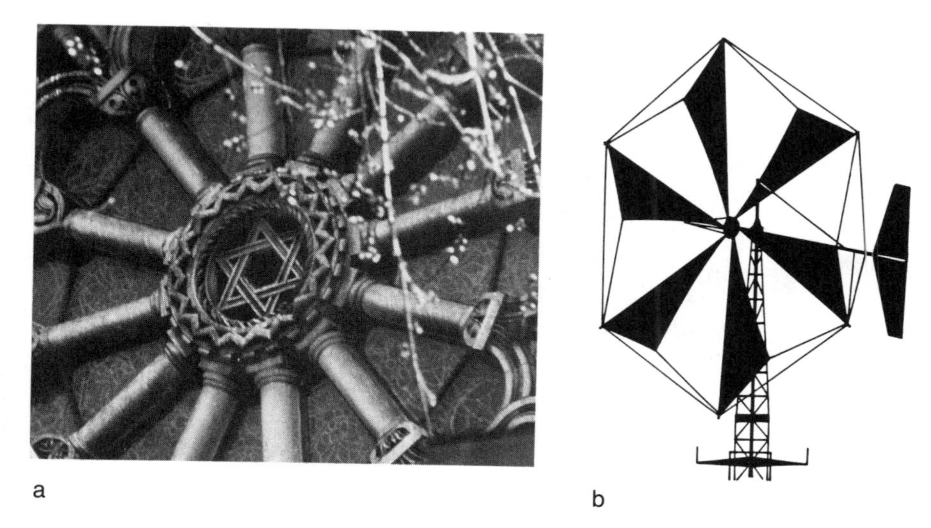

Figure 2-21. Examples of sixfold rotational symmetry: (a) Star of David, New York; photograph by the authors; (b) six-blade windmill.

Figure 2-22. Examples of rotational symmetry: (a) Hawaiian flower, $n = 7$; (b) Hawaiian flower, $n = 8$; Photographs by the authors. (*Continued on next page*)

c

d

Figure 2-22. (*Continued*) (c) seedpod of the Autograph Tree (Hawaii), $n = 9$; (d) turbine wheel in Trondheim, Norway, $n = 21$. Photographs by the authors.

2.3 COMBINED SYMMETRIES

The symmetry plane and the rotation axis are symmetry elements. If a figure has a symmetry element, it is symmetrical. If it has no symmetry element, it is asymmetrical. Even an asymmetrical figure has a onefold rotation axis, or, actually, an infinity of onefold rotation axes.

The application of a symmetry element is a *symmetry operation*. The symmetry elements are the *symmetry operators*. The consequence of a symmetry operation is a *symmetry transformation*. Strict definitions refer to geometrical symmetry and will serve us as guidelines only. They will be followed qualitatively in our discussion of primarily nongeometrical symmetries, according to the ideas presented in Chapter 1.

So far, symmetries with *either* a symmetry plane *or* a rotation axis have been discussed. These symmetry elements may also be combined. The simplest case occurs when the symmetry planes include a rotation axis.

2.3.1 A Rotation Axis with Intersecting Symmetry Planes

The dot between n and m in the label $n{\cdot}m$ indicates that the rotation axis is in the symmetry plane. This combination of a rotation axis and a symmetry plane produces further symmetry planes. Their total number will be n as a consequence of the application of the n-fold rotational symmetry to the symmetry plane. The complete set of symmetry operations of a figure is its symmetry group.

Figure 2-23 shows two flowers. The periwinkle (*Vinca minor*) has fourfold rotational symmetry and no symmetry plane. The Norwegian tulip has threefold rotational symmetry with the axis of rotation in a symmetry plane. The threefold rotation axis will, of course, rotate not only the flower but any other symmetry element, in this case the symmetry plane, as well. The 120°

a

b

Figure 2-23. (a) *Vinca minor* and Norwegian tulip. Photographs by the authors; (b) Stone carving along Via Appia Antica in Rome. Photograph by the authors.

rotations will generate altogether three symmetry planes, and these planes will make an angle of 60° with each other. The lower part of Figure 2-23 shows an ancient stone carving along Via Appia Antica in Rome depicting two flowers with the same symmetries as the Vinca minor and the Norwegian tulip.

Some primitive organisms are shown in Figure 2-24, after Häckel [2-13]. They all have fivefold rotation axes, and some of them have intersecting (vertical) symmetry planes as well. The symmetry class of the starfish in the

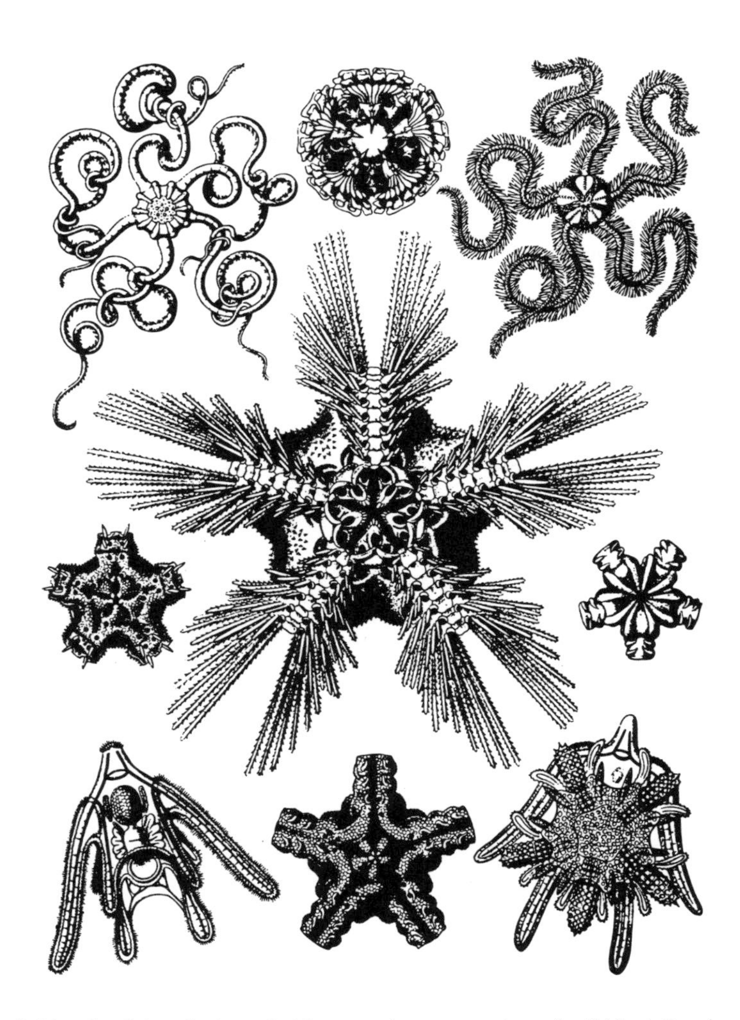

Figure 2-24. Starfish and other primitive organisms possessing a fivefold rotational symmetry axis, which may or may not have symmetry planes intersecting it. From Häckel [2-13].

middle, for example, is $5 \cdot m$. This starfish consists of ten congruent parts, with each pair related by a symmetry plane. The whole starfish is unchanged either by $360°/5 = 72°$ rotation around the rotation axis or by mirror reflection through the symmetry planes which intersect at an angle of 36°. Fivefold symmetry with fivefold rotation and coinciding mirror reflection is quite common among fruits and flowers. On the other hand, this symmetry is conspicuously absent in the world of crystals as will be discussed in more detail later.

Examples of $n \cdot m$ symmetries are shown in Figure 2-25. It is a much

Figure 2-25. Examples of $n \cdot m$ symmetries: $n = 3$ Hawaiian flower (a); $n = 4$ Eiffel Tower, Paris, from below (b); $n = 5$, Pentagonal star (c); and Hawaiian flower (d); Photographs by the authors. (*Continued on next page*)

Figure 2-25. (*Continued*) $n = 6$, Korean beam-end decoration (e); $n = 8$, street lamp (f); $n > 10$, rosetta on the Notre Dame Cathedral, Paris (g); and cupola of the Hungarian Parliament, Budapest (h). Photographs by the authors.

favored symmetry for builders of important buildings, as demonstrated by the cupolas of churches and state houses, etc.

2.3.2 A Rotation Axis with Intersecting Symmetry Planes and a Perpendicular Symmetry Plane

The combination of symmetries considered in this section is labeled $m{\cdot}n{:}m$, and it is characteristic of highly symmetrical objects. Accordingly, their shapes are relatively simple. As seen in Figure 2-26, some fundamental shapes have $m{\cdot}n{:}m$ symmetries. Examples include the square prism, $m{\cdot}4{:}m$, the pentagonal prism, $m{\cdot}5{:}m$, the trigonal bipyramid, $m{\cdot}3{:}m$, the square bipyramid, $m{\cdot}4{:}m$, and the bicone, the cylinder, and the ellipsoid, all having $m{\cdot}\infty{:}m$ symmetry. One of the most beautiful and most common examples of this symmetry is the $m{\cdot}6{:}m$ symmetry of snowflakes.

2.3.3 Snowflakes

The magnificent hexagonal symmetry of snow crystals, the virtually endless variety of their shapes, and their natural beauty make them outstanding examples of symmetry. The fascination in the shape and symmetry of snowflakes goes far beyond the scientific interest in their formation, variety, and properties. The morphology of snowflakes is determined by their internal structures and the external conditions of their formation. The mechanism of snowflake formation has been the subject of considerable research efforts. It is well known that the internal hexagonal arrangement of water molecules

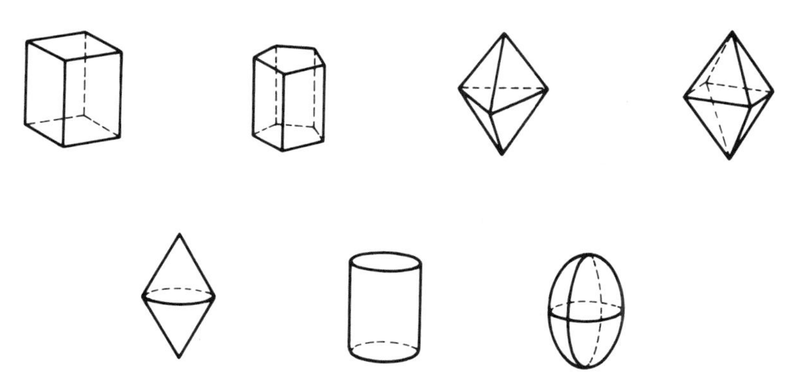

Figure 2-26. Examples of $m{\cdot}n{:}m$ symmetry: prisms, bipyramids, bicone, cylinder, and ellipsoid.

produced by hydrogen bonds is responsible for the hexagonal symmetry of snowflakes. However, this does not explain the countless number of different shapes of snowflakes and, furthermore, why even the smallest variations from the basic underlying shape of a snowflake are repeated in all six directions. Perfection and diversity of shape are illustrated by Figure 2-27.

As the really puzzling questions concerning snowflakes are related to their morphology rather than to their internal structures, these questions will be discussed at some length in the present section. The process of solidification of fluids into crystals has been simulated using mathematical models. The investigation of the relative stability of various shapes is especially rewarding [2-15]. These simulations showed that crystals with sharp tips grew rapidly and had high stability, while crystals with fat shapes grew slowly and were less stable. However, when these slowly growing shapes were slightly perturbed, they tended to split into sharp, rapidly growing tips. This observation led to the hypothesis of the so-called *points of marginal stability*.

According to Langer's marginal stability model [2-15], the snow crystal may start with a relatively stable shape. The crystal may, however, be easily destabilized by a small perturbation. A rapid process of crystallization from the surrounding water vapor ensues. The rapid growth gradually transforms the

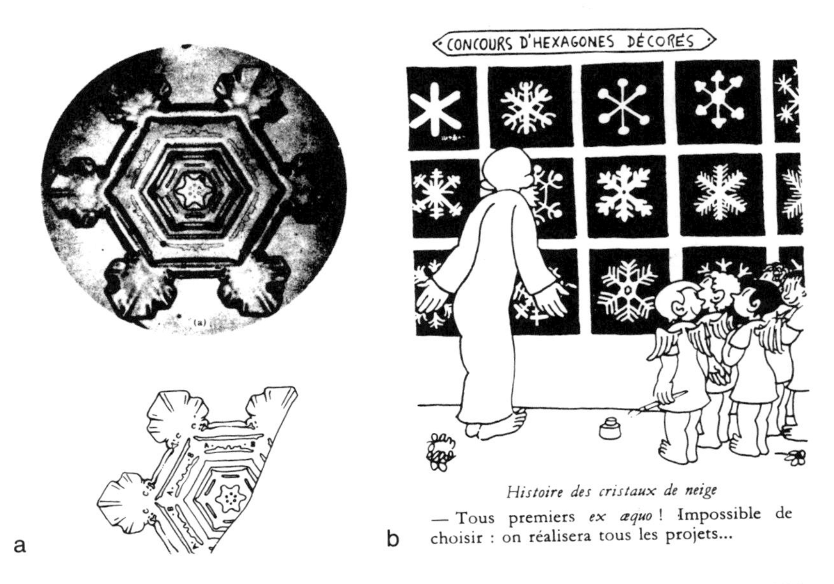

Figure 2-27. Perfection and diversity of shape: (a) photomicrograph and sketch from Nakaya [2-14]; (b) creation of the great variety of snowflake shapes. From Jean Effel, *La Création du Monde*. Reproduced by permission; Copyright Mme Jean Effel and Agence Hoffman, Paris.

crystal into another semistable shape. A subsequent perturbation may then occur, resulting again in a new direction of growth with a different rate. The marginal stability of the snowflake makes the growing crystal very sensitive to even slight changes in its microenvironment.

The uniqueness of snowflakes may be related to the marginal stability. The ice starts crystallizing in a flat sixfold pattern of water crystals so that it is growing in six equivalent directions. As the ice is quickly solidifying, latent heat is released and flows between the growing six bulges. The released latent heat retards the growth in the areas between these bulges. This model accounts for the dendritic or treelike growth. Both the minute differences in the conditions of two growing crystals and their marginal stability make them develop differently: "Something that is almost unstable, will be very susceptible to changes, and will respond in a large way to a small force" [2-15]. At each step of growth, slightly new microenvironmental conditions are encountered, causing new variations in the branches. However, it is assumed that each of the six branches will encounter exactly the same microenvironmental conditions, hence their almost exact likeness.

The marginal stability model is attractive in its explanation of the great variety of snowflake shapes. It is somewhat less convincing in explaining the repetitiveness of the minute variations in all six directions since the microenvironmental changes may occur also across the snowflakes themselves and not only between the spaces assigned to different snowflakes.

In order to explain the morphological symmetry of the dendritic snow crystals, McLachlan [2-16] suggested a mechanism about three decades ago which has not yet been seriously challenged. This author posed the very question already mentioned above: "How does one branch of the crystal know what the other branches are doing during growth?" McLachlan noted that the kind of regularity encountered among snowflakes is not uncommon among flowers and blossoms or among sea animals, in which hormones and nerves coordinate the development of the *living* organisms.

McLachlan's explanation [2-16] for the coordination of the growth among the six branches of a snow crystal is based on the existence of thermal and acoustical standing waves in the crystal. As the snowflake grows by deposition of water molecules upon a small nucleus, it undergoes thermal vibrations at temperatures between 250 and 273 K. The water molecules strike and bounce off the nucleus, and those which stay add to the growth. Branching occurs at points with high concentration of water molecules. If the starting ice nucleus has the hexagonal shape shown in Figure 2-28a and the conditions favor dendritic growth, then the six corners would be receiving more molecules and would be releasing more heat of crystallization than the flat portions. The dendritic development evolving from this situation is shown in Figure 2-28b. The next stage in the development of a snowflake is the production of a new set

Figure 2-28. (a)–(f) McLachlan's [2-16] representation of the coordinated growth of the six branches of a snowflake based on his standing wave theory. The original photographs were from Bentley's collection [2-17].

of equally spaced dendritic branches determined by the modes of vibration along the spines of the flake. The long spines of Figure 2-28c are thought to be particular molecular arrays which correspond to the ice structure. The molecules are vibrating, and the energy distribution between the modes of vibration is influenced by the boundary conditions. When one of the spines becomes "heavily loaded" at some point, then nodes are induced along this spine. These nodes will eject dendritic branches that are equally spaced as indicated in Figure 2-28d–f. The question of how the standing waves in one of the six branches are coupled with those in the other branches is answered by considering the torque about an axis through the intersection point. This torque transmits the same frequencies and induces the same nodes in all the branches. Thus, McLachlan asserts that the dendritic development is identical in all six branches and is independent of the particular branch in which the change in the conditions occurred.

During the past decade intensive research has continued into the mechanism of snowflake formation (see, e.g., Refs. [2-18]–[2-22]). This research encompasses the broader question of dendritic crystal growth. New approaches, such as fractal models, and copious use of computer simulation have greatly facilitated these attempts. However, these investigations assume two-dimensional dendrites whereas actual ice dendrites have three-dimensional patterns. As Furukawa and Shimada [2-21] noted, it cannot be concluded "that a full understanding of the pattern formation of ice crystals has been established." It is truly intriguing, in the words of Kobayashi [2-22], "how such complex patterns can be formed by systems which seem to be too simple to yield them." It is also fascinating how dendritic growth penetrates even chemical synthetic work, witnessed by the development of *dendrimer chemistry* of ever-increasing complexity [2-23] and illustrated by Figure 2-29.

Returning to the snowflakes, an eloquent description of their beauty and symmetry is given by Thomas Mann in *The Magic Mountain* [2-1]:

> Indeed, the little soundless flakes were coming down more quickly as he stood. Hans Castorp put out his arm and let some of them to rest on his sleeve; he viewed them with the knowing eye of the nature-lover. They looked mere shapeless morsels; but he had more than once had their like under his good lens, and was aware of the exquisite precision of form displayed by these little jewels, insignia, orders, agraffes—no jeweller, however skilled, could do finer more minute work. Yes, he thought, there was a difference, after all, between this light, soft, white powder he trod with his skis, that weighed down the trees, and covered the open spaces, a difference between it and the sand on the beaches at home, to which he had likened it. For this powder was not made of tiny grains of stone; but of myriads of tiniest drops of water which in freezing had darted together in symmetrical variation—parts, then, of the same inorganic substance which was the source of protoplasm, of plant life, of the human body. And among these myriads of enchanting little stars, in their hidden splendour that was too small for man's naked eye to see, there was not one like unto another and endless inventiveness governed the development and unthinkable differentiation of one and the same basic scheme, the equilateral, equiangular hexagon. Yet each, in itself—this was the uncanny, the anti-organic, the life-denying character of them all—each of them was absolutely symmetrical, icily regular in form. They were too regular, as substance adapted to life never was to this degree—the

living principle shuddered at this perfect precision, found it deathly, the very marrow of death—Hans Castorp felt he understood now the reason why the builders of antiquity purposely and secretly introduced minute variation from absolute symmetry in their columnar structures.

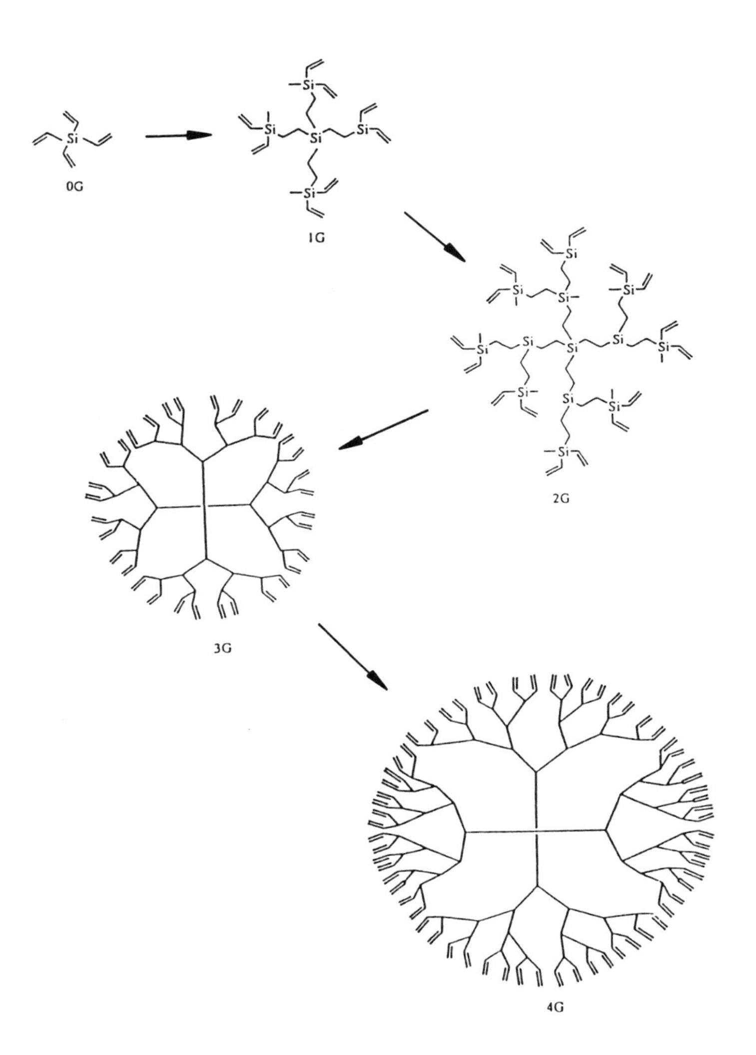

Figure 2-29. Examples of dendrimer chemistry, after Tomalia and Durst [2-23]. Used by permission. Copyright (1993) Springer-Verlag.

The coldness and lifelessness of too much symmetry is as beautifully expressed by Thomas Mann as the beauty of the hexagonal symmetry of the snow crystal. Michael Polányi [2-24] remarked that an environment that was perfectly ordered was not a suitable human habitat. Crystallographers Fedorov and Bernal simply stated, "Crystallization is death" [2-25].

Human interest in snowflakes has a long history. The oldest known recorded statement on snowflake forms dates back to the second century B.C. and comes from China according to Needham and Lu Gwei-Djen [2-26]. "Flowers of plants and trees are generally five-pointed, but those of snow are always six-pointed" . . . was stated as early as 135 B.C. Six was a symbolic number for water in many classical Chinese writings. The contrast between five-pointed plant structures and six-pointed snowflakes became a literary commonplace in subsequent centuries. Of several other relevant citations collected by Needham and Lu Gwei-Djen [2-26], another is reproduced here, from a statement by a physician from 1189:

> The reason why double-kernelled peaches and apricots are harmful to people is that the flowers of these trees are properly speaking five-petalled yet if they develop with sixfold (symmetry), twinning will occur. Plants and trees all have the fivefold pattern; only the yellow-berry and snowflake crystals are hexagonal. This is one of the principles of Yin and Yang. So if double-kernelled peaches and apricots with an (aberrant) sixfold (symmetry) are harmful, it is because these trees have lost their standard rule.

The examination of snowflake shapes and their comparison with other shapes has apparently been a great achievement in East Asia. The involvement of yin and yang amply demonstrates how much importance was given to these studies. As a forerunner of the modern investigations of the correlation between snowflake shapes and environmental (i.e., meteorological) conditions, the following passage from the thirteenth century is cited [2-26]:

> The Yin embracing the Yang gives hail, the Yang embracing the Yin gives sleet. When snow gets six-pointedness, it becomes snow crystals. When hail gets three-pointedness, it becomes solid. This is the sort of difference that arises from Yin and Yang.

The first known sketches of snowflakes from Europe in the sixteenth century did not reflect their hexagonal shape. Johannes Kepler was the first in Europe to recognize the hexagonal symmetry of the snowflakes as he described it in his Latin tractate entitled *The Six-Cornered Snowflake* [2-2], published in 1611. By this time, Kepler had already discovered the first two laws of planetary motion and thus found the true celestial geometry when he turned his attention to the snowflakes. He considered their perfect form and, for the first time, sought the origin of shape and symmetry in the internal structure. The relationship between crystal habit and the internal structure will be discussed in the chapter on crystals (Chapter 9).

Descartes observed and recorded the shapes of snow crystals. Some of his sketches from 1635 are reproduced in Figure 2-30, after Nakaya [2-14]. As these were the first drawings of hexagonal snowflakes recorded, it was quite an achievement that even rare versions such as those composed of a hexagonal column with plane crystals developed at both ends could be found among them. More such important contributions in this field [2-27], among them Hooke's observations using his microscope, occurred in the seventeenth century. Branching in snow crystals has also been recorded by several investigators. Among the later works, Scoresby's observations and sketches are especially important [2-28]. Figure 2-31 reproduces some of them. Scoresby, who later became an Arctic scientist, made these drawings in his log book in 1806 at the age of 16 while he was on a voyage with his father to the Greenland whale fisheries. A few years after the publication of Scoresby's work (1820), the Japanese Doi communicated a series of excellent sketches, some of which are reproduced in Figure 2-32.

There are two fundamental books containing collections of snowflake pictures available today as a result of photomicrography. Bentley [2-17] devoted his lifetime to taking photomicrographs of snow crystals and collected at least 6000 of them. About half of them appeared in his book coauthored with Humphreys [2-17]. This most well known book on snowflakes is probably unsurpassable. Bentley's photomicrographs have been reproduced innumerable

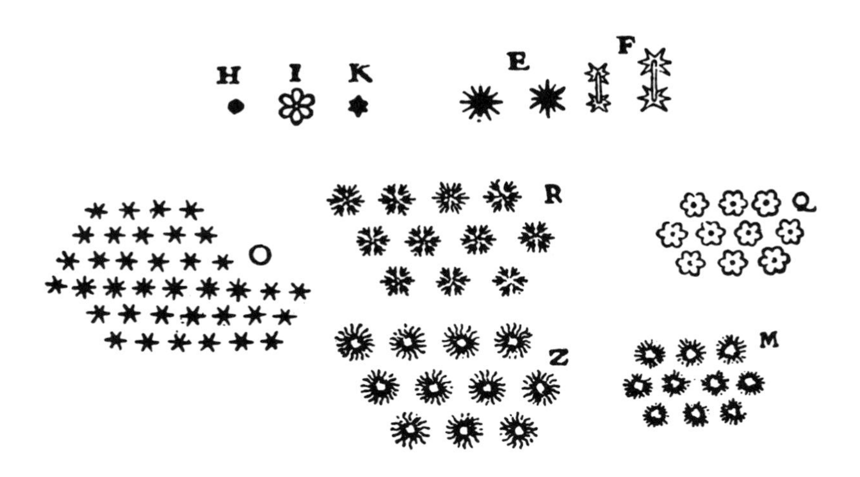

Figure 2-30. Snow crystals by Descartes from 1635 after Nakaya [2-14].

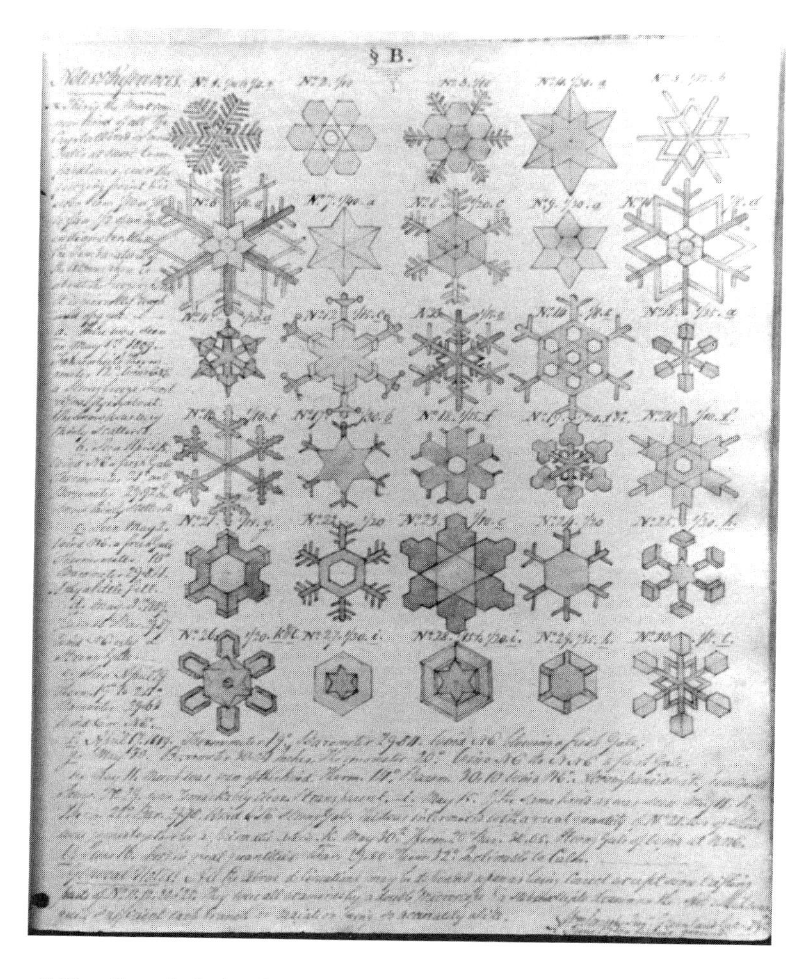

Figure 2-31. Scoresby's sketches of snowflakes from his log book (1806), after Stamp and Stamp [2-28]. Reproduced by permission.

times in various places—sometimes without indicating the source. Some characteristic examples of snowflakes from this collection are shown in Figure 2-33.

The other outstanding contribution is Nakaya's [2-14]. He recorded the naturally occurring snow crystals, classified them, and investigated their mass, speed of fall, electrical properties, frequency of occurrence, and so on. In addition, Nakaya and co-workers developed methods of producing snow

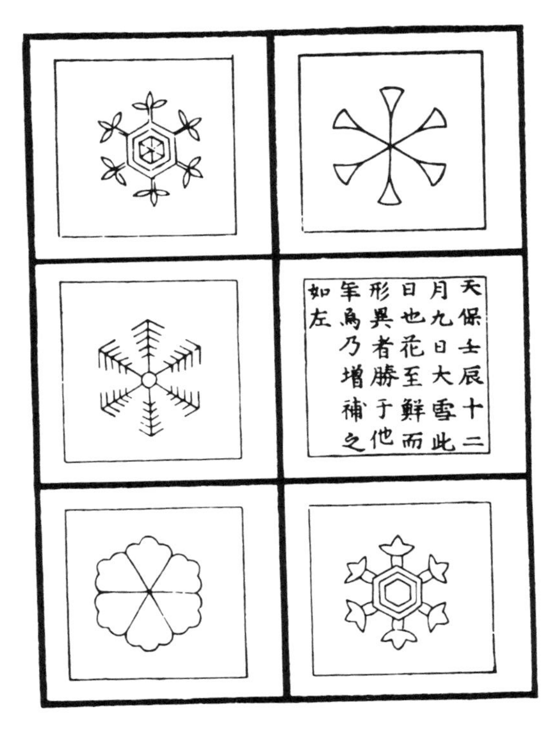

Figure 2-32. Snow crystals from Sekka Zusetsu of Doi (from 1832), after Nakaya [2-14].

crystals artificially. There is a statue (Figure 2-34) on the campus of Hokkaido University in Sapporo honoring Nakaya and commemorating the first artificial snowflake in 1936. Nakaya and co-workers succeeded in determining the conditions of formation of all different types of snowflakes.

The major part of the general classification of snow crystals by Nakaya is given in Figure 2-35 and Table 2-1. The hexagonal plane crystals are the most common and the best known.

Nakaya made important contributions to observing not only the perfect or near perfect symmetries of the snow crystals but also distortions from hexagonal symmetry. Of course, the atomic arrangement is always hexagonal, but the morphology or crystal habit may be less than perfectly regular hexagonal. Nakaya called such crystals *malformed* and stated that these asymmetric crystals may be more common than the symmetric ones. Of course, the question of symmetry is a matter of degree. Even the snowflakes which are

Figure 2-33. Snowflake photomicrographs by Bentley, after Bentley and Humphreys [2-17].

Figure 2-34. Statue, commemorating the production of the first artificial snowflake by Nakaya, on the campus of Hokkaido University, Sapporo, Japan. Photograph by the authors.

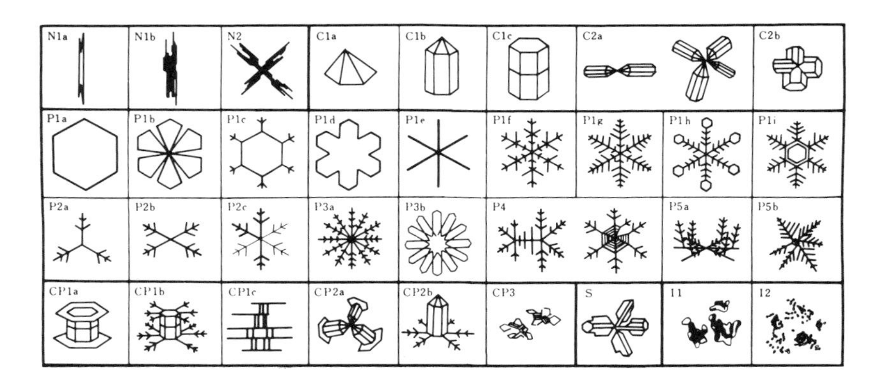

Figure 2-35. From Nakaya's general classification of snow crystals [2-14].

Table 2-1. Part of Nakaya's General Classification of Snow Crystals[a]

Main groups	Subgroups	Types
Needle (N)	1. Simple	a. Elementary needle
		b. Bundle of needles
	2. Combination	
Columnar (C)	1. Simple	a. Pyramid
		b. Bullet
		c. Hexagonal
	2. Combination	a. Bullets
		b. Columns
Plane (P)	1. Regular, developed in plane	a. Simple plate
		b. Branches in sector form
		c. Plate with simple extensions
		d. Broad branches
		e. Simple stellar form
		f. Ordinary dendritic form
		g. Fernlike
		h. Stellar form with plates at ends
		i. Plate with dendritic extensions
	2. Irregular number of branches	a. Three-branched
		b. Four-branched
		c. Others
	3. Twelve branches	a. Fernlike
		b. Broad branches
	4. Malformed	Many varieties
	5. Spatial assemblage of plane branches	a. Spatial hexagonal
		b. Radiating
Column/plane combinations (CP)	1. Column with plane at both ends	a. Column with plates
		b. Column with dendrites
	2. Bullets with plates	a. Bullets with plates
		b. Bullets with dendrites
	3. Irregular	
Columnar with extended side planes (S)		
Irregular snow particles (I)	1. Ice	
	2. Rimed	
	3. Miscellaneous	

[a]After Nakaya [2-14]; cf. Figure 2-35.

considered to be the most symmetrical may reveal slight differences in their branches when examined closely.

2.4 INVERSION

What is the symmetry of the 1,2-dibromo-1,2-dichloroethane molecule, as shown in Figure 2-36? There is obviously no symmetry plane and no rotation axis. However, any two atoms of the same kind are related by a line connecting them and going through the midpoint of the central bond. This midpoint is the only symmetry element of this molecule, and it is called the symmetry center or *inversion point*. The application of this symmetry element interchanges the atoms, or more generally, any two points located at the same distance from the center along the line going through the center. This interchange is called *inversion*. The notation of inversion symmetry is i.

An inversion may also be represented as the consecutive application of two simple symmetry elements, namely, a twofold rotation and mirror reflec-

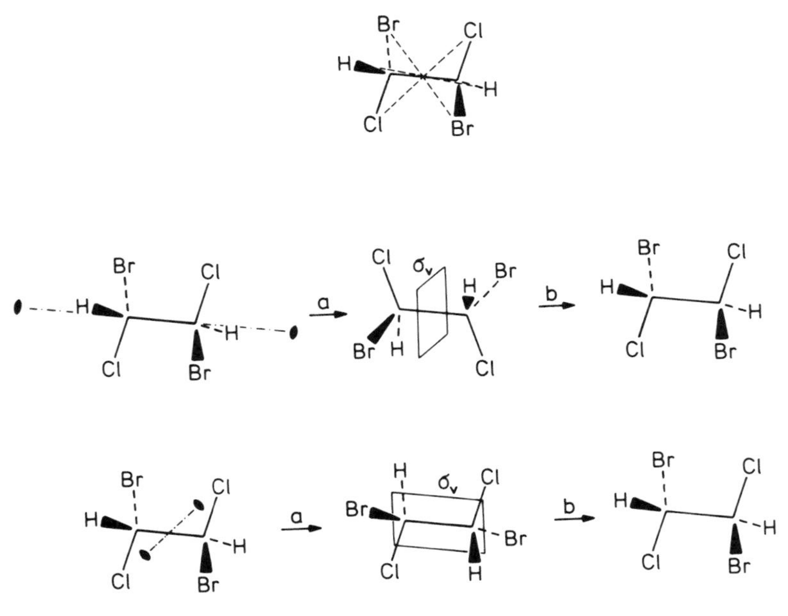

Figure 2-36. The 1,2-dibromo-1,2-dichloroethane molecule. Its center of symmetry is the midpoint of the C–C bond. An inversion is equivalent to the consecutive application of a twofold rotation axis and a reflection.

tion, or vice versa. For the molecule of Figure 2-36, this could be described, for example, in the following way: (a) rotate the molecule by 180° about the C–C bond as the rotation axis and (b) apply a symmetry plane perpendicular to and bisecting the C–C bond; or (a) apply a twofold rotation axis perpendicular to the ClCCCl plane and going through the midpoint of the C–C bond and then (b) apply a mirror plane coinciding with the ClCCCl plane. These operations are indicated in Figure 2-36, and in both examples the results are invariant to the order in which the two operations are performed.

The parallelepiped of Figure 2-37 is a typical example of an object possessing a center of symmetry. Each apex, edge, and face has its corresponding one through the inversion center. If there is any direction of a line or a segment of a face, the symmetry center will invert that direction, and the counterpart line or face is obtained.

The sphere is a highly symmetrical object which possesses a center of symmetry. Conjugate locations on the surface of a sphere are related by an inversion through the center of symmetry. The geographical consequences of such an inversion are emphasized in a newspaper article on New Zealand by James Reston in his "Letter from Wellington. Search for End of the Rainbow" [2-29]:

> Nothing is quite the same here. Summer is from December to March. It is warmer in the North Island and colder in the South Island. The people drive on the left rather than on the right. Even the sky is different—dark blue velvet with stars of the Southern Cross—and the fish love the hooks

Madrid, Spain, corresponds approximately to Wellington, New Zealand, by inversion.

The notation of the symmetry center or inversion center is $\bar{1}$ while the corresponding combined application of twofold rotation and mirror reflection

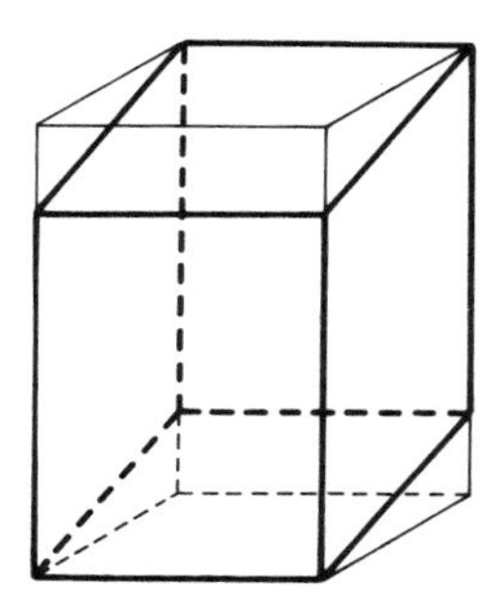

Figure 2-37. Parallelepiped: Illustration of an object with a center of symmetry.

may also be considered to be just one symmetry transformation. The symmetry element is called a mirror-rotation symmetry axis of the second order, or twofold mirror-rotation symmetry axis, and it is labeled $\tilde{2}$. Thus $\bar{1} \equiv \tilde{2}$.

The twofold mirror-rotation axis is the simplest among the mirror-rotation axes. The object shown in Figure 2-38a has a fourfold mirror-rotation axis. It was prepared from a square shape with an obliquely inscribed square. The emerging corners are bent alternately up and down. The object obtained in this way has a twofold rotation axis perpendicular to the square plane and intersecting its midpoint. Moreover, a 90° rotation about the rotation axis plus a reflection through the square plane also brings the object into coincidence with itself. This combined operation is determined by a fourfold mirror-rotation axis, labeled $\tilde{4}$. Generally speaking, a $2n$-fold mirror-rotation axis consists of the following operations: a rotation by $(360/2n)°$ and a reflection through the plane perpendicular to the rotation axis. Another example, a sixfold mirror-rotation axis, $\tilde{6}$, is shown in Figure 2-38b. It should be noted that only mirror-rotation axes with an even order ($2n$) can be present in the objects shown in Figure 2-38.

The symmetry of the snowflake involves this type of mirror-rotation axis. The snowflake obviously has a center of symmetry. The symmetry class $m \cdot 6{:}m$ contains a center of symmetry at the intersection of the sixfold rotation axis and the perpendicular symmetry plane. In general, for all *$m \cdot n{:}m$ symmetry classes with n* even, the point of intersection of the n-fold rotation axis and the perpendicular symmetry plane is also a center of symmetry. When n is odd in an $m \cdot n{:}m$ symmetry class, however, there is no center of symmetry present.

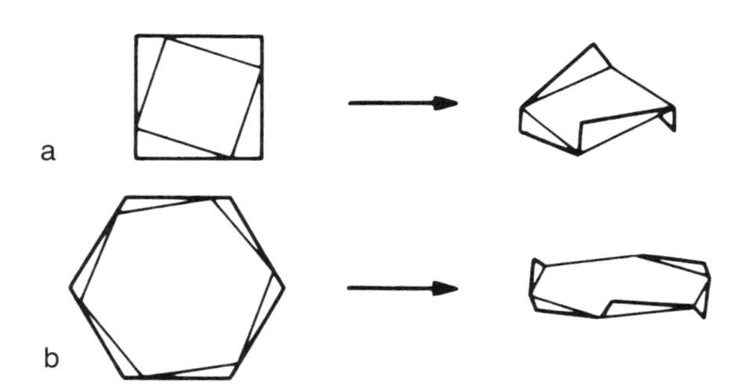

Figure 2-38. (a) Example of fourfold mirror-rotation symmetry. (b) Example of sixfold mirror-rotation symmetry.

2.5 SINGULAR POINT AND TRANSLATIONAL SYMMETRY

The midpoint of a square is unique, there is no other point equivalent to it (Figure 2-39). It is called a singular point. A corner of the same square is not singular. The symmetry transformations of the square reproduce it, and there are altogether four equivalent corner points of the square. An arbitrarily chosen point in a square will have seven other equivalent points because of the symmetry transformations of the square. Altogether there will be eight equivalent points. However, if the chosen point coincides with one of the corners of the square, there will only be four equivalent points. The same argument applies if the point happens to be on one of the symmetry axes of the square. The multiplicity of a corner point of the square or any point on a symmetry axis is two. The product of the number of *equivalent points* and *multiplicity* is constant (viz., eight for the square). Finally, if the chosen point coincides with the midpoint of the square, the number of equivalent points will be one, and the multiplicity will be eight.

In an asymmetric figure, each point is singular and the multiplicity of each point is one.

The symmetry classes characterizing figures or objects which have at least one singular point are called point groups. The center of the circular pattern of the pavement in Figure 2-40a is a singular point. Another pattern is displayed by the pavement in Figure 2-40b, consisting of identical arcs. If it is supposed that this pavement is a fragment of an infinitely large one, there is no

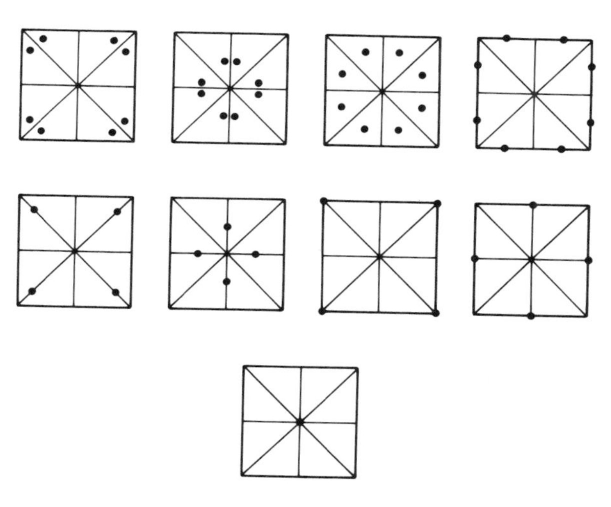

Figure 2-39. The singular point and the multiplicity of points of a square.

Figure 2-40. Italian pavements: (a) The system of concentric circles has point-group symmetry; (b) the pattern of arcs, if extended to infinity, has space-group symmetry. Photographs by the authors.

Table 2-2. Dimensionality (m) and Periodicity (n) of Symmetry Groups G_n^m after Engelhardt[a]

Periodicity / Dimensionality	$n = 0$, no periodicity	$n = 1$, periodicity in one direction	$n = 2$, periodicity in two directions	$n = 3$, periodicity in three directions
$m = 0$, dimensionless	G_0^0			
$m = 1$, one-dimensional	G_0^1	G_1^1		
$m = 2$, two-dimensional	G_0^2	G_1^2	G_2^2	
$m = 3$, three-dimensional	G_0^3	G_1^3	G_2^3	G_3^3

[a]Reference [2-30].

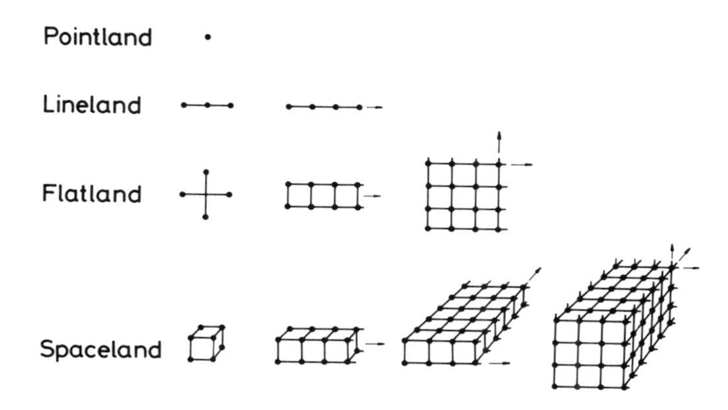

Figure 2-41. Dimensionality and periodicity in point groups and space groups. This figure is consistent with Table 2-2.

singular point in it. Assuming an infinite extent for this pavement pattern is natural because of its periodicity. The absence of a singular point leads to regularity expressed in infinite repetition, which characterizes translational symmetry. This kind of symmetry precludes the presence of singular points though it does not preclude the presence of a singular line or plane. The symmetry classes characterizing entities with translational symmetry are called space groups. One-dimensional space groups describe the symmetries involving infinite repetition or periodicity in one direction, two-dimensional space groups those involving periodicity in two directions, and three-dimensional space groups those involving periodicity in all three directions. Table 2-2 and Figure 2-41 summarize the possible cases in terms of dimensionality and periodicity. The nomenclature is somewhat inconsistent but has some relationship to Abbott's classic *Flatland* [2-31].

2.6 POLARITY

A line is polar if its two directions can be distinguished, and a plane is polar if its two surfaces are not equivalent. This defnition of polarity has, of course, nothing to do with charge separation. A polar line has a "head" and a "tail," and a polar plane has a "front" and a "back." A vertical line on the surface of the Earth is polar with respect to gravity, and a sheet of paper with one of its sides painted is polar with respect to its color.

An axis is polar if its two ends are not brought into coincidence by the symmetry transformations of the symmetry group of its figure. An analogous definition applies to the two sides of a polar plane.

If a symmetry group includes a center of symmetry, polarity is excluded. It has already been seen (cf., e.g., Figure 2-37) that in a centrosymmetric figure a directed line or segment of a face changes direction by inversion. In the case of the absence of a center of symmetry, there will be at least one directed line or face which is not accompanied by parallel counterparts reversed in direction.

The significance of polar axes can be demonstrated, for example, in crystal morphology. Curtin and Paul [2-32] have summarized the chemical consequences of the polar axis in organic crystal chemistry. A few examples will be mentioned here following Curtin and Paul. Figure 2-42a shows two centrosymmetric acetanilide crystals. The faces occur in parallel pairs in both habits. On the other hand, the *p*-chloroacetanilide crystal shown in Figure 2-42b is noncentrosymmetric, and some of the faces occur without parallel ones at the opposite end of the crystal. This crystal has a polar axis parallel to its long direction.

The morphological symmetry differences between the acetanilide and *p*-chloroacetanilide crystals originate from their internal structures. The acet-

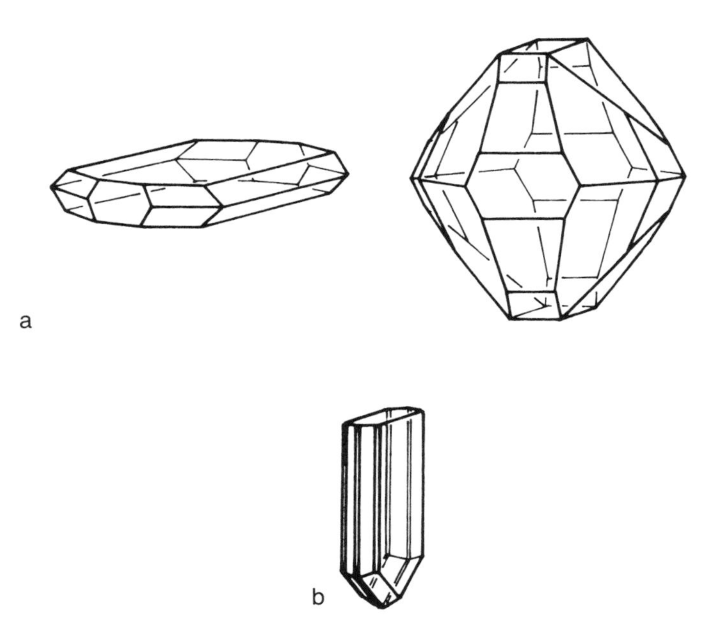

Figure 2-42. Crystals from Groth's *Chemische Kristallographie* [2-33]: (a) Centrosymmetri rhombic bipyramidal acetanilide; (b) noncentrosymmetric rhombic pyramidal *p*-chloro acetanilide.

anilide molecules appear in pairs, and the two molecules in each pair are related by an inversion center (Figure 2-43). On the other hand, the p-chloroacetanilide molecules are all aligned in one direction. The molecular arrangements in the crystal are shown in Figures 2-44a and b.

Even very simple structures may form polar crystals. For example, in a polar crystal composed of diatomic molecules AB, the molecular axis will be oriented more along the polar direction of the crystal than perpendicular to it. Furthermore, as there is an ABAB . . . array in the crystal, it is required that the spacings between the atom A and the two adjacent atoms B be unequal in order to have a polar axis present:

A B A B A B . . .

Curtin and Paul characterized this situation from the point of view of a submicroscopic traveler proceeding along this array of atoms. The observer is able to determine the direction of travel thanks to the difference in spacings. The distance is always longer from atom B to atom A and shorter from atom A to the next atom B in one direction whereas the reverse is true in the opposite direction.

It is not required that a molecule possess a large dipole moment in order to be suitable for building polar crystal habits. Curtin and Paul cite the nearly "nonpolar" 1-*tert*-butyl-4-methylbenzene molecule, which crystallizes in a polar habit with one end of the crystal being formed by the methyl groups and the other end by the *tert*-butyl groups. It is not fully understood why some classes of substances prefer to form polar crystals while others with similar potentials do not. Aromatic compounds with certain functional groups (e.g., an amino group) more often form polar crystals than do such compounds with other groups (e.g., carboxyl group). *Meta*-disubstituted benzene derivatives crystallize more often in a polar habit than do *ortho* and *para* derivatives. Sometimes, the molecular polar axis is oriented almost perpendicular to the crystal polar axis, and only a small component of the molecular polarity contributes to the crystal polarity.

Crystal polarity may have important consequences for the chemical

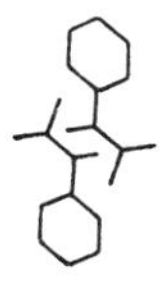

Figure 2-43. Two acetanilide molecules related by inversion.

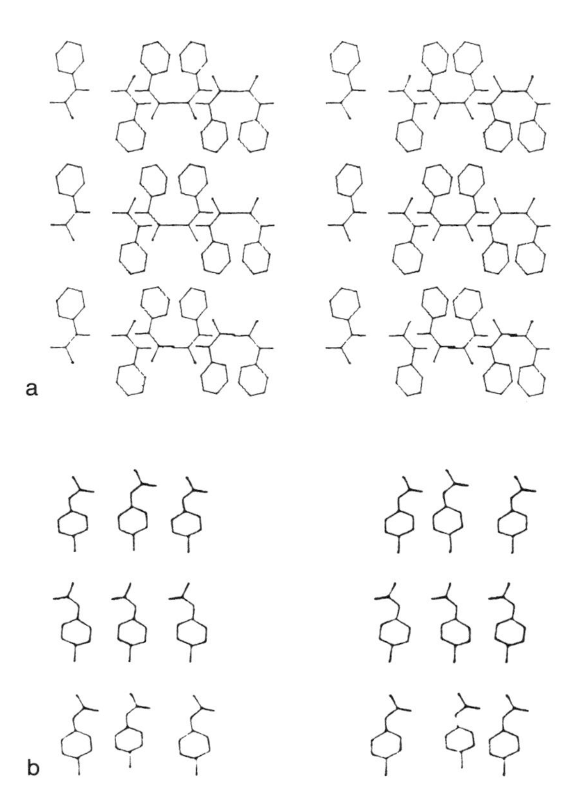

Figure 2-44. (a) The centrosymmetric arrangement of acetanilide molecules in the crystal, resulting in a centrosymmetric crystal habit. Reprinted with permission from Ref. [2-32]. Copyright (1981) Americal Chemical Society; (b) The head-to-tail alignment of *p*-acetanilide molecules in the crystal, resulting in the occurrence of a polar axis in the crystal habit. Reprinted with permission from Ref. [2-32]. Copyright (1981) American Chemical Society.

behavior. In solid/gas reactions, for example, crystal polarity may be a source of considerable anisotropy.

There are also important physical properties characterizing polar crystals, such as pyroelectricity and piezoelectricity and others [2-34]. The primitive cell of a pyroelectric crystal possesses a dipole moment. The separation of the centers of the positive and negative charges changes upon heating. In this process the two charges migrate to the two ends of the polar axis. Piezoelectricity is the separation of the positive and negative charges upon expansion/compression of the crystal. Both pyroelectricity and piezoelectricity have practical uses.

2.7 CHIRALITY

There are many objects, both animate and inanimate, which have no symmetry planes but which occur in pairs related by a symmetry plane and whose mirror images cannot be superposed. Figure 2-45 shows a building decoration, a detail from Bach's *The Art of the Fugue*, a pair of molecules, and a pair of crystals. The simplest chiral molecules are those in which a carbon atom is surrounded by four different ligands—atoms or groups of atoms—at the vertices of a tetrahedron. All the naturally occurring amino acids are chiral, except glycine.

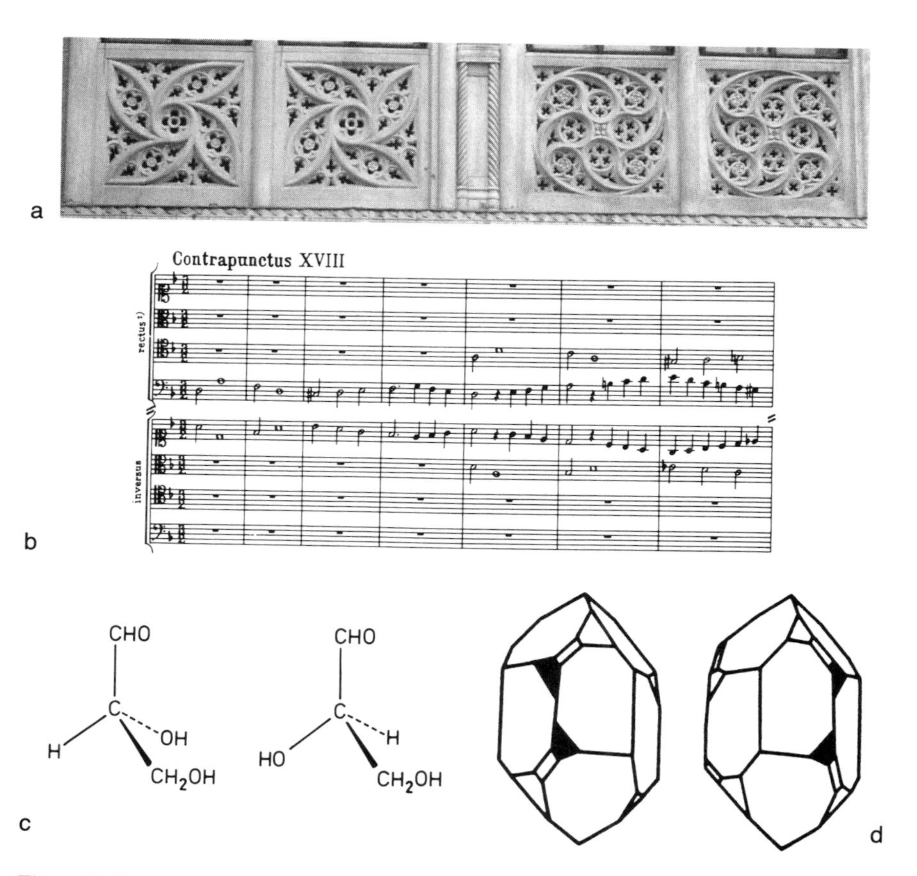

Figure 2-45. Illustrations of chiral pairs: (a) Decorations whose motifs (of fourfold rotational symmetry) are each other's mirror images; photographs by the authors; (b) J. S. Bach, *Die Kunst der Fuge, Contrapunctus XVIII*, detail; (c) glyceraldehyde molecules; (d) quartz crystals.

W. H. Thomson, Lord Kelvin, wrote [2-35]: "I call any geometrical figure or group of points 'chiral,' and say it has chirality, if its image in a plane mirror, ideally realized, cannot be brought into coincidence with itself." He called forms of the same sense *homochiral* and forms of the opposite sense *heterochiral*. The most common example of a heterochiral form is hands. Indeed, the word chirality itself comes from the Greek word for hand. Figures 2-46 and 2-47 show some heterochiral and homochiral pairs of hands.

A chiral object and its mirror image are enantiomorphous, and they are

Figure 2-46. Heterochiral pairs of hands: (a) Tombstone in the Jewish cemetery, Prague; photograph by the authors; (b) Albrecht Dürer's *Praying Hands* on the cover of the German magazine *Der Spiegel*, June 15, 1992; reproduced by permission; (c) Buddha in Tokyo; photograph by the authors; (d) United Nations stamp. (*Continued on next page*)

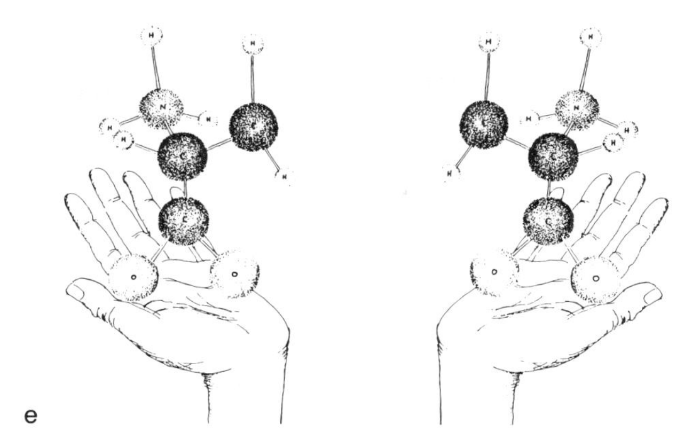

Figure 2-46. (*Continued*) (e) heterochiral pair of hands and models of a heterochiral pair of amino acid molecules [2-36]; reproduced by permission from R. N. Bracewell.

Figure 2-47. Homochiral pairs of hands: (a) Cover of the German magazine *Der Spiegel*, May 18, 1992; reproduced by permission; (b) U.S. stamp; (c) logo with SOS distress sign at a Swiss railway station; photograph by the authors.

Figure 2-48. Louis Pasteur's bust in front of the Pasteur Institute, Paris. Photograph by the authors.

each other's enantiomorphs. Louis Pasteur (Figure 2-48) first suggested that molecules can be chiral. In his famous experiment in 1848, he recrystallized a salt of tartaric acid and obtained two kinds of small crystals which were mirror images of each other, as shown by Pasteur's models in Figure 2-49. The two kinds of crystals had the same chemical composition but differed in their optical activity. One was levo-active (L), and the other was dextro-active (D). Since the true absolute configuration of molecules could not be determined at the time, an arbitrary convention was applied, which, luckily, proved to coincide with reality. If a molecule or a crystal is chiral, it is necessarily optically active. The converse is, however, not true. There are, in fact

Figure 2-49. Pasteur's models of enantiomeric crystals in the Pasteur Institute, Paris. Photographs by the authors.

nonenantiomorphous symmetry classes of crystals which may exhibit optical activity.

Whyte [2-37] extended the definition of chirality as follows: "Three-dimensional forms (point arrangements, structures, displacements, and other processes) which possess non-superposable mirror images are called 'chiral'." A chiral process consists of successive states, all of which are chiral. The two main classes of chiral forms are screws and skews. Screws may be conical or cylindrical and are ordered with respect to a line. Examples of the latter are the left-handed and right-handed helices in Figure 2-50. The skews, on the other hand, are ordered around their center. Examples are chiral molecules having point-group symmetry.

From the point of view of molecules, or crystals, left and right are intrinsically equivalent. An interesting overview of the left/right problem in science has been given by Gardner [2-39]. Distinguishing between left and right has also considerable social, political, and psychological connotations. For example, left-handedness in children is viewed with varying degrees of tolerance in different parts of the world. Figure 2-51a shows a classroom at the University of Connecticut with different (homochiral and heterochiral) chairs

Figure 2-50. Left-handed and right-handed helix decorations from Zagorsk, Russia [2-38]. Photograph by the authors.

a

b

Figure 2-51. Classrooms with heterochiral and homochiral chairs: (a) Chairs for both the right-handed *and* left-handed students; (b) Older chairs, all for right-handed students only. Photographs by the authors.

to accommodate both the right-handed and the left-handed students. Older classrooms at the same university have chairs for the right-handed only (Figure 2-51b).

2.7.1 Asymmetry and Dissymmetry

Symmetry operations of the first kind and of the second kind are sometimes distinguished in the literature (cf. Ref. [2-40]). Operations of the first kind are sometimes also called even-numbered operations. For example, the identity operation is equivalent to two consecutive reflections from a symmetry plane. It is an even-numbered operation, an operation of the first kind. Simple rotations are also operations of the first kind. Mirror rotation leads to figures consisting of right-handed and left-handed components and therefore is an operation of the second kind. Simple reflection is also an operation of the second kind as it may be considered as a mirror rotation about a onefold axis. A simple reflection is related to the existence of two enantiomorphic components in a figure. Figure 2-52 illustrates these distinctions by a series of simple sketches after Shubnikov [2-40]. In accordance with the above description, chirality is sometimes defined as the absence of symmetry elements of the second kind.

Sometimes, the terms asymmetry, dissymmetry, and antisymmetry are confused in the literature although the scientific meaning of these terms is in complete conformity with the etymology of these words. Asymmetry means the complete absence of symmetry, dissymmetry means the derangement of symmetry, and antisymmetry means the symmetry of opposites (see Section 4.6).

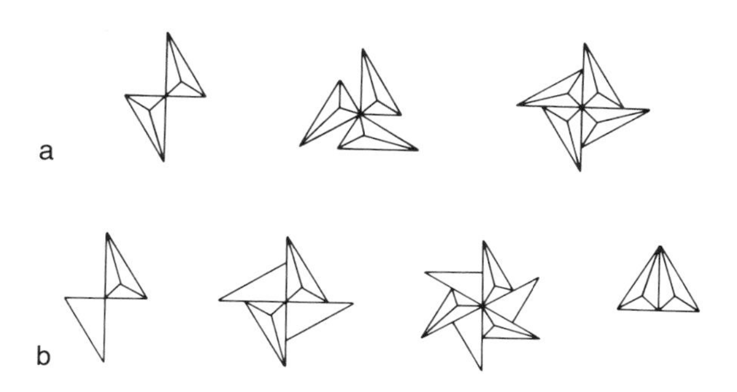

Figure 2-52. Examples of symmetry operations of the first kind (a) and of the second kind (b), after Shubnikov [2-40].

Pasteur used "dissymmetry" for the first time as he designated the absence of elements of symmetry of the second kind in a figure. Accordingly, dissymmetry did not exclude elements of symmetry of the first kind. Pierre Curie suggested an even broader application of this term. He called a crystal dissymmetric in the case of the absence of those elements of symmetry upon which depends the existence of one or another physical property in that crystal. In Pierre Curie's original words [2-41], "Dissymmetry creates the phenomenon." Namely a phenomenon exists and is observable due to dissymmetry, i.e., due to the absence of some symmetry elements from the system. Finally, Shubnikov [2-40] called dissymmetry the *falling out* of one or another element of symmetry from a *given* group. He argued that to speak of the absence of elements of symmetry makes sense only when these symmetry elements are present in some other structures.

Thus, from the point of view of chirality any asymmetric figure is chiral, but asymmetry is not a necessary condition for chirality. All dissymmetric figures are also chiral if dissymmetry means the absence of symmetry elements of the second kind. In this sense, dissymmetry is synonymous with chirality.

An assembly of molecules may be achiral for one of two reasons. Either all the molecules present are achiral or the two kinds of enantiomorphs are present in equal amounts. Chemical reactions between achiral molecules lead to achiral products. Either all product molecules will be achiral or the two kinds of chiral molecules will be produced in equal amounts. Chiral crystals may sometimes be obtained from achiral solutions. When this happens, the two enantiomorphs will be obtained in (roughly) equal numbers, as was observed by Pasteur. Quartz crystals are an inorganic example of chirality (Figure 2-45d). Roughly equal numbers of left-handed and right-handed crystals are obtained from the achiral silica melt.

Incidentally, Pierre Curie's teachings on symmetry are probably not so widely known as they should be, considering their fundamental and general importance. The fact that his works on symmetry were characterized by extreme brevity may have contributed to this. Marie Curie and Aleksei V. Shubnikov [2-42, 2-43] have considerably facilitated the dissemination of Curie's teachings. Our discussion also relies on their works. There is also a critical and fascinating discussion of Pierre Curie's symmetry teachings by Stewart and Golubitsky [2-44].

Pierre Curie's above-quoted statement concerning the role of dissymmetry in "creating" a phenomenon is part of a broader formulation. It states that in every phenomenon there may be elements of symmetry compatible with, though not required by, its existence. What is necessary is that certain elements of symmetry shall not exist. In other words, it is the absence of certain symmetry elements that is a necessary condition for the phenomenon to exist.

Another important statement of Pierre Curie's is that when several phenomena are superposed in the same system, the dissymmetries are added together. As a result, only those symmetry elements which were common to each phenomenon will be characteristic of the system.

Finally, concerning the symmetry relationships of causes and effects, Marie Curie [2-42] formulated the following principles from Pierre Curie's teachings. (1) "When certain causes produce certain effects, the elements of symmetry in the causes ought to reappear in the effects produced"; (2) "When certain effects reveal a certain dissymmetry, this dissymmetry should be apparent in the causes which have given them birth"; however, (3) "The converse of these two statements does not hold . . . [and] the effects produced can be more symmetrical than their causes."

2.7.2 Relevance to Origin of Life

The situation with respect to living organisms is unique. Living organisms contain a large number of chiral constituents, but only L-amino acids are present in proteins and only D-nucleotides are present in nucleic acids. This happens in spite of the fact that the energy of both enantiomorphs is equal and their formation has equal probability in an achiral environment. However, only one of the two occurs in nature, and the particular enantiomorphs involved in life processes are the same in humans, animals, plants, and microorganisms. The origin of this phenomenon is a great puzzle which, according to Prelog [2-45], may be regarded as a problem of molecular theology.

This problem has long fascinated those interested in the molecular basis of the origin of life (e.g., Refs. [2-46], [2-47]). There are in fact two questions. One is why do all the amino acids in a protein have the same L-configuration, or why do all the components of a nucleic acid, that is, all its nucleotides, have the same D-configuration? The other question, the more intriguing one, is why does that particular configuration happen to be L for the amino acids and why does it happen to be D for nucleotides in all living organisms? This second question seems to be impossible to answer satisfactorily at the present time.

According to Prelog [2-45], a possible explanation is that the creation of living matter was an extremely improbable event, which occurred only once. We may then suppose that if there are living forms similar to ours on a distant planet, their molecular structures may be the mirror images of the corresponding molecular structures on the earth. We know of no structural reason at the molecular level for living organisms to prefer one type of chirality. (There may be reasons at the atomic nuclear level. The violation of parity at the nuclear level has already been referred to in Chapter 1.) Of course, once the selection is made, the consequences of this selection must be examined in relation to the first question. The fact remains, however, that chirality is intimately associated with life. This means that at least dissymmetry and possibly asymmetry are basic characteristics of living matter.

Although Pasteur believed that there is a sharp gap between vital and nonliving processes, he attributed the asymmetry of living matter to the asymmetry of the structure of the universe and not to a vital force. Pasteur himself wrote that he was inclined to think that life, as it appears to us, must be a product of the dissymmetry of the universe (see Ref. [2-48]).

Concerning the first question, Orgel [2-46] suggests that we compare the structure of DNA to a spiral staircase. The regular DNA right-handed double helix is composed of D-nucleotides. On the other hand, if a DNA double helix were synthesized from L-nucleotides, it would be left-handed. These two helices can be visualized as right-handed and left-handed spiral staircases, respectively. Both structures can perform useful functions. A DNA double helix containing both D- and L-nucleotides, however, could not form a truly helical structure at all since its handedness would be changing. Just consider the analogous spiral staircase that Orgel suggested as shown in Figure 2-53.

If each component of a complex system is replaced by its mirror image, the mirror image of the original system is obtained. However, if only *some* components of the complex system are replaced by their mirror images, a chaotic system emerges. Chemical systems that are perfect mirror images of each other behave identically, whereas systems in which only some, but not all, components have been replaced by their mirror images have quite different chemical properties. If, for example, a naturally occurring enzyme made up of L-amino acids synthesizes a D-nucleotide, then the corresponding artificial enzyme obtained from D-amino acids would synthesize the L-nucleotide. On the other hand, a corresponding polypeptide containing both D- and L-amino acids would probably lack the enzymic activity.

Recently, the first enzymatically active D-protein has been synthesized

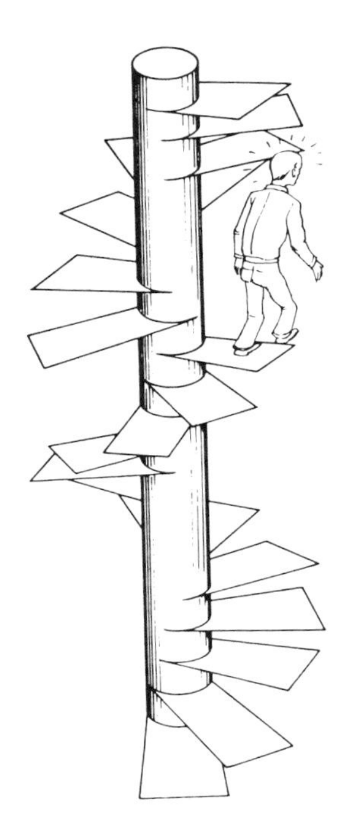

Figure 2-53. A spiral staircase which changes its chirality [2-46]. Reproduced by permission from L. E. Orgel.

[2-49]. There are many potential applications, both therapeutic and nontherapeutic, that may open up with such progress [2-50]. It has been known for some time that the two enantiomers of drugs and pesticides may have vastly different responses in a living organism. Natural products extracted from plants and animals are enantiomerically pure while the synthesized ones are obtained in a 1:1 ratio of the enantiomers. In some cases, the twin of the one exerting the beneficial action is harmless. In other cases, however, the drug molecule has an "evil twin" [2-51]. A tragic example was the thalidomide case, in Europe, in which the right-handed molecule was a sedative and the left-handed one caused birth defects. Other examples include one enantiomer of ethambutol fighting tuberculosis with its evil twin causing blindness, and one enantiomer of naproxen reducing arthritic inflammation with its evil twin poisoning the liver. Bitter and sweet asparagine are represented by structural formulas in Figure 2-54.

Ibuprofen is a lucky case in which the twin of the enantiomer that provides the curing is converted to the beneficial version by the body.

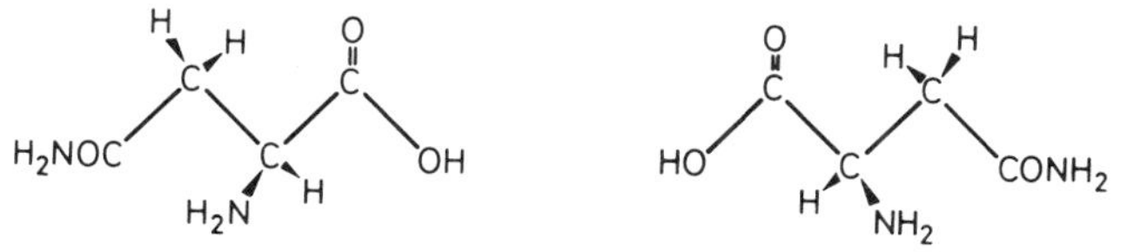

Figure 2-54. Bitter and sweet asparagine represented by structural formulas [2-52].

Even when the twin is harmless, it represents waste and a potential pollutant. Thus, a lot of efforts are directed toward producing enantiomerically pure drugs and pesticides. The techniques of asymmetric synthesis (see, e.g., Refs. [2-53] and [2-54]), based on the strategy of employing chiral catalysts (see, e.g., Ref. [2-55]), are used to this end. One of the fascinating possibilities is to produce sweets from chiral sugars of the enantiomer that would not be capable of contributing to obesity yet would retain the taste of the other enantiomer. Chiral separation and purity is an increasingly important question. Worldwide sales of enantiopure drugs topped $35 billion in 1993 and are expected to reach about $40 billion in 1997 [2-56]. There is a rapidly growing literature on the subject, with even special journals exclusively dedicated to this topic. Production of enantiomerically pure substances has also become a topic in investment reports and the daily press.

2.7.3 *La coupe du roi*

Among the many chemical processes in which chirality/achirality relationships may be important are the fragmentation of some molecules and the reverse process of the association of molecular fragments. Such fragmentation and association can be considered generally and not just for molecules. The usual cases are those in which an achiral object is bisected into achiral or heterochiral halves. On the other hand, if an achiral object can be bisected into two homochiral halves, it cannot be bisected into two heterochiral ones. A relatively simple case is the tessellation of planar achiral figures into achiral, heterochiral, and homochiral segments. Some examples are shown in Figure 2-55. For a detailed discussion, see Ref. [2-7].

Anet *et al.* [2-57] have cited a French parlor trick called *la coupe du roi*—or *the royal section*—in which an apple is bisected into two homochiral halves, as shown in Figure 2-56. An apple can be easily bisected into two achiral halves. On the other hand, it is impossible to bisect an apple into two heterochiral halves. Two heterochiral halves, however, can be obtained from two apples, both cut into two homochiral halves in the opposite sense (see Figure 2-56).

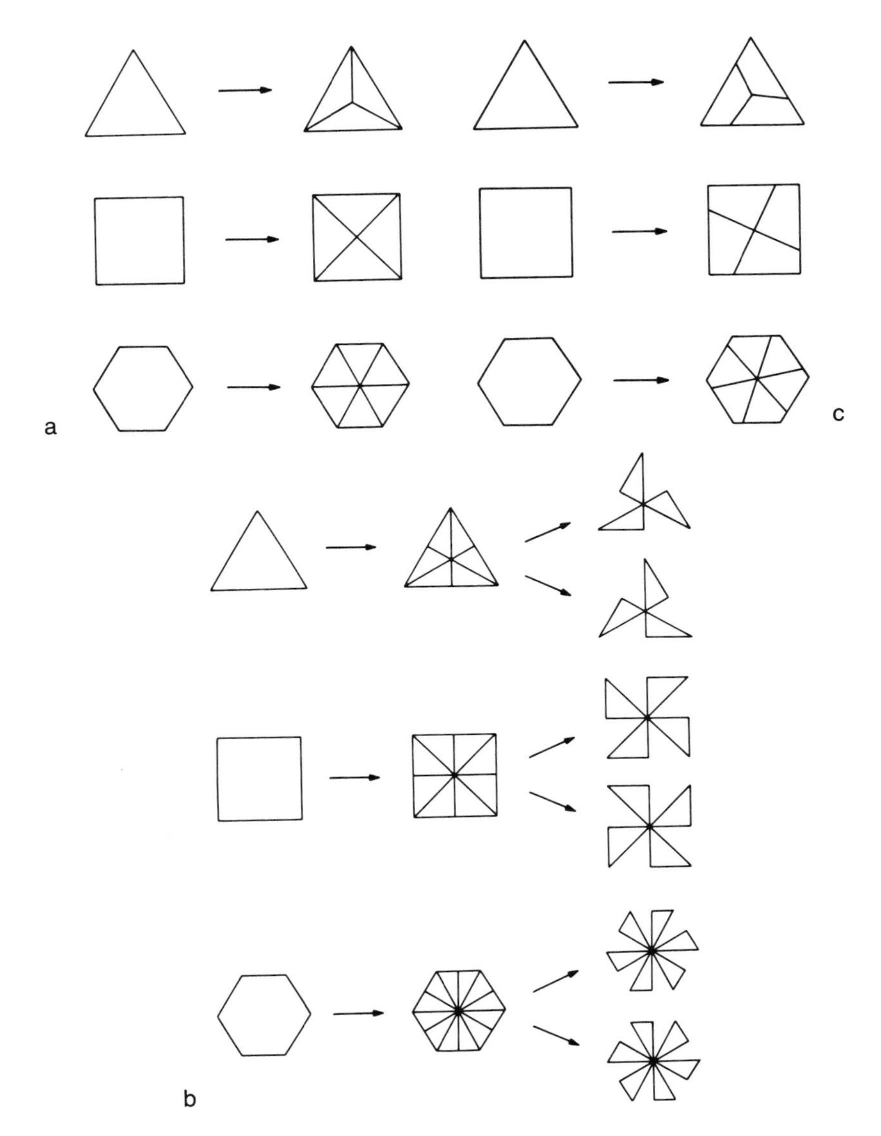

Figure 2-55. Dissection of planar achiral figures into achiral (a), heterochiral (b), and homochiral segments (c): some examples.

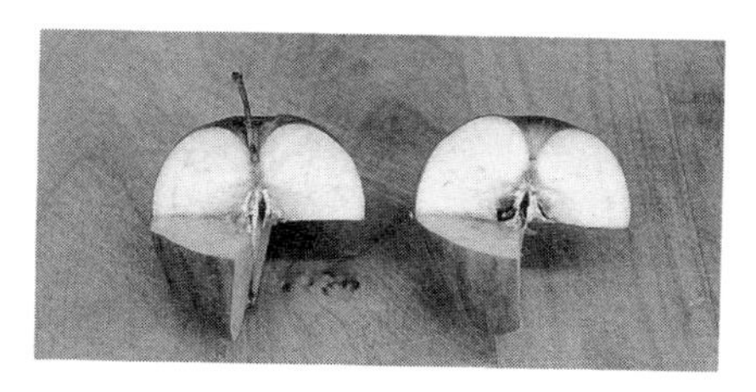

Figure 2-56. The French parlor trick *la coupe du roi*, after Anet *et al.* [2-57]. An apple can be cut into two homochiral halves in two ways which are enantiomorphous to each other. An apple cannot be cut into two heterochiral halves. Two heterochiral halves originating from two different apples cannot be combined into one apple.

In *la coupe du roi*, two vertical half cuts are made through the apple—one from the top to the equator, and another, perpendicularly, from the bottom to the equator. In addition, two nonadjacent quarter cuts are made along the equator. If all this is properly done, the apple should separate into two homochiral halves as seen in Figure 2-56.

The first chemical analog of *la coupe du roi* was demonstrated by Cinquini *et al.* [2-58] by bisecting the achiral molecule *cis*-3,7-dimethyl-1,5-cyclooctanedione into homochiral halves, viz., 2-methyl-1,4-butanediol. The reaction sequence is depicted in Figure 2-57 after Cinquini *et al.* [2-58], who painstakingly documented the analogy with the pomaceous model. Only examples of the *reverse coupe du roi* had been known prior to the work of Cinquini *et al.* Thus, Anet *et al.* [2-57] had reported the synthesis of chiral 4-(bromomethyl)-6-(mercaptomethyl)[2.2]metacyclophane. They then showed that two homochiral molecules can be combined to form an achiral dimer as shown in and illustrated by Figure 2-58.

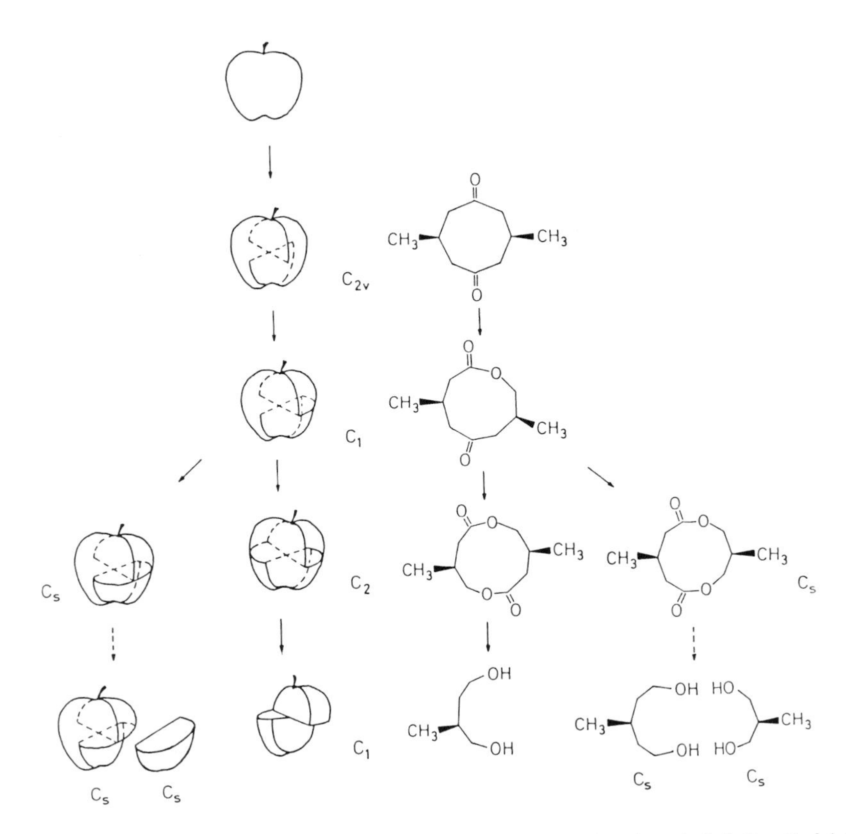

Figure 2-57. *La coupe du roi* and the reaction sequence transforming *cis*-3,7-dimethyl-1,5-cyclooctanedione into 2-methyl-1,4-butanediol. After Cinquini *et al*. [2-58]. Used by permission. Copyright (1988) American Chemical Society.

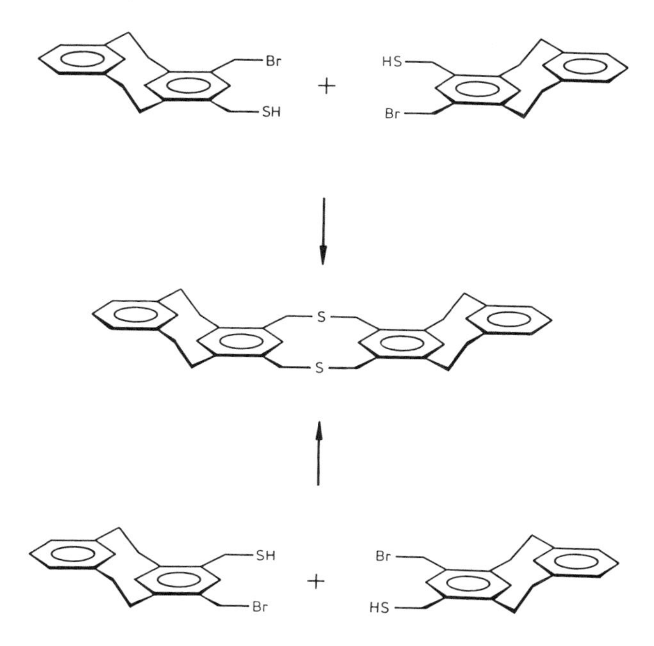

Figure 2-58. *Reverse coupe du roi* and the formation of a dimer from two homochiral 4-(bromo-methyl)-6-(mercaptomethyl)[2.2]metacyclophane molecules. After Anet *et al.* [2-57]. Used by permission. Copyright (1983) American Chemical Society.

2.8 POLYHEDRA

"A convex polyhedron is said to be *regular* if its faces are regular and equal, while its vertices are all surrounded alike" [2-59]. A polyhedron is convex if every dihedral angle is less than 180°. The dihedral angle is the angle formed by two polygons joined along a common edge.

There are only five regular convex polyhedra, a very small number indeed. The regular convex polyhedra are called *Platonic solids* because they constituted an important part of Plato's natural philosophy. They are the tetrahedron, cube (hexahedron), octahedron, dodecahedron, and icosahedron. The faces are regular polygons, either regular triangles, regular pentagons, or squares.

A *regular polygon* has equal interior angles and equal sides. Figure 2-59 presents a regular triangle, a regular quadrangle (i.e., a square), a regular pentagon, and so on. The circle is obtained in the limit as the number of sides approaches infinity. The regular polygons have an n-fold rotational symmetry

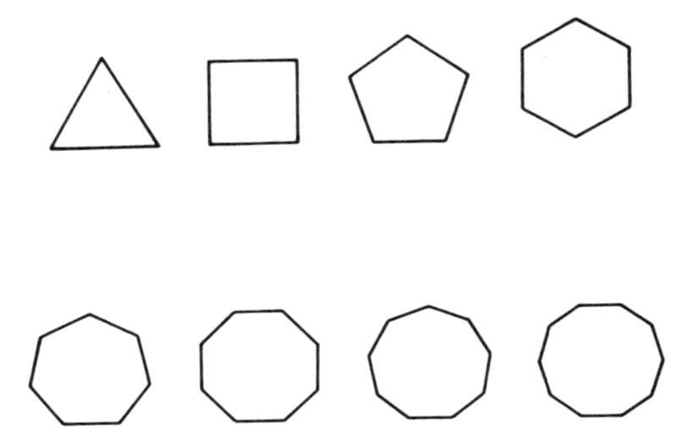

Figure 2-59. Regular polygons.

axis perpendicular to their plane and going through their midpoint. Here n is 1, 2, 3, . . . up to infinity for the circle.

The five regular polyhedra are shown in Figure 2-60. Their characteristic parameters are given in Table 2-3. Figure 2-61 reproduces an East German stamp with Euler and his equation, $V - E + F = 2$, where V, E, and F are the number of vertices, edges, and faces. The equation is valid for polyhedra having any kind of polygonal faces. According to Weyl [2-9], the existence of

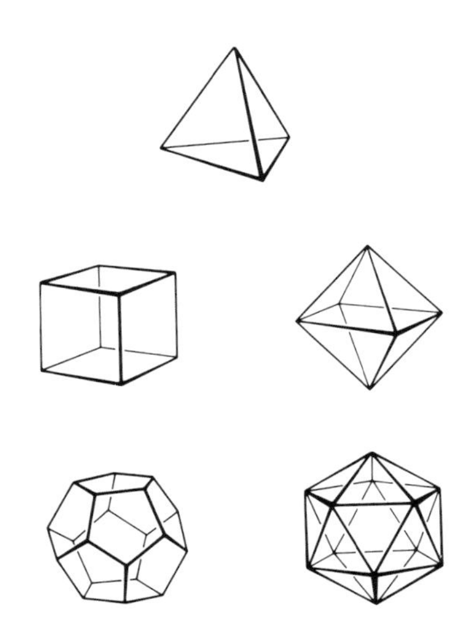

Figure 2-60. The five Platonic solids.

Table 2-3. Characteristics of the Regular Polyhedra

Name	Polygon	Number of faces	Vertex figure	Number of vertices	Number of edges
Tetrahedron	3	4	3	4	6
Cube	4	6	3	8	12
Octahedron	3	8	4	6	12
Dodecahedron	5	12	3	20	30
Icosahedron	3	20	5	12	30

the tetrahedron, cube, and octahedron is a fairly trivial geometric fact. On the other hand, he considered the discovery of the regular dodecahedron and the regular icosahedron "one of the most beautiful and singular discoveries made in the whole history of mathematics." However, to ask who first constructed the regular polyhedra is, according to Coxeter [2-59], like asking who first used fire.

Many primitive organisms have the shape of the pentagonal dodecahedron. As will be seen later, it is not possible to have crystal structures having this symmetry. Belov [2-60] suggested that the pentagonal symmetry of primitive organisms represents their defense against crystallization. Several radiolarians of different shapes from Häckel's book [2-13] are shown in Figure 2-62. Artistic representations of regular polyhedra are shown in Figure 2-63.

Figure 2-64 shows Kepler and his planetary model based on the regular solids [2-61]. According to this model, the greatest distance of one planet from the sun stands in a fixed ratio to the least distance of the next outer planet from the sun. There are five ratios describing the distances of the six planets that were known to Kepler. A regular solid can be interposed between two adjacent planets so that the inner planet, when at its greatest distance from the sun, lays

Figure 2-61. Euler and his equation $e - k + f = 2$, corresponding to $V - E + F = 2$, where the German *e* (*E*cke), *k* (*K*ante), and *f* (*F*läche) correspond to vertex (*V*), edge (*E*), and face (*F*), respectively.

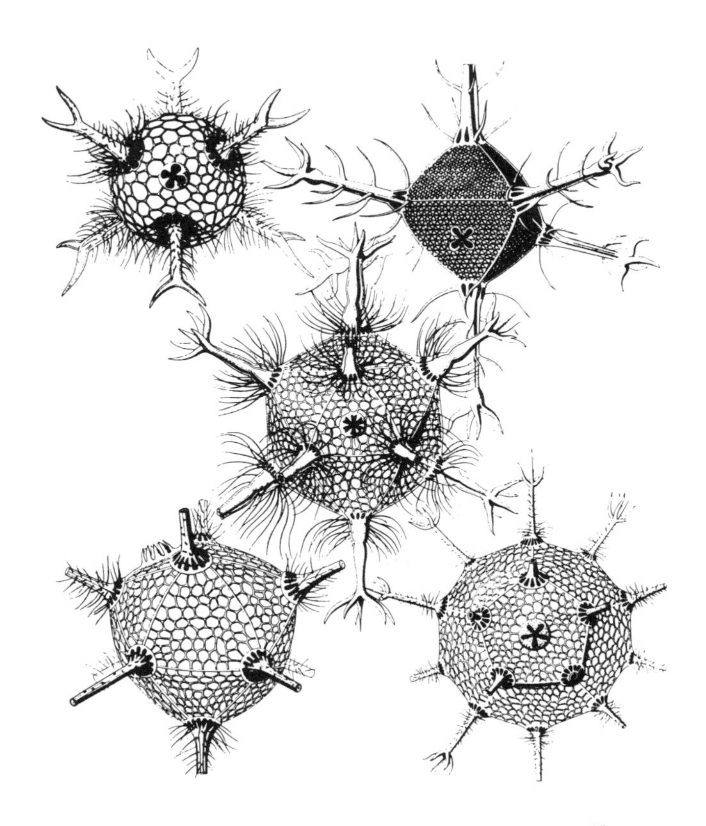

Figure 2-62. Radiolarians from Häckel's book [2-13].

on the inscribed sphere of the solid, while the outer planet, when at its least distance, lays on the circumscribed sphere.

Arthur Koestler in *The Sleepwalkers* [2-62] called this planetary model "a false inspiration, a supreme hoax of the Socratic *daimon*, . . .". However, the planetary model, which is also a densest packing model, probably represents Kepler's best attempt at attaining a unifed view of his work both in astronomy and in what we call today crystallography.

There are excellent monographs on regular figures, two of which are especially noteworthy [2-59, 2-63]. The Platonic solids have very high symmetries and one especially important common characteristic. None of the rotational symmetry axes of the regular polyhedra is unique, but each axis is associated with several axes equivalent to itself. The five regular solids can be classifed into three symmetry classes:

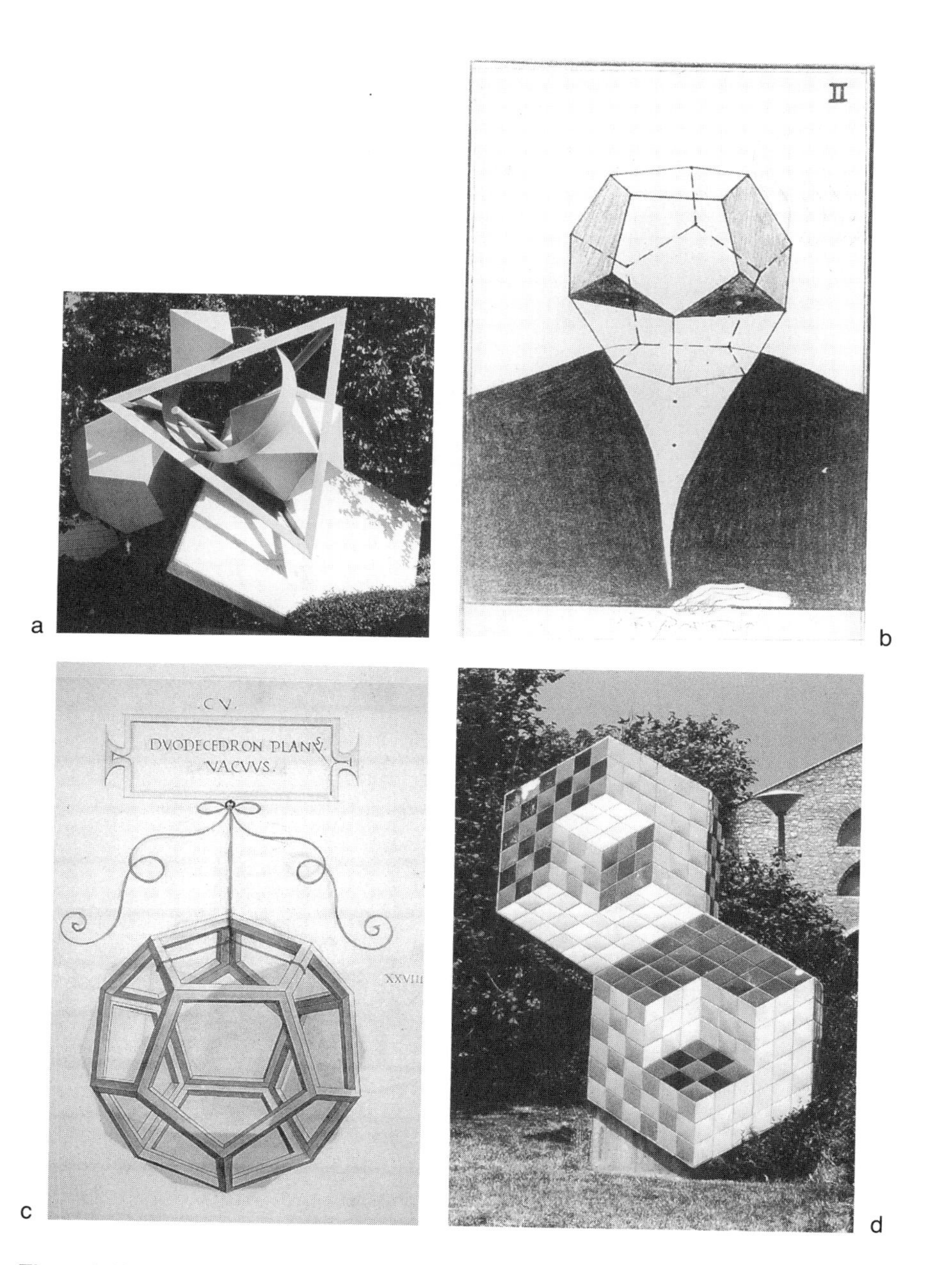

Figure 2-63. Artistic representations of regular polyhedra: (a) Sculpture in the garden of Tel Aviv University; photograph by the authors; (b) pentagonal dodecahedron by Horst Janssen, *ChriStall-Knecht* (crystal slave); reproduced by permission; (c) Leonardo da Vinci's dodecahedron drawn for Luca Pacioli's *De Divina Proportione*; (d) sculpture by Victor Vasarely in Pécs, Hungary; photograph by the authors.

Figure 2-64. Johannes Kepler on a Hungarian stamp and a detail of his planetary model based on the regular solids [2-61].

Tetrahedron	$3/2{\cdot}m = 3/\tilde{4}$
Cube and octahedron	$3/4{\cdot}m = \tilde{6}/4$
Dodecahedron and icosahedron	$3/5{\cdot}m = 3/\widetilde{10}$

It is equivalent to describe the symmetry class of the tetrahedron as $3/2{\cdot}m$ or $3/\tilde{4}$. The skew line between two axes means that they are not orthogonal. The symbol $3/2{\cdot}m$ denotes a threefold axis and a twofold axis which are not perpendicular and a symmetry plane which includes these axes. These three symmetry elements are indicated in Figure 2-65. The symmetry class $3/2{\cdot}m$ is equivalent to a combination of a threefold axis and a fourfold mirror-rotation axis. In both cases the threefold axes connect one of the vertices of the tetrahedron with the midpoint of the opposite face. The fourfold mirror-rotation axes coincide with the twofold axes. The presence of the fourfold mirror-rotation axis is easily seen if the tetrahedron is rotated by a quarter rotation about a twofold axis and is then reflected by a symmetry plane perpendicular to this axis. The symmetry operations chosen as basic will then generate the remaining symmetry elements. Thus, the two descriptions are equivalent.

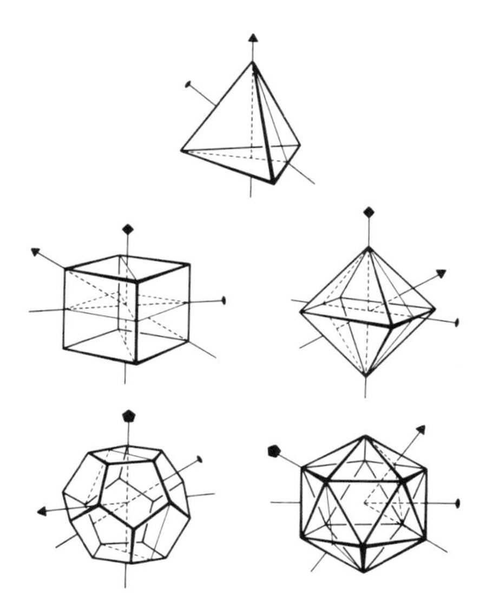

Figure 2-65. Characteristic symmetry elements of the Platonic solids.

Characteristic symmetry elements of the cube are shown in Figure 2-65. Three different symmetry planes go through the center of the cube parallel to its faces. Furthermore, six symmetry planes connect the opposite edges and also diagonally bisect the faces. The fourfold rotation axes connect the midpoints of opposite faces. The sixfold mirror-rotation axes coincide with threefold rotation axes. They connect opposite vertices and are located along the body diagonals. The symbol $\tilde{6}/4$ does not directly indicate the symmetry planes connecting the midpoints of opposite edges, the twofold rotation axes, or the center of symmetry. These latter elements are generated by the others. The presence of a center of symmetry is well seen by the fact that each face and edge of the cube has its parallel counterpart. The tetrahedron, on the other hand, has no center of symmetry.

The octahedron is in the same symmetry class as the cube. The antiparallel character of the octahedron faces is especially conspicuous. As seen in Figure 2-65, the fourfold symmetry axes go through the vertices, the threefold axes go through the face midpoints, and the twofold axes go through the edge midpoints.

The pentagonal dodecahedron and the icosahedron are in the same symmetry class. The fivefold, threefold, and twofold rotation axes intersect the midpoints of faces, the vertices, and the edges of the dodecahedron, respectively (Figure 2-65). On the other hand, the corresponding axes intersect the vertices and the midpoints of faces and edges of the icosahedron (Figure 2-65).

Consequently, the five regular polyhedra exhibit a dual relationship as regards their faces and vertex figures. The tetrahedron is self-dual (Table 2-3).

If the definition of regular polyhedra is not restricted to convex figures, their number rises from five to nine. The additional four are depicted in Figure 2-66 (for more information, see, e.g., Refs. [2-59] and [2-63]–[2-66]). They are called by the common name of regular star polyhedra. One of them, viz., the great stellated dodecahedron, is illustrated by the decoration at the top of the Sacristy of St. Peter's Basilica in Vatican City, and another, the small stellated dodecahedron, by an ordinary lamp in Figure 2-67.

The *sphere* deserves special mention. It is one of the simplest possible figures and, accordingly, one with high and complicated symmetry. It has an infinite number of rotation axes with infinite order. All of them coincide with body diagonals going through the midpoint of the sphere. The midpoint, which is also a singular point, is the center of symmetry of the sphere. The following symmetry elements may be chosen as basic ones: two infinite order rotation axes which are not perpendicular plus one symmetry plane. Therefore, the symmetry class of the sphere is $\infty/\infty \cdot m$. Concerning the symmetry of the sphere, Kepes [2-67] quotes Copernicus:

> The spherical is the form of all forms most perfect, having need of no articulation; and the spherical is the form of greatest volumetric capacity, best able to contain and circumscribe all else; and all the separated parts of the world—I mean the sun, the moon, and the stars—are observed to have spherical form; and all things tend to limit themselves under this form—as appears in drops of water and other liquids whenever of themselves they tend to limit themselves.

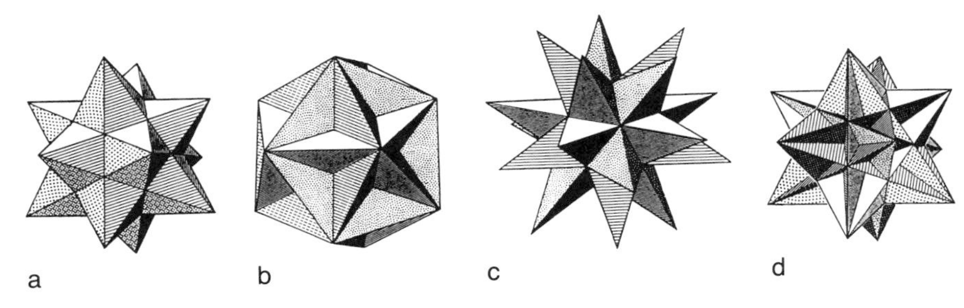

Figure 2-66. The four regular star polyhedra: (a) Small stellated dodecahedron; (b) the great dodecahedron; (c) great stellated dodecahedron; (d) the great icosahedron. From H. M. Cundy and A. P. Rollett [2-64]. Used by permission of Oxford University Press.

a

b

Figure 2-67. Examples of star polyhedra: (a) Great stellated dodecahedron as decoration at the top of the Sacristy of St. Peter's Basilica, Vatican City; (b) small stellated dodecahedron as a lamp in an Italian home, Bologna. Photographs by the authors.

> So no one may doubt that the spherical is the form of the world, the divine body.

Artistic appearances of spheres are shown in Figure 2-68.

In addition to the regular polyhedra, there are various families of polyhedra with decreased degrees of regularity [2-59, 2-63–2-66]. The so-called *semiregular* or *Archimedean* polyhedra are similar to the Platonic polyhedra in that all their faces are regular and all their vertices are congruent. However, the polygons of their faces are not all of the same kind. The thirteen semiregular polyhedra are listed in Table 2-4, and some of them are also shown in Figure 2-69. Table 2-4 also enumerates their rotation axes.

The simplest semiregular polyhedra are obtained by symmetrically shav-

a

Figure 2-68. Artistic expressions of the sphere: (a) In front of the World Trade Center, New York; (b) sculpture by J.-B. Carpeaux in Paris. Photographs by the authors.

Table 2-4. The Thirteen Semiregular Polyhedra

		Number of			Number of rotation axes			
No.	Name	Faces	Vertices	Edges	2-fold	3-fold	4-fold	5-fold
1	Truncated tetrahedron[a]	8	12	18	3	4	0	0
2	Truncated cube[a]	14	24	36	6	4	3	0
3	Truncated octahedron[a]	14	24	36	6	4	3	0
4	Cuboctahedron[b]	14	12	24	6	4	3	0
5	Truncated cuboctahedron	26	48	72	6	4	3	0
6	Rhombicuboctahedron	26	24	48	6	4	3	0
7	Snub cube	38	24	60	6	4	3	0
8	Truncated dodecahedron[a]	32	60	90	15	10	0	6
9	Icosidodecahedron[b]	32	30	60	15	10	0	6
10	Truncated icosahedron[a]	32	60	90	15	10	0	6
11	Truncated icosidodecahedron	62	120	180	15	10	0	6
12	Rhombicosidodecahedron	62	60	120	15	10	0	6
13	Snub dodecahedron	92	60	150	15	10	0	6

[a]Truncated regular polyhedron.
[b]Quasiregular polyhedron.

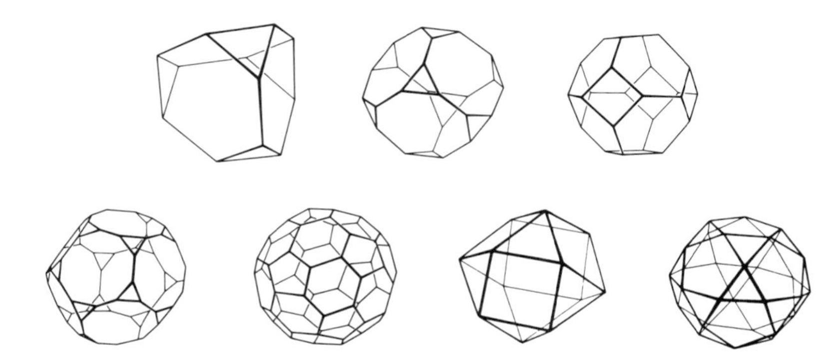

Figure 2-69. Some of the semiregular polyhedra: the so-called truncated regular polyhedra and quasiregular polyhedra.

ing off the corners of the regular solids. They are the truncated regular polyhedra and are marked with a superscript *a* in Table 2-4. One of them is the truncated icosahedron, the shape of the buckminsterfullerene molecule. Two semiregular polyhedra are classified as so-called quasiregular polyhedra. They have two kinds of faces, and each face of one kind is entirely surrounded by faces of the other kind. They are marked with a superscript *b* in Table 2-4. All these seven semiregular polyhedra are shown in Figure 2-69. The remaining six semiregular polyhedra may be derived from the other semiregular polyhedra. The structures of *zeolites*, aluminosilicates, are rich in polyhedral shapes, including the channels and cavities they form (see, e.g., Ref. [2-68]). One of the most common zeolites is sodalite, $Na_6[Al_6Si_6O_{24}]\cdot 2NaCl$, whose name refers to its sodium content. The sodalite unit itself is represented by a truncated octahedron in Figure 2-70a, where the line drawing ignores the oxygen atoms and each line represents T . . . T (T = Al, Si). The three remaining models of Figure 2-70 (b, c, and d) represent different modes of linkages between the sodalite units. It is especially interesting to see the different cavities formed by different modes of linkage [2-69]. Some other examples of semiregular polyhedra are shown in Figure 2-71.

The prisms and antiprisms are also important polyhedron families. A prism has two congruent and parallel faces, and they are joined by a set of parallelograms. An antiprism also has two congruent and parallel faces, but they are joined by a set of triangles. There is an infinite number of prisms and antiprisms, and some of them are shown in Figure 2-72. A prism or an antiprism is semiregular if all its faces are regular polygons. A cube can be considered a square prism, and an octahedron can be considered a triangular antiprism.

There are additional polyhedra which are important in discussing molecular geometries and crystal structures.

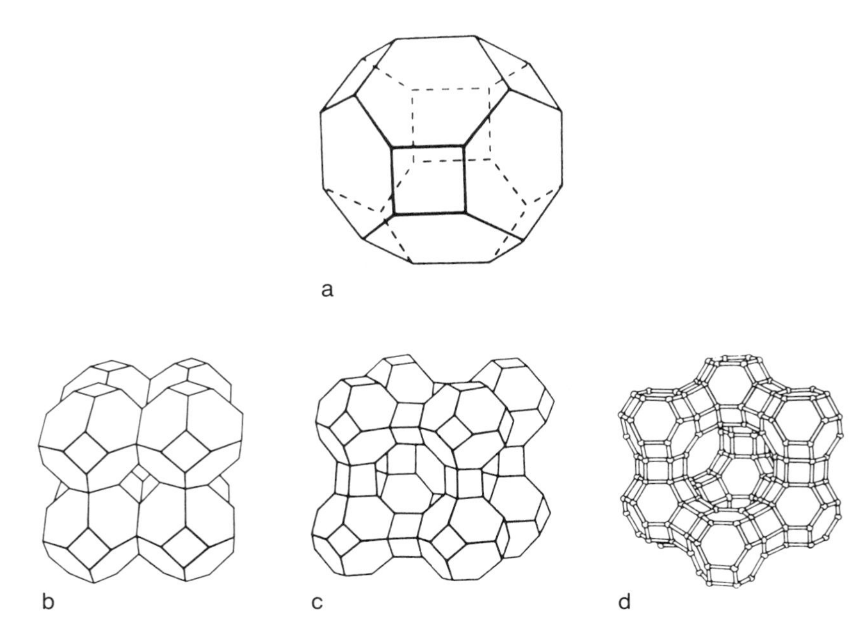

Figure 2-70. Zeolite structures: the shape and various modes of linkage of the sodalite units after Beagley and Titiloye [2-69]. Reproduced by permission. The line drawings ignore th oxygen atoms and represent T . . . T (T = Al, Si); (a) Sodalite unit; (b) and (c) sodalite unit linked through double 4-rings; (d) Sodalite units linked through double 6-rings. In this mode double lines represent T . . . T.

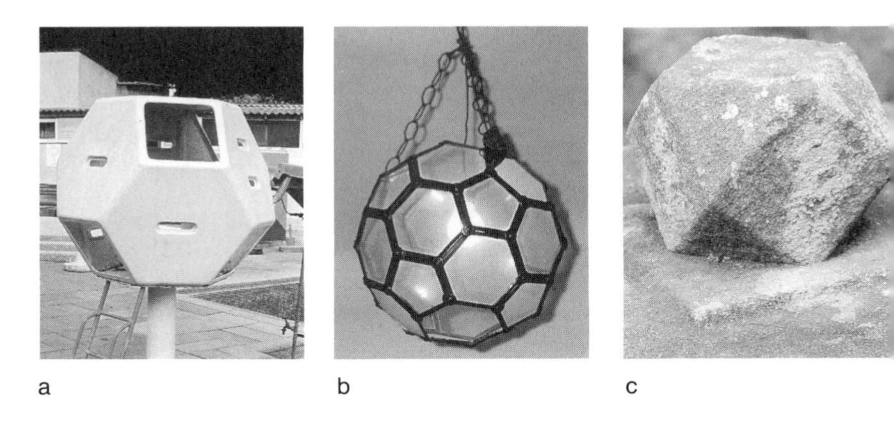

Figure 2-71. Examples of semiregular polyhedra: (a) Truncated octahedron in a Tel Avi playground; (b) truncated icosahedron as a lamp in an Italian home, Bologna; (c) cuboctahedro as a top decoration of a garden lantern in Kyoto, Japan. Photographs by the authors.

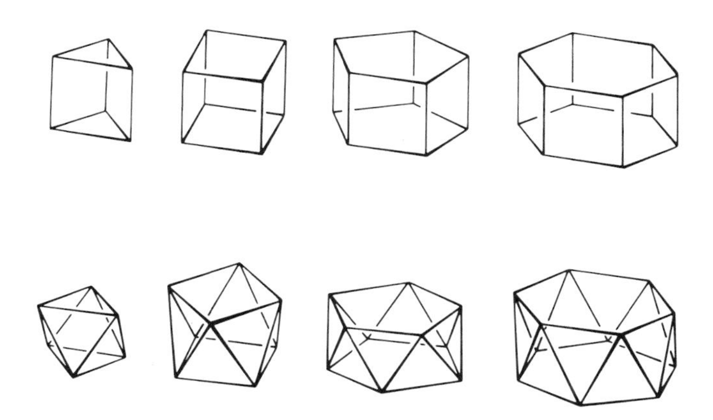

Figure 2-72. Prisms and antiprisms.

REFERENCES

[2-1] T. Mann, *The Magic Mountain*. The cited passage on p. 21 is in French both in the original German and its English translation. The English translation cited in our text was kindly provided by Dr. Jack M. Davis, Professor of English, The University of Connecticut, 1984.

[2-2] J. Kepler, *Strena, seu De Nive Sexangula* (1611); English translation, *The Six-Cornered Snowflake*, Clarendon Press, Oxford (1966).

[2-3] I. I. Shafranovskii, Kepler's crystallographic ideas and his tract "The Six-Cornered Snowflake," in *Kepler, Four Hundred Years* (A. Beer and P. Beer, eds.), Proceedings of conferences held in honor of Johannes Kepler, *Vistas Astron*. **18** (1975).

[2-4] H. A. Sackeim, R. C. Gur, and M. C. Saucy, *Science* **202**, 434 (1978).

[2-5] S. P. Springer and G. Deutsch, *Left Brain, Right Brain*, W. Freeman & Co., San Francisco (1981).

[2-6] J. B. Hellige, *Hemispheric Asymmetry. What's Right and What's Left*, Harvard University Press, Cambridge, Massachusetts (1993).

[2-7] A. V. Shubnikov and V. A. Koptsik, *Symmetry in Science and Art*, Plenum Press, New York (1974.); Russian original, *Simmetriya v nauke i isskustve*, Nauka, Moscow (1972).

[2-8] R. Jastrow, *Red Giants and White Dwarfs*, W. W. Norton, New York (1979).

[2-9] H. Weyl, *Symmetry*, Princeton University Press, Princeton, New Jersey (1952).

[2-10] L. R. H. Appleton, *American Indian Design and Decoration*, Dover Publications, New York (1971).

[2-11] W. Herzfielde, *John Heartfield. Leben und Werk dargestellt von seinem Bruder*, 3rd revised and expanded ed., VEB Verlag der Kunst, Dresden (1976).

[2-12] M. Ickis, *The Standard Book of Quilt Making and Collecting*, Dover Publications, New York (1959).

[2-13] E. Häckel, *Kunstformen der Natur*, Vols. 1–10, Verlag des Bibliographischen Instituts, Leipzig (1899–1904).

[2-14] U. Nakaya, *Snow*, Iwanami-Shoten Publishing Co., Tokyo (1938) [in Japanese].

[2-15] G. Taubes, *Discover* **1984** (January), 75.

[2-16] D. McLachlan, *Proc. Natl. Acad. Sci. U.S.A.* **43**, 143 (1957).
[2-17] W. A. Bentley and W. J. Humphreys, *Snow Crystals*, McGraw-Hill, New York (1931).
[2-18] J. Nittmann and H. E. Stanley, *Nature* **321**, 663 (1986).
[2-19] S. Kai (ed.), *Pattern Formation in Complex Dissipative Systems*, World Scientific, Singapore (1992).
[2-20] S. Ohta and H. Honjo, in *Santa Fe Institute's Studies in the Sciences of Complexity*, Addison-Wesley Publishing Co. in press.
[2-21] Y. Furukawa and W. Shimada, *J. Cryst. Growth* **128**, 234 (1993).
[2-22] R. Kobayashi, *Physica D* **63**, 410 (1993).
[2-23] D. A. Tomalia and H. D. Durst, *Top. Curr. Chem.* **165**, 193 (1993).
[2-24] Attributed to M. Polányi. Private communication from Professor W. Jim Neidhardt, New Jersey Institute of Technology, 1984.
[2-25] A. L. Mackay, *Jugosl. Cent. Kristallogr.* **10**, 5 (1975); A. L. Mackay, personal communication, 1982.
[2-26] J. Needham and Lu Gwei-Djen, *Whether* **16**, 319 (1961).
[2-27] G. Hellmann, *Schneekrystalle*, Berlin (1893).
[2-28] T. Stamp and C. Stamp, *William Scoresby; Arctic Scientist*, Caedmon of Whitby (1976).
[2-29] J. Reston, *International Herald Tribune*, May 7, 1981, p. 4.
[2-30] W. Engelhardt, *Mathematischer Unterricht* **9**(2), 49 (1963).
[2-31] E. A. Abbott, *Flatland. A Romance of Many Dimensions*, Barnes and Noble Books, 5th ed., New York (1983).
[2-32] D. Y. Curtin and I. C. Paul, *Chem. Rev.* **81**, 525 (1981).
[2-33] P. Groth, *Chemische Kristallographie*, 5 vols., Verlag von Wilhelm Engelmann, Leipzig (1906–1919).
[2-34] G. N. Desiraju, *Crystal Engineering: The Design of Organic Solids*, Elsevier, Amsterdam (1989).
[2-35] Lord Kelvin, *Baltimore Lectures*, C. J. Clay and Sons, London (1904).
[2-36] R. N. Bracewell, *The Galactic Club. Intelligent Life in the Outer Space*, W. H. Freeman & Co., San Francisco (1975).
[2-37] L. L. Whyte, *Leonardo* **8**, 245 (1975); *Nature* **182**, 198 (1958).
[2-38] I. Hargittai and M. Hargittai, *Symmetry: A Unifying Concept*, Shelter Publications, Bolinas, California (1994).
[2-39] M. Gardner, *The New Ambidextrous Universe. Symmetry and Asymmetry from Mirror Reflections to Superstrings*, 3rd revised ed., W. H. Freeman and Co., New York (1990).
[2-40] A. V. Shubnikov, *Simmetriya i antisimmetriya konechnykh figur*, Izd. Akad. Nauk SSSR, Moscow (1951).
[2-41] "C'est la dissymétrie qui crée le phénomène"; P Curie, *J. Phys. Paris* **3**, 393 (1894).
[2-42] M. Curie, *Pierre Curie*, The Macmillan Company, New York (1929).
[2-43] A. V. Shubnikov, in *Shubnikov Centennial Papers* (I. Hargittai and B. K. Vainshtein, eds.), *Crystal Symmetries* p. 357, Pergamon Press, Oxford (1988). [Originally appeared in Russian in *Usp. Fiz. Nauk* **59**, 591 (1956).]
[2-44] I. Stewart and M. Golubitsky, *Fearful Symmetry. Is God a Geometer?* Blackwell, Oxford (1992).
[2-45] V. Prelog, *Science* **193**, 17 (1976).
[2-46] L. E. Orgel, *The Origins of Life: Molecules and Natural Selection*, John Wiley & Sons, New York (1973).
[2-47] J. D. Bernal, *The Origin of Life*, The World Publishing Co., Cleveland (1967).
[2-48] J. B. S. Haldane, *Nature* **185**, 87 (1960), citing L. Pasteur, *C. R. Acad. Sci. Paris*, June 1, 1874.
[2-49] R. C. Milton, S. C. F. Milton, and S. B. H. Kent, *Science* **256**, 1445 (1992).

[2-50] G. Jung, *Angew. Chem. Int. Ed. Engl.* **31**, 1457 (1992).
[2-51] K. F. Schmidt, *Sci. News* **143**, 348 (1993).
[2-52] R. Janoschek (ed.), *Chirality—From Weak Bosons to the Alpha-Helix*, Springer-Verlag, Berlin (1991).
[2-53] M. Nógrádi, *Stereoselective Synthesis. A Practical Approach*, VCH, Weinheim (1995).
[2-54] R. A. Aitken and S. N. Kilényi (eds.), *Asymmetric Synthesis*, Blackie Academic & Professional, London (1992).
[2-55] H. Brunner and W. Zettlmeier, *Handbook of Enantioselective Catalysis*, Vols. I and II, VCH, Weinheim (1993).
[2-56] S. C. Stinson, *Chem. Eng. News*, September 19, 1994, p. 38.
[2-57] F. A. L. Anet, S. S. Miura, J. Siegel, and K. Mislow, *J. Am. Chem. Soc.* **105**, 1419 (1983).
[2-58] M. Cinquini, F. Cozzi, F. Sannicoló and A. Sironi, *J. Am. Chem. Soc.* **110**, 4363 (1988).
[2-59] H. S. M. Coxeter, *Regular Polytopes*, 3rd ed., Dover Publications, New York (1973).
[2-60] N. V. Belov, *Ocherki po strukturnoi mineralogii*, Nedra, Moscow (1976).
[2-61] J. Kepler, *Mysterium cosmographicum* (1595).
[2-62] A. Koestler, *The Sleepwalkers*, The Universal Library, Grosset and Dunlap, New York (1963), p. 252.
[2-63] L. Fejes Tóth, *Regular Figures*, Pergamon Press, New York (1964).
[2-64] H. M. Cundy and A. P. Rollett, *Mathematical Models*, Clarendon Press, Oxford (1961).
[2-65] M. J. Wenninger, *Polyhedron Models*, Cambridge University Press, New York (1971).
[2-66] P. Pearce and S. Pearce, *Polyhedra Primer*, Van Nostrand Reinhold, New York (1978).
[2-67] N. Copernicus, *De Revolutionibus Orbium Caelestium* (1543), as cited in G. Kepes, *The New Landscape in Art and Science*, Theobald & Co., Chicago (1956).
[2-68] W. M. Meier and D. H. Olson, *Atlas of Zeolite Structure Types*, 3rd revised ed., Butterworth-Heinemann, London (1992).
[2-69] B. Beagley and J. O. Titiloye, *Struct. Chem.* **3**, 429 (1992).

Chapter 3

Molecules: Shape and Geometry

A molecule is not simply a collection of its constituent atoms. It is kept together by interactions among these atoms. Thus, for some purposes it is better to consider the molecule as consisting of the nuclei of its constituent atoms and its electron density distribution. Generally, it is the geometry and symmetry of the arrangement of the atomic nuclei that is considered to be the geometry and symmetry of the molecule itself.

Molecules are finite figures with at least one singular point in their symmetry description. Thus, point groups are applicable to them. There is no inherent limitation on the available symmetries for molecules. On the other hand, severe restrictions apply to the symmetries of crystals, as will be seen later (Section 9.3). In fact, molecules occupy a more fundamental level in the hierarchy of structures than do crystals. Many crystals themselves are built from molecules (see, Section 9.6).

Molecules in the gas phase are considered to be *free*. They are so far apart that they are unperturbed by interactions with other molecules. On the other hand, intermolecular interactions may occur between the molecules in condensed phases, i.e., in liquids, melts, amorphous solids, or crystals. In the present discussion, all molecules will be assumed to be unperturbed by their environment, regardless of the phase or state of matter in which they exist.

Molecules are never motionless. They are performing vibrations all the time. In addition, gaseous molecules, and also molecules in liquids, are performing rotational and translational motion as well. Molecular vibrations constitute relative displacements of the atomic nuclei with respect to their equilibrium positions and occur in all phases, including the crystalline state, and even at the lowest possible temperatures. The magnitude of molecular vibrations is relatively large, amounting to several percent of the internuclear distances. Typically, there are about 10^{12} to 10^{14} vibrations per second.

Symmetry considerations are fundamental in any description of molecular vibrations, as will be seen later in detail (Chapter 5). First, however, we will discuss the molecular symmetries, ignoring entirely the motion of the molecules. Various molecular symmetries will be illustrated by examples from outside chemistry. A simple model will also be discussed to gain some insight into the origins of the various shapes and symmetries in the world of molecules. Our considerations will be restricted, however, to relatively simple and thus rather symmetrical systems. The importance and consequences of intramolecular motion involving relatively large amplitudes will be commented upon in the final section of this chapter.

3.1 FORMULAS, ISOMERS

The empirical formula of a chemical compound expresses its composition. For example, $C_2H_4O_2$ indicates that the molecule consists of two carbon, four hydrogen, and two oxygen atoms. This formulation, however, provides no information on the order in which these atoms are linked. This empirical formula may correspond to methyl formate (**3-1**), acetic acid (**3-2**), and glycolaldehyde (**3-3**). Only the structural formulas for these compounds, shown below, distinguish among them. This is called structural isomerism.

$HCOOCH_3$ CH_3COOH $HCOCH_2OH$

3-1 **3-2** **3-3**

Although these molecules, as a whole, are not symmetric, some of thei component parts may be symmetrical. They possess what is called loca symmetry. Similar atomic groups in different molecules often have simila geometries, and thus similar local symmetries. The structural formulas revea considerable information about these local symmetries, or at least their sim ilarities and differences in various molecules. The above simplified structura formulas are especially useful in this respect. This approach is widely applica ble in organic chemistry, where relatively few kinds of atoms build a enormous number of different molecules. A far greater diversity of structura peculiarities is characteristic for inorganic compounds.

The symbol for the carbon atom occurs twice in all three simplified structural formulas above, a fact that indicates differences in the structural positions of these carbon atoms. The same argument applies to the oxygen atoms. On the other hand, three hydrogens are equivalent in both methyl formate and acetic acid, with the fourth being different in the two molecules. There are three different types of hydrogen positions in glycolaldehyde.

Molecules with the same formula but in which the distances between corresponding atoms are not all the same are called structural isomers (Figure 3-1). They are of two types. If their atomic connectivities are the same, they are diastereomers, and if their atomic connectivities are different, they are constitutional isomers. Some diastereomers can become superimposable by rotation about a bond, and they are called rotational isomers. Depending on the magnitude of the barrier to rotation, geometrical isomers (high barrier) and conformers (low barrier) are distinguished.

Identical molecules have the same formula, the same atomic connectivity, and the same distances between corresponding atoms. In addition, they are

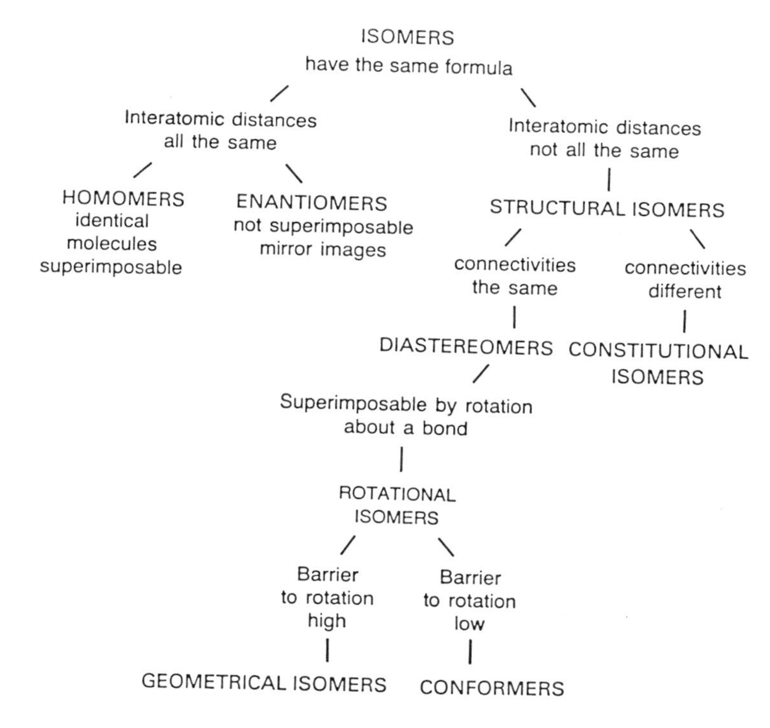

Figure 3-1. The hierarchy of isomers.

superimposable (homomers). Enantiomers have the same formula, the same atomic connectivity, and the same distances between corresponding atoms, but they are not superimposable; instead, they are mirror images of each other (cf. Section 2.7 on chirality).

3.2 ROTATIONAL ISOMERISM

The four-atom chain is the simplest system for which rotational isomerism is possible. It is shown in Figure 3-2. Rotational isomers, or *conformers*, are various forms of the same molecule related by rotation around a bond as axis. The various rotational forms of a molecule are described by the same empirical formula and by the same structural formula. Only the relative positions of the two bonds (or groups of atoms) at the two ends of the rotation axis are changed. The molecular point groups for various rotational isomers may be entirely different.

Rotational isomers can be conveniently represented by so-called projection formulas in which the two bonds (or groups of atoms) at the two ends are projected onto a plane which is perpendicular to the central bond. This plane is denoted by a circle whose center coincides with the projection of the rotation axis. The bonds in front of this plane are drawn as originating from the center. The bonds behind this plane, i.e., the bonds from the other end of the rotation axis, are drawn as originating from the perimeter of the circle.

The drawings by Degas *End of the Arabesque* and *Seated Dancer Adjusting Her Shoes* may be looked at as illustrations of the staggered and eclipsed conformations of $A_2B–BC_2$ molecules. They are shown in Figure 3-3a. Their projection-like representations are given in Figure 3-3b, while the conformers of the molecules are depicted in Figure 3-3c. Degas' drawings are also helpful in understanding the representation for the rotational isomer described above. The projections in Figure 3-3 represent views along the B–B bond, i.e., the dancer's body. The plane bisecting the B–B bond is shown by the circle, and it corresponds to the dancer's skirt. The dancer's arms and

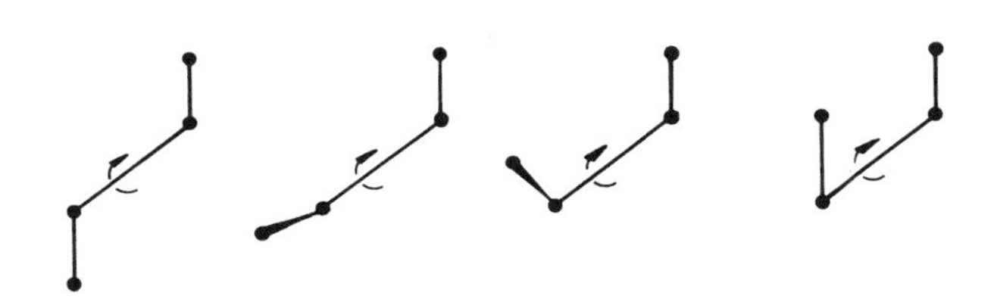

Figure 3-2. Rotational isomerism of a four-atom chain.

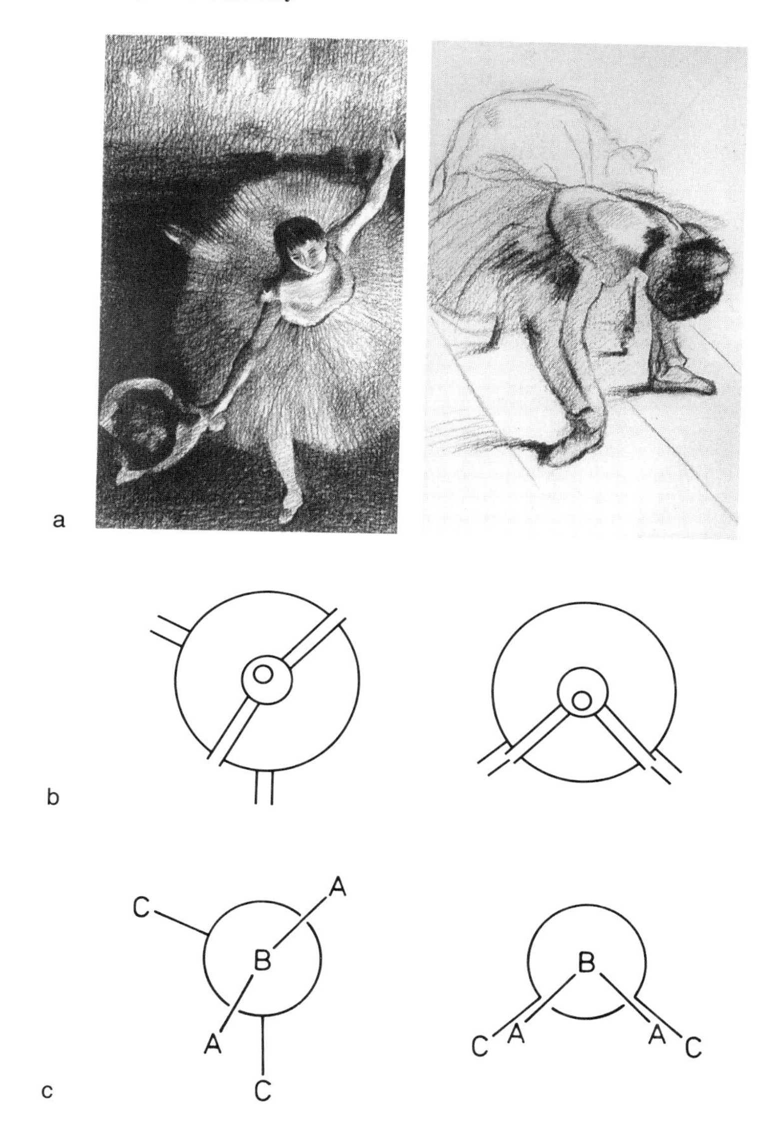

Figure 3-3. Illustration for the projectional representation of rotational isomers [3-1]. (a) Left: A drawing after Degas' *End of the Arabesque* by Ferenc Lantos. Right: A drawing after Degas' *Seated Dancer Adjusting Her Shoes* by Ferenc Lantos. Full-color reproductions of the original drawings are available in editions of Degas' work. The original drawings are in the Louvre, Musée de l'Impressionisme, Paris, and in the Hermitage, St. Petersburg, respectively. (b) Contour drawings of the dancers. (c) Staggered and eclipsed rotational isomers of the A_2BBC_2 molecule by Newman projections representing view along the B–B bond.

legs refer to the bonds B–A and B–C, respectively. Incidentally, the bouquet in the right hand of the dancer in the staggered conformation may be viewed as a different substituent.

Two important cases in rotational isomerism may be distinguished by considering the nature of the central bond. When it is a double bond, rotation of one form into another is hindered by a very high potential barrier. This barrier may be so high that the two rotational isomers will be stable enough to make their physical separation possible. An example is 1,2-dichloroethylene (**3-4**).

cis *trans*

3-4

The symmetry of the *cis* isomer is characterized by two mutually perpendicular mirror planes generating also a twofold rotational axis. This symmetry class is labeled *mm*. An equivalent notation is C_{2v}, as will be seen in the next section. The *trans* isomer has one twofold rotation axis with a perpendicular symmetry plane. Its symmetry class is $2/m$ (C_{2h}).

Rotational isomerism relative to a single bond is illustrated by ethane and 1,2-dichloroethane in Figure 3-4. During a complete rotation of one methyl

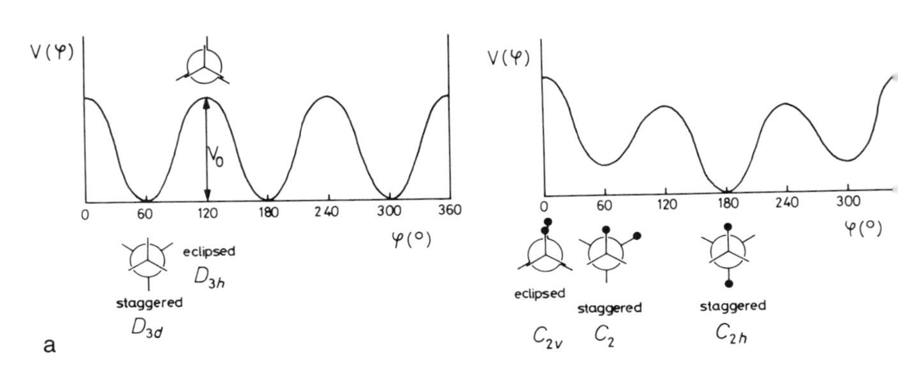

Figure 3-4. Potential energy functions for rotation about a single bond; φ is the angle of rotation. (a) Ethane, $H_3C–CH_3$. There are two different symmetrical forms. Both the staggered form with D_{3d} symmetry and the eclipsed form with D_{3h} symmetry occur three times in a complete rotational circuit. (b) 1,2-Dichloroethane, $ClH_2C–CH_2Cl$. There is no other symmetrical form in the region between the two symmetrical staggered forms shown. Only partial eclipsing can occur here because of insufficient symmetry (cf. Ref. [3-2]). The eclipsed form with C_{2v} symmetry and the staggered form with C_{2h} symmetry occur once, while the staggered form with C_2 symmetry occurs twice in a complete rotational circuit.

group around the C–C bond relative to the other methyl group, the ethane molecule appears three times in the stable staggered form and three times in the unstable eclipsed form. As all the hydrogen atoms of one methyl group are equivalent, the three energy minima are equivalent, and so are the three energy maxima, as seen in Figure 3-4a. The situation becomes more complicated when the three ligands bonded to the carbon atoms are not the same. This is seen for 1,2-dichloroethane in Figure 3-4b. There are three highly symmetrical forms. Of these, two are staggered, with C_{2h} and C_2 symmetry, respectively. The third is an eclipsed form with C_{2v} symmetry. This form has Cl/Cl and H/H eclipsing. There is no other fully eclipsed form because of insufficient symmetry [3-2].

Figure 3-4 shows only the symmetrical conformers by projection formulas. The symmetrical forms always belong to extreme energies, either minima or maxima. The barriers to internal rotation in the potential energy functions depicted in Figure 3-4 are about 10 kJ/mol. Typical barriers for systems where the double bonds would be considered to be the "rotational axis" may be as much as 30 times greater than those for systems with single bonds.

3.3 SYMMETRY NOTATIONS

So far, the so-called International or Hermann–Mauguin symmetry notations have been used in the descriptions in this text. Another, older system by Schoenflies is generally used, however, to describe the molecular point-group symmetries. This notation has been given in parentheses in the preceding section. The Schoenflies notation has the advantage of succinct expression for even complicated symmetry classes combining various symmetry elements. The two systems are compiled in Table 3-1 (see, e.g., Ref. [3-3]) for a selected set of symmetry classes. The set includes all point-group symmetries in the world of crystals, which are restricted to 32 classes. The reasons for and significance of these restrictions will be discussed later in the chapter on crystals (Section 9.3). There are no restrictions on the point-group symmetries for individual molecules, and a few additional, so-called *limiting*, classes are also listed in Table 3-1.

The Schoenflies notation for rotation axes is C_n, and for mirror-rotation axes the notation is S_{2n}, where n is the order of the rotation. The symbol i refers to the center of symmetry (cf. Section 2.4). Symmetry planes are labeled σ; σ_v is a vertical plane, which always coincides with the rotation axis with an order of two or higher, and σ_h is a horizontal plane, which is always perpendicular to the rotation axis when it has an order of two or higher.

Point-group symmetries not listed in Table 3-1 may easily be assigned the appropriate Schoenflies notation by analogy. Thus, for example, C_{5v}, C_{5h},

Table 3-1. Symmetry Notations of the Crystallographic and a Few Limiting Groups

Hermann–Mauguin	Schoenflies	Hermann–Mauguin	Schoenflies
Crystallographic groups			
1	C_1	$\bar{3}m$	D_{3d}
$\bar{1}$	C_i	$\bar{6}$	C_{3h}
m	C_s	6	C_6
2	C_2	$6/m$	C_{6h}
$2/m$	C_{2h}	$\bar{6}m2$	D_{3h}
mm	C_{2v}	$6mm$	C_{6v}
222	D_2	622	D_6
mmm	D_{2h}	$6/mmm$	D_{6h}
4	C_4	23	T
$\bar{4}$	S_4	$m\bar{3}$	T_h
$4/m$	C_{4h}	$\bar{4}3m$	T_d
$4mm$	C_{4v}	432	O
$\bar{4}2m$	D_{2d}	$m3m$	O_h
422	D_4	Limiting groups	
$4/mmm$	D_{4h}	∞	C_∞
3	C_3	$\infty 2$	D_∞
$\bar{3}$	S_6	∞/m	$C_{\infty h}$
$3m$	C_{3v}	∞mm	$C_{\infty v}$
32	D_3	∞/mm	$D_{\infty h}$

C_7, C_8, etc., can be established. Such symmetries may well occur among real molecules.

3.4 ESTABLISHING THE MOLECULAR POINT GROUP

Figure 3-5 shows a possible scheme for establishing the molecular point group (cf. Refs. [3-4] and [3-5]). The symmetry of most molecules may be reliably established by this scheme.

First, an examination is carried out to ascertain whether the molecule belongs to some "special" group. If the molecule is linear, it may have a perpendicular symmetry plane ($D_{\infty h}$) or it may not have one ($C_{\infty v}$). Very high symmetries are easy to recognize. Each of the groups T, T_h, T_d, O, and O_h has four threefold rotation axes. Both icosahedral I and I_h groups require ten threefold rotation axes and six fivefold rotation axes. The molecules belonging to these groups have a central tetrahedron, octahedron, cube, or icosahedron.

If the molecule does not belong to one of these "special" groups, a systematic approach is followed. Firstly, the possible presence of rotation axes in the molecule is checked. If there is no rotation axis, then it is determined

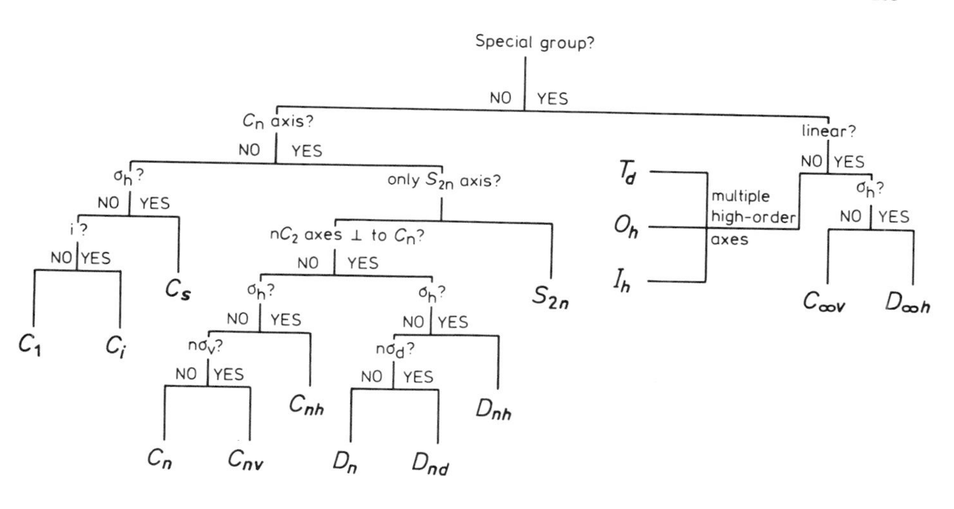

Figure 3-5. Scheme for establishing the molecular point groups (cf. Refs. [3-4] and [3-5]).

whether there is a symmetry plane (C_s). In the absence of rotational axes and mirror planes, there may only be a center of symmetry (C_i), or there may be no symmetry element at all (C_1). If the molecule has rotation axes, it may have a mirror-rotation axis with even-number order (S_{2n}) coinciding with the rotation axis. For S_4, there will be a coinciding C_2; for S_6, a coinciding C_3; and for S_8, both C_2 and C_4.

In any case the search is for the highest order C_n axis. Then it is ascertained whether there are n C_2 axes present perpendicular to the C_n axis. If such C_2 axes are present, then there is D symmetry. If in addition to D symmetry there is a σ_n plane, the point group is D_{nh}, while if there are n symmetry planes (σ_d) bisecting the twofold axes, the point group is D_{nd}. If there are no symmetry planes in a molecule with D symmetry, the point group is D_n.

Finally, if no C_2 axes perpendicular to C_n are present, then the lowest symmetry will be C_n; when a perpendicular symmetry plane is present, the point group will be C_{nh}, and when there are n coinciding symmetry planes, it will be C_{nv}.

3.5 EXAMPLES

In this section, actual molecular structures are shown for the various point groups along with occasional examples from outside chemistry. The Schoenflies notation is used and the characteristic symmetry elements are enumerated.

C_1. There are no symmetry elements except the onefold rotation axis, or identity, of course. C_1 symmetry is *asymmetry*. Some examples are shown in Figure 3-6.

$C_2, C_3, C_4, C_5, C_6, \ldots, C_n$. One twofold, threefold, fourfold, fivefold, and sixfold rotation axis, respectively, and this series can be continued by analogy. C_n has one n-fold rotation axis. Examples: Figure 3-7.

C_i. Center of symmetry. Example: Figure 3-8.

C_s. One symmetry plane. Examples: Figure 3-9.

S_4. One fourfold mirror-rotation axis. Example: Figure 3-10a.

S_6. One sixfold mirror-rotation axis, which is, of course, equivalent to one threefold rotation axis plus a center of symmetry. Example: Figure 3-10b.

$C_{2h}, C_{3h}, \ldots, C_{nh}$. One twofold, threefold, . . ., n-fold rotation axis with a symmetry plane perpendicular to it. Examples: Figure 3-11.

C_{2v}. Two perpendicular symmetry planes whose crossing line is a twofold rotation axis. Examples: Figure 3-12a.

C_{3v}. One threefold rotation axis with three symmetry planes which include the rotation axis. The angle is 60° between two symmetry planes. Examples: Figure 3-12b.

C_{4v}. One fourfold rotation axis with four symmetry planes which include the rotation axis. The four planes are grouped in two nonequivalent pairs. One pair is rotated relative to the other pair by 45°. The angle between the two planes within each pair is 90°. Examples: Figure 3-12c.

$C_{5v}, C_{6v}, \ldots, C_{nv}$. This series can be continued by analogy. When n is even, there are two sets of symmetry planes. One set is rotated relative to the other set by $(180/n)°$. The angle between the planes within each set is $(360/n)°$.

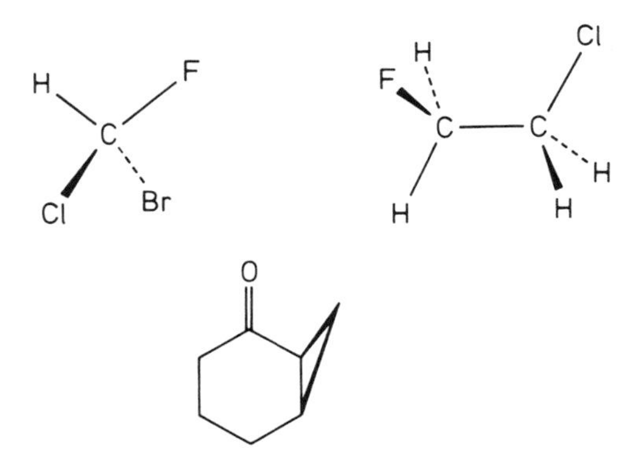

Figure 3-6. Examples with C_1 symmetry: no symmetry elements except the onefold rotation axis (C_1 symmetry is asymmetry).

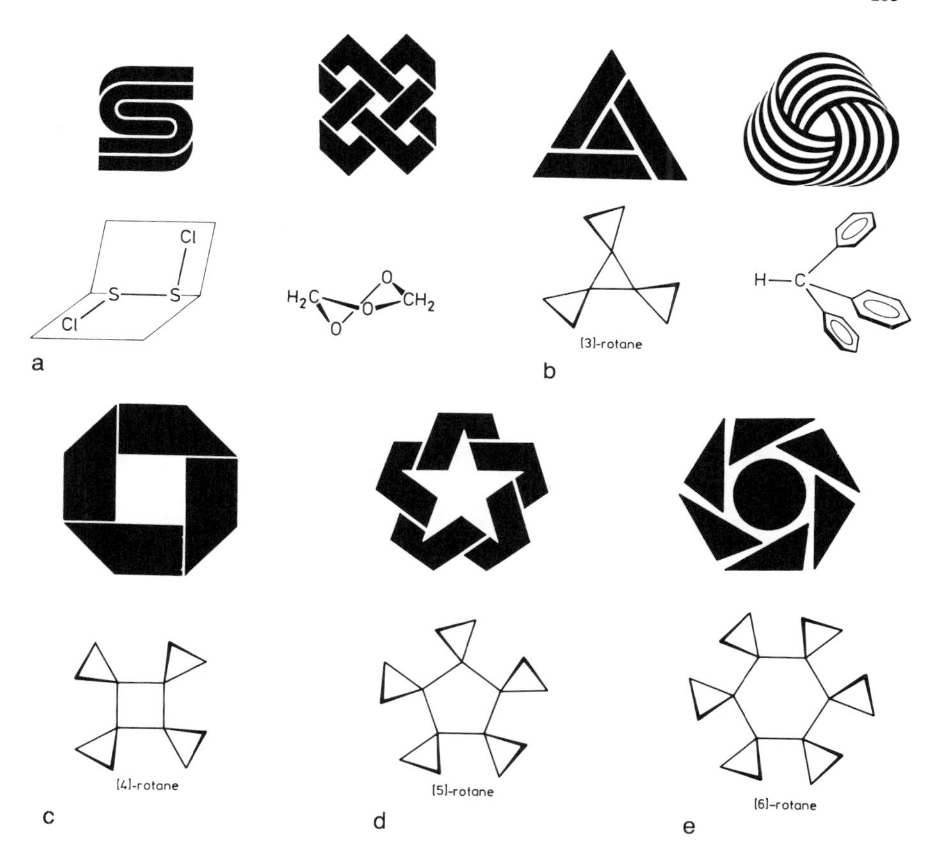

Figure 3-7. Some logos and molecules illustrating C_n symmetries. (a) C_2. Logos of Security First National Bank, California (left) and United Banks of Colorado (right). (b) C_3. Logos of Pittsburgh National Bank (left) and Woolmark (right). (c) C_4. Logo of Chase Manhattan Bank. (d) C_5. Logo of First American National Bank, Tennessee. (e) C_6. Logo of Crocker Bank. Logos are from Ref. [3-6].

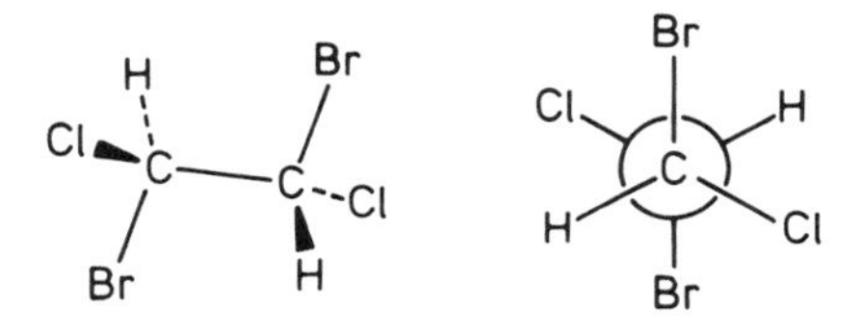

Figure 3-8. Example with C_i symmetry.

Figure 3-9. Examples with C_s symmetry. The pictures show the tail of a whale, off Plymouth Massachusetts, a leaf, and the Flatiron building in New York City. Photographs by the authors

a H_3C O Si O CH_3 H_3C O O CH_3

b $[(CH_2)_2CH]_3CC[CH(CH_2)_2]_3$

Figure 3-10. Examples with mirror-rotation axes. (a) S_4 symmetry. (b) S_6 symmetry.

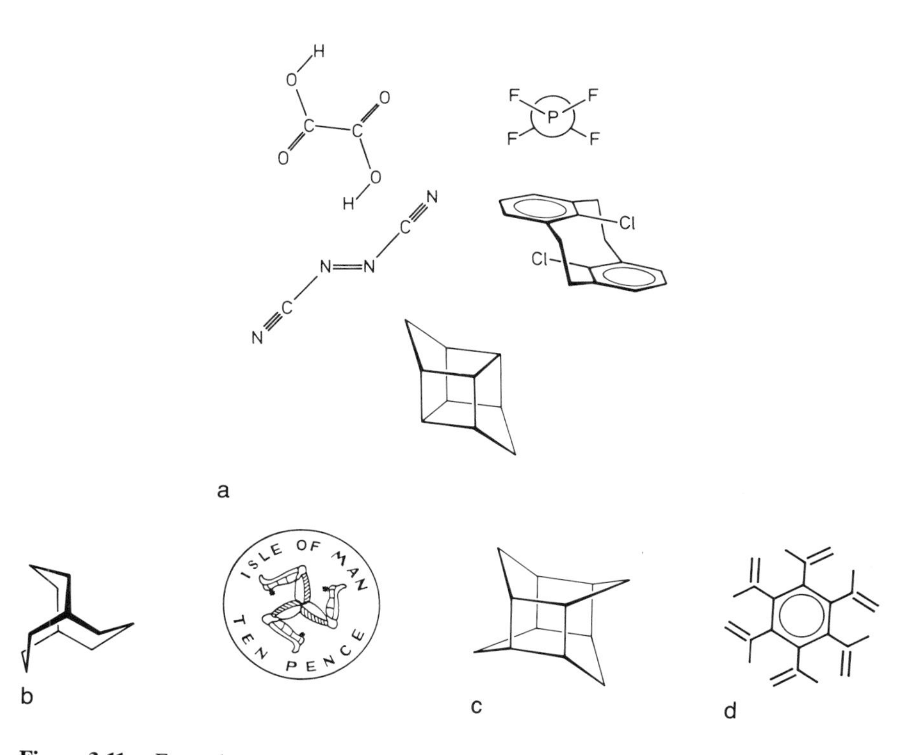

Figure 3-11. Examples with rotation axis and perpendicular symmetry plane. (a) C_{2h} symmetry. (b) C_{3h} symmetry. The molecule bicyclo[3.3.3]undecane is also called "manxane." It has C_{3h} symmetry indeed. The Isle of Man coin shows a one-sided rosette whose symmetry is only C_3. (c) C_{4h} symmetry. (d) C_{6h} symmetry.

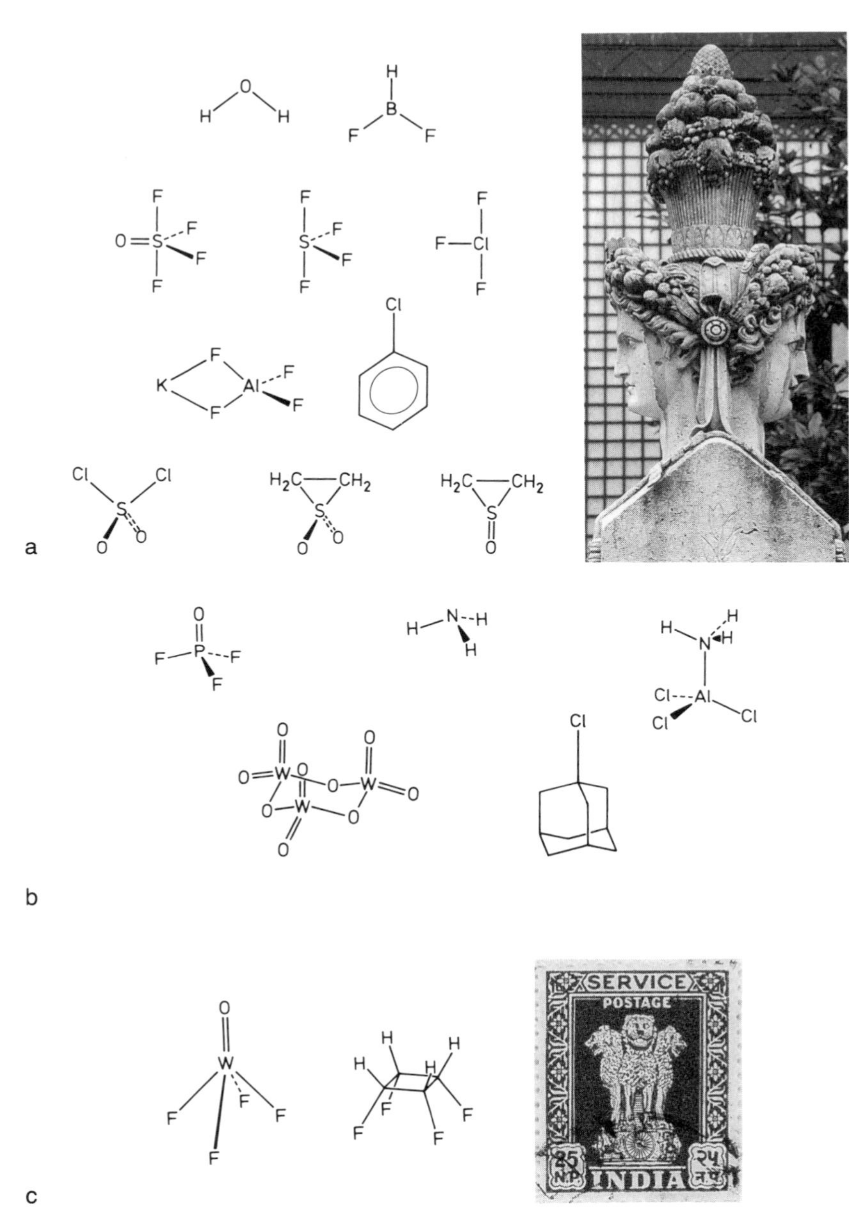

Figure 3-12. Examples with rotation axis and symmetry planes containing the rotation axis. (a) C_{2v} symmetry. The examples include a sculpture in Paris; photograph by the authors. (b) C_{3v} symmetry. (c) C_{4v} symmetry. The examples include an Indian stamp. (*Continued on next page*)

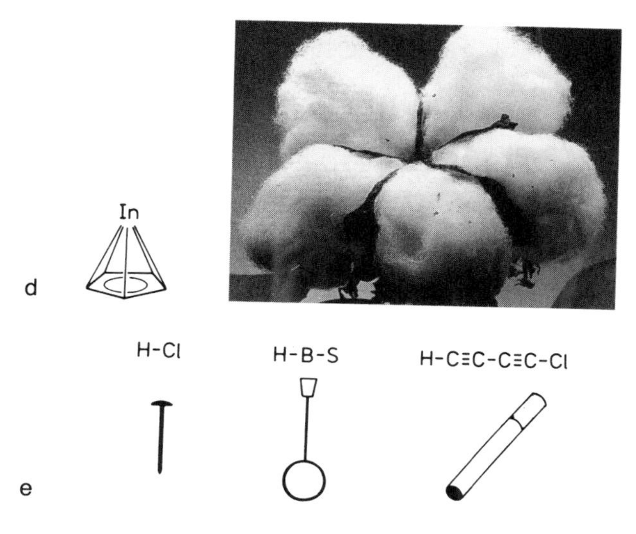

Figure 3-12. (*Continued*) (d) C_{5v} symmetry. The examples include a cotton plant. (e) $C_{\infty v}$ symmetry.

When n is odd, the angle between the symmetry planes is $(180/n)°$. Examples: Figure 3-12d.

$C_{\infty v}$. One infinite-fold rotation axis with an infinite number of symmetry planes which include the rotation axis. Examples: Figure 3-12e.

D_2. Three mutually perpendicular twofold rotation axes. Example: Figure 3-13a.

D_3. One threefold rotation axis and three twofold rotation axes perpendicular to the threefold axis. The twofold axes are at 120°, so the minimum angle between two such axes is 60°. Examples: Figure 3-13b.

D_4. One fourfold rotation axis and four twofold rotation axes which are perpendicular to the fourfold axis. The four axes are grouped in two nonequivalent pairs. One pair is rotated relative to the other pair by 45°. The angle between the two axes within each pair is 90°.

D_5, D_6, D_7, . . ., D_n. This series can be continued by analogy. It is characterized by one n-fold rotation axis and n twofold rotation axes perpendicular to the n-fold axis.

D_{2d}. Three mutually perpendicular twofold rotation axes and two symmetry planes. The planes include one of the three rotation axes and bisect the angle between the other two. Examples: Figure 3-14a.

D_{3d}. One threefold rotation axis with three twofold rotation axes perpen-

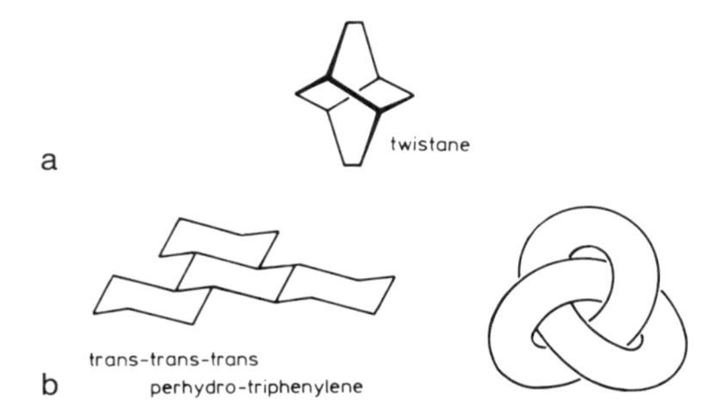

Figure 3-13. Examples with D_n symmetries. (a) D_2. (b) D_3.

dicular to it and three symmetry planes. The angle between the twofold axes is 60°. The symmetry planes include the threefold axis and bisect the angles between the twofold axes. Examples: Figure 3-14b.

$D_{4d}, D_{5d}, D_{6d}, D_{7d}, \ldots, D_{nd}$. One fourfold rotation axis with four twofold rotation axes perpendicular to it and four symmetry planes. The angle between the twofold axes is 45°. The symmetry planes include the fourfold axis and bisect the angles between the twofold axes. The series can be continued by analogy. Examples: Figures 3-14c and 3-14d.

D_{2h}. Three mutually perpendicular symmetry planes. Their three crossing lines are three twofold rotation axes, and their crossing point is a center of symmetry. Examples: Figure 3-15a.

D_{3h}. One threefold rotation axis, three symmetry planes (at 60°) which contain the threefold axis, and another symmetry plane perpendicular to the threefold axis. Examples: Figure 3-15b.

D_{4h}. One fourfold axis, one symmetry plane perpendicular to it, and four symmetry planes which include the fourfold axis. The four planes make two pairs. One pair is rotated relative to the other pair by 45°. The two planes in each pair are perpendicular to each other. Examples: Figure 3-15c.

D_{5h}. One fivefold rotation axis, one symmetry plane perpendicular to it, and five symmetry planes which include the fivefold rotation axis. The angle between the adjacent five planes is 36°. Examples: Figure 3-15d.

D_{6h}. One sixfold rotation axis, one symmetry plane perpendicular to it, and six symmetry planes which include the sixfold axis. The six planes are grouped in two sets. One set is rotated relative to the other set by 30°. The angle between the planes within each set is 60°. Examples: Figure 3-15e.

D_{nh}. The series can be continued by analogy. There will be one n-fold rotation axis, one symmetry plane perpendicular to it, and n symmetry planes which include the n-fold axis. When n is even, there are two sets of symmetry planes. One set is rotated relative to the other set by $(180/n)°$. The angle

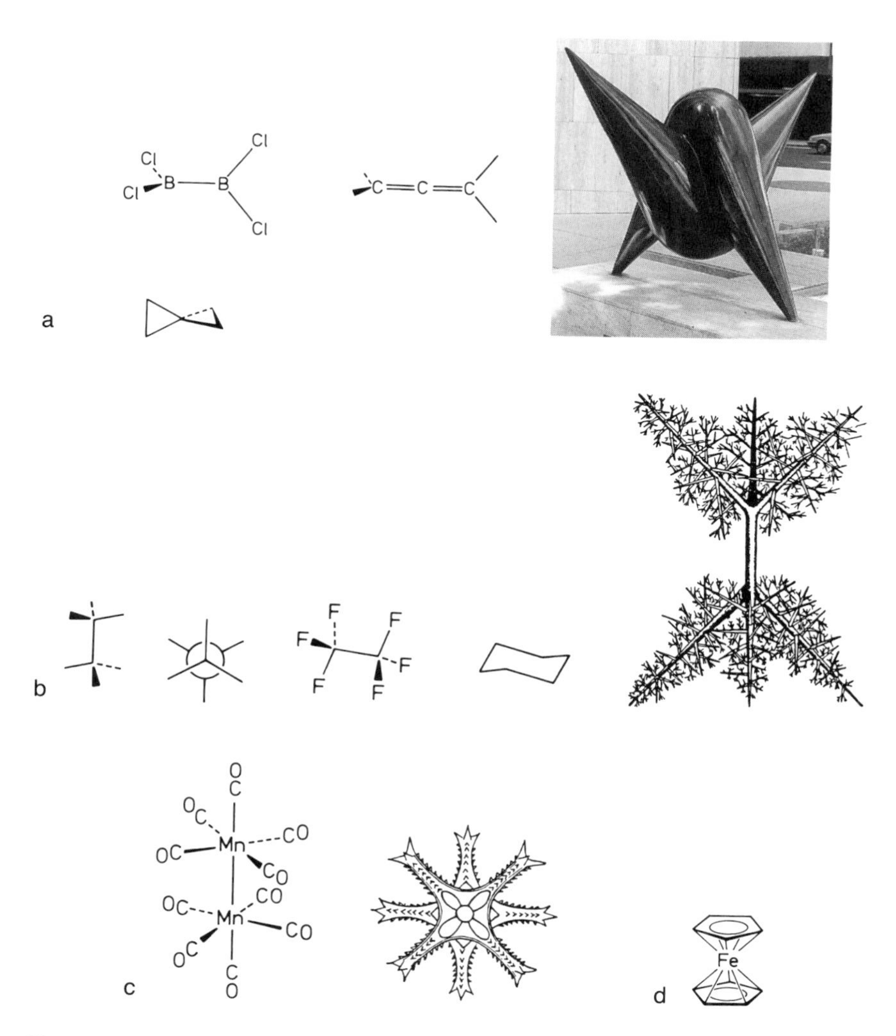

Figure 3-14. Examples with D_{nd} symmetries. (a) D_{2d}. The examples include a sculpture in Honolulu, Hawaii; photograph by the authors. (b) D_{3d}. The drawing of the radiolarian is from Ref. [3-7]. (c) D_{4d}. The drawing of the plant is from Ref. [3-7]. (d) D_{5d}.

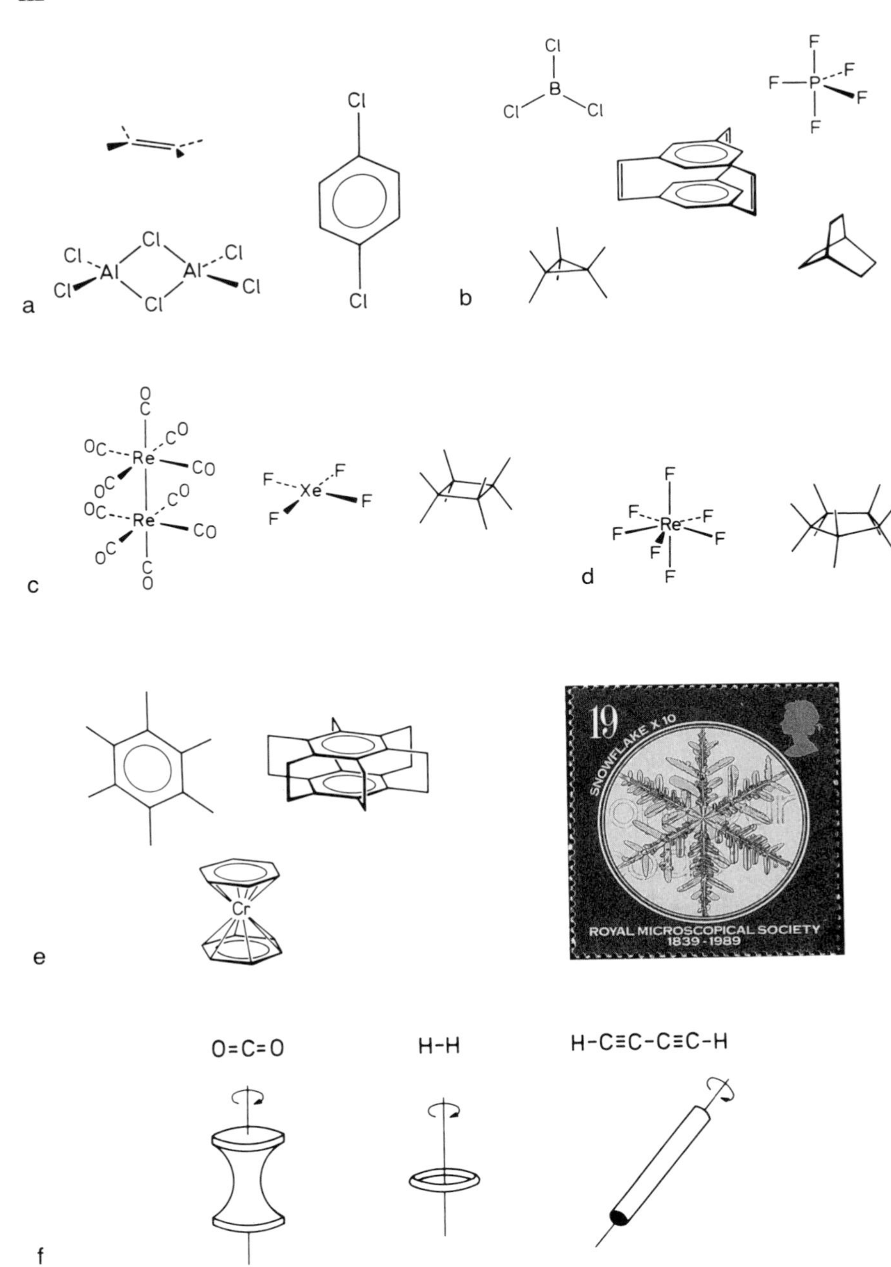

Figure 3-15. Examples with D_{nh} symmetries. (a) D_{2h}. (b) D_{3h}. (c) D_{4h}. (d) D_{5h}. (e) D_{6h}. The examples include a snowflake on a British stamp. (f) $D_{\infty h}$.

between the planes within each set is $(360/n)°$. When n is odd, the angle between the symmetry planes is $(180/n)°$.

$D_{\infty h}$. One ∞-fold axis and a symmetry plane perpendicular to it. Of course, there are also an infinite number of symmetry planes which include the ∞-fold rotation axis. Examples: Figure 3-15f.

T. Three mutually perpendicular twofold rotation axes and four threefold rotation axes. The threefold axes all go through a vertex of a tetrahedron and the midpoint of the opposite face center. The twofold axes connect the midpoints of opposite edges of this tetrahedron. Examples: Figure 3-16a.

T_d. In addition to the symmetry elements of symmetry T, there are six symmetry planes, each pair of them being mutually perpendicular. All of these symmetry planes contain two threefold axes. Examples: Figure 3-16b.

T_h. In addition to the symmetry elements of symmetry T, there is a center of symmetry which introduces also three symmetry planes perpendicular to the twofold axes. Example: Figure 3-16c.

O_h. Three mutually perpendicular fourfold rotation axes and four threefold rotation axes, which are tilted with respect to the fourfold axes in a uniform manner, and a center of symmetry. Examples: Figure 3-17.

I_h. The most characteristic feature of this point group is the presence of six fivefold rotation axes. Examples: Figure 3-18.

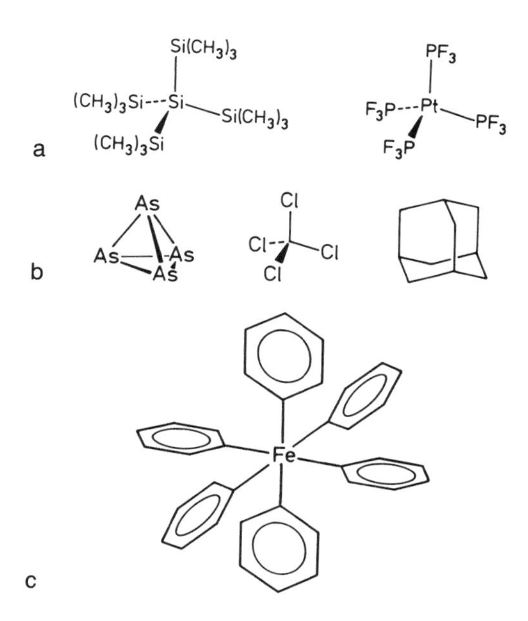

Figure 3-16. Examples with T (a), T_d (b), and (c) T_h symmetry.

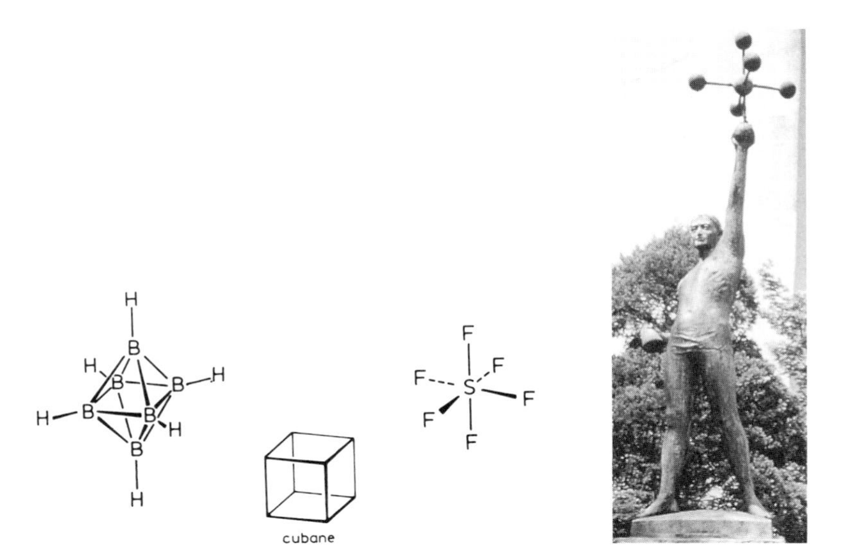

Figure 3-17. Examples with O_h symmetry. The sculpture, in Seoul, shows a chemist holding an octahedral molecular structure and is by Eui Soon Choi; photograph by the authors.

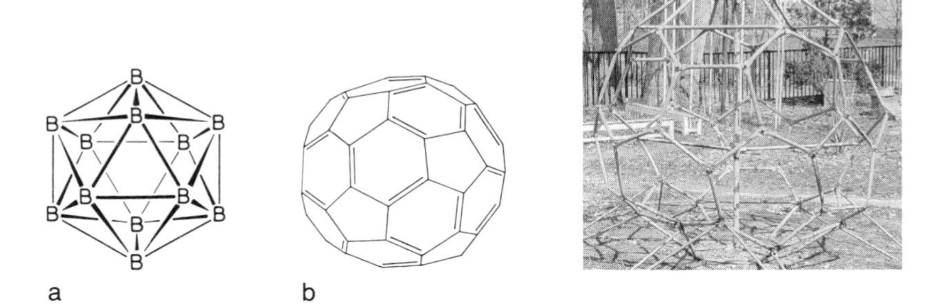

Figure 3-18. Examples with I_h symmetry: (a) The regular icosahedral boron skeleton of the $B_{12}H_{12}^{2-}$ ion; (b) truncated icosahedral structure of buckminsterfullerene, C_{60}, and of a climber in Sapporo, Japan. Photograph by the authors.

3.6 CONSEQUENCES OF SUBSTITUTION

A tetrahedral AX_4 molecule, for example, methane, CH_4, has the point group of the regular tetrahedron, T_d. Gradual substitution of the X ligands by B ligands leads to less symmetrical tetrahedral configurations (Figure 3-19a), until complete substitution is accomplished.

If each consecutive substitution introduces a new kind of ligand, then the symmetry will continue to decrease. This is shown for the tetrahedral case in Figure 3-19b. As the sites of all X ligands are equivalent in each of these configurations, the symmetry changes accompanying the substitution are determined *a priori*.

Let us consider now an octahedral AX_6 molecule, for example, sulfur hexafluoride, SF_6, which has the symmetry of the regular octahedron, O_h. Substitution of an X ligand by a B ligand results in an AX_5B molecule whose symmetry is again determined *a priori* to be C_{4v}. The substitution of a second X ligand by another ligand B may lead to alternative structures as the sites of the five X ligands after the first substitution are no longer equivalent. The symmetry variations in this substitution process are illustrated in Figure 3-20. A yet larger variety is obtained if each consecutive substitution introduces a new kind of ligand.

Another example among fundamental structures is the benzene geometry, D_{6h}. Gradual substitution of an increasing number of hydrogens by ligands X results in the symmetry variations illustrated in Figure 3-21. As regards the

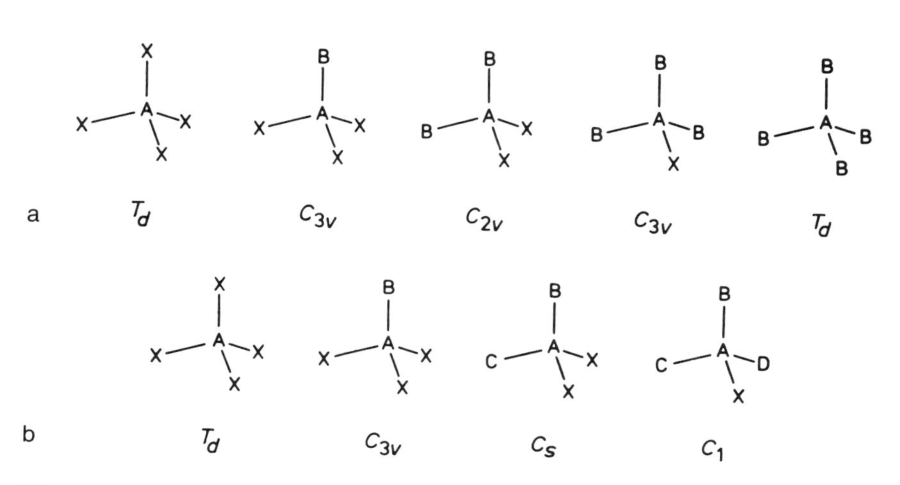

Figure 3-19. Substitution in a tetrahedral AX_4 molecule: (a) Gradual substitution of the ligands X by ligands B; (b) substitution of the ligands X by different ligands.

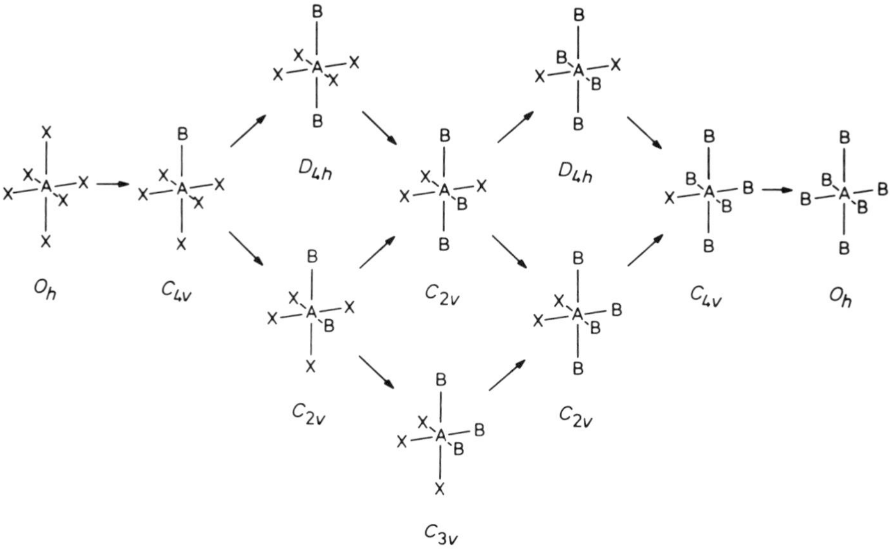

Figure 3-20. Gradual substitution of the ligands X in an octahedral AX_6 molecule by ligands B.

molecular point group, the monosubstituted and the pentasubstituted derivatives are equivalent. All derivatives can be grouped in such pairs with each of the trisubstituted benzenes constituting a pair by itself. Again, only the simplest case is considered here, with one kind of ligand used in all substituted positions. The decrease in the symmetry of the molecular point group for the substituted derivatives occurs because of the presence of the substituent ligands. It does not presuppose a change in the hexagonal symmetry of the benzene ring itself. Modern structure analyses have determined, however, that an appreciable deformation of the ring from regularity may also take place, depending on the nature of the substituents. The largest deformation usually occurs at the so-called *ipso* angle adjacent to the substituent. According to the general observation, electronegative substituents tend to compress the ring while electropositive substituents elongate it [3-8].

Complex formation usually implies the association of molecules or other species which may also exist separately in chemically nonextreme conditions. Complex formation often has important consequences on the shapes and symmetries of the constituent molecules [3-9]. The $H_3N{\cdot}AlCl_3$ donor–acceptor complex [3-9], for example, has a triangular antiprismatic shape with C_{3v} symmetry as seen in Figure 3-22. The symmetry of the donor part (NH_3) remains unchanged in the complex, and the geometrical changes are relatively

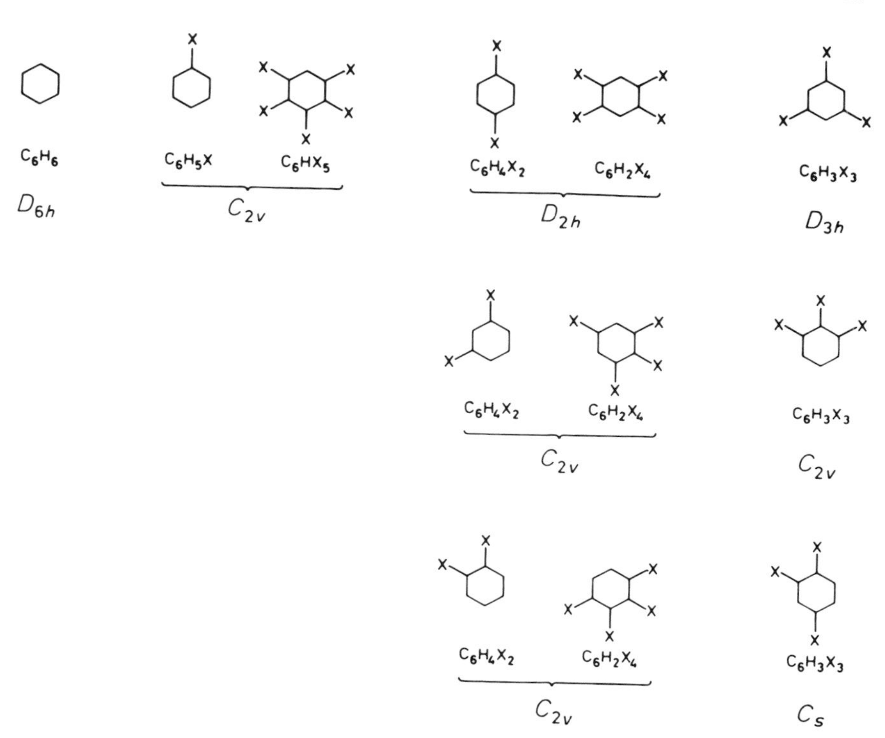

Figure 3-21. The symmetries of benzene and its $C_6H_nX_{6-n}$ derivatives.

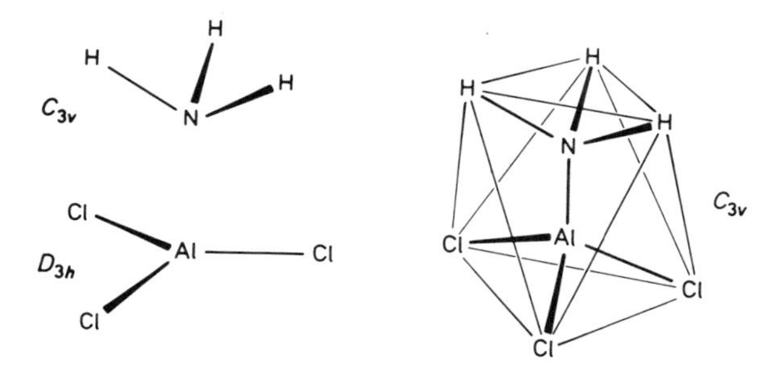

Figure 3-22. The uncomplexed ammonia and aluminum trichloride molecules and the triangular antiprismatic shape of the $H_3N \cdot AlCl_3$ donor–acceptor complex.

small. On the other hand, there are more drastic geometrical changes in the acceptor part ($AlCl_3$) due to loss of coplanarity of the four atoms, and this results in a reduction in the point group. However, the structural change in the acceptor part may also be viewed as if the complex formation completes the tetrahedral configuration around the central atoms in the component molecules. The nitrogen configuration may be considered to be tetrahedral already in ammonia, with the lone pair of electrons being the fourth ligand. For aluminum, it is indeed the complexation that makes the tetrahedral configuration complete. Coordination molecules often demonstrate the utility of polyhedra in describing molecular shapes, symmetries, and geometries. Of course, such description may be useful for many other classes of compounds as well.

3.7 POLYHEDRAL MOLECULAR GEOMETRIES

In the Preface to the third edition of his *Regular Polytopes* [3-10], the great geometer H. S. M. Coxeter called attention to the icosahedral structure of a boron compound in which twelve boron atoms are arranged like the vertices of an icosahedron. It had been widely believed that there would be no inanimate occurrence of an icosahedron, or of a regular dodecahedron either.

In 1982 the synthesis and properties of a new polycyclic $C_{20}H_{20}$ hydrocarbon, dodecahedrane, were reported [3-11]. The 20 carbon atoms of this molecule are arranged like the vertices of a regular dodecahedron. When in the early sixties Schultz [3-12] discussed the topology of the polyhedrane and prismane molecules (*vide infra*), at that time it was in terms of a geometrical diversion rather than true-life chemistry. Since then it has become real chemistry.

It should be reemphasized that the above high-symmetry examples refer to isolated molecules and not to crystal structures. Crystallography has, of course, been one of the main domains where the importance of polyhedra has been long recognized, together with some limitations which forbid the occurrence of regular *pentagonal* figures in crystals. Polyhedra are not less important in the world of molecules, where the limitations existing in crystals do not apply.

In the first edition of *Regular Polytopes* [3-10], Coxeter stated, "the chief reason for studying regular polyhedra is still the same as in the times of the Pythagoreans, namely, that their symmetrical shapes appeal to one's artistic sense." The success of modern molecular chemistry does not diminish the validity of this statement. On the contrary. There is no doubt that aesthetic appeal has much contributed to the rapid development of what could be termed polyhedral chemistry. One of the pioneers in the area of polyhedral borane chemistry, Earl Muetterties, movingly described [3-13] his attraction to the chemistry of boron hydrides, comparing it to Escher's devotion to periodic drawings [3-14]. Muetterties' words are quoted here*:

> When I retrace my early attraction to boron hydride chemistry, Escher's poetic introspections strike a familiar note. As a student intrigued by early descriptions of the extraordinary hydrides, I had not the prescience to see the future synthesis developments nor did I have then a scientific appreciation of symmetry, symmetry operations, and group theory. Nevertheless, some inner force also seemed to drive me but in the direction of boron hydride chemistry. In my initial synthesis efforts, I was not the master of these molecules; they seemed to have destinies unperturbed by my then amateurish tactics. Later as the developments in polyhedral borane chemistry were evident on the horizon, I found my general outlook changed in a characteristic fashion. For example, my doodling, an inevitable activity of mine during meetings, changed from characters of nondescript form to polyhedra, fused polyhedra and graphs.
>
> I (and others, my own discoveries were not unique nor were they the first) was profoundly impressed by the ubiquitous character of the three-center relationship in bonding (e.g., the boranes) and nonbonding situations. I found a singular uniformity in geometric relationships throughout organic, inorganic, and organometallic chemistry: The favored geometry in coordination compounds, boron hydrides, and metal clusters is the polyhedron that has all faces equilateral or near equilateral triangles . . .

The polyhedral description of molecular geometries is, of course, generally applicable as these geometries are spatial constructions. To emphasize that even planar or linear molecules are also included, the term polytopal could be used rather than polyhedral. The real utility of the polyhedral description is for molecules possessing a certain amount of symmetry. Because of this and also because of the introductory character of our discussion, only molecules with relatively high symmetries will be mentioned.

The polyhedral description may be useful for widely different systems. Thus, for example, both the tetraarsene, As_4, and the methane, CH_4, molecules have tetrahedral shapes (Figure 3-23) and T_d symmetry. However, there is an important difference in their structures. In the As_4 molecule all the four constituent nuclei are located at the vertices of a regular tetrahedron, and all the edges of this tetrahedron are chemical bonds between the As atoms. In the methane molecule, there is a central carbon atom, and four chemical bonds are

*From Ref. [3-13], p. 98. The passage is quoted with permission from Academic Press.

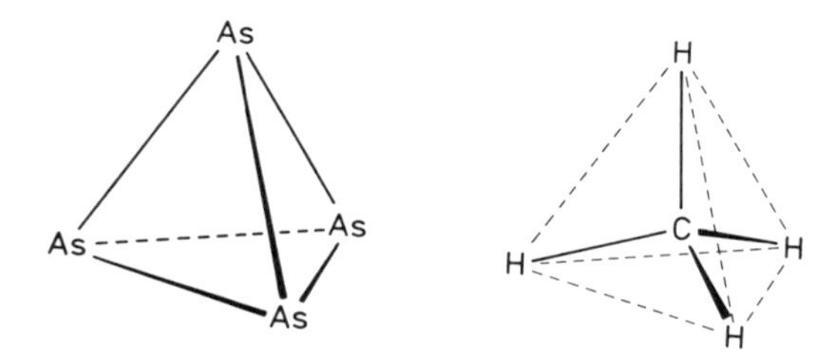

Figure 3-23. The molecular shapes of As_4 and CH_4.

directed from it to the four vertices of a regular tetrahedron, where the four protons are located. The edges are not chemical bonds.

The As_4 and CH_4 molecules are clear-cut examples of the two distinctly different arrangements. However, these distinctions are not always so unambiguous. An interesting example is the structure of zirconium borohydride, $Zr(BH_4)_4$. Two independent studies [3-15, 3-16] described its structure by the same polyhedral configuration, while they differed in the assignment of the chemical bonds (Figure 3-24). The most important difference between the two interpretations concerns the linkage between the central zirconium atom and the four boron atoms situated at the four vertices of a regular tetrahedron. According to one interpretation [3-15], there are four Zr–B bonds in the

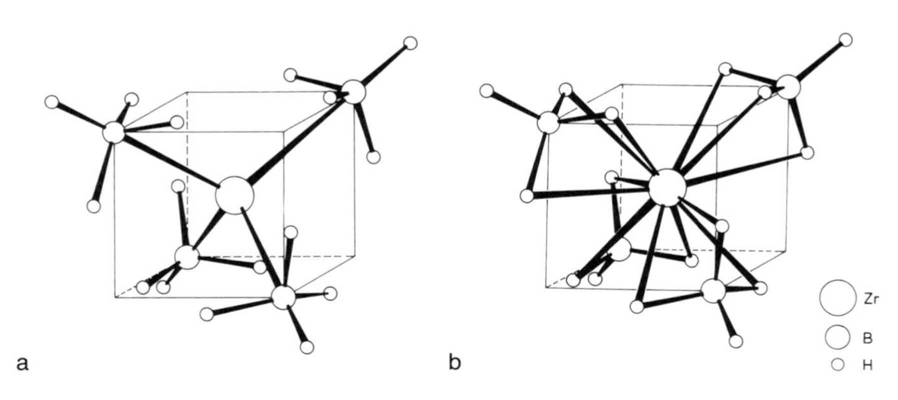

Figure 3-24. The molecular configuration of zirconium borohydride, $Zr(BH_4)_4$, in two interpretations but described by the same polyhedral shape. (a) According to one interpretation [3-15], the zirconium atom is directly bonded to the four tetrahedrally arranged boron atoms. (b) According to another interpretation [3-16], the zirconium and the tetrahedrally arranged boron atoms are not bonded directly. Their linkage is established by four times three hydrogen bridges.

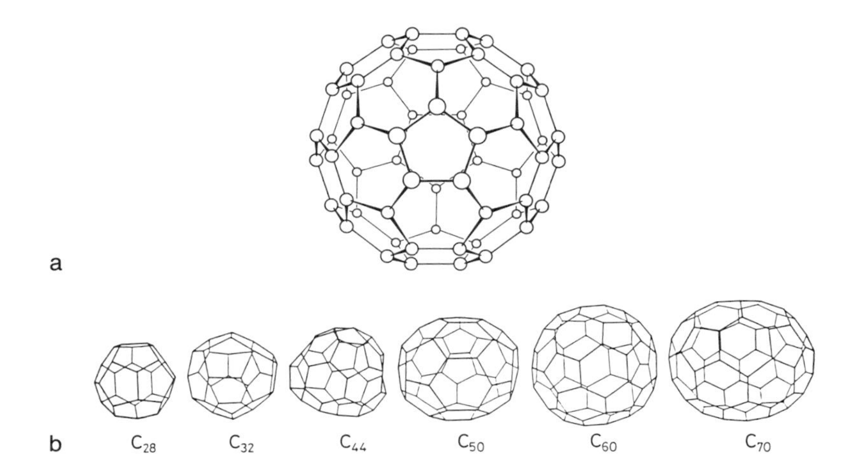

Figure 3-25. (a) Buckminsterfullerene, "the roundest, most symmetrical large molecule found so far" [3-19]. (b) A few members of the fullerene family.

tetrahedral arrangement. On the other hand, according to the other interpretation [3-16], there is no direct Zr–B bond, but each boron atom is linked to the zirconium atom by three hydrogen bridges. Zirconium borohydride is one of the interesting metal borohydrides whose molecular geometries have presented a challenge to the structural chemist [3-9].

Polyhedral molecular geometries have made even the mass media recently with the discovery of buckminsterfullerene [3-17], C_{60} (Figure 3-25), and especially with the whole new chemistry of the fullerenes (see, e.g., Ref. [3-18]).* Buckminsterfullerene was named "Molecule of the Year" in the December 20, 1991, issue of *Science* magazine [3-19] while it was only the first runner-up the previous year [3-20]. Although even a runner-up status is of the highest prestige, in 1990 even the structural formula was drawn erroneously (cf. Ref. [3-21]), and buckminsterfullerene was referred to as a "distant cousin" of diamond [3-20]. By December, 1991, all this had changed, and the *Science* editorial [3-22] stated,

> Part of the exhilaration of the fullerenes is the shock that an old reliable friend, the carbon atom, has for all these years been hiding a secret life-style. We were all familiar with the charming versatility of carbon, the backbone of organic chemistry, and its

*There is also a whole new journal devoted to this new class of compounds, *Fullerene Science and Technology* (T. Braun, ed.), Marcel Dekker, New York.

> infinite variation in aromatic and aliphatic chemistry, but when you got it naked, we believed it existed in two well-known forms, diamond and graphite. The finding that it could exist in a shockingly new structure unleashes tantalizing new experimental and theoretical ideas.

Then it added something that certainly carried a flavor of the broadest possibl implications:

> Perhaps the least surprising might be that improving life through science is a path that would see all the citizens of the world holding hands like carbon atoms in C_{60} and like them, welcoming any newcomer, no matter how different his or her skills or challenges.

Figure 3-25 shows a series of fullerenes, the C_{20} molecule being the smalles fullerene molecule.

3.7.1 Boron Hydride Cages

The boron hydrides are one of most beautiful classes of polyhedra compounds. Its representatives range from the simplest to the most compli cated systems. Our description here is purely phenomenological. Only in passing is reference made to the relationship between the characteristi polyhedral cage arrangements of the boron hydrides and the peculiarities o multicenter bonding (see, e.g., Refs. [3-23–3-26]).

All faces of the boron hydride polyhedra are equilateral or nearly equi lateral triangles. Those boron hydrides that have a complete polyhedral shap are called *closo* boranes (the Greek *closo* meaning closed). One of the mos symmetrical, and, accordingly, most stable, polyhedral boranes is th $B_{12}H_{12}^{2-}$ ion. Its regular icosahedral configuration is shown in Figure 3-26a The structural systematics of $B_nH_n^{2-}$ *closo* boranes and related $C_2B_{n-2}H$ *closo* carboranes are presented in Table 3-2, after Muetterties [3-13]. In carboranes some of the boron sites are taken by carbon atoms.

Another structural class of the boron hydrides is the so-called quasi-*closo* boranes. They are related to the *closo* boranes by removing a framework aton from the latter and adding in its stead a pair of electrons. Thus, one of th polyhedron framework sites is taken by an electron pair.

There are boron hydrides in which one or more of the polyhedral sites i truly removed. Figure 3-26b shows the systematics of borane polyhedra fragments as obtained from *closo* boranes, after Williams [3-27] and Rudolp [3-24]. All the faces of the polyhedral skeletons are triangular, and thus th polyhedra may be termed deltahedra and the derived fragments deltahedral The starting deltahedra are the tetrahedron, the trigonal bipyramid, the octa

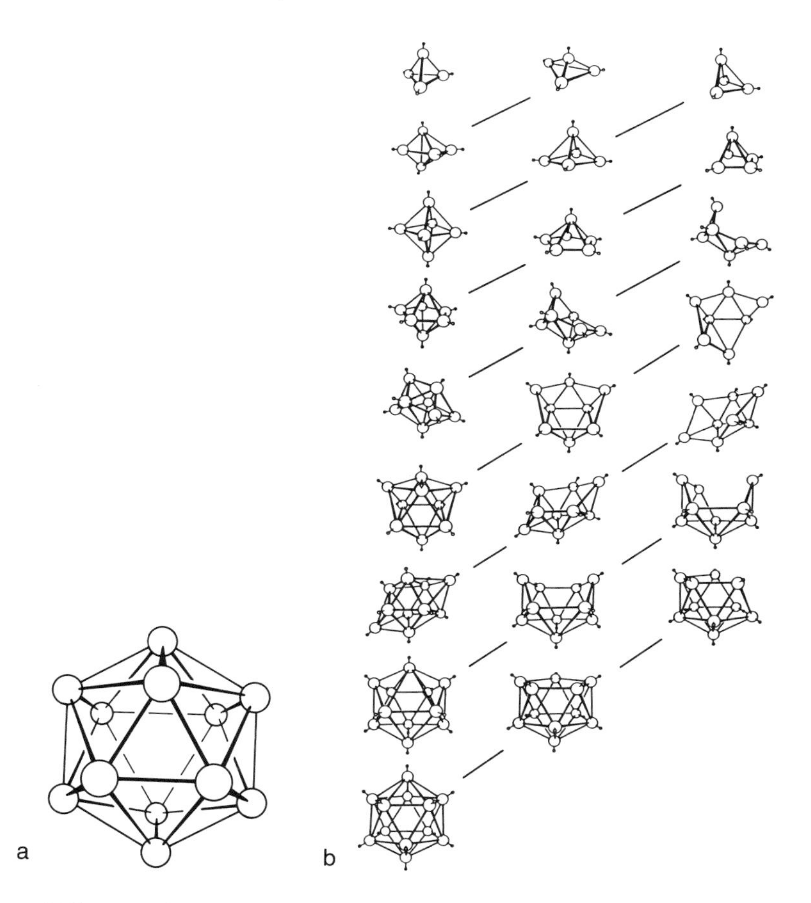

Figure 3-26. (a) The regular icosahedral boron skeleton of $B_{12}H_{12}^{2-}$. (b) *Closo*, *nido*, and *arachno* boranes. The genetic relationships are indicated by diagnoal lines. After Williams [3-27] and Rudolph [3-24]. Reprinted with permission from Ref. [3-24]. Copyright (1976) American Chemical Society.

hedron, the pentagonal bipyramid, the bisdisphenoid, the symmetrically tricapped trigonal prism, the bicapped square antiprism, the octadecahedron, and the icosahedron. The geometrical systematics have been recently updated [3-28].

A *nido* (nestlike) boron hydride is derived from a *closo* borane by the removal of one skeleton atom. If the starting *closo* borane is not a regular polyhedron, then the atom removed is the one at a vertex with the highest connectivity. An *arachno* (weblike) boron hydride is derived from a *closo*

Table 3-2. Structural Systematics of $B_nH_n^{2-}$ *closo* Boranes and $C_2B_{n-2}H_n$ *closo* Carboranes[a]

Polyhedron and point group	Borane	Dicarbaborane
Tetrahedron, T_d	$(B_4Cl_4)^b$	—
Trigonal bipyramid, D_{3h}	—	$C_2B_3H_5$
Octahedron, O_h	$B_6H_6^{2-}$	$C_2B_4H_6$
Pentagonal bipyramid, D_{5h}	$B_7H_7^{2-}$	$C_2B_5H_7$
Dodecahedron (triangulated), D_{2d}	$B_8H_8^{2-}$	$C_2B_6H_8$
Tricapped trigonal prism, D_{3h}	$B_9H_9^{2-}$	$C_2B_7H_9$
Bicapped square antiprism, D_{4d}	$B_{10}H_{10}^{2-}$	$C_2B_8H_{10}$
Octadecahedron, C_{2v}	$B_{11}H_{11}^{2-}$	$C_2B_9H_{11}$
Icosahedron, I_h	$B_{12}H_{12}^{2-}$	$C_2B_{10}H_{12}$

[a] After Muetterties [3-13].
[b] B_4H_4 not known.

borane by the removal of two adjacent skeleton atoms. If the starting *closo* borane is not a regular polyhedron, then, again, one of the two atoms removed is at a vertex with the highest connectivity. Complete *nido* and *arachno* structures are shown in Figure 3-27 together with the starting boranes [3-13]. The fragmented structures are completed by a number of bridging and terminal hydrogens. The above examples are, of course, from among the simplest boranes and their derivatives.

3.7.2 Polycyclic Hydrocarbons

Some fundamental polyhedral shapes are found among polycyclic hydrocarbons. The bond arrangements around the carbon atoms in such configurations may be far from the energetically most advantageous, causing *strain* in these structures [3-29]. The strain may be so large as to render particular arrangements too unstable to exist under any reasonable conditions. On the other hand, the fundamental character of these shapes and their high symmetry and aesthetic appeal make them an attractive and challenging "playground" to the organic chemist [3-30]. Incidentally, these substances have also great practical importance as they are building blocks for such natural products as steroids, alkaloids, vitamins, carbohydrides, and antibiotics.

Tetrahedrane, $(CH)_4$, would be the simplest regular polyhedral polycyclic hydrocarbon (Figure 3-28a). However, since it has such a high strain energy and provides easy access to attacking reagents, its preparation may not be possible. Its derivative, tetra-*tert*-butyltetrahedrane (Figure 3-28b), however, has been prepared [3-31]. This compound is amazingly stable, perhaps because the substituents help "clasp" the molecule together.

The next Platonic solid is the cube, and the corresponding polycyclic

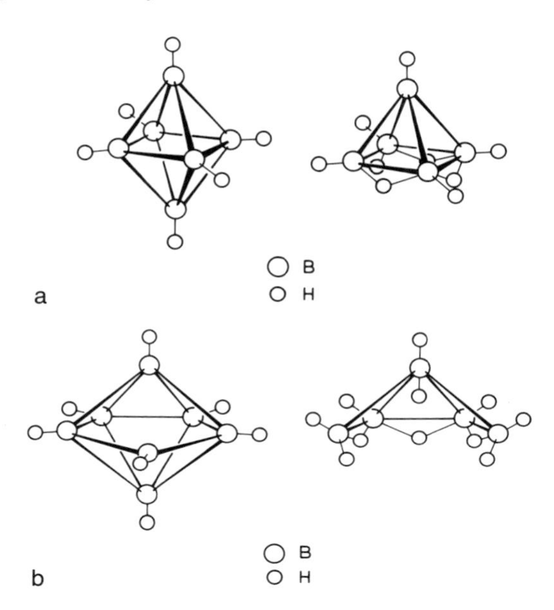

Figure 3-27. Examples of *closo*/*nido* and *closo*/*arachno* structural relationships: (a) *closo*-$B_6H_6^{2-}$ and *nido*-B_5H_9: (b) *closo*-$B_7H_7^{2-}$ and *arachno*-B_5H_{11}. After Muetterties [3-13].

hydrocarbon, cubane, $(CH)_8$ (Figure 3-28c), has been known for some time [3-32]. The strain energy of the C–C bonds in cubane is among the highest known. It is unstable thermodynamically but stable kinetically, like a "rock" [3-33]. The preparation of dodecahedrane, $(CH)_{20}$ (Figure 3-28d), by Paquette and co-workers [3-11] followed a prediction almost two decades before, by Schultz [3-12], concerning possible hydrocarbon polyhedranes:

> Dodecahedrane is the one substance of the series with almost ideal geometry, physically the molecule is practically a miniature ball bearing! One would expect the substance to have a low viscosity, a high melting point but low boiling point, high thermal stability, a very simple infrared spectrum and perhaps an aromatic-like p.m.r. spectrum. Chemically one might expect a relatively easy (for an aliphatic hydrocarbon) removal of a tertiary proton from the molecule, for the negative charge thus deposited on the molecule could be accommodated on any one of the twenty completely equivalent carbon atoms, the carbanion being stabilized by a "rolling charge" effect that delocalizes the extra electron.

Incidentally, the simplest fullerene is dodecahedrene, C_{20}.

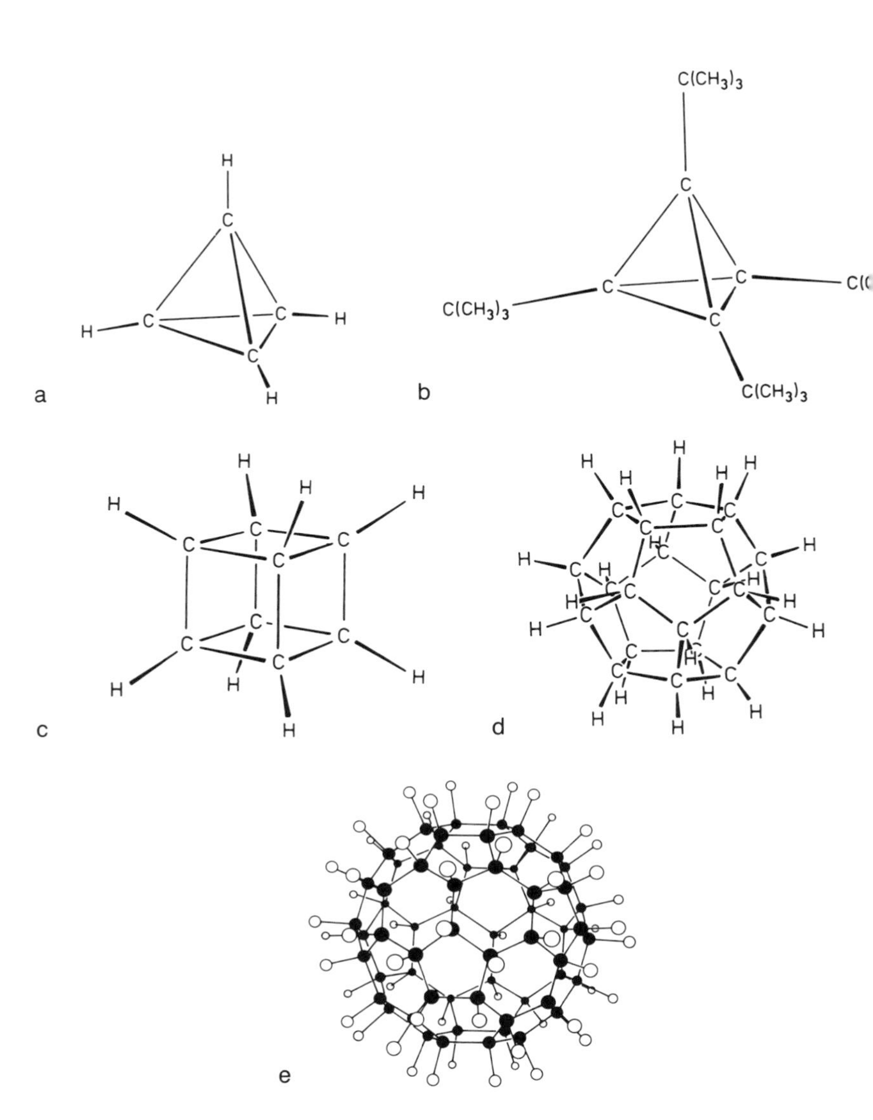

Figure 3-28. Polyhedrane molecules. (a) Tetrahedrane, $(CH)_4$. It has very high strain ene and has not (yet?) been prepared. (b) Tetra-*tert*-butyltetrahedrane, $\{C[C(CH_3)_3]\}_4$, [3-31]. Cubane, $(CH)_8$, [3-32]. (d) Dodecahedrane, $(CH)_{20}$, [3-11]. (e) $C_{60}H_{60}$ [3-34], not yet prepa

(*Continued on next page*)

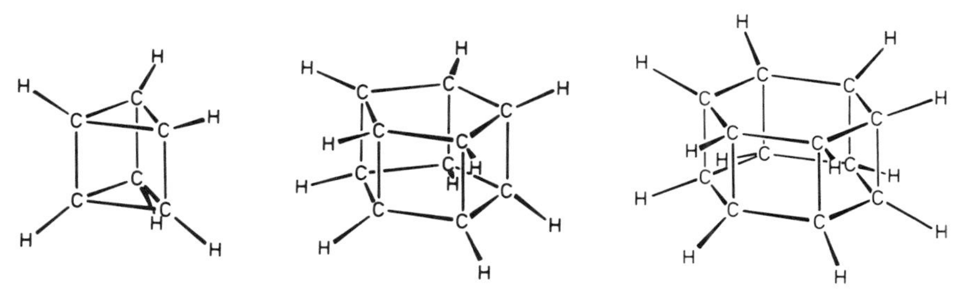

Figure 3-28. (*Continued*) (f) Triprismane, C_6H_6, pentaprismane, $C_{10}H_{10}$, and hexaprismane, $C_{12}H_{12}$ (not yet prepared).

In the $(CH)_n$ convex polyhedral hydrocarbon series, each carbon atom is bonded to three other carbon atoms. The fourth bond is directed externally to a hydrogen atom. Around the all-carbon polyhedron, there is thus a similar polyhedron whose vertices are protons. The edges of the all-carbon polyhedron are carbon–carbon chemical bonds, while the edges of the larger all-proton polyhedron do not correspond to any chemical bonds. This kind of arrangement of the polycyclic hydrocarbons is not possible for the remaining two Platonic solids. There are four bonds meeting at the vertices of the octahedron and five at the vertices of the icosahedron. For similar reasons, only 7 of the 13 Archimedean polyhedra can be considered in the $(CH)_n$ polyhedral series. One of them is "fuzzyball," or $C_{60}H_{60}$, a predicted form of fully hydrogenated buckminsterfullerene (Figure 3-28e) [3-34]. Table 3-3 presents some characteristics of the polyhedranes, after Schultz [3-12]. It also indicates which of the hydrocarbon polyhedranes have already been synthesized, as of 1994.

The cubane molecule may also be considered and called tetraprismane (cf. Figs. 2-72 and 3-28c). It may be described as composed of eight identical methine units arranged at the corners of a regular tetragonal prism with O_h symmetry and bound into two parallel four-membered rings conjoined by four four-membered rings. Triprismane, $(CH)_6$ [3-35] has D_{3h} symmetry and pentaprismane, $(CH)_{10}$ [3-36], has D_{5h} symmetry. Triprismane, pentaprismane, and hexaprismane, $C_{12}H_{12}$ (not yet prepared), are shown in Figure 3-28f. The quest for a synthesis of pentaprismane is a long story with a happy ending [3-36]. Hexaprismane, $(CH)_{12}$, which is the face-to-face dimer of benzene, has not yet been prepared. Table 3-4 presents some characteristic geometric information on the hydrocarbon prismane molecules, after Schultz [3-12]. The description of the general n-prismane is that it is composed of $2n$ identical methine units arranged at the corners of a regular prism with D_{nh}

Table 3-3. Characterization of Polyhedrane Molecules[a]

Name	Formula	Geometry and number of faces (all regular)	Face angles	Has been prepared?[b]
Tetrahedrane[c]	$(CH)_4$	Triangle, 4	60°	No
Cubane[d]	$(CH)_8$	Square, 6	90°	Yes
Dodecahedrane[e]	$(CH)_{20}$	Pentagon, 12	108°	Yes
Truncated tetrahedrane	$(CH)_{12}$	Triangle, 4 Hexagon, 4	60°	No
Truncated octahedrane	$(CH)_{24}$	Square, 6 Hexagon, 8	90° 120°	No
Truncated cubane	$(CH)_{24}$	Triangle, 8 Octagon, 6	60° 135°	No
Truncated cuboctahedrane	$(CH)_{48}$	Square, 12 Hexagon, 8 Octagon, 6	90° 120° 135°	No
Truncated icosahedrane[f]	$(CH)_{60}$	Pentagon, 12 Hexagon, 20	108° 120°	No
Truncated dodecahedrane	$(CH)_{60}$	Triangle, 20 Decagon, 12	60° 144°	No
Truncated icosidodecahedrane	$(CH)_{120}$	Square, 30 Hexagon, 20 Decagon, 12	90° 120° 144°	No

[a] After Schultz [3-12].
[b] As of 1994.
[c] Figure 3-28a.
[d] Figure 3-28c.
[e] Figure 3-28d.
[f] Figure 3-28e.

symmetry and bound into two parallel n-membered rings conjoined by n four-membered rings.

Incidentally, the *regular* prisms and the *regular* antiprisms are also semiregular, i.e., Archimedean, solids. Moreover, the second prism, in its most symmetrical configuration, is a regular solid, the cube; and the first antiprism, in its most symmetrical configuration, is also a regular solid, the octahedron.

Only a few highly symmetrical structures have been mentioned above. The varieties become virtually endless if one reaches beyond the most symmet-

Table 3-4. Characterization of Prismane Molecules[a]

Name	Formula	Geometry and number of faces	Face angles	Has been prepared?[b]
Tripismane	C_6H_6	Triangle, 2 Square, 3	60° 90°	Yes
Tetrapismane (cubane)	C_8H_8	Square, 6	90°	Yes
Pentaprismane	$C_{10}H_{10}$	Pentagon, 2 Square, 5	108° 90°	Yes
Hexaprismane	$C_{12}H_{12}$	Hexagon, 2 Square, 6	120° 90°	No
Heptaprismane	$C_{14}H_{14}$	Heptagon, 2 Square, 7	128°34′ 90°	No
n-Prismane	$C_{2n}H_{2n}$	*n*-gon, 2 Square, *n*	—[c] 90°	

[a]After Schultz [3-12].
[b]As of 1994.
[c]Approaches 180° as *n* increases.

rical convex polyhedral shapes. For example, the number of possible isomers is 5,291 for the tetracyclic structures of the $C_{12}H_{18}$ hydrocarbons with 12 skeletal carbon atoms [3-27]. Of all these geometric possibilities, however, only a few are stable [3-37]. One is *iceane*, shown in Figure 3-29 [3-38]. The molecule may be visualized as two chair cyclohexanes connected to each other by three axial bonds. Alternatively, the molecule may be viewed as consisting of three fused boat cyclohexanes. The trivial name iceane had been proposed for this molecule by Fieser [3-39] almost a decade before its preparation [3-38]. As Fieser was considering the arrangement of the water molecules in the ice crystal (Figure 3-29), he noticed three *vertical* hexagons with boat conformations. The emerging *horizontal* $(H_2O)_6$ units possess three equatorial hydrogen atoms and three equatorial hydrogen bonds available for horizontal building. Fieser [3-39] further noted that this structure "suggests the possible existence of a hydrocarbon of analogous conformation of the formula $C_{12}H_{18}$, which might be named 'iceane.' The model indicates a stable strain-free structure analogous to adamantane and twistane. 'Iceane' thus presents a challenging target for synthesis." Within a decade the challenge was met [3-38].

There is a close relationship between the adamantane, $C_{10}H_{16}$, molecule and the diamond crystal. The Greek work *adamant* means diamond and diamond has been termed the "infinite adamantylogue to adamantane" [3-40]. While iceane has D_{3h} symmetry, adamantane has T_d. This high symmetry can

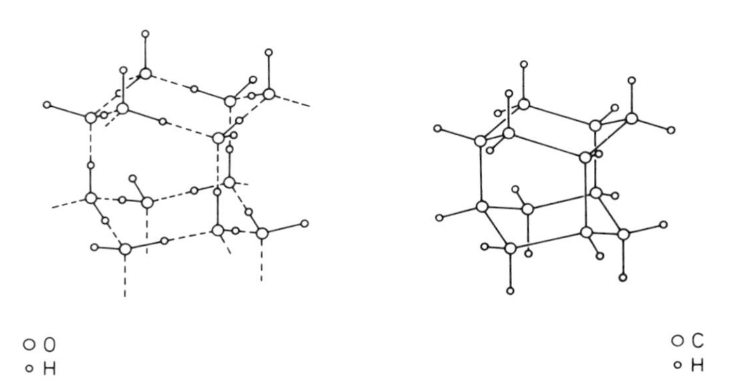

Figure 3-29. Ice crystal structure and the iceane hydrocarbon molecule, after Fieser [3-39] and Cupas and Hodakowski [3-38], respectively.

be clearly seen when the configuration of adamantane is described by four imaginary cubes packed one inside the other, two of which are shown in Figure 3-30a [3-41]. The Czechoslovakian stamp in Figure 3-30a pays tribute to the discovery of adamantane [3-42]. Similar structures are found among inorganic compounds where, by analogy to adamantane, $(CH)_4(CH_2)_6$, the general formula is A_4B_6. Here A may be, e.g., P, As, Sb, PO, or PN [3-43], as illustrated in Figure 3-30b.

Adamantane molecules may be imagined to join at vertices, at edges, or even at faces. Examples are shown in Figure 3-31; most of them, however, have not yet been synthesized (for references, see Ref. [3-29]).

3.7.3 Structures with Central Atom

Adamantane is sometimes regarded as the cage analog of methane while diamantane and triamantane are regarded as the analogs of ethane and propane. Methane has, of course, a tetrahedral structure with the point group of the regular tetrahedron, T_d. Important structures may be derived by joining two tetrahedra, or, for example, two octahedra, at a common vertex, edge, or face as shown in Figure 3-32. Ethane, $H_3C{-}CH_3$, ethylene, $H_2C{=}CH_2$, and acetylene, $HC{\equiv}CH$, may be derived formally from joined tetrahedra in such a way. The analogy with the joining tetrahedra is even more obvious in some metal halide structures with halogen bridges [3-48]. Thus, for example, the $Al_2Cl_7^-$ ion may be considered as two aluminum tetrachloride tetrahedra joined at a common vertex, or the Al_2Cl_6 molecule may be looked at as two such tetrahedra joined at a common edge. These examples are shown in Figure 3-33.

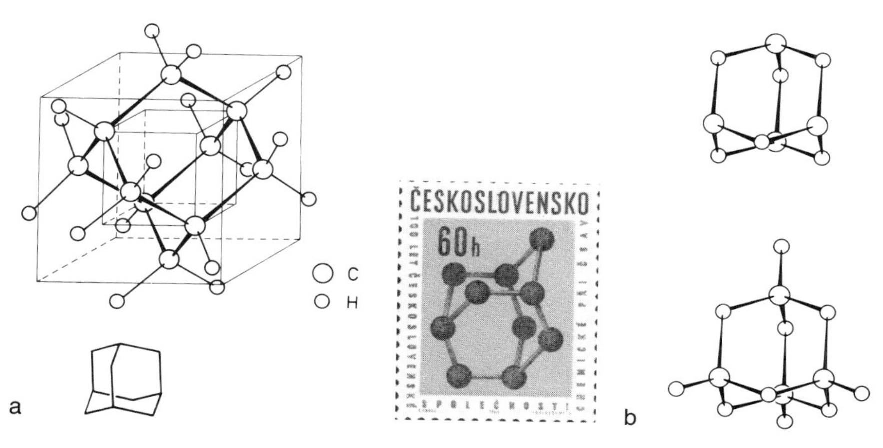

Figure 3-30. (a) Adamantane, $C_{10}H_{16}$ or $(CH)_4(CH_2)_6$, in three representations. (b) Inorganic adamantane analogs: P_4O_6, $(PO)_4O_6$.

In mixed-halogen complexes, such as potassium tetrafluoroaluminate, $KAlF_4$ [3-49], there is also a tetrahedral metal coordination. In fact, the regular or nearly regular tetrahedral tetrafluoroaluminate part of the molecule is an especially well defined structural unit. It is relatively rigid, whereas the position of the potassium atom around the AlF_4 tetrahedron is rather loose. The most plausible model for this molecule is also shown in Figure 3-33 [3-49]. The $KAlF_4$ molecule is merely a representative from a large class of com-

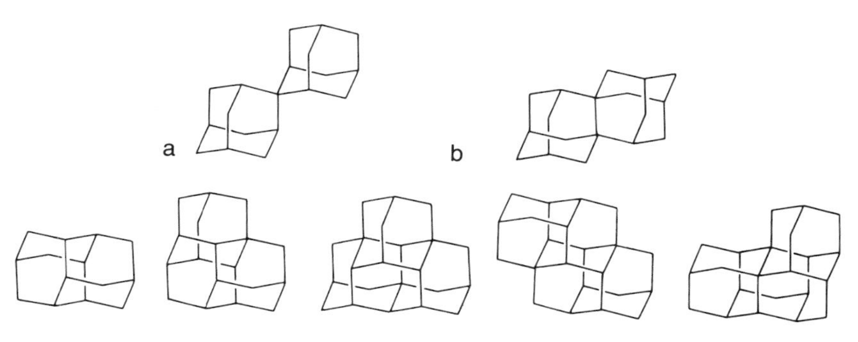

Figure 3-31. Joined adamantanes: (a) At vertices, [1]diadamantane [3-44]; (b) at edges, [2]diadamantane [3-45]; (c) at faces, diamantane (congressane) [3-40], triamantane [3-46], and three isomers of tetramantane [3-47]: "iso," C_{3v}; "anti," C_{2h}; and "skew," C_2.

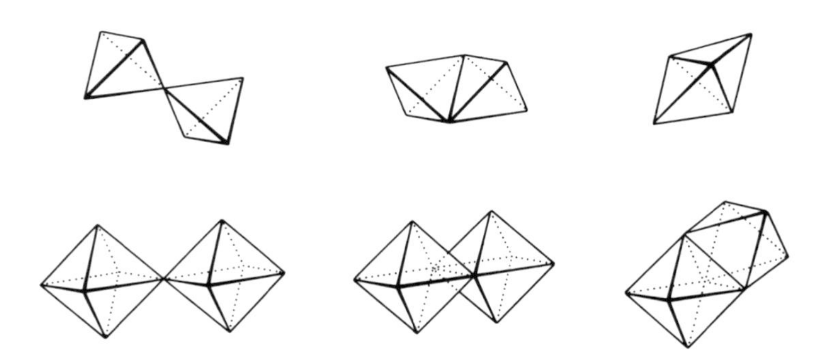

Figure 3-32. Joined tetrahedra and octahedra.

pounds with great practical importance: the mixed halides have greatly enhanced volatility compared with the individual metal halides.

For tetralithiotetrahedrane, $(CLi)_4$, the structure with the lithium atoms above the faces of the carbon tetrahedron was found in calculations to be more stable than that with the lithium atoms above the vertices [3-50] (Figure 3-34).

The prismatic cyclopentadienyl and benzene complexes of transition metals (see, e.g., Ref. [3-9]) are reminiscent of the polycyclic hydrocarbon prismanes. Figure 3-35 shows ferrocene, $(C_5H_5)_2Fe$, for which both the barrier to rotation and the free energy difference between the prismatic (eclipsed) and antiprismatic (staggered) conformations are very small [3-51]. Figure 3-35 presents also a prismatic model with D_{6h} symmetry for dibenzenechromium, $(C_6H_6)_2Cr$.

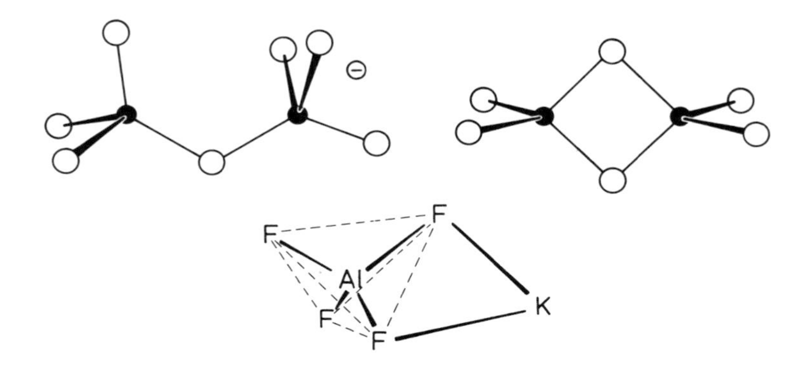

Figure 3-33. The configurations of the $Al_2Cl_7^-$ ion and Al_2Cl_6 and $KAlF_4$ molecules.

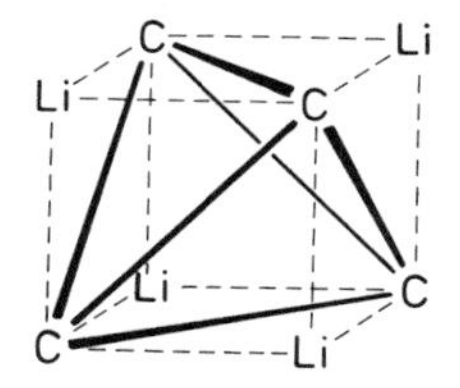

Figure 3-34. Model of the $(CLi)_4$ molecule [3-50].

Molecules with multiple bonds between metal atoms often have structures with beautiful and highly symmetrical polyhedral shapes [3-52]. The square prismatic $[Re_2Cl_8]^{2-}$ ion [3-53], shown in Figure 3-36, played an important role in the history of the discovery of metal–metal multiple bonds. Figure 3-37a shows another molecular model with a metal–metal multiple bond. Its shape is similar to the paddles that propel riverboats. There is then a whole class of hydrocarbons called paddlanes [3-56], and one of their representatives is shown in Figure 3-37b.

3.7.4 Regularities in Nonbonded Distances

The structure of the ONF_3 molecule (Figure 3-38) can be looked at as a regular tetrahedron formed by three fluorines and one oxygen. The nonbonded F · · · F and F · · · O distances representing the lengths of the edges of a tetrahedron are equal within the experimental errors of their determination [3-57], as shown in Figure 3-38. The bond lengths and bond angles are also given. The molecule has C_{3v} symmetry, and the central nitrogen atom is obviously not in the center of the essentially regular tetrahedron of its ligands.

In some molecular geometries, the so-called intramolecular 1,3 separations are remarkably constant. The "1,3" label refers to the interactions between two atoms in the molecule which are separated by a third atom. The

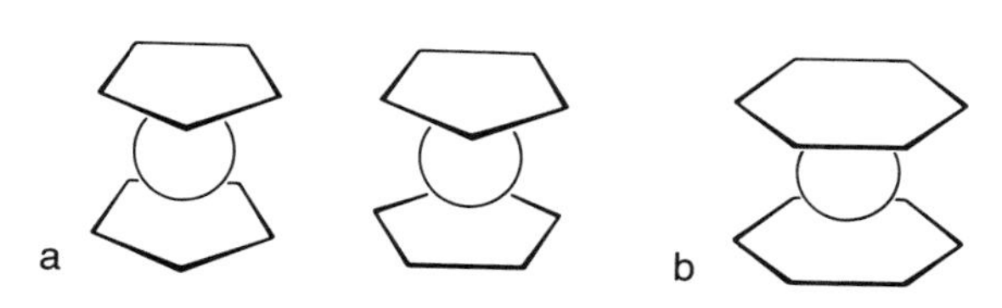

Figure 3-35. (a) Ferrocene: prismatic (D_{5h}) and antiprismatic (D_{5d}) models. (b) Dibenzenechromium: prismatic model (D_{6h}).

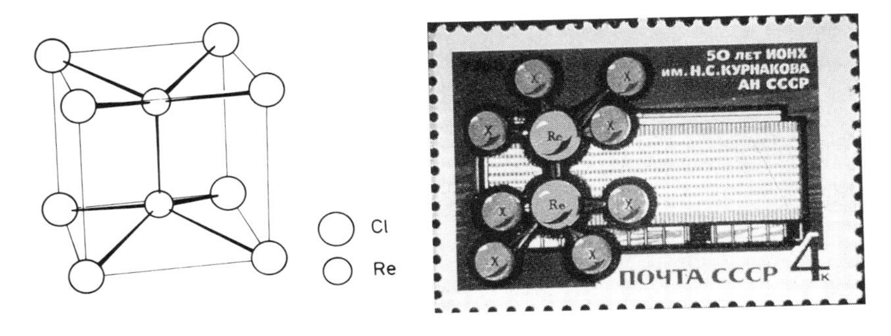

Figure 3-36. The square prismatic structure of the $[Re_2Cl_8]^{2-}$ ion, which played a historic role in the discovery of metal–metal multiple bonds ([3-54], cf. Ref. [3-53]), and a Soviet stamp with the same structure. On the stamp, the building in the background is the Moscow research institute where the first such structure was obtained [3-54].

near equality of the nonbonded distances in the ONF_3 molecule is a special case. What is usually observed is the constancy of a certain 1,3 nonbonded distance throughout a series of related molecules. Significantly, this constancy of 1,3 distances may be accompanied by considerable changes in the bond lengths and bond angles within the three-atom group. The intramolecular 1,3 interactions have also been called intramolecular van der Waals interactions, and Bartell [3-58] postulated a set of intramolecular nonbonded 1,3 radii. These 1,3 non-bonded radii are intermediate in value between the corresponding covalent radii and "traditional" van der Waals radii. All these values are compiled for some elements in Table 3-5.

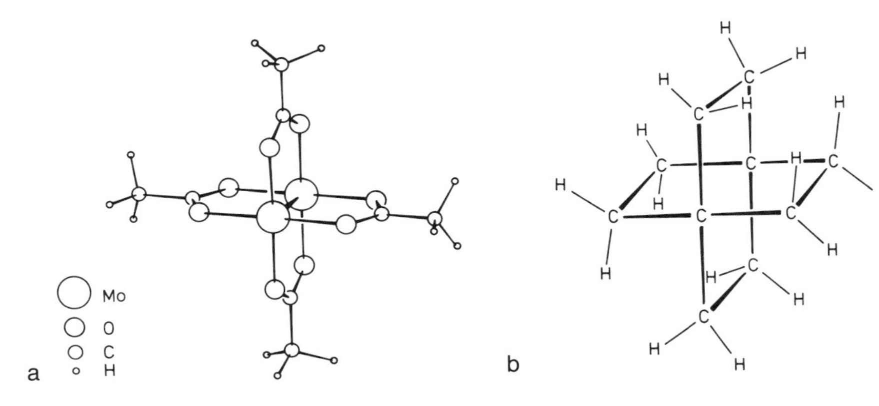

Figure 3-37. (a) Dimolybdenum tetraacetate, $Mo_2(O_2CCH_3)_4$ [3-55]. (b) [2.2.2.2]Paddlan not yet prepared.

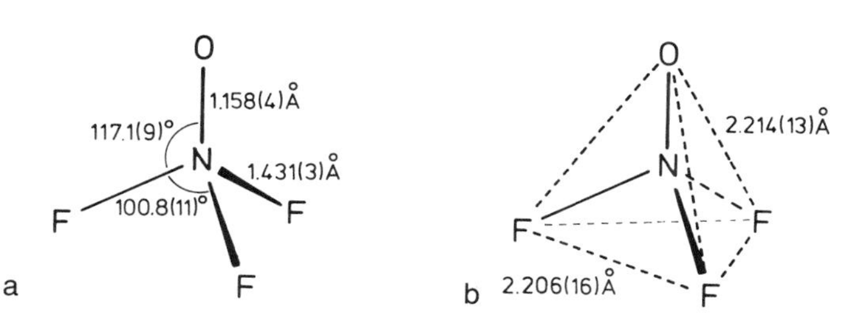

Figure 3-38. The molecular geometry of ONF_3 [3-57]. (a) Bond lengths and bond angles; (b) nonbonded distances.

Figure 3-39 shows some structural peculiarities which originally prompted Bartell [3-61] to recognize the importance of the intramolecular nonbonded interactions. It was an interesting observation that the three outer carbon atoms in $H_2C{=}C(CH_3)_2$ were arranged as if they were at the corners of an approximately equilateral triangle, as shown in Figure 3-39a. Since the central carbon atom in this arrangement is obviously not in the center of the triangle, the bond angle between the bulky methyl groups is smaller than the ideal 120°. In the other example, in Figure 3-39b, the C–C bond lengthening is related to the increasing number of nonbonded interactions.

Table 3-5. Covalent, 1,3 Intramolecular Nonbonded, and van der Waals Radii of Some Elements

Element	Covalent radius[a] (Å)	1,3 Intramolecular nonbonded radius[b] (Å)	van der Waals radius[a] (Å)
B	0.817	1.33	
C	0.772	1.25	
N	0.70	1.14	1.5
O	0.66	1.13	1.40
F	0.64	1.08	1.35
Al	1.202	1.66	
Si	1.17	1.55	
P	1.10	1.45	1.9
S	1.04	1.45	1.85
Cl	0.99	1.44	1.80
Ga	1.26	1.72	
Ge	1.22	1.58	
As	1.21	1.61	2.0
Se	1.17	1.58	2.00
Br	1.14	1.59	1.95

[a]After Pauling [3-59].
[b]After Bartell [3-58] and Glidewell [3-60].

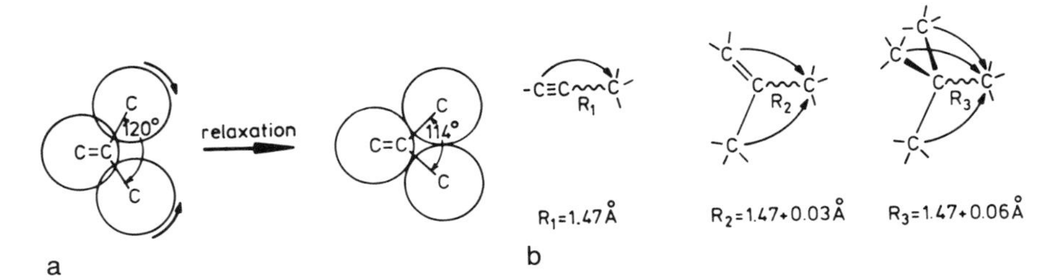

Figure 3-39. Geometrical consequences of nonbonded interactions, after Bartell [3-61]. (a) The three outer carbon atoms of $H_2C{=}C(CH_3)_2$ are at the corners of an approximately equilateral triangle, leading to a relaxation of the bond angle between the methyl groups. (b) Considerations of nonbonded interactions in the interpretation of the C–C single bond length changes in a series of molecules.

Of course, the 1,3 intramolecular nonbonded radii (Table 3-5) are purely empirical, but so are the other kinds of radii. Thus, the 1,3 nonbonded radii may be updated from time to time (see, e.g., Ref. [3-60]).

The F · · · F nonbonded distances have been observed to be remarkably constant in trifluoromethyl derivatives at 2.16 Å [3-62]. Similarly, the O · · · O nonbonded distances in XSO_2Y sulfones have been observed to be remarkably constant at 2.48 Å [3-63] in a relatively large series of compounds. At the same time the S=O bond lengths vary by up to 0.05 Å and the O=S=O bond angles by up to 5° depending on the nature of the X and Y ligands. The geometrical variations in the sulfone series could be visualized (Figure 3-40a) as if the two oxygen ligands were firmly attached to two of the four vertices of the ligand tetrahedron around the sulfur atom, and this central atom were moving along the bisector of the OSO angle depending on the X and Y ligands [3-63]. The sulfuric acid, H_2SO_4 or $(HO)SO_2(OH)$, molecule has its four oxygens around the sulfur at the vertices of a nearly regular tetrahedron (Figure 3-40b). Compared with the differences in the various OSO angles (up to 20°) and in the two kinds of SO bonds (up to 0.15 Å), the greatest difference among the six O · · · O nonbonded distances is only 0.07 Å [3-64].

The alkali sulfate molecules used to appear in old textbooks with the following structural formula:

Na—O   O
     S
Na—O   O

3-5

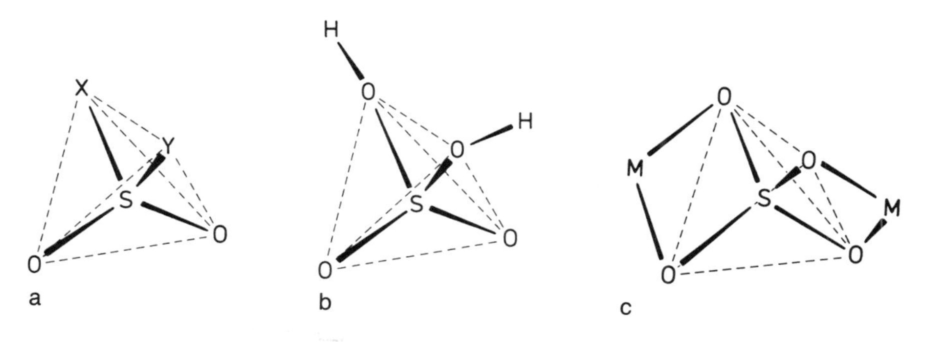

Figure 3-40. The configurations of XSO_2Y sulfone molecules (a), the sulfuric acid molecule (b), and (c) alkali sulfate molecules, M_2SO_4.

However, the SO_4 groups have nearly regular tetrahedral configuration in such molecules. The metal atoms are located on axes perpendicular to the edges of the SO_4 tetrahedron. Thus, this structure is bicyclic as shown in Figure 3-40c.

3.7.5 The VSEPR Model

Numerous examples of molecular structures have been introduced in the preceding sections. They are all confirmed by modern experiments and/or calculations. We would like to know, however, not only *what* is the structure of a molecule and its symmetry, but also, *why* a certain structure with a certain symmetry is realized.

It has been a long-standing goal in chemistry to determine the shape and measure the size of molecules, and also to calculate these properties. Today, quantum chemistry is capable of determining the molecular structure, at least for relatively simple molecules, starting from the mere knowledge of the atomic composition, and without using any empirical information. Such calculations are called *ab initio*. The primary results from these calculations are, however, wave functions and energies, which may also be considered "raw measurements," similar to some experimental data. At the same time there is a desire to understand molecular structures in simple terms—such as, for example, the localized chemical bond—that have proved so useful to chemists' thinking. There is a need for a bridge between the measurements and calculations, on one hand, and simple qualitative ideas, on the other hand. There are several qualitative models for molecular structure that serve this purpose well. These models can explain, for example, why the methane molecule is regular tetrahedral, T_d, why ammonia is pyramidal, C_{3v}, why water is bent, C_{2v}, and why the xenon tetrafluoride molecule is square planar, D_{4h}. It is also important to understand why seemingly analogous molecules such as OPF_3 and $OClF_3$ have so different symmetries, the former C_{3v}, and the latter C_s, as seen in Figure 3-41.

The structure of a series of the simplest AX_n type molecules will be examined in terms of one of these useful and successful qualitative models. A is the central atom, the X's are the ligands, and not necessarily all n ligands are the same.

Qualitative models simplify. They usually consider only a few, if not just one, of the many effects which are obviously present and are interacting in a most complex way. The measure of the success of a qualitative model is in its ability to create consistent patterns for interpreting individual structures and structural variations in a series of molecules and, above all, in its ability to correctly predict the structures of molecules, not yet studied or not even yet prepared.

One of the simplest models is based on the following postulate [3-65]: *The geometry of the molecule is determined by the repulsions among the electron pairs in the valence shell of its central atom*. The valence shell of an atom may have bonding pairs and other electron pairs that do not participate in bonding and belong to this atom alone. The latter are called unshared or lone pairs of electrons. The above postulate emphasizes the importance of *both* bonding pairs and lone pairs in establishing the molecular geometry. The model is appropriately called the valence shell electron pair repulsion or VSEPR model. The bond configuration around atom A in the molecule AX_n is such that the electron pairs of the valence shell are at maximum distances from each other. Thus, the situation may be visualized in such a way that the electron pairs occupy well-defined parts of the space around the central atom, corresponding to the concept of localized molecular orbitals.

If it is assumed that the valence shell of the central atom retains its spherical symmetry in the molecule, then the electron pairs will be at equal distances from the nucleus of the central atom. In this case the arrangements at

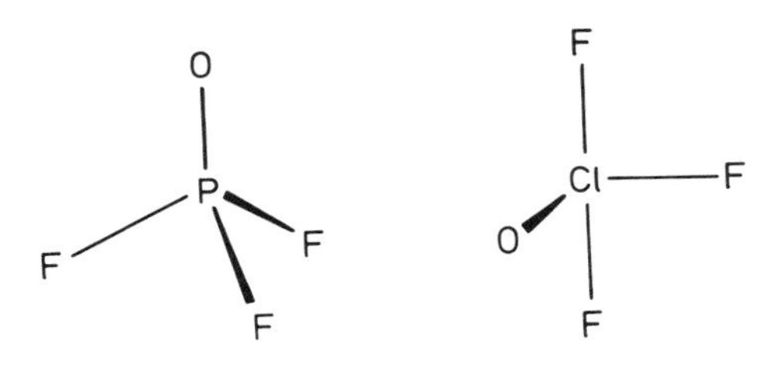

Figure 3-41. Molecular configuration of OPF_3 (C_{3v}) and $OClF_3$ (C_s).

which the distances among the electron pairs are at maximum will be those listed in Table 3-6. If the electron pairs are represented by points on the surface of a sphere, then the shapes shown in Figure 3-42 are obtained by connecting these points. Of the three polyhedra shown in Figure 3-42, only two are regular, viz., the tetrahedron and the octahedron. The trigonal bipyramid is not a regular polyhedron; although its six faces are equivalent, its edges and vertices are not. Incidentally, the trigonal bipyramid is not a unique solution to the five-point problem. Another, and only slightly less advantageous, arrangement is the square pyramidal configuration.

The repulsions considered in the VSEPR model may be expressed by the potential energy terms

$$V_{ij} = k/r_{ij}^{n}$$

where k is a constant, r_{ij} is the distance between the points i and j, and the exponent n is large for strong, or "hard," repulsion interactions and small for weak, or "soft," repulsion interactions, and is generally much larger than it would be for simple electrostatic coulomb interactions. Indeed, when n is larger than 3, the results become rather insensitive to the value of n. That is very fortunate because n is not really known. This insensitivity to the choice of n is what provides the wide applicability of the VSEPR model.

3.7.5.1 Analogies

It is easy to demonstrate the three-dimensional consequences of the VSEPR model in reality. We need only to blow up a few balloons that children play with. If groups of two, three, four, five, and six balloons, respectively, are connected at the ends near their openings, the resulting arrangements are those shown in Figure 3-43. Obviously, the space requirements of the various groups

Table 3-6. Arrangements of Two to Six Electron Pairs That Maximize Their Distances Apart

Number of electron pairs in the valence shell	Arrangement
2	Linear
3	Equilateral triangle
4	Tetrahedron
5	Trigonal bipyramid
6	Octahedron

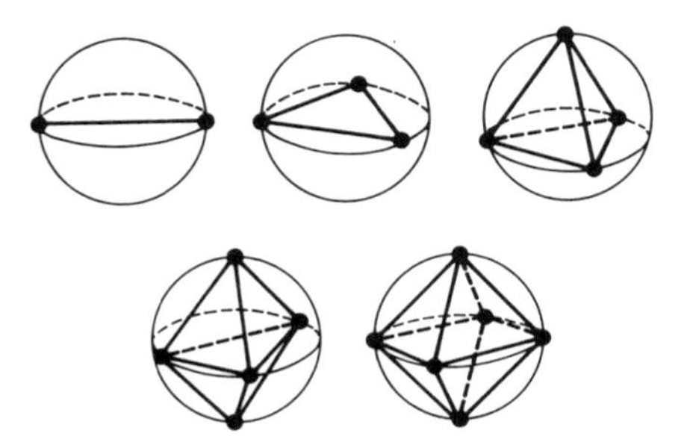

Figure 3-42. Molecular shapes from a points-on-the-sphere model.

of balloons, acting as mutual repulsions, determine the shapes and symmetries of these assemblies. The balloons here play the role of the electron pairs of the valence shell.

Another beautiful analogy with the VSEPR model, and one found directly in nature, is demonstrated in Figure 3-44. These are hard-shell fruits growing together. The small clusters of walnuts, for example, have exactly the same arrangements for two, three, four, and five walnuts in assemblies as predicted by the VSEPR model or as those shown by the balloons. The walnuts are required to accommodate themselves to each other's company and find the arrangements that are most advantageous considering the space requirements of all. Incidentally, the balloons and the walnuts may be considered as "soft" and "hard" objects, with weak and strong interactions, respectively.

3.7.5.2 Molecular Shapes

Using the VSEPR model, it is simple to predict the shape and symmetry of a molecule from the *total* number of bonding pairs, n, and lone pairs, m, of

Figure 3-43. Shapes of groups of balloons.

Figure 3-44. (a) Walnut clusters. Photographs by the authors. (b) Chestnut clusters. Photograph by Dr. Anna Rita Campanelli, University of Rome. (c) Ho'awa (endemic Hawaiian tree). Photograph by the authors.

electrons in the valence shell of its central atom. The molecule may then be written as AX_nE_m, where E denotes a lone pair of electrons. Only a few examples will be described here for illustration. For a comprehensive coverage, see, e.g., Ref. [3-65].

First, we shall consider the methane molecule, shown in the second row of Figure 3-45, together with ammonia and water. Originally, there were four electrons in the carbon valence shell, and these formed four C–H bonds, with the four hydrogens contributing the other four electrons. Thus, methane is represented as AX_4 and its symmetry is, accordingly, regular tetrahedral. In ammonia, originally there were five electrons in the nitrogen valence shell, and the formation of the three N–H bonds added three more. With the three bonding pairs and one lone pair in the nitrogen valence shell, ammonia may be

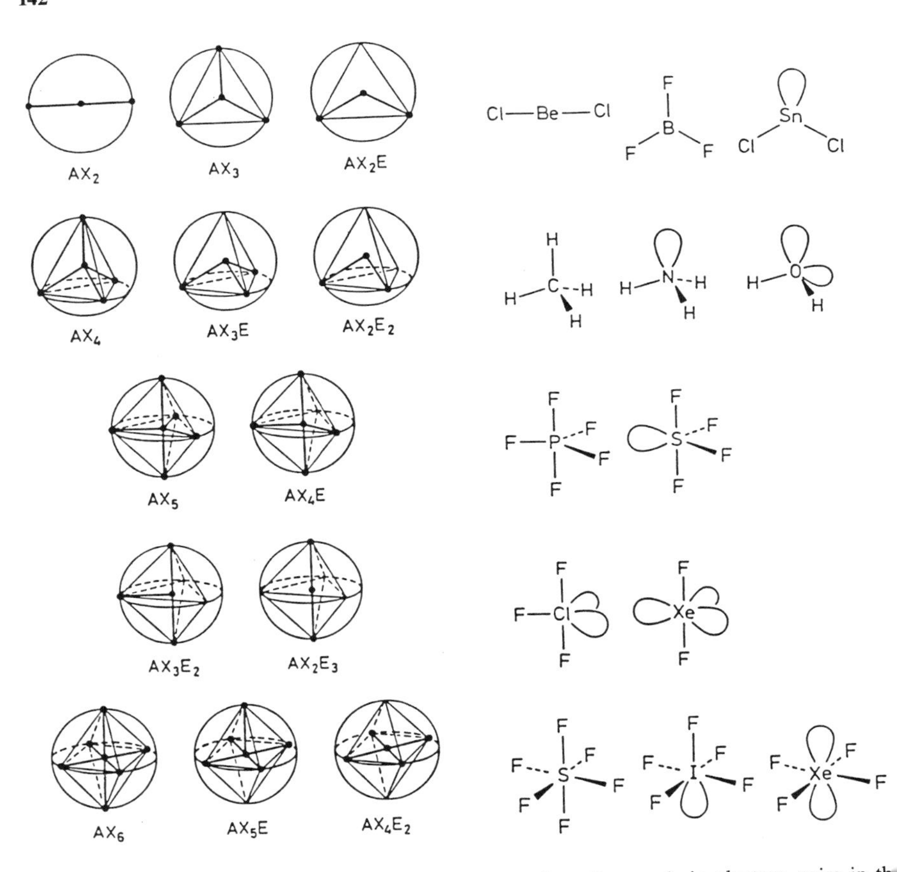

Figure 3-45. Bond configurations with two, three, four, five, and six electron pairs in th valence shell of the central atom, after Ref. [3-65].

written as AX_3E, and, accordingly, the arrangement of the molecule is relate to a tetrahedron. However, only in three of its four directions do we find bond and consequently ligands, while in the fourth there is a lone pair of electron Hence, a pyramidal geometry is found for the ammonia molecule. The be configuration of the water molecule can be similarly deduced.

In order to establish the total number of electron pairs in the valence shel the number of electrons originally present and the number of bonds forme need to be considered. A summary of molecular shapes based on the arrang ments of two to six valence shell electron pairs is shown in Figure 3-45.

The molecular shape to a large extent determines the bond angles. Thu the bond angle X–A–X is 180° in the linear AX_2 molecule, 120° in the trigon planar AX_3 molecule, and 109°28′ in the tetrahedral AX_4 molecule. The arrangements shown in Figure 3-45 correspond to the assumption that the strengths of the repulsions from all electron pairs are equal. In reality, however, the space requirements and, accordingly, the strengths of the repulsions from various electron pairs may be different depending on various circumstances as described in the following three subrules [3-65]:

1. A lone pair, E, in the valence shell of the central atom has a greater space requirement in the vicinity of the central atom than does a bonding pair. Thus, a lone pair exercises a stronger repulsion towards the neighboring electron pairs than does a bonding pair, b. The repulsion strengths weaken in the following order:

$$E/E > E/b > b/b$$

   This order is well illustrated by the various angles in sulfur difluoride in Figure 3-46 as determined by *ab initio* molecular orbital calculations [3-66]. This is also why, for example, the bond angles H–N–H of ammonia, 106.7° [3-67], are smaller than the ideal tetrahedral value, 109.5°. Unless stated otherwise, the parameters in the present discussion are taken from Ref. [3-67].
2. Multiple bonds, b_m, have greater space requirements than do single bonds and thus exercise stronger repulsions toward the neighboring electron pairs than do single bonds. The repulsion strengths weaken in the following order:

$$b_m/b_m > b_m/b > b/b$$

   A consequence of this is that the bond angles will be larger between multiple bonds than between single bonds. The structure of dimethyl sulfate provides a good example as shown in Figure 3-47. This molecule has three different types of OSO bond angles, and they decrease in the following order:

$$S{=}O/S{=}O > S{=}O/S{-}O > S{-}O/S{-}O$$

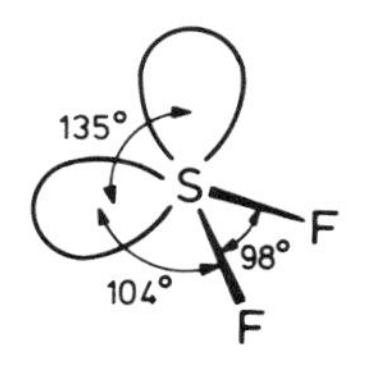

Figure 3-46. The angles of sulfur difluoride as determined by *ab initio* molecular orbital calculations [3-66].

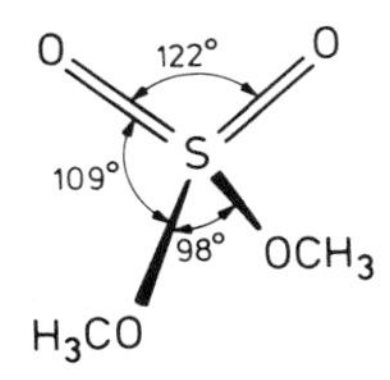

Figure 3-47. The three different kinds of oxygen–sulfur–oxygen bond angles in the dimethyl sulfate molecule as determined by electron diffraction [3-68].

3. A more electronegative ligand decreases the electron density in the vicinity of the central atom as compared with a less electronegative ligand. Accordingly, the bond to a less electronegative ligand, b_X, has a greater space requirement than the bond to a more electronegative ligand, b_Y. The repulsion strengths then weaken in the following order:

$$b_X/b_X > b_X/b_Y > b_Y/b_Y$$

Consequently, the bond angles are smaller for more electronegative ligands than for less electronegative ligands. An example of this effect can be seen in a comparison of sulfur difluoride (98°) and sulfur dichloride (103°).

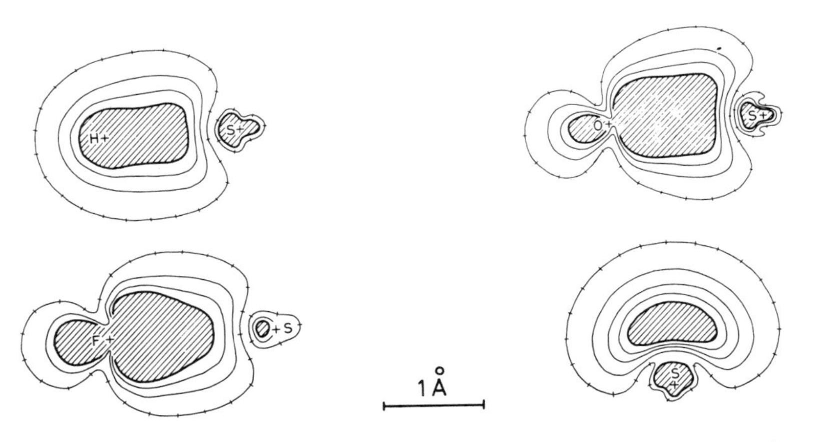

Figure 3-48. Localized molecular orbitals, represented by contour lines denoting electron densities of 0.02, 0.04, 0.06, etc. electron/bohr^3 from theoretical calculations [3-66], for the S–H, S–F, and S=O bonds and the lone pair on sulfur.

It is interesting to compare the implications expressed by these subrules with the depiction of some localized molecular orbitals in Figure 3-48, after Schmiedekamp *et al.* [3-66]. The lone pair of electrons occupies more space than do the bonding pairs in the vicinity of the central atom. Also, a bond to a more electronegative ligand such as fluorine occupies less space in the vicinity of the central atom than does a bond to a less electronegative ligand such as hydrogen. Finally, a double bond occupies more space than a single bond. The angular ranges of the corresponding contours in the electron density plots are all in good qualitative agreement with the postulates of the VSEPR model.

The VSEPR model has a fourth subrule that concerns the relative availability of space in the valence shell:

4. There is less space available in a completely filled valence shell than in a partially filled valence shell. Accordingly, the repulsions are stronger and the possibility for angular changes are smaller in the filled valence shell than in the partially filled one. Thus, for example, the bond angles of ammonia (107°) are closer to the ideal tetrahedral value than are those of phosphine (94°).

Thus, the differences in the electron pair repulsions may account for the bond angle variations in various series of molecules. The question now arises as to whether these differences have any effect on the *symmetry choice* of the molecules. In the four-electron-pair systems, the differences in the electron pair repulsions have a decisive role in the sense that the AX_4, EBX_3, and E_2CX_2 molecules have T_d, C_{3v}, and C_{2v} symmetries, respectively. Within each series, however, the symmetry is preserved regardless of the changes in the ligand electronegativities. For example, only the bond angles change in the molecules EBX_3, and EBY_3; the symmetry remains the same.

Ligand electronegativity changes may have decisive effects, however, on the symmetry choices of various bipyramidal systems, of which the trigonal bipyramidal configuration is the simplest.

When five electron pairs are present in the valence shell of the central atom, the trigonal bipyramidal configuration is usually found, although a tetragonal pyramidal arrangement cannot be excluded in some cases. Even intermediate arrangements between these two may appear to be the most stable in some special structures. The trigonal bipyramidal configuration with an equilateral triangle in the equatorial plane has D_{3h} symmetry while the square pyramidal configuration has C_{4v} symmetry. The intermediate arrangements have C_{2v} symmetry or nearly so. Indeed, rearrangements often occur in trigonal bipyramidal structures performing low-frequency large-amplitude motion. Such rearrangements will be illustrated later.

The positions in the D_{3h} trigonal bipyramid are generally not equivalent, and the axial ligand position is further away from the central atom than the equatorial one. This has no effect on the symmetry of the AX_5 structures, and this is comforting from the point of view of the applicability of the VSEPR model in establishing the point-group symmetries of such molecules.

On the other hand, when there is inequality among the electron pairs, the differences in the axial and equatorial positions do have importance for symmetry considerations. The PF_5 molecule, as an AX_5 system, unambiguously shows D_{3h} symmetry in its trigonal bipyramidal configuration. However, the prediction of the symmetry of the SF_4 molecule, which may be written as AX_4E, is less obvious. For SF_4 the problem is, where will the lone pair of electrons occur?

An axial position in the trigonal bipyramidal arrangement has three nearest neighbors at 90° away and one more neighbor at 180°. For an equatorial position there are two nearest neighbors at 90° and two further ones at 120°. As the closest electron pairs exercise by far the strongest repulsion, the axial positions are affected more than the equatorial ones. In agreement with this reasoning, the axial bonds are usually found to be longer than the equatorial ones. If there is a lone pair of electrons with a relatively large space requirement, it should be found in the more advantageous equatorial position. Accordingly, the SF_4 structure has C_{2v} symmetry, as does the ClF_3 molecule, which is of the AX_3E_2 type. Finally, the XeF_2 molecule is AX_2E_3 with all three lone pairs in the equatorial plane; hence, its symmetry is $D_{\infty h}$. All these structures are depicted in Figure 3-45.

By similar reasoning, the VSEPR model predicts that a double bond will also occupy an equatorial position. Thus, the point group may easily be established for the molecules $O{=}SF_4$, (C_{2v}), $O{=}ClF_3$ (C_s), XeO_3F_2 (D_{3h}), and XeO_2F_2 (C_{2v}). We note the C_s symmetry for the $OClF_3$ molecule (cf. Figure 3-41) as a consequence of the bipyramidal geometry with both the Cl=O double bond and the lone pair in the equatorial plane. The molecule OPF_3 (cf. Figure 3-41) is only seemingly analogous. There is no lone pair in the phosphorus valence shell, and thus the molecule has a distorted tetrahedral bond configuration. The P=O double bond is along the threefold axis, and the point group is C_{3v}, like that for ammonia.

Lone pairs and/or double bonds replaced single bonds in the above examples. Similar considerations are applicable when only ligand electronegativity changes take place. A typical example is demonstrated by a comparison of the structures of PF_2Cl_3 and PF_3Cl_2. The chlorine atoms are less electronegative ligands than the fluorines, and they will be in equatorial positions in *both* structures, as seen in Figure 3-49. The point groups are C_{2v} for PF_3Cl_2 and D_{3h} for PF_2Cl_3 [3-69]. Were the chlorines in the axial positions in PF_3Cl_2, this molecule would also have the much higher symmetry D_{3h}.

All six electron pairs are equivalent in the AX_6 molecule and so the symmetry is unambiguously O_h. An example is SF_6. The IF_5 molecule, how-

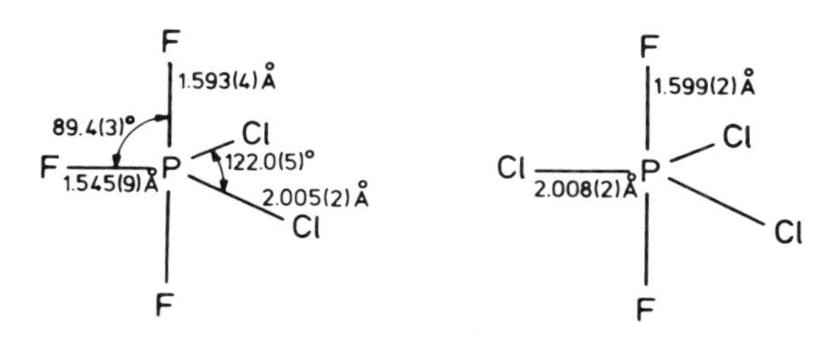

Figure 3-49. The molecular structures of PF_3Cl_2 and PF_2Cl_3 are not analogous: the chlorine ligands occupy equatorial positions in both cases.

ever, corresponds to AX_5E, and its square pyramidal configuration has C_{4v} symmetry. There is no question here as to the preferred position for the lone pair, as any of the six equivalent sites may be selected. When, however, a second lone pair is introduced, then the favored arrangement is that in which the two lone pairs find themselves at the maximum distance apart. Thus, for XeF_4, i.e., AX_4E_2, the bond configuration is square planar, point group D_{4h}. These structures are depicted in Figure 3-45.

The difficulties encountered in the discussion of the five- electron-pair valence shells are intensified in the case of the seven-electron-pair case. Here again the ligand arrangements are less favorable than for the nearest coordination neighbors, i.e., six and eight. It is not possible to arrange seven equivalent points in a regular polyhedron, while the number of nonisomorphic polyhedra with seven vertices is large, viz., 34 [3-70]. A few of them are shown in Figure 3-50. No single one of them is distinguished, however, from the others on the basis of relative stability. There may be quite rapid rearrangements among the various configurations. One of the early successes of the VSEPR model was

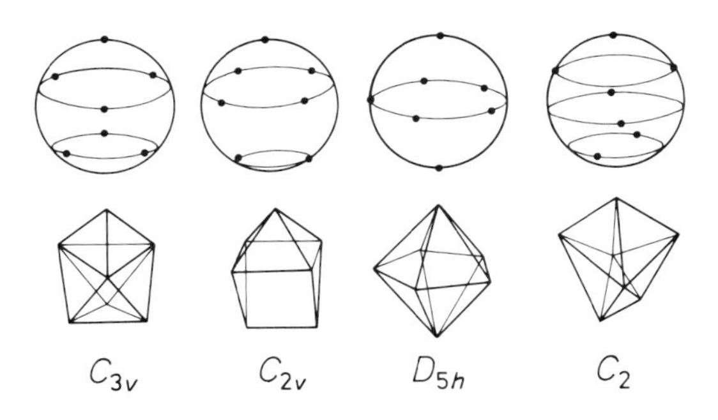

Figure 3-50. A sample of configurations for seven electron pairs in the valence shell.

that it correctly predicted a nonregular structure for XeF_6 by considering it as a seven-coordination case, AX_6E.

Numerous examples, a wealth of structural data, and detailed considerations on the potential and limitations of the applicability of the VSEPR model are given in a recent monograph [3-65].

3.7.5.3 Historical Remarks

The simplicity of the VSEPR model is one of its primary strengths. In addition, the model provides a continuity in the development of the qualitative ideas about the nature of the chemical bond and its correlation with molecular structure. Abegg's octet rule (see, e.g., Ref. [3-71]) and Lewis's theory of the shared electron pair [3-72] may be considered as direct forerunners of the model.

Lewis's cubical atom [3-72] deserves special mention. It was instrumental in shaping the concept of the shared electron pair. It also permitted a resolution of the apparent contradiction between the two distinctly different bonding types, viz., the shared electron pair and the ionic electron-transfer bond. In terms of Lewis's theory, the two bonding types could be looked at as mere limiting cases. Lewis's cubical atoms are illustrated in Figure 3-51. They are also noteworthy as an example of a certainly useful though not necessarily correct application of a polyhedral model.

Sidgwick and Powell [3-73] were first to correlate the number of electron pairs in the valence shell of a central atom and its bond configuration. Then Gillespie and Nyholm [3-74] introduced allowances for the difference between the effects of bonding pairs and lone pairs and applied the model to large classes of inorganic compounds.

There have been attempts to provide quantum-mechanical foundations for the VSEPR model. These attempts have developed along two lines. One is concerned with assigning a rigorous theoretical basis to the model, primarily involving the Pauli exclusion principle, to the extent that it was even suggested that the application of the model be named "Pauli mechanics" [3-75]. The other line is the numerous quantum-chemical calculations (e.g., Refs. [3-66] and [3-76]) which have already produced a large amount of structural data consistent with the VSEPR model, demonstrating that it indeed captures some important effects determining the structures of molecules. It has also been shown that while the total electron density distribution of a molecule does not provide any evidence for the localized electron pairs, the charge concentrations obtained by deriving the second derivative of this distribution parallel the features of these localized pairs [3-77]. This may be considered as supporting evidence, or even a physical basis, for the VSEPR model. We would stress, however, that the VSEPR model is a qualitative tool, and, as such, it overemphasizes some effects and ignores many others. Its simplicity, wide appli-

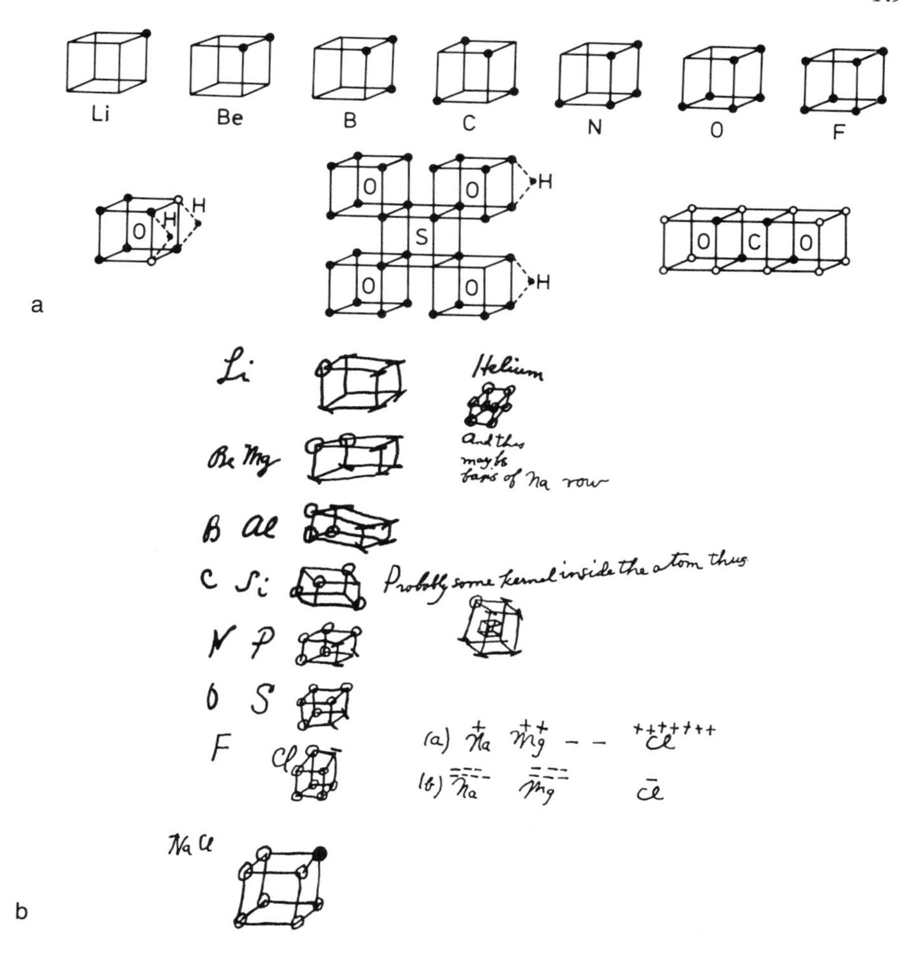

Figure 3-51. (a) Lewis's cubical atoms and some molecules built from such atoms (cf. Ref. [3-72]). (b) Lewis's original sketches, after Ref. [3-72].

cability, and predictive power have been repeatedly demonstrated, making it useful in both research and education.

3.7.6 Consequences of Intramolecular Motion

Imagine the merry-go-round (Figure 3-52a) revolving and one of the wooden horses getting lifted and, upon its returning to the ground level, the next horse is lifted and so on. In addition to the real revolution of the whole

Figure 3-52. (a) Merry-go-round (Bologna, Italy, photograph by the authors). (b) Henri Matisse, *Dance*. The Hermitage, St. Petersburg. Reproduced by permission.

circle, the vertical motion is transmitted from horse to horse; this can be considered pseudorotation. If we take a picture of the merry-go-round in operation and the exposure is long enough, there will be a blurred image of all the horses in the elevated position in addition to the ground circle. With a very sensitive film, however, the exposure may be reduced so that we get a picture of a single horse being lifted. Another fitting analogy may be the *Dance* by Henri Matisse (Figure 3-52b). Let us imagine the following choreography for this dance: one of the dancers jumps and is thus out of the plane of the other four. As soon as this dancer returns into the plane of the others, it is now the role of the next to jump, and so on. The exchange of roles from one dancer to another throughout the five-member group is so quick that if we take a normal photograph, we will have a blurred picture of the five dancers. However, if we have a very sensitive film, we may be able to use such a short exposure that a well-defined configuration of the dancers at a particular moment can be identified.

The above descriptions simulate well the *pseudorotation* of the cyclopentane molecule, although on a different time scale. The cyclopentane, $(CH_2)_5$, molecule has a special degree of freedom when the out-of-plane carbon atom exchanges roles with one of its two neighboring carbon atoms (and their hydrogen ligands). This is equivalent to a rotation by $2\pi/5$ about the axis perpendicular to the plane of the four in-plane carbons (Figure 3-53) (see, e.g., Ref. [3-78]). All three examples emphasize the importance of the *relationship* between the time scale of motion and the time scale of measurement. This relationship must be taken into account when making a conclusion about the symmetry of a moving structure.

In discussing molecular structure, an extreme approach is to disregard intramolecular motion and to consider the molecule to be motionless. A completely rigid molecule is a hypothetical state corresponding to the minimum position of the potential energy function for the molecule. Such a motionless structure has an important and well-defined physical meaning and is called the *equilibrium structure*. It is this equilibrium structure that emerges from quantum-chemical calculations. On the other hand, real molecules are never motionless, not even at temperatures approaching 0 K. Furthermore, the various physical measurement techniques determine the structures of real molecules. As our discussion of the merry-go-round and Matisse's *Dance* illustrated, the relationship between the lifetime of the configuration under investigation and the time scale of the investigating technique is of crucial importance.

Large-amplitude, low-frequency intramolecular vibrations may lower the molecular symmetry of the average structure from the higher symmetry of the

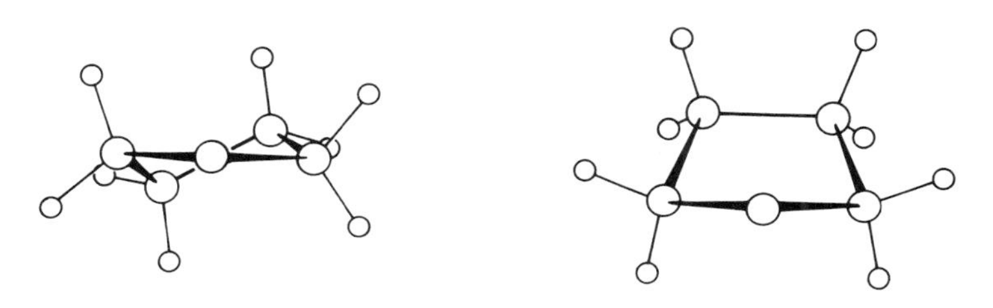

Figure 3-53. Pseudorotation of the cyclopentane molecule.

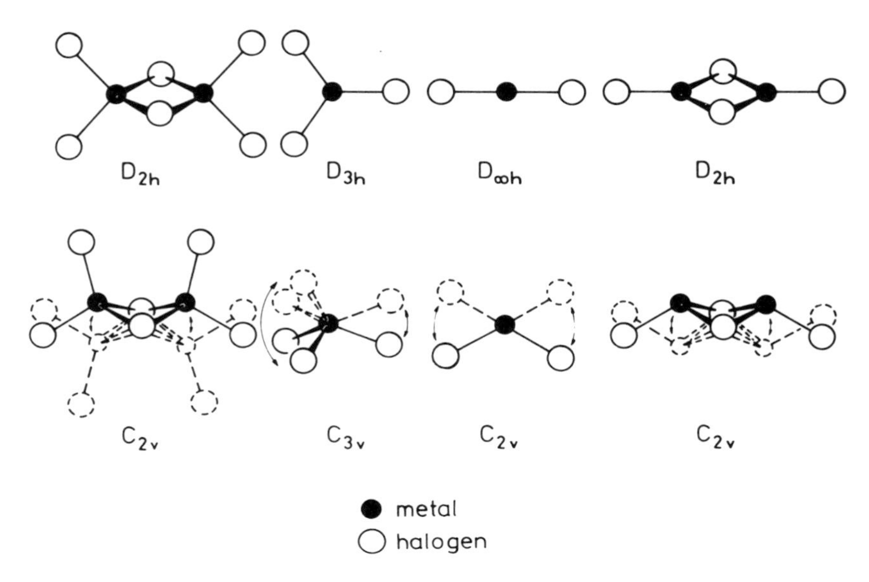

Figure 3-54. Equilibrium versus average structures of metal halide molecules with low-frequency, large-amplitude deformation vibrations (filled circles, metal; open circles, halogen).

equilibrium structure. Some examples from metal halide molecules are shown in Figure 3-54.

If we determine the average interatomic distances of symmetric triatomic molecules, for example, the emerging geometry will always be bent, regardless of whether the equilibrium structure is linear or bent, because of the consequences of bending vibrations (Figure 3-55a). In order to distinguish between linear and truly bent molecules, the potential energy function describing the bending motion must be scrutinized [3-79]. The bending potential energy functions of $ZnCl_2$ and $SrBr_2$ are shown in Figure 3-55b; $\rho_e = 0°$ corresponds to the linear configuration. The minimum of the potential energy function appears at $\rho_e = 0°$ for both molecules. It is also seen though that the minimum is much more shallow for $SrBr_2$ than for $ZnCl_2$. Figure 3-55c shows the bending potential energy functions of $SiBr_2$ and, again, of $SrBr_2$. The relatively high barrier at $\rho_e = 0°$ for $SiBr_2$ indicates an unambiguously bent configuration. Further enlarging the scale reveals a small barrier at $\rho_e = 0°$ for $SrBr_2$, so small that it lies below the level of the ground vibrational state. Such structures are called quasilinear.

A rapid interconversion of the nuclei takes place in the bullvalene molecule under very mild conditions in fluid media. This process involves

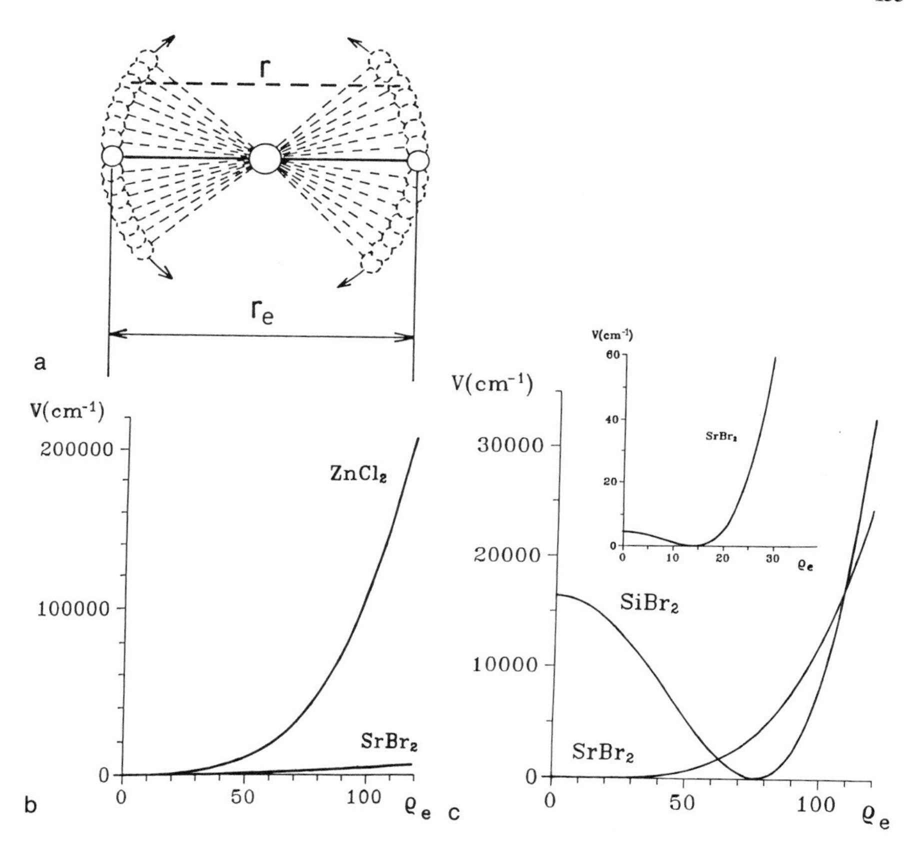

Figure 3-55. (a) Even a linear triatomic molecule appears bent due to its bending vibrations ($r < r_e$). Comparison of bending potential functions makes it possible to distinguish between linear and truly bent molecules [3-79]. (b) Linear models of $ZnCl_2$ and $SrBr_2$. (c) Bent models of $SrBr_2$ and $SiBr_2$.

making and breaking bonds, but this is accompanied by very small shifts in the nuclear positions. The molecular formula is $(CH)_{10}$, and the carbon skeleton is shown in Figure 3-56a. There are only four different kinds of carbon positions (and hydrogen positions, accordingly), and all four positions are being interconverted simultaneously [3-80]. Hypostrophene is another $(CH)_{10}$ hydrocarbon. Its trivial name was chosen to reflect its behavior [3-81]. The Greek *hypostrophe* means turning about, a recurrence. The molecule is ceaselessly undergoing the intramolecular rearrangements indicated in Figure 3-56b. The atoms have a complete time-averaged equivalence yet hypostrophene could not be converted into pentaprismane (cf. Ref. [3-36]).

Permutational isomerism among inorganic substances was discovered by R. S. Berry [3-82] for trigonal bipyramidal structures. Although the trigonal bipyramid and the square pyramid have very different symmetries, D_{3h} versus C_{4v}, they easily interconvert by means of bending vibrations as is illustrated in Figure 3-57. The possible change in the potential energy during this structural reorganization is also shown. The permutational isomerism of an AX_5 molecule, e.g., PF_5, is easy to visualize as the two axial ligands replace two of the three equatorial ones, while the third equatorial ligand becomes the axial ligand in the transitional square pyramidal structure. The rearrangements quickly follow one another without any position being constant for any significant time period. The C_{4v} form originates from a D_{3h} structure and yields then again to another D_{3h} form. A somewhat similar pathway was established [3-83] for the $(CH_3)_2NPF_4$ molecule, in which the dimethylamine group is permanently locked in an equatorial position whereas the fluorines exchange in pairs all the time.

The structure of the $(CH_3)_2NPF_4$ molecule and its investigation by nuclear magnetic resonance (NMR) spectroscopy is also a good example demonstrating the importance of the relationship between the lifetime of a configuration and the time scale of the investigating technique [3-83]. The ^{31}P NMR spectra of $(CH_3)_2NPF_4$ at *low temperatures* provide evidence of two different kinds of

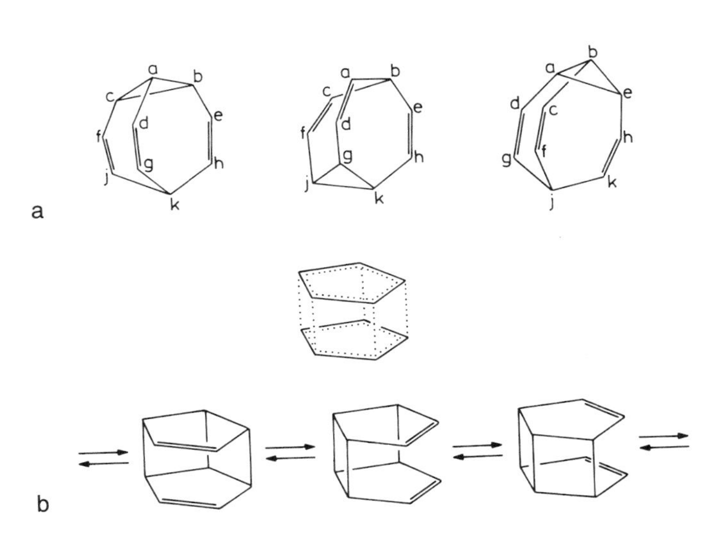

Figure 3-56. (a) The interconversion of bullvalene [3-80] and (b) hypostrophene [3-81].

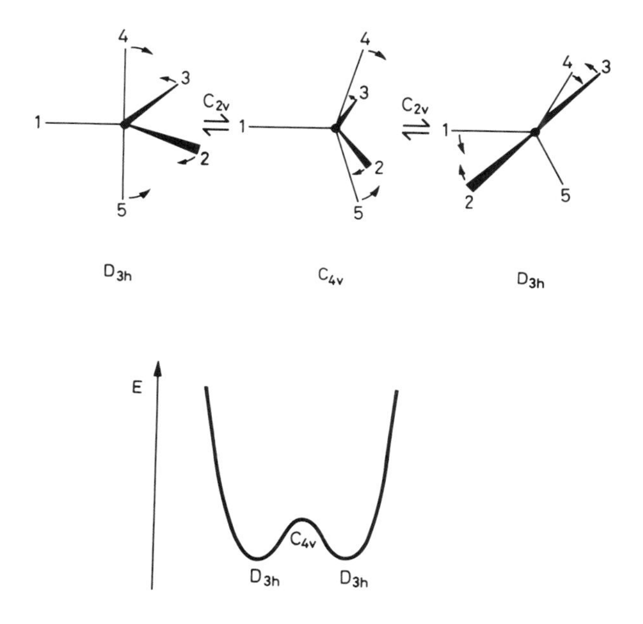

Figure 3-57. Berry pseudorotation of PF_5-type molecules [3-84].

P–F bond in this molecule, viz., axial and equatorial. At low temperatures the interconversion is slow, and the lifetimes of the fluorines in the axial and equatorial positions are much greater than the interaction time for producing the spectrum so the two kinds of P–F bond give separate resonances in the spectrum. At *higher temperatures* the intramolecular exchange of the fluorine positions accelerates, and the lifetimes of the fluorines in the axial and equatorial positions decrease. As the interaction time needed to produce the spectrum remains the same, the spectrum becomes simpler, and the nonequivalent fluorines are no longer distinguished. Since the time scale of NMR spectroscopy is commensurable with the lifetimes of separate configurations in intramolecular motion, different molecular shapes may be observed at different temperatures. Other techniques utilize interactions on different time scales. Thus, for example, the time scale of electron diffraction is several orders of magnitude smaller, and, accordingly, the two different fluorine positions will always be distinguished in an electron diffraction analysis.

Iodine heptafluoride, IF_7, has a pentagonal bipyramidal structure of at least approximately D_{5h} symmetry [3-84]. Its dynamic behavior has been described by pseudorotation.

The rearrangement that characterizes the PF_5 molecule also describes well the permutation of the atomic nuclei in five-atom polyhedral boron skeletons in borane molecules [3-85].

Lipscomb [3-86] has elaborated a general concept for the rearrangements of polyhedral boranes. According to this concept, two common triangulated faces are stretched to a square face in the borane polyhedra. There is an intermediate polyhedral structure with square faces. In the final step of the rearrangement, the intermediate configuration may revert to the original polyhedron with no net change, but it may as well turn into a different arrangement. The arrangement has rectangular faces with an orthogonal linkage with respect to the bonding situation in the original polyhedron [3-85]. This is illustrated in Figure 3-58. There are many practical examples, among which is the rearrangements of dicarba-*closo*-dodecaboranes, illustrated in Figure 3-59. There are three isomers of this beautiful carborane molecule:

1,2-dicarba-*closo*-dodecaborane, or *o*-$C_2B_{10}H_{12}$,
1,7-dicarba-*closo*-dodecaborane, or *m*-$C_2B_{10}H_{12}$, and
1,12-dicarba-*closo*-dodecaborane, or *p*-$C_2B_{10}H_{12}$.

Whereas the ortho isomer easily transforms into the meta isomer in agreement with the above-mentioned model, the para isomer is obtained only under more drastic conditions and only in a small amount ([3-86]; see also Refs. [3-88] and [3-89]). A similar model has been proposed [3-90] for the so-called carbonyl scrambling mechanism in molecules like $Co_4(CO)_{12}$, $Rh_4(CO)_{12}$, and $Ir_4(CO)_{12}$.

Incidentally, the carbonyl ligands can have several modes of coordination, viz., terminal and a variety of bridging possibilities. Rapid interconversion between the different coordination modes is possible, even in the solid state [3-91]. The above-mentioned metal-carbonyl molecules belong to a large class of compounds whose general formula is $M_m(CO)_n$, where M is a transition

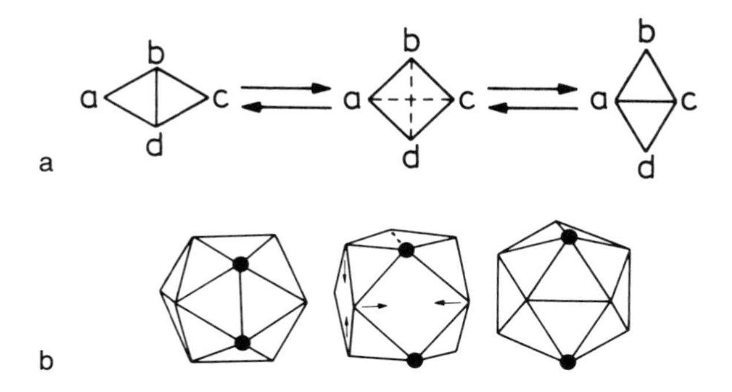

Figure 3-58. (a) The Lipscomb model of the rearrangement in polyhedral boranes [3-86]. (b) An example of icosahedron/cuboctahedron/icosahedron rearrangement [3-86].

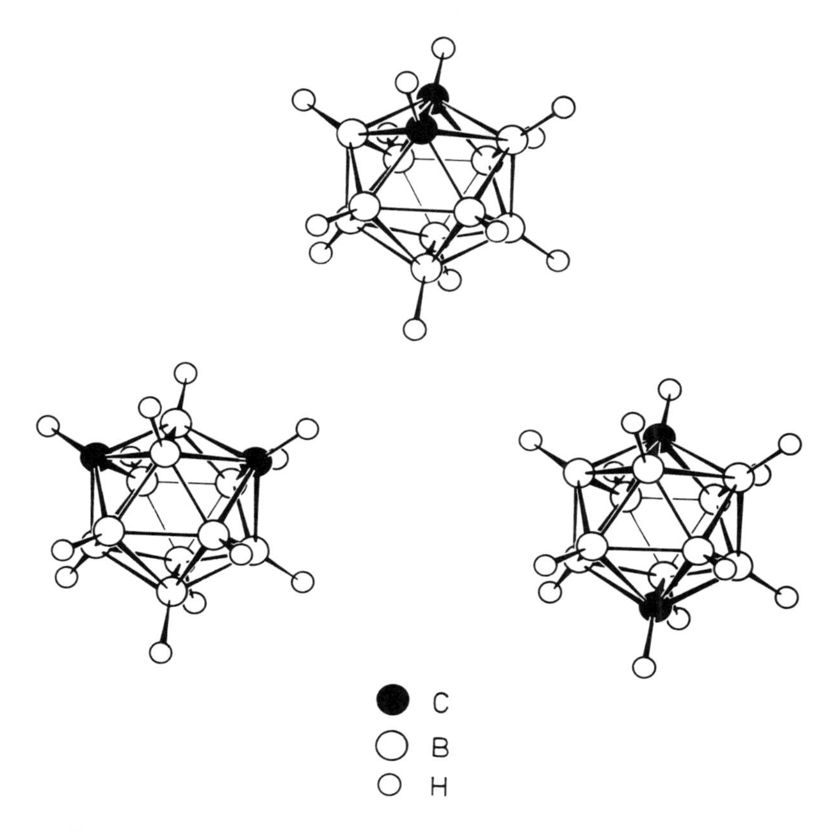

Figure 3-59. Structures of *o*-, *m*-, and *p*-dicarba-*closo*-dodecaborane [3-87].

metal. The usually small *m*-atomic metal cluster polyhedron is enveloped by another polyhedron whose vertices are occupied by the carbonyl oxygens [3-92]. An attractive example is the structure of $[Co_6(CO)_{14}]^{4-}$, in which the octahedral metal cluster has six terminal and eight triply bridging carbonyl groups, as shown in Figure 3-60a. This structure may also be represented by an omnicapped cube enveloping an octahedron as shown in Figure 3-60b, after Ref. [3-92]. These models are reminiscent of another model in which, also, polyhedra were enveloping other polyhedra. That model was Kepler's planetary system [3-93] cited in Figure 2-64.

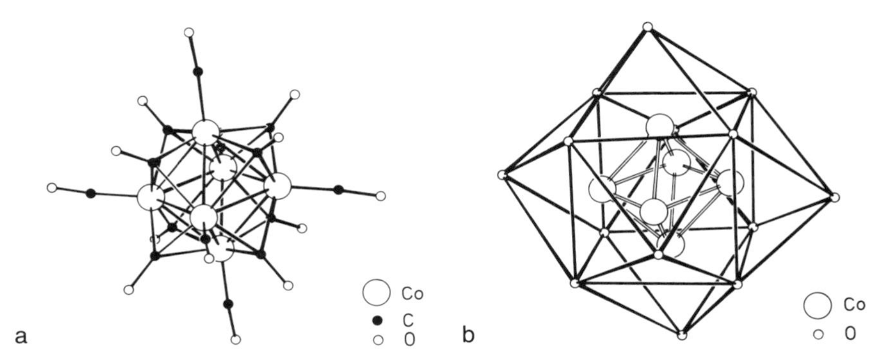

Figure 3-60. The structure of $[Co_6(CO)_{14}]^{4-}$ in two representations, after Benfield and Johnson [3-92]. (a) The octahedron of the cobalt cluster possesses six terminal and eight triply bridging carbonyl groups. (b) An omnicapped cube of the carbonyl oxygens envelopes the cobalt octahedron.

REFERENCES

[3-1] I. Hargittai, *J. Chem. Educ.* **60**, 94 (1983).
[3-2] W. D. Hounshell, D. A. Dougherty, and K. Mislow, *J. Am. Chem. Soc.* **100**, 3149 (1978).
[3-3] N. F. M. Henry and K. Lonsdale (eds.), *International Tables for X-ray Crystallography*, Vol. I., *Symmetry Groups*, Kynoch Press, Birmingham, England (1969).
[3-4] F. A. Cotton, *Chemical Applications of Group Theory*, 3rd ed., Wiley-Interscience, New York (1990).
[3-5] M. Orchin and H. H. Jaffe, *J. Chem. Educ.* **47**, 372 (1970).
[3-6] B. Baer Capitman, *American Trademark Design*, Dover Publications, New York (1976).
[3-7] K. L. Wolf and R. Wolff; *Symmetrie*, Böhlau-Verlag, Münster/Cologne (1956).
[3-8] A. Domenicano, in *Accurate Molecular Structures. Their Determination and Importance*, A. Domenicano and I. Hargittai, (eds.), Oxford University Press, Oxford (1992).
[3-9] M. Hargittai and I. Hargittai, *The Molecular Geometries of Coordination Compounds in the Vapour Phase*, Akadémiai Kiadó, Budapest, and Elsevier, Amsterdam (1977).
[3-10] H. S. M. Coxeter, *Regular Polytopes*, 3rd ed., Dover Publications, New York (1973).
[3-11] R. J. Ternansky, D. W. Balogh, and L. A. Paquette, *J. Am. Chem. Soc.* **104**, 4503 (1982).
[3-12] H. P. Schultz, *J. Org. Chem.* **30**, 1361 (1965).
[3-13] E. L. Muetterties, in *Boron Hydride Chemistry*, (E. L. Muetterties, ed.), Academic Press, New York (1975).
[3-14] C. H. MacGillavry, *Symmetry Aspects of M. C. Escher's Periodic Drawings*, Bohn, Scheltema and Holkema, Utrecht (1976).
[3-15] V. P. Spiridonov and G. I. Mamaeva, *Zh. Strukt. Khim.* **10**, 113 (1969).
[3-16] V. Plato and K. Hedberg, *Inorg. Chem.* **10**, 590 (1970).
[3-17] H. W. Kroto, J. R. Heath, S. O'Brien, R. F. Curl, and R. E. Smalley, *Nature* **318**, 162 (1985).
[3-18] H. W. Kroto and D. R. M. Walton (eds.), *The Fullerenes: New Horizons for the Chemistry, Physics and Astrophysics of Carbon*, Cambridge University Press, Cambridge (1993).

[3-19] *Science* **254**, December 20 (1991).
[3-20] *Science* **250**, December 12 (1990).
[3-21] H. Hart, *Science* **251**, 1162 (1991).
[3-22] D. E. Koshland, *Science* **254**, 1705 (1991).
[3-23] W. N. Lipscomb, in *Boron Hydride Chemistry* (E. L. Muetterties, ed.), Academic Press, New York (1975).
[3-24] R. W Rudolph, *Acc. Chem. Res.* **9**, 446 (1976).
[3-25] K. Wade, in *Electron Deficient Boron and Carbon Clusters* (G. A. Olah, K. Wade, and R. E. Williams, eds.), John Wiley & Sons, New York (1991).
[3-26] S. O. Kang and L. G. Sneddon, in *Electron Deficient Boron and Carbon Clusters* (G. A. Olah, K. Wade, and R. E. Williams, eds.), John Wiley & Sons, New York (1991).
[3-27] R. E. Williams, *Inorg. Chem.* **10**, 210 (1971).
[3-28] R. E. Williams, in *Electron Deficient Boron and Carbon Clusters* (G. A. Olah, K. Wade, and R. E. Williams, eds.), John Wiley & Sons, New York (1991).
[3-29] A. Greenberg and J. F Liebman, *Strained Organic Molecules*, Academic Press, New York (1978).
[3-30] L. N. Ferguson, *J. Chem. Educ.* **46**, 404 (1969).
[3-31] G. Maier, S. Pfriem, U. Schafer, and R. Matush, *Angew. Chem. Int. Ed. Engl.* **17**, 520 (1978).
[3-32] P. E. Eaton and T. J. Cole, *J. Am. Chem. Soc.* **86**, 3157 (1964); see also, P. E. Eaton, *Angew. Chem. Int. Ed. Engl.* **31**, 1421 (1992).
[3-33] A. P. Marchand, *Chem. Int.*, to be published.
[3-34] R. F Curl and R. E. Smalley, *Sci. Am.* **1991** (October), 32.
[3-35] T. J. Katz and N. Acton, *J. Am. Chem. Soc.* **95**, 2738 (1973); V. Ramamurthy and T. J. Katz, *Nouv. J. Chim.* **1**, 363 (1977).
[3-36] P. E. Eaton, Y. S. Or, and S. J. Branca, *J. Am. Chem. Soc.* **103**, 2134 (1981).
[3-37] D. Farcasin, E. Wiskott, E. Osawa, W. Thielecke, E. M. Engler, J. Slutsky, and P. v. R. Schleyer, *J. Am. Chem. Soc.* **96**, 4669 (1974).
[3-38] C. A. Cupas and L. Hodakowski, *J. Am. Chem. Soc.* **96**, 4668 (1974).
[3-39] L. F Fieser, *J. Chem. Educ.* **42**, 408 (1965).
[3-40] C. Cupas, P v. R. Schleyer, and D. J. Trecker, *J. Am. Chem. Soc.* **87**, 917 (1965); T. M. Gund, E. Osawa, V. Z. Williams, and P v. R. Schleyer, *J. Org. Chem.* **39**, 2979 (1974).
[3-41] I. Hargittai and K. Hedberg, in *Molecular Structures and Vibrations* (S. J. Cyvin, ed.), Elsevier, Amsterdam (1972).
[3-42] S. Landa, *Chem. Listy* **27**, 415 (1933).
[3-43] O. J. Scherer, *Angew. Chem. Int. Ed. Engl.* **31**, 170 (1992).
[3-44] E. Boelema, J. Strating, and H. Wynberg, *Tetrahedron Lett.* **1972**, 1175; W. D. Graham and P. v. R. Schleyer, *Tetrahedron Lett.* **1972**, 1179.
[3-45] W. D. Graham, P v. R. Schleyer, E. W. Hagaman, and E. Wenkert, *J. Am. Chem. Soc.* **95**, 5785 (1973).
[3-46] W. Z. Williams, P v. R. Schleyer, G. J. Gleicher, and L. B. Rodewald, *J. Am. Chem. Soc.* **88**, 3862 (1966).
[3-47] W. Burns, T. R. B. Mitchell, M. A. McKervey, J. J. Rooney, G. Ferguson, and P. Roberts, *J. Chem. Soc., Chem. Commun.* **1976**, 893.
[3-48] M. Hargittai, *Kém. Közlem.* **50**, 371 (1978); **50**, 489 (1978).
[3-49] E. Vajda, I. Hargittai, and J. Tremmel, *Inorg. Chim. Acta* **25**, L143 (1977).
[3-50] G. Rauscher, T. Clark, D. Poppinger, and P v. R. Schleyer, *Angew. Chem. Int. Ed. Engl.* **17**, 276 (1978).
[3-51] A. Haaland and J. E. Nilsson, *Acta Chem. Scand.* **22**, 2653 (1968).

[3-52] F. A. Cotton and R. A. Walton, *Multiple Bonds between Metal Atoms*, 2nd ed., Clarendon Press, Oxford (1993).
[3-53] F. A. Cotton and C. B. Harris, *Inorg. Chem.* **4**, 330 (1965).
[3-54] V. G. Kuznetsov and P. A. Koz'min, *Zh. Strukt. Khim.* **4**, 55 (1963).
[3-55] M. H. Kelly and M. Fink, *J. Chem. Phys.* **76**, 1407 (1982).
[3-56] E. H. Hahn, H. Bohm, and D. Ginsburg, *Tetrahedron Lett.* **507**, (1973).
[3-57] V. Plato, W. D. Hartford, and K. Hedberg, *J. Chem. Phys.* **53**, 3488 (1970).
[3-58] L. S. Bartell, *J. Chem. Phys.* **32**, 827 (1960).
[3-59] L. Pauling, *The Nature of the Chemical Bond*, 3rd ed., Cornell University Press, Ithaca, New York (1960).
[3-60] C. Glidewell, *Inorg. Chim. Acta* **20**, 113 (1976).
[3-61] L. S. Bartell, *J. Chem. Educ.* **45**, 754 (1968).
[3-62] I. Hargittai, *J. Mol. Struct.* **54**, 287 (1979).
[3-63] I. Hargittai, *The Structure of Volatile Sulphur Compounds*, Reidel, Dordrecht (1985).
[3-64] R. L. Kuczkowski, R. D. Suenram, and F. J. Lovas, *J. Am. Chem. Soc.* **103**, 2561 (1981).
[3-65] R. J. Gillespie and I. Hargittai, *The VSEPR Model of Molecular Geometry*, Allyn and Bacon, Boston (1991).
[3-66] A. Schmiedekamp, D. W. J. Cruickshank, S. Skaarup, P. Pulay, I. Hargittai, and J. E. Boggs, *J. Am. Chem. Soc.* **101**, 2002 (1979).
[3-67] *Landolt-Börnstein, Numerical Data and Functional Relationships in Science and Technology (New Series)*, Vols. II/7, II/15, II/21; *Structure Data of Free Polyatomic Molecules*, Springer-Verlag, Berlin (1976, 1987, 1992).
[3-68] J. Brunvoll, O. Exner, and I. Hargittai, *J. Mol. Struct.* **73**, 99 (1981).
[3-69] C. Macho, R. Minkwitz, J. Rohmann, B. Steger, V. Wölfel, and H. Oberhammer, *Inorg. Chem.* **25**, 2828 (1986).
[3-70] R. Hoffmann, B. F. Beier, E. L. Muetterties, and A. Rossi, *Inorg. Chem.* **16**, 511 (1977).
[3-71] W. B. Jensen, *J. Chem. Educ.* **61**, 191 (1984).
[3-72] G. N. Lewis, *J. Am. Chem. Soc.* **38**, 762 (1916); G. N. Lewis, *Valence and the Structure of Atoms and Molecules*, Chemical Catalog Co., New York (1923).
[3-73] N. V. Sidgwick and H. M. Powell, *Proc. R. Soc. London, Ser A* **176**, 153 (1940).
[3-74] R. J. Gillespie and R. S. Nyholm, *Quart. Rev. Chem. Soc.* **11**, 339 (1957).
[3-75] L. S. Bartell, *Kém. Közlem.* **43**, 497 (1975).
[3-76] R. F. W. Bader, P. J. MacDougall, and C. D. H. Lau, *J. Am. Chem. Soc.* **106**, 1594 (1984).
[3-77] R. F. W. Bader, *Atoms and Molecules: A Quantum Theory*, Oxford University Press, Oxford (1990).
[3-78] R. S. Berry, in *Quantum Dynamics of Molecules. The New Experimental Challenge to Theorists* (R. G. Wooley, ed.), Plenum Press, New York (1980).
[3-79] M. Hargittai and I. Hargittai, in *Structures and Conformations of Non-Rigid Molecules* (J. Laane, M. Dakkouri, B. van der Veken, and H. Oberhammer, eds.), pp. 465–489, NATO ASI Series C: Mathematical and Physical Sciences, Vol. 410, Kluwer, Dordrecht (1993).
[3-80] W. v. E. Doering and W. R. Roth, *Tetrahedron* **19**, 720 (1963); G. Schroeder, *Angew. Chem. Int. Ed. Engl.* **2**, 481 (1963); M. Saunders, *Tetrahedron Lett.* **1963**, 1699.
[3-81] J. S. McKennis, L. Brener, J. S. Ward, and R. Pettit, *J. Am. Chem. Soc.* **93**, 4957 (1971).
[3-82] R. S. Berry, *J. Chem. Phys.* **32**, 933 (1960).
[3-83] G. M. Whitesides and H. L. Mitchell, *J. Am. Chem. Soc.* **91**, 5384 (1969).
[3-84] L. S. Bartell, M. J. Rothman, and A. Gavezzotti, *J. Chem. Phys.* **76**, 4136 (1982) and references therein; K. O. Christe, E. C. Curtis, and D. A. Dixon, *J. Am. Chem. Soc.* **115**, 1520 (1993).

[3-85] E. L. Muetterties and W. H. Knoth, *Polyhedral Boranes*, Marcel Dekker, New York (1968).
[3-86] W. N. Lipscomb, *Science* **153**, 373 (1966).
[3-87] R. K. Bohn and M. D. Bohn, *Inorg. Chem.* **10**, 350 (1971).
[3-88] D. M. P. Mingcs and D. J. Wales, in *Electron Deficient Boron and Carbon Clusters* (G. A. Olah, K. Wade, and R. E. Williams, eds.), John Wiley & Sons, New York (1991).
[3-89] D. J. Wales, *J. Am. Chem. Soc.* **115**, 1557 (1993).
[3-90] B. F. G. Johnson and R. E. Benfield, *J. Chem. Soc., Dalton Trans.* **1978**, 1554.
[3-91] B. E. Hanson, M. J. Sullivan, and R. J. Davis, *J. Am. Chem. Soc.* **106**, 251 (1984).
[3-92] R. E. Benfield and B. F. G. Johnson, *J. Chem. Soc., Dalton Trans.* **1980**, 1743.
[3-93] J. Kepler, *Mysterium cosmographicum* (1595).

Chapter 4

Helpful Mathematical Tools

4.1 GROUPS

So far, our discussion has been nonmathematical. Ignoring mathematics, however, does not necessarily make things easier. Group theory is the mathematical apparatus for describing symmetry operations. It facilitates the understanding and the use of symmetries. It may not even be possible to successfully attack some complex problems without the use of group theory. Besides, groups are fascinating.

This introductory chapter gives the reader the tools necessary to understand the next three chapters, in which molecular vibrations, electronic structure, and chemical reactions are discussed. Further reading is recommended for broader knowledge of the subject [4-1–4-7].

A mathematical group is a very general idea. It is a special case when the elements of the group are symmetry operations. When the symmetries of molecules are characterized by Schoenflies symbols, for example, C_{2v}, C_{3v} or C_{2h}, these symbols represent well-defined groups of symmetry operations. Let us consider first the C_{2v} point group. It consists of a twofold rotation, C_2, and two reflections through mutually perpendicular symmetry planes, σ_v and σ'_v, whose intersection coincides with the rotation axis. All the corresponding elements are shown in Figure 4-1. One more operation can be added to these, called the identity operation, E. Its application leaves the molecule unchanged. The set of the operations C_2, σ_v, σ'_v, and E together make a mathematical group.

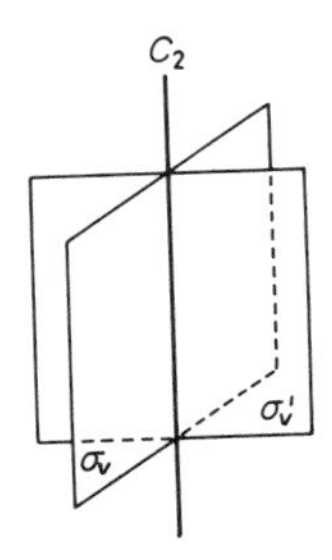

Figure 4-1. Symmetry operations in the C_{2v} point group.

A mathematical group is a set of elements related by certain rules. They will be illustrated on the symmetry operations.

1. *The product of any two elements of a group is also an element of the group* The product here means consecutive application of the elements rathe than common multiplication. Thus, for example, the product $\sigma_v \cdot C_2$ mean that first a twofold rotation is applied to an operand* and then reflection i applied to the new operand. Let us perform these operations on the atomi positions of a sulfuryl chloride molecule as is shown in Figure 4-2a. Th same final result is obtained by simply applying the symmetry plane σ_v', a is also shown in Figure 4-2b. Thus,

$$\sigma_v \cdot C_2 = \sigma_v'$$

The products of the elements in a group are generally not commutativ That means that the result of the consecutive application of the symmetr operations depends on the order in which they are applied. This is why it so important to read the multiplication sign as "preceded by." Figure 4- gives an example for the ammonia molecule, which belongs to the C point group. Depending on whether the C_3 operation is applied first ar then the σ_v'' or vice versa, the effect is different. There are some groups f which multiplication is commutative; they are called *Abelian groups*. Th C_{2v} point group is an example. Thus, in Figure 4-2a we could get the san result by first applying the σ_v reflection and then the twofold rotation.

2. *One element in the group must commute with all other elements in th group and leave them unchanged*. This is the *identity element*. Thus,

$$E \cdot X = X \cdot E = X$$

*Shortly, we shall use a wide range of operands related to molecular structure.

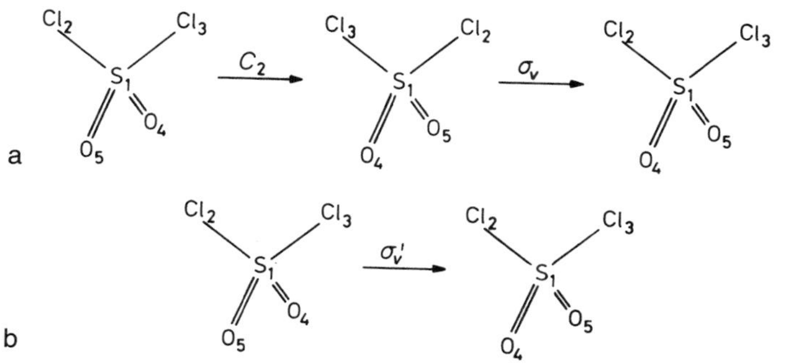

a

b

Figure 4-2. (a) Consecutive application of two symmetry operations, C_2 and σ_v, to the nuclear positions of the SO_2Cl_2 molecule. (b) Application of σ_v' to SO_2Cl_2.

3. *The products of the elements in a group are always associative.* That means that if there is a consecutive application of several symmetry operations, their application may be grouped in any way without changing the final result as long as the order of application remains the same. Thus, for example,

$$C_2 \cdot \sigma_v \cdot \sigma_v' = C_2 \cdot (\sigma_v \cdot \sigma_v') = (C_2 \cdot \sigma_v) \cdot \sigma_v'$$

4. *For each element in a group, there is an inverse or reciprocal operation which is also an element of the group* and satisfies the following condition:

$$X \cdot X^{-1} = X^{-1} \cdot X = E$$

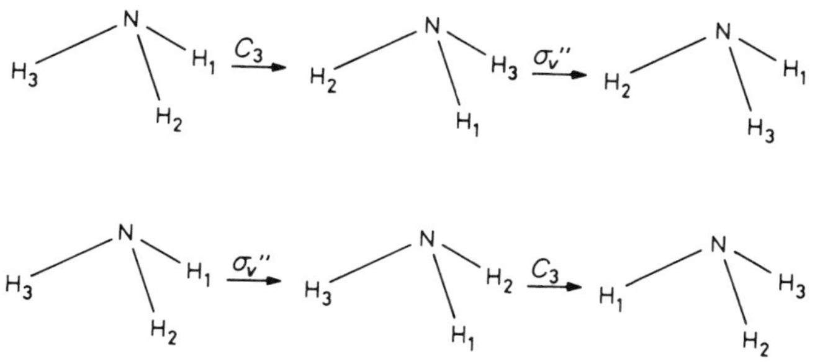

Figure 4-3. Illustration for the noncommutative character of the symmetry operations.

For example,

$$C_2 \cdot C_2^{-1} = C_2^{-1} \cdot C_2 = E$$

or

$$\sigma_v \cdot \sigma_v^{-1} = \sigma_v^{-1} \cdot \sigma_v = E$$

The symmetry operation corresponding to an inverse operation can be found in group multiplication tables. These tables contain the products of the elements of a group. An example is shown in Table 4-1, for the C_{2v} point group. Here each element of the group, that is, each symmetry operation, is listed only once in the initial row at the top and in the initial column at the far left. In forming the product of any two elements, one belonging to the row and the other to the column, the order of the application of the elements is strictly defined. First, the element in the top row is applied, followed by the element in the far left column. The result is found at the intersection of the corresponding column and row. Any one of the results is also a symmetry operation belonging to the C_{2v} point group. In fact, each row and each column in the field of the results is a rearranged list of the initial operations, but no two rows or two columns may be identical. From the C_{2v} multiplication table, it is seen that the inverse operation of C_2 is C_2, since their intersection is E; similarly, the inverse operation of σ_v is σ_v in this group.

The multiplication table of the C_{3v} point group is compiled in Table 4-2. Here,

$$C_3 \cdot C_3 = C_3^2$$

means two successive applications of the threefold rotation. Applying it once yields a 120° rotation, while C_3^2 corresponds to a 240° rotation altogether.

Table 4-1. Group Multiplication Table for the C_{2v} Point Group

C_{2v}	E	C_2	σ_v	σ_v'
E	E	C_2	σ_v	σ_v'
C_2	C_2	E	σ_v'	σ_v
σ_v	σ_v	σ_v'	E	C_2
σ_v'	σ_v'	σ_v	C_2	E

Table 4-2. Group Multiplication Table for the C_{3v} Point Group

C_{3v}	E	C_3	C_3^2	σ_v	σ_v'	σ_v''
E	E	C_3	C_3^2	σ_v	σ_v'	σ_v''
C_3	C_3	C_3^2	E	σ_v''	σ_v	σ_v'
C_3^2	C_3^2	E	C_3	σ_v'	σ_v''	σ_v
σ_v	σ_v	σ_v'	σ_v''	E	C_3	C_3^2
σ_v'	σ_v'	σ_v''	σ_v	C_3^2	E	C_3
σ_v''	σ_v''	σ_v	σ_v'	C_3	C_3^2	E

Accordingly, for example, the meaning of C_5^2 is a rotation by $2{\cdot}(360°/5) = 144°$.

The number of elements in a group is called the *order of the group*. Its conventional symbol is h. The group multiplication tables show that $h = 4$ for the C_{2v} point group and $h = 6$ for C_{3v}.

A group may be divided into two kinds of subunits: subgroups and classes. A *subgroup* is a smaller group within a group that still possesses the four fundamental properties of a group. The identity operation, E, is always a subgroup by itself, and it is also a member of all other possible subgroups.

A *class* is a complete set of elements, in our case symmetry operations, of the group that are conjugate to one another. Elements A and B of a group are *conjugates* if there is some group element, Z, for which

$$B = Z^{-1}{\cdot}A{\cdot}Z$$

Designating a conjugate B to a symmetry operation A is called a *similarity transformation*. B is a similarity transform of A by Z, or, in other words, A and B are conjugates. Elements belong to one class if they are conjugate to one another. The inverse operation can be applied with the aid of the multiplication table and rule 4 given above,

$$Z^{-1}{\cdot}Z = Z{\cdot}Z^{-1} = E$$

To find out what operations belong to the same class within a group, all possible similarity transformations in the group have to be performed. Let us work this out for the C_{3v} point group and begin with the identity operation. Since E commutes with any other elements Z (see under rule 2 above), we have

$$Z^{-1}{\cdot}E{\cdot}Z = Z^{-1}{\cdot}Z{\cdot}E = E{\cdot}E = E$$

for all elements in the class. Consequently, E is not conjugate with any other element, and it always forms a class by itself. This is true for all other point groups as well.

Consider now σ_v:

$$
\begin{aligned}
&E^{-1}\cdot(\sigma_v\cdot E) = E^{-1}\cdot\sigma_v = \sigma_v \\
&C_3^{-1}\cdot(\sigma_v\cdot C_3) = C_3^{-1}\cdot\sigma_v' = C_3^2\cdot\,\sigma_v' = \sigma_v'' \\
&(C_3^2)^{-1}\cdot(\sigma_v\cdot C_3^2) = (C_3^2)^{-1}\cdot\sigma_v'' = C_3\cdot\sigma_v'' = \sigma_v' \\
&\sigma_v^{-1}\cdot(\sigma_v\cdot\sigma_v) = \sigma_v^{-1}\cdot E = \sigma_v\cdot E = \sigma_v \\
&\sigma_v'^{-1}\cdot(\sigma_v\cdot\sigma_v') = \sigma_v'^{-1}\cdot C_3 = \sigma_v'\cdot C_3 = \sigma_v'' \\
&\sigma_v''^{-1}\cdot(\sigma_v\cdot\sigma_v'') = \sigma_v''^{-1}\cdot C_3^2 = \sigma_v''\cdot C_3^2 = \sigma_v'
\end{aligned}
$$

We have performed all possible similarity transformations for the operation σ_v. As a result, it is seen that the three operations expressing vertical mirror symmetry belong to the same class. We could reach the same conclusion by similarity transformations on either of the other two σ_v operations.

Next let us examine C_3:

$$
\begin{aligned}
&E^{-1}\cdot(C_3\cdot E) = E^{-1}\cdot C_3 = E\cdot C_3 = C_3 \\
&C_3{}^{-1}\cdot(C_3\cdot C_3) = C_3{}^{-1}\cdot C_3^2 = C_3^2\cdot C_3^2 = C_3 \\
&(C_3^2)^{-1}\cdot(C_3\cdot C_3^2) = (C_3^2)^{-1}\cdot E = C_3\cdot E = C_3 \\
&\sigma_v{}^{-1}\cdot(C_3\cdot\sigma_v) = \sigma_v{}^{-1}\cdot\sigma_v'' = \sigma_v\cdot\sigma_v'' = C_3^2 \\
&\sigma_v'^{-1}\cdot(C_3\cdot\sigma_v') = \sigma_v'^{-1}\cdot\sigma_v = \sigma_v'\cdot\sigma_v = C_3^2 \\
&\sigma_v''^{-1}\cdot(C_3\cdot\sigma_v'') = \sigma_v''^{-1}\cdot\sigma_v' = \sigma_v''\cdot\sigma_v' = C_3^2
\end{aligned}
$$

According to these transformations, C_3 and C_3^2 are conjugates and thus belong to the same class.

The *order of a class* is defined as the number of elements in the class. For example, the order of the class of the reflection operations in C_{3v} is 3, and the order of the class of the rotation operations is 2. The order of a class, or a subgroup, is an integral divisor of the order of the group.

The mathematical handling of the symmetry operations is done by means of matrices.

4.2 MATRICES

A matrix is a rectangular array of numbers, or symbols for numbers. These elements are put between square brackets. A numerical example of a matrix is shown below:

$$\begin{bmatrix} 3 & 1 & 0 & 2 \\ 5 & 7 & 0 & -3 \\ 0 & 0 & -2 & 1 \end{bmatrix}$$

Generally, a matrix has m rows and n columns:

$$\begin{bmatrix} a_{11} & a_{12} & \cdots & a_{1n} \\ a_{21} & a_{22} & \cdots & a_{2n} \\ a_{31} & a_{32} & \cdots & a_{3n} \\ \cdot & & & \cdot \\ \cdot & & & \cdot \\ \cdot & & & \cdot \\ a_{m1} & a_{m2} & \cdots & a_{mn} \end{bmatrix}$$

The above matrix may be represented by a capital letter A. Another notation is $[a_{ij}]$. The symbol a_{ij} represents the matrix element standing in the ith row and the jth column. The number of rows is m, and the number of columns is n, and $1 \leqslant i \leqslant m$ and $1 \leqslant j \leqslant n$.

There are some special matrices that are important for our discussion. A *square matrix* has equal numbers of rows and columns. According to the general notation, a matrix $[a_{ij}]$ is a square matrix if $m = n$. The *dimension* of a square matrix is the number of its rows or columns.

A special square matrix is the *unit matrix*, in which all elements along the top-left-to-bottom-right diagonal are 1 and all the other elements are zero. The short notation for a unit matrix is E. Some unit matrices are presented here:

$$\begin{bmatrix} 1 & 0 \\ 0 & 1 \end{bmatrix} \quad \begin{bmatrix} 1 & 0 & 0 \\ 0 & 1 & 0 \\ 0 & 0 & 1 \end{bmatrix} \quad \begin{bmatrix} 1 & 0 & 0 & 0 & 0 \\ 0 & 1 & 0 & 0 & 0 \\ 0 & 0 & 1 & 0 & 0 \\ 0 & 0 & 0 & 1 & 0 \\ 0 & 0 & 0 & 0 & 1 \end{bmatrix}$$

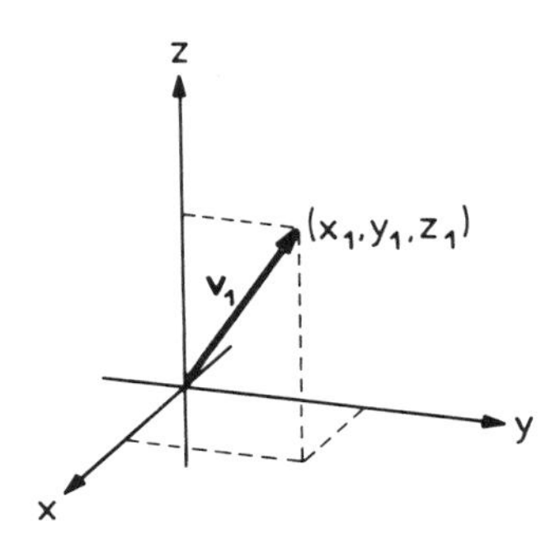

Figure 4-4. Representation of a vector in three-dimensional space.

A *column matrix* consists of only one column. Column matrices are used to represent vectors. A *vector* is characterized by its length and direction. A vector in three-dimensional space is shown in Figure 4-4. If one end of the vector is at the origin of the Cartesian coordinate system, then the three coordinates of its other end fully describe the vector. These three Cartesian coordinates can be written as a column matrix:

$$\begin{bmatrix} x_1 \\ y_1 \\ z_1 \end{bmatrix}$$

Thus, this column matrix represents the vector.

While column matrices are used to represent vectors, square matrices are used to *represent symmetry operations*. Performing a symmetry operation on a vector is actually a geometrical transformation. How can these geometrical transformations be translated into matrix "language"? Consider a specific example and see how the symmetry operations of the C_s symmetry group can be applied to the vector of Figure 4-4. For a matrix representation, we first write (or usually just imagine) the coordinates of the original vector in the top row and the coordinates of the vector resulting from the symmetry operation in the left-hand column:

$$\begin{array}{lc} & \begin{matrix} x_1 & y_1 & z_1 \end{matrix} \quad \leftarrow \text{original vector} \\ \text{resultant vector} \begin{matrix} x_1' \\ y_1' \\ z_1' \end{matrix} & \begin{bmatrix} \quad & \quad & \quad \\ & & \\ & & \end{bmatrix} \end{array}$$

Then we examine the effect of the symmetry operation in detail. If a coordinate is transformed into itself, 1 is placed into the intersection position, and if it is transformed into its negative self, −1 is put into the intersection position. Both these positions will be *along* the diagonal of the matrix. If a coordinate is transformed into another coordinate or into the negative of this other coordinate, 1 or −1 is placed into the intersection position, respectively. These intersection positions will be *off* the matrix diagonal.

There are two symmetry operations in the C_s point group, E and σ_h. The identity operation, E, does not change the position of the vector so it can be represented by a unit matrix:

$$\begin{array}{c} \\ x_1' \\ y_1' \\ z_1' \end{array} \begin{array}{c} \begin{array}{ccc} x_1 & y_1 & z_1 \end{array} \\ \begin{bmatrix} 1 & 0 & 0 \\ 0 & 1 & 0 \\ 0 & 0 & 1 \end{bmatrix} \end{array} \cdot \begin{bmatrix} x_1 \\ y_1 \\ z_1 \end{bmatrix} = \begin{bmatrix} x_1 \\ y_1 \\ z_1 \end{bmatrix}$$

Accordingly,

$$E \cdot \mathbf{v}_1 = \mathbf{v}_1$$

If the matrix elements are a_{ij} and the vector components are b_j, then the components of the product vector c_i are given by

$$c_i = \sum_j a_{ij} \cdot b_j$$

To get the first member of the resulting matrix, all the elements of the first row of the square matrix are multiplied by the consecutive members of the column matrix and then added together. To get the second member, the same procedure is followed with the second row of the square matrix, and so on, as shown below:

$$\begin{bmatrix} 1 & 0 & 0 \\ 0 & 1 & 0 \\ 0 & 0 & 1 \end{bmatrix} \cdot \begin{bmatrix} x_1 \\ y_1 \\ z_1 \end{bmatrix} = \begin{bmatrix} 1 \cdot x_1 + 0 \cdot y_1 + 0 \cdot z_1 \\ 0 \cdot x_1 + 1 \cdot y_1 + 0 \cdot z_1 \\ 0 \cdot x_1 + 0 \cdot y_1 + 1 \cdot z_1 \end{bmatrix} = \begin{bmatrix} x_1 \\ y_1 \\ z_1 \end{bmatrix}$$

The other symmetry operation of the C_s point group is the horizontal reflection (see Figure 4-5). In matrix language this operation can be written as follows:

$$\begin{array}{c} \\ x_1' \\ y_1' \\ z_1' \end{array} \begin{array}{c} \begin{array}{ccc} x_1 & y_1 & z_1 \end{array} \\ \begin{bmatrix} 1 & 0 & 0 \\ 0 & 1 & 0 \\ 0 & 0 & -1 \end{bmatrix} \end{array} \cdot \begin{bmatrix} x_1 \\ y_1 \\ z_1 \end{bmatrix} = \begin{bmatrix} 1\cdot x_1 + 0\cdot y_1 + 0\cdot z_1 \\ 0\cdot x_1 + 1\cdot y_1 + 0\cdot z_1 \\ 0\cdot x_1 + 0\cdot y_1 + (-1)\cdot z_1 \end{bmatrix} = \begin{bmatrix} x_1 \\ y_1 \\ -z_1 \end{bmatrix}$$

$$E \cdot \mathbf{v}_1 = \mathbf{v}_2$$

It often happens that the coordinates are not transformed simply into each other by a symmetry operation. Trigonometric relations must be used to express, for instance, the consequences of threefold rotation.

Figure 4-6 illustrates a vector rotated by an angle α in the xy plane. The coordinates of the rotated vector are related to the coordinates of the original vector in the following way (β is an auxiliary angle shown in Figure 4-6, and the rotation is clockwise):

$$x_1 = r\cdot\cos\beta \quad \text{and} \quad y_1 = r\cdot\sin\beta \tag{4-1}$$

$$x_2 = r\cdot\cos(\alpha - \beta) \quad \text{and} \quad y_2 = -r\cdot\sin(\alpha - \beta) \tag{4-2}$$

Utilizing the trigonometric expressions:

$$\cos(\alpha - \beta) = \cos\alpha\cdot\cos\beta + \sin\alpha\cdot\sin\beta \tag{4-3a}$$

$$\sin(\alpha - \beta) = \sin\alpha\cdot\cos\beta - \cos\alpha\cdot\sin\beta \tag{4-3b}$$

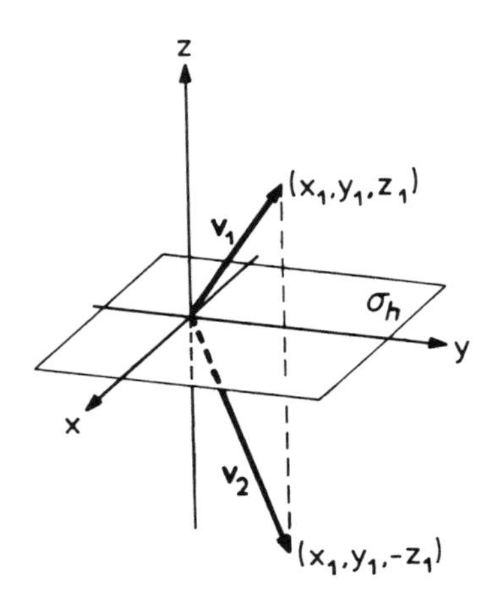

Figure 4-5. Reflection of a vector by a horizontal mirror plane.

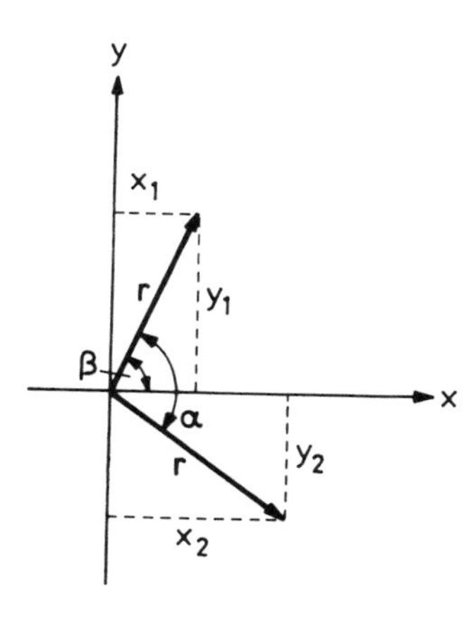

Figure 4-6. Rotation of a vector by an angle α in the *xy* plane.

and substituting Eqs. (4-3) and (4-1) into Eq. (4-2), we get:

$$x_2 = r\cdot\cos\alpha\cdot\cos\beta + r\cdot\sin\alpha\cdot\sin\beta = x_1\cdot\cos\alpha + y_1\cdot\sin\alpha \qquad \text{(4-4a)}$$

$$y_2 = -r\cdot\sin\alpha\cdot\cos\beta + r\cdot\cos\alpha\cdot\sin\beta = -x_1\cdot\sin\alpha + y_1\cdot\cos\alpha \qquad \text{(4-4b)}$$

or, in matrix formulation:

$$\begin{bmatrix} \cos\alpha & \sin\alpha \\ -\sin\alpha & \cos\alpha \end{bmatrix} \cdot \begin{bmatrix} x_1 \\ y_1 \end{bmatrix} = \begin{bmatrix} x_2 \\ y_2 \end{bmatrix}$$

The square matrix above is the matrix representation of a rotation through an angle α.

Since matrices can be used to represent symmetry operations, the set of matrices representing all symmetry operations of a point group will be a representation of that group. Moreover, if a set of matrices forms a representation of a symmetry group, it will obey all the rules of a mathematical group. It will also obey the group multiplication table. Let the SO_2Cl_2 molecule serve as an example again. This molecule belongs to the C_{2v} point group, and some of its symmetry operations have already been illustrated in Figure 4-2. To construct the corresponding matrices, the same procedure can be applied as used before with a vector. The original nuclear positions of the molecule can be written at the top row, and the nuclear positions resulting from the symmetry operation at the far left column.

There are four operations in the C_{2v} point group. E leaves the molecule unchanged, so the corresponding matrix will be a unit matrix:

$$E = \begin{array}{c} \\ S_1' \\ Cl_2' \\ Cl_3' \\ O_4' \\ O_5' \end{array} \begin{array}{c} \begin{array}{ccccc} S_1 & Cl_2 & Cl_3 & O_4 & O_5 \end{array} \\ \begin{bmatrix} 1 & 0 & 0 & 0 & 0 \\ 0 & 1 & 0 & 0 & 0 \\ 0 & 0 & 1 & 0 & 0 \\ 0 & 0 & 0 & 1 & 0 \\ 0 & 0 & 0 & 0 & 1 \end{bmatrix} \end{array}$$

The twofold rotation changes the positions of the two chlorine atoms and also the positions of the two oxygen atoms. The sulfur atom remains in place.

$$C_2 = \begin{array}{c} \\ S_1' \\ Cl_2' \\ Cl_3' \\ O_4' \\ O_5' \end{array} \begin{array}{c} \begin{array}{ccccc} S_1 & Cl_2 & Cl_3 & O_4 & O_5 \end{array} \\ \begin{bmatrix} 1 & 0 & 0 & 0 & 0 \\ 0 & 0 & 1 & 0 & 0 \\ 0 & 1 & 0 & 0 & 0 \\ 0 & 0 & 0 & 0 & 1 \\ 0 & 0 & 0 & 1 & 0 \end{bmatrix} \end{array}$$

The σ_v operation changes the positions of the two chlorines and leaves the other three atoms in place (the auxiliary top row and left-hand column will no longer be indicated):

$$\sigma_v = \begin{bmatrix} 1 & 0 & 0 & 0 & 0 \\ 0 & 0 & 1 & 0 & 0 \\ 0 & 1 & 0 & 0 & 0 \\ 0 & 0 & 0 & 1 & 0 \\ 0 & 0 & 0 & 0 & 1 \end{bmatrix}$$

Finally, σ_v' changes the positions of the two oxygen atoms and leaves the sulfur and the two chlorines in their original positions:

$$\sigma_v' = \begin{bmatrix} 1 & 0 & 0 & 0 & 0 \\ 0 & 1 & 0 & 0 & 0 \\ 0 & 0 & 1 & 0 & 0 \\ 0 & 0 & 0 & 0 & 1 \\ 0 & 0 & 0 & 1 & 0 \end{bmatrix}$$

Since each of these four 5×5 matrices represents one of the symmetry operations of the C_{2v} point group, the set of these four 5×5 matrices will be a representation of this group. They will also obey the C_{2v} multiplication table. As was shown in Figure 4-2,

$$\sigma_v \cdot C_2 = \sigma_v'$$

The corresponding matrix representations are the following:

$$\underset{\sigma_v}{\begin{bmatrix} 1 & 0 & 0 & 0 & 0 \\ 0 & 0 & 1 & 0 & 0 \\ 0 & 1 & 0 & 0 & 0 \\ 0 & 0 & 0 & 1 & 0 \\ 0 & 0 & 0 & 0 & 1 \end{bmatrix}} \cdot \underset{C_2}{\begin{bmatrix} 1 & 0 & 0 & 0 & 0 \\ 0 & 0 & 1 & 0 & 0 \\ 0 & 1 & 0 & 0 & 0 \\ 0 & 0 & 0 & 0 & 1 \\ 0 & 0 & 0 & 1 & 0 \end{bmatrix}} =$$

$$\begin{bmatrix} 1\cdot1 + 0\cdot0 + 0\cdot0 + 0\cdot0 + 0\cdot0 & 1\cdot0 + 0\cdot0 + 0\cdot1 + 0\cdot0 + 0\cdot0 \ldots \\ 0\cdot1 + 0\cdot0 + 1\cdot0 + 0\cdot0 + 0\cdot0 & 0\cdot0 + 0\cdot0 + 1\cdot1 + 0\cdot0 + 0\cdot0 \ldots \\ 0\cdot1 + 1\cdot0 + 0\cdot0 + 0\cdot0 + 0\cdot0 & 0\cdot0 + 1\cdot0 + 0\cdot1 + 0\cdot0 + 0\cdot0 \ldots \\ 0\cdot1 + 0\cdot0 + 0\cdot0 + 1\cdot0 + 0\cdot0 & 0\cdot0 + 0\cdot0 + 0\cdot1 + 1\cdot0 + 0\cdot0 \ldots \\ 0\cdot1 + 0\cdot0 + 0\cdot0 + 0\cdot0 + 1\cdot0 & 0\cdot0 + 0\cdot0 + 0\cdot1 + 0\cdot0 + 1\cdot0 \ldots \end{bmatrix} =$$

$$\begin{bmatrix} 1 & 0 & 0 & 0 & 0 \\ 0 & 1 & 0 & 0 & 0 \\ 0 & 0 & 1 & 0 & 0 \\ 0 & 0 & 0 & 0 & 1 \\ 0 & 0 & 0 & 1 & 0 \end{bmatrix}$$

$$\sigma_v'$$

The multiplication is shown here in detail only for the first two columns of the resulting matrix. The elements of the product matrix are given by

$$c_{ik} = \sum_j a_{ij} \cdot b_{jk}$$

To get the first member of the first row, all elements of the first row of the first matrix are multiplied by the corresponding elements of the first column of the second matrix and the results are added. To get the second member of the first row, all elements of the first row of the first matrix are multiplied by the corresponding members of the second column of the second matrix and the results are added, and so on. To get the second-row members, the same procedure is repeated with the second-row members of the first matrix, and so on. It is also possible to visualize the second matrix as a series of column matrices and then consider the multiplication of each of these column matrices, one by one, by the first matrix.

4.3 REPRESENTATION OF GROUPS

Any collection of quantities (or symbols) which obey the multiplication table of a group is a *representation* of that group [4-2]. These quantities are the matrices in our examples showing how certain characteristics of a molecule behave under the symmetry operations of the group. The symmetry operations may be applied to various characteristics or descriptions of the molecule. The particular description to which the symmetry operations are applied forms the *basis* for a representation of the group. Generally speaking, any set of algebraic functions or vectors may be the basis for a representation of a group [4-1]. Our choice of a suitable basis depends on the particular problem we are studying. After choosing the basis set, the task is to construct the matrices which transform the basis or its components according to each symmetry operation. The most common basis sets in chemical applications are summarized in Section 4.11. Some of them will be used in the following discussion. Let us now work out the representation of a point group for a very simple basis. We will choose just the *changes*, Δr_1 and Δr_2, of the two N–H bond lengths of the diimide molecule, N_2H_2 (**4-1**).

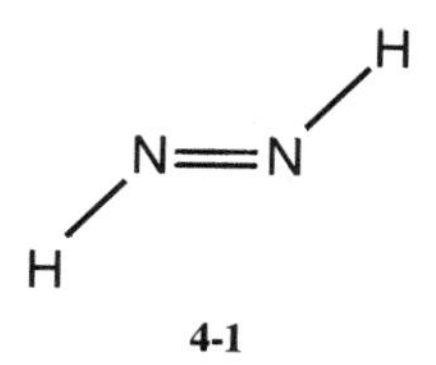

4-1

These two vectors may be used in the description of the stretching vibrations of the molecule. The molecular symmetry is C_{2h}. Figure 4-7 helps to visualize the effects of the symmetry operations of this group on the selected basis. There are four symmetry operations in the C_{2h} point group, E, C_2, i, and σ_h. E leaves the basis unchanged, so the corresponding matrix representation is a unit matrix:

$$E \cdot \begin{bmatrix} \Delta r_1 \\ \Delta r_2 \end{bmatrix} = \begin{bmatrix} 1 & 0 \\ 0 & 1 \end{bmatrix} \cdot \begin{bmatrix} \Delta r_1 \\ \Delta r_2 \end{bmatrix}$$

Figure 4-7. The four symmetry operations of the C_{2h} point group applied to the two N–H bond length changes of the HNNH molecule.

Both C_2 and i interchange the two vectors; i.e., Δr_1 "goes into" Δr_2 and vice versa:

$$C_2 \cdot \begin{bmatrix} \Delta r_1 \\ \Delta r_2 \end{bmatrix} = \begin{bmatrix} 0 & 1 \\ 1 & 0 \end{bmatrix} \cdot \begin{bmatrix} \Delta r_1 \\ \Delta r_2 \end{bmatrix}$$

$$i \cdot \begin{bmatrix} \Delta r_1 \\ \Delta r_2 \end{bmatrix} = \begin{bmatrix} 0 & 1 \\ 1 & 0 \end{bmatrix} \cdot \begin{bmatrix} \Delta r_1 \\ \Delta r_2 \end{bmatrix}$$

Finally, σ_h leaves the molecule unchanged:

$$\sigma_h \cdot \begin{bmatrix} \Delta r_1 \\ \Delta r_2 \end{bmatrix} = \begin{bmatrix} 1 & 0 \\ 0 & 1 \end{bmatrix} \cdot \begin{bmatrix} \Delta r_1 \\ \Delta r_2 \end{bmatrix}$$

With this basis the representation consists of four 2×2 matrices.

Let us take now a more complicated basis, and consider all the nuclear coordinates of HNNH shown in Figure 4-8a. These are the so-called Cartesian displacement vectors and will be discussed in Chapter 5 on molecular vibrations. Let us find the matrix representation of the σ_h operation (see Figure 4-8b). The horizontal mirror plane leaves all x and y coordinates unchanged while all z coordinates will "go" into their negative selves. In matrix notation this is expressed in the following way:

$$\sigma_h \cdot \begin{bmatrix} x_1 \\ y_1 \\ z_1 \\ x_2 \\ y_2 \\ z_2 \\ x_3 \\ y_3 \\ z_3 \\ x_4 \\ y_4 \\ z_4 \end{bmatrix} = \begin{bmatrix} 1 & 0 & 0 & 0 & 0 & 0 & 0 & 0 & 0 & 0 & 0 & 0 \\ 0 & 1 & 0 & 0 & 0 & 0 & 0 & 0 & 0 & 0 & 0 & 0 \\ 0 & 0 & -1 & 0 & 0 & 0 & 0 & 0 & 0 & 0 & 0 & 0 \\ 0 & 0 & 0 & 1 & 0 & 0 & 0 & 0 & 0 & 0 & 0 & 0 \\ 0 & 0 & 0 & 0 & 1 & 0 & 0 & 0 & 0 & 0 & 0 & 0 \\ 0 & 0 & 0 & 0 & 0 & -1 & 0 & 0 & 0 & 0 & 0 & 0 \\ 0 & 0 & 0 & 0 & 0 & 0 & 1 & 0 & 0 & 0 & 0 & 0 \\ 0 & 0 & 0 & 0 & 0 & 0 & 0 & 1 & 0 & 0 & 0 & 0 \\ 0 & 0 & 0 & 0 & 0 & 0 & 0 & 0 & -1 & 0 & 0 & 0 \\ 0 & 0 & 0 & 0 & 0 & 0 & 0 & 0 & 0 & 1 & 0 & 0 \\ 0 & 0 & 0 & 0 & 0 & 0 & 0 & 0 & 0 & 0 & 1 & 0 \\ 0 & 0 & 0 & 0 & 0 & 0 & 0 & 0 & 0 & 0 & 0 & -1 \end{bmatrix} \cdot \begin{bmatrix} x_1 \\ y_1 \\ z_1 \\ x_2 \\ y_2 \\ z_2 \\ x_3 \\ y_3 \\ z_3 \\ x_4 \\ y_4 \\ z_4 \end{bmatrix}$$

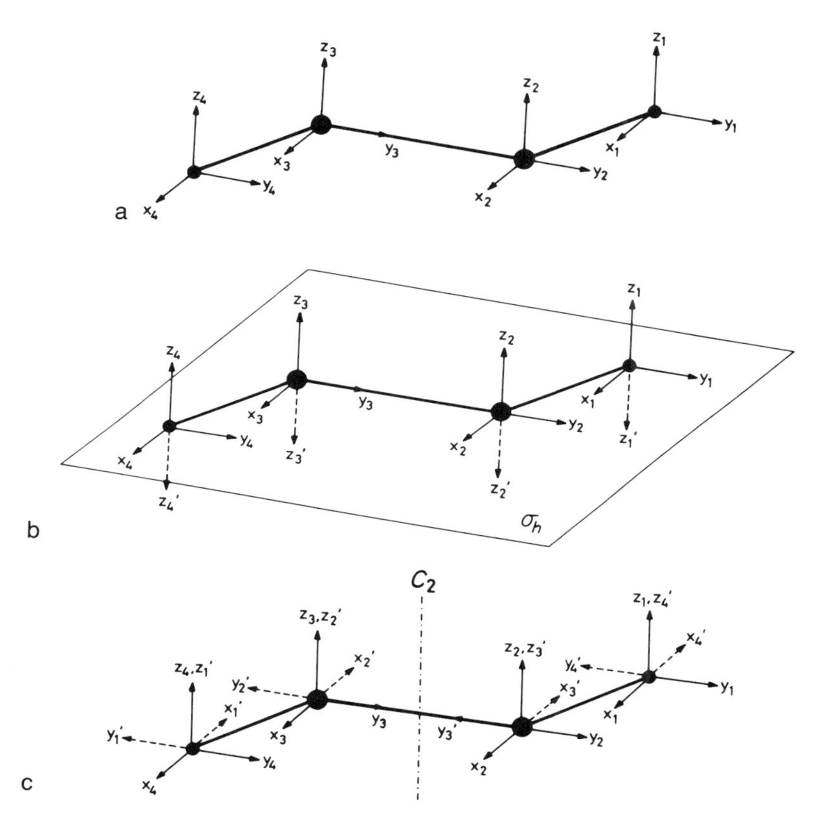

Figure 4-8. (a) Cartesian coordinates as basis for a representation; (b) the effect of σ_h; (c) the effect of C_2.

Take one more operation, the C_2 rotation (Figure 4-8c). This operation introduces the following changes:

x_1, y_1, and z_1 to $-x_4$, $-y_4$, and z_4,
x_2, y_2, and z_2 to $-x_3$, $-y_3$, and z_3,
x_3, y_3, and z_3 to $-x_2$, $-y_2$, and z_2, and
x_4, y_4, and z_4 to $-x_1$, $-y_1$, and z_1.

In matrix notation:

$$C_2 \cdot \begin{bmatrix} x_1 \\ y_1 \\ z_1 \\ x_2 \\ y_2 \\ z_2 \\ x_3 \\ y_3 \\ z_3 \\ x_4 \\ y_4 \\ z_4 \end{bmatrix} = \begin{bmatrix} 0 & 0 & 0 & 0 & 0 & 0 & 0 & 0 & 0 & -1 & 0 & 0 \\ 0 & 0 & 0 & 0 & 0 & 0 & 0 & 0 & 0 & 0 & -1 & 0 \\ 0 & 0 & 0 & 0 & 0 & 0 & 0 & 0 & 0 & 0 & 0 & 1 \\ 0 & 0 & 0 & 0 & 0 & 0 & -1 & 0 & 0 & 0 & 0 & 0 \\ 0 & 0 & 0 & 0 & 0 & 0 & 0 & -1 & 0 & 0 & 0 & 0 \\ 0 & 0 & 0 & 0 & 0 & 0 & 0 & 0 & 1 & 0 & 0 & 0 \\ 0 & 0 & 0 & -1 & 0 & 0 & 0 & 0 & 0 & 0 & 0 & 0 \\ 0 & 0 & 0 & 0 & -1 & 0 & 0 & 0 & 0 & 0 & 0 & 0 \\ 0 & 0 & 0 & 0 & 0 & 1 & 0 & 0 & 0 & 0 & 0 & 0 \\ -1 & 0 & 0 & 0 & 0 & 0 & 0 & 0 & 0 & 0 & 0 & 0 \\ 0 & -1 & 0 & 0 & 0 & 0 & 0 & 0 & 0 & 0 & 0 & 0 \\ 0 & 0 & 1 & 0 & 0 & 0 & 0 & 0 & 0 & 0 & 0 & 0 \end{bmatrix} \cdot \begin{bmatrix} x_1 \\ y_1 \\ z_1 \\ x_2 \\ y_2 \\ z_2 \\ x_3 \\ y_3 \\ z_3 \\ x_4 \\ y_4 \\ z_4 \end{bmatrix}$$

Considering all four symmetry operations of the C_{2h} point group, the complete representation of the displacement coordinates of HNNH as basis consists of four 12×12 matrices. Working with such big matrices is awkward and time-consuming. Fortunately, they can be simplifed. We shall not go into the details of how this is done since only the easiest and quickest methods utilizing matrix representations will be used in the next chapters. We shall merely outline the procedure leading from the big unpleasant representations of symmetry operations to simpler tools [4-1]. With the help of suitable similarity transformations, matrices can be turned into so-called *block-diagonal matrices*. A block-diagonal matrix has nonzero values only in square blocks along the diagonal from the top left to the bottom right. The merits of block-diagonal matrices are best illustrated in their multiplication. Suppose, for example, that two 5×5 matrices are to be multiplied, as follows:

$$\begin{bmatrix} 2 & 3 & 0 & 0 & 0 \\ 1 & 2 & 0 & 0 & 0 \\ 0 & 0 & 1 & 1 & 0 \\ 0 & 0 & 1 & 1 & 0 \\ 0 & 0 & 0 & 0 & 2 \end{bmatrix} \cdot \begin{bmatrix} 1 & 2 & 0 & 0 & 0 \\ 2 & 1 & 0 & 0 & 0 \\ 0 & 0 & 2 & 2 & 0 \\ 0 & 0 & 1 & 2 & 0 \\ 0 & 0 & 0 & 0 & 1 \end{bmatrix} = \begin{bmatrix} 8 & 7 & 0 & 0 & 0 \\ 5 & 4 & 0 & 0 & 0 \\ 0 & 0 & 3 & 4 & 0 \\ 0 & 0 & 3 & 4 & 0 \\ 0 & 0 & 0 & 0 & 2 \end{bmatrix}$$

The determination of the first row is already quite complicated:

$$\begin{aligned} 2\cdot1 + 3\cdot2 + 0\cdot0 + 0\cdot0 + 0\cdot0 &= 8 \\ 2\cdot2 + 3\cdot1 + 0\cdot0 + 0\cdot0 + 0\cdot0 &= 7 \\ 2\cdot0 + 3\cdot0 + 0\cdot2 + 0\cdot1 + 0\cdot0 &= 0 \\ 2\cdot0 + 3\cdot0 + 0\cdot2 + 0\cdot2 + 0\cdot0 &= 0 \\ 2\cdot0 + 3\cdot0 + 0\cdot0 + 0\cdot0 + 0\cdot1 &= 0 \end{aligned}$$

Notice that the product of two equally block-diagonalized matrices—such as those two above—is another similarly block-diagonalized matrix. It is especially important that this resulting matrix can be obtained simply by multiplying the corresponding individual blocks of the original matrices. Check this on the above example:

$$\begin{bmatrix} 2 & 3 \\ 1 & 2 \end{bmatrix} \cdot \begin{bmatrix} 1 & 2 \\ 2 & 1 \end{bmatrix} = \begin{bmatrix} 2\cdot1 + 3\cdot2 & 2\cdot2 + 3\cdot1 \\ 1\cdot1 + 2\cdot2 & 1\cdot2 + 2\cdot1 \end{bmatrix} = \begin{bmatrix} 8 & 7 \\ 5 & 4 \end{bmatrix}$$

$$\begin{bmatrix} 1 & 1 \\ 1 & 1 \end{bmatrix} \cdot \begin{bmatrix} 2 & 2 \\ 1 & 2 \end{bmatrix} = \begin{bmatrix} 1\cdot2 + 1\cdot1 & 1\cdot2 + 1\cdot2 \\ 1\cdot2 + 1\cdot1 & 1\cdot2 + 1\cdot2 \end{bmatrix} = \begin{bmatrix} 3 & 4 \\ 3 & 4 \end{bmatrix}$$

$$\begin{bmatrix} 2 \end{bmatrix} \cdot \begin{bmatrix} 1 \end{bmatrix} = \begin{bmatrix} 2 \end{bmatrix}$$

Generally, if two matrices A and B can be transformed by similarity transformation into identically shaped block-diagonalized matrices, their product matrix C will also have the same block-diagonal form:

$$\begin{bmatrix} A_1 & & \\ & A_2 & \\ & & A_3 \end{bmatrix} \cdot \begin{bmatrix} B_1 & & \\ & B_2 & \\ & & B_3 \end{bmatrix} = \begin{bmatrix} C_1 & & \\ & C_2 & \\ & & C_3 \end{bmatrix}$$

The multiplication will also be valid for the individual blocks:

$$\begin{aligned} A_1\cdot B_1 &= C_1 \\ A_2\cdot B_2 &= C_2 \\ A_3\cdot B_3 &= C_3 \end{aligned}$$

Since the blocks themselves will obey the same multiplication table that the big matrices do, each block will be a new representation for an operation of the group. Thus, if the above A and B matrices are representations for the respective symmetry operations σ_v and σ'_v in the C_{2v} point group, so will be the matrices A_1, A_2, and A_3 and B_1, B_2, and B_3, respectively. The C_{2v} multiplication table (Table 4-1) shows that

$$\sigma_v \cdot \sigma'_v = C_2$$

and, accordingly, not only the big C matrix but also the small matrices C_1, C_2, and C_3 will be representations of the C_2 operation. This way the big matrices *reduce* into smaller ones which are more convenient to handle. Let us suppose that the above big matrices A, B, and C together with the E matrix constitute a representation for the C_{2v} point group. This is called then a *reducible representation* of the group, indicating that it is possible to find a similarity transformation that reduces all its matrices into new ones with smaller dimension. If this is repeated until it is no longer possible to find a similarity transformation to reduce simultaneously all the matrices of a representation into smaller ones, we call this representation *irreducible*. Suppose now that in the example above the small matrices along the diagonals of the big ones cannot be reduced further by a similarity transformation. In this case each set of the small matrices will be an irreducible representation of the C_{2v} point group. The set of A_1, B_1, C_1, and E_1 will be an irreducible representation, so will be the set of A_2, B_2, C_2, and E_2, and yet another irreducible representation will be the set of A_3, B_3, C_3, and E_3. Thus, the reducible representation was reduced to three irreducible representations. Since the symmetry operations can be applied to all kinds of bases for a molecule, there may be countless numbers of reducible representations. The important thing is that all these representations reduce into a *small and finite number* of irreducible representations for practically all point groups. These irreducible representations, often called *symmetry species*, are then used in many areas of chemistry to describe symmetry properties.

4.4 THE CHARACTER OF A REPRESENTATION

Considering the sizes of the initial matrices, using irreducible representations is a great improvement. Fortunately, even further simplification is possible. Instead of working with irreducible representations, we can simply use their *characters*. The utility of this approach will be amply demonstrated later. The *character* (or trace) *of a matrix* is the sum of its diagonal elements. For the following matrix

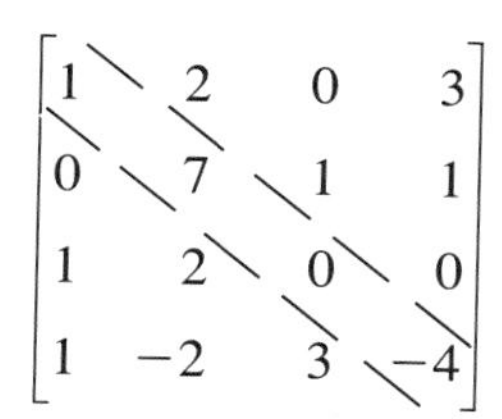

the character is

$$1 + 7 + 0 + (-4) = 4$$

Since a representation—reducible or irreducible—is a set of matrices corresponding to all symmetry operations of a group, the representation can be described by the set of characters of all these matrices. For the simple basis of Δr_1 and Δr_2 used before for the HNNH molecule in the C_{2h} point group, the representation consisted of four 2×2 matrices:

characters

$$E = \begin{bmatrix} 1 & 0 \\ 0 & 1 \end{bmatrix} \qquad 1 + 1 = 2$$

$$C_2 = \begin{bmatrix} 0 & 1 \\ 1 & 0 \end{bmatrix} \qquad 0 + 0 = 0$$

$$i = \begin{bmatrix} 0 & 1 \\ 1 & 0 \end{bmatrix} \qquad 0 + 0 = 0$$

$$\sigma_h = \begin{bmatrix} 1 & 0 \\ 0 & 1 \end{bmatrix} \qquad 1 + 1 = 2$$

Thus, the characters of this representation are

$$2 \quad 0 \quad 0 \quad 2$$

We do not know yet, however, whether this representation is reducible or irreducible. To answer this question, first we have to know the characters of the irreducible representations of the C_{2h} point group.

4.5 CHARACTER TABLES AND PROPERTIES OF IRREDUCIBLE REPRESENTATIONS

The characters of irreducible representations are collected in so-called *character tables*. We shall not discuss here how to find the characters of a given irreducible representation. The character tables are always available in textbooks and handbooks, and some of them are also given in the subsequent chapters of this book. Table 4-3 shows the character table for the C_{2h} point group. The top row contains the complete set of symmetry operations of this group. The left column shows, for the time being, some temporary names. Γ is the generally used label for the representations. The main body of the character table contains the characters themselves. Thus, each row constitutes the characters of an irreducible representation, and the number of rows gives us the number of irreducible representations of the particular point group. The irreducible representations have some important and useful properties:

1. *The sum of the squares of the dimensions of all irreducible representations in a group is equal to the order of the group*. The dimension of an irreducible representation is simply the dimension of any of its matrices, which is the number of rows or columns of the matrix. Since the identity operation always leaves the molecules unchanged, its representation is a unit matrix. The character of a unit matrix is equal to the number of rows or columns of that matrix, as is demonstrated below:

Table 4-3. A Preliminary Character Table for the C_{2h} Point Group

C_{2h}	E	C_2	i	σ_h
Γ_1	1	1	1	1
Γ_2	1	−1	1	−1
Γ_3	1	1	−1	−1
Γ_4	1	−1	−1	1

$$E = \begin{bmatrix} 1 & 0 & 0 \\ 0 & 1 & 0 \\ 0 & 0 & 1 \end{bmatrix} \qquad \text{character} = 1 + 1 + 1 = 3$$

$$E = \begin{bmatrix} 1 & 0 \\ 0 & 1 \end{bmatrix} \qquad \text{character} = 1 + 1 = 2$$

$$E = \begin{bmatrix} 1 \end{bmatrix} \qquad \text{character} = 1$$

From this it follows that the *character under E is always the dimension of the given irreducible representation*. The one-dimensional representations are nondegenerate, and the two- or higher-dimensional representations are degenerate. The meaning of degeneracy will be discussed in Chapter 6.

2. *The sum of the squares of the absolute values of characters of any irreducible representation in a group is equal to the order of the group.*
3. *The sum of the products of the corresponding characters (or one character with the conjugate of another in the case of imaginary characters) of any two different irreducible representations of the same group is zero.*
4. *The characters of all matrices belonging to operations in the same class are identical in a given irreducible representation.*
5. *The number of irreducible representations of a group is equal to the number of classes of that group.*

Let us check these rules on the C_{2h} character table given above. All four irreducible representations have 1 as their character under E, so all of them are one-dimensional. Applying rule 1,

$$1^2 + 1^2 + 1^2 + 1^2 = 4$$

This is, indeed, the order of the group since there are four symmetry operations in C_{2h}. Let us check rule 2 with the Γ_2 representation:

$$1^2 + (-1)^2 + 1^2 + (-1)^2 = 4$$

Table 4-4. A Preliminary Character Table for the C_{3v} Point Group

C_{3v}	E	C_3	C_3^2	σ_v	σ_v'	σ_v''
Γ_1	1	1	1	1	1	1
Γ_2	1	1	1	−1	−1	−1
Γ_3	2	−1	−1	0	0	0

This is, again, the order of the group. Let us form the sum of the products of Γ_3 and Γ_4 according to rule 3:

$$1 \cdot 1 + 1 \cdot (-1) + (-1) \cdot (-1) + (-1) \cdot 1 = 0$$

Since all four symmetry elements in C_{2h} stand by themselves, rule 4 cannot be checked with this point group. Finally, the number of irreducible representations is four just as is the number of classes, according to rule 5.

Table 4-4 shows a preliminary character table for the C_{3v} point group. The complete set of symmetry operations is listed in the upper row. Clearly, some of them must belong to the same class since the number of irreducible representations is 3 and the number of symmetry operations is 6. A closer look at this table reveals that the characters of all irreducible representations are equal in C_3 and C_3^2 and also in σ_v, σ_v', and σ_v'', respectively. Thus, according to rule 4 C_3 and C_3^2 form one class, and σ_v, σ_v', and σ_v'' together form another class.

A complete character table is given in Table 4-5 for the C_{3v} point group. The classes of symmetry operations are listed in the upper row, together with the number of operations in each class. Thus, it is clear from looking at this character table that there are two operations in the class of threefold rotations and three in the class of vertical reflections. The identity operation, E, always forms a class by itself, and the same is true for the inversion operation, i (which is, however, not present in the C_{3v} point group). The number of classes in C_{3v} is 3; this is also the number of irreducible representations, satisfying rule 5 as well.

Table 4-5. Complete Character Table for the C_{3v} Point Group

C_{3v}	E	$2C_3$	$3\sigma_v$		
A_1	1	1	1	z	$x^2 + y^2$, z^2
A_2	1	1	−1	R_z	
E	2	−1	0	(x, y) (R_x, R_y)	$(x^2 - y^2, xy)$ (xz, yz)

Table 4-6. Symbols for Irreducible Representations of Finite Groups

Dimension of representation	Character under: E	C_n	i	σ_h	C_2[a] or σ_v	Symbol(s)
1	1	1				A
	1	−1				B
2	2					E
3	3					T
			1			A_g B_g E_g T_g
			−1			A_u B_u E_u T_u
				1		A' B'
				−1		A'' B''
					1	A_1 B_1
					−1	A_2 B_2

[a] C_2 axis perpendicular to the principal axis.

Consider now the symbols used for the names of the irreducible representations. These are the so-called Mulliken symbols, and their meaning is described below, along with other Mulliken symbols collected in Table 4-6.

Letters A and B are used for one-dimensional irreducible representations, depending on whether they are symmetric or antisymmetric with respect to rotation around the principal axis of the point group. Antisymmetric behavior here means changing sign or direction.* The character for a symmetric representation is $+1$, and this is designated by the letter A. An antisymmetric behavior is represented by the letter B and has -1 character. E is the symbol† for two-dimensional, and T (sometimes F) the symbol for three-dimensional representations. The subscripts g and u indicate whether the representation is symmetric or antisymmetric with respect to inversion. The German *gerade* means even, and *ungerade* means odd. The superscripts $'$ and $''$ are used for irreducible representations which are symmetric and antisymmetric with respect to a horizontal mirror plane, respectively. The subscripts 1 and 2 with A and B refer to symmetric (1) and antisymmetric (2) behavior with respect to either a C_2 axis perpendicular to the principal axis or, in its absence, a vertical mirror plane. The meaning of subscripts 1 and 2 with E and T is more complicated and will not be discussed here. The character tables of the infinite groups, $C_{\infty v}$ and $D_{\infty h}$, use Greek rather than Latin letters: Σ stands for one-

*Antisymmetry will be discussed in Section 4.6.

†Not to be confused with the symbol of the identity operation, which is also E.

dimensional representations, and Π, Δ, Φ, etc., for two-dimensional representations.

It is always possible to find a behavior that remains unchanged under any of the symmetry operations of the given point group. Thus, there is always an irreducible representation which has only +1 characters. This is the *totally symmetric irreducible representation*, and it is always the first one in any character table.

The character tables usually consist of four main areas (sometimes three if the last two are merged), as is seen in Table 4-5 for the C_{3v} and in Table 4-7 for the C_{2h} group. The first area contains the symbol of the group (in the upper left corner) and the Mulliken symbols referring to the dimensionality of the representations and their relationship to various symmetry operations. The second area contains the classes of symmetry operations (in the upper row) and the characters of the irreducible representations of the group.

The third and fourth areas of the character table contain some chemically important basis functions for the group. The third area contains six symbols: x, y, z, R_x, R_y, and R_z. The first three are the Cartesian coordinates that we have already used as bases for a representation of the C_{2h} point group. The symbols R_x, R_y, and R_z stand for rotations around the x, y, and z axes, respectively. A popular toy, the spinning top, is helpful in visualizing the consequences of symmetry operations on rotation. Let us work out the characters for rotation around the z axis in the C_{3v} point group (Figure 4-9a). Obviously, the identity operation leaves the rotating spinning top unchanged (character 1). So does the rotation around the same axis since the rotational symmetry axis is indistinguishable from the axis of rotation of the toy. The corresponding character is again 1. Now place a mirror next to the rotating toy (Figure 4-9b). Irrespective of the position of the mirror, the rotation of the mirror image will always have the opposite direction with respect to the real rotation. Accordingly, the character will be −1.

Thus, the characters of the rotation around the z axis in the C_{3v} point group will be:

$$1 \quad 1 \quad -1$$

Table 4-7. C_{2h} Character Table

C_{2h}	E	C_2	i	σ_h		
A_g	1	1	1	1	R_z	x^2, y^2, x^2, xy
B_g	1	−1	1	−1	R_x, R_y	xz, yz
A_u	1	1	−1	−1	z	
B_u	1	−1	−1	1	x, y	

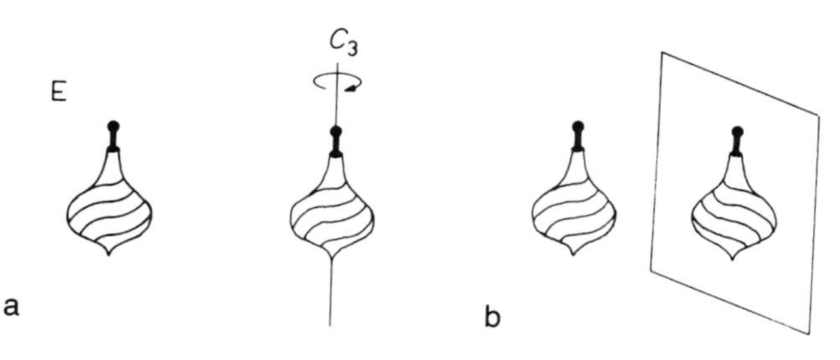

Figure 4-9. (a) Applying the identity and the C_3 operation to a rotating spinning top. (b) Illustration of the effect of mirror planes on the rotating spinning top.

Indeed, R_z belongs to the irreducible representation A_2 in the C_{3v} character table. In other words, R_z *transforms* as A_2, or, it *forms* a *basis* for A_2.

The fourth area of the character table contains all the squares and binary products of the coordinates according to their behavior under the symmetry operations. All the coordinates and their products listed in the third and fourth areas of the character table are important basis functions. They have the same symmetry properties as the atomic orbitals under the same names; z corresponds to p_z, $x^2 - y^2$ to $d_{x^2-y^2}$, and so on. We shall meet them again in the discussion of the properties of atomic orbitals.

The term antisymmetry has occurred several times above, and it is a whole new idea in our discussion. It is again a point where chemistry and other fields meet in a uniquely important symmetry concept.

4.6 ANTISYMMETRY

Antisymmetry is the symmetry of opposites [4-8]. "Operations of antisymmetry transform objects possessing two possible values of a given property from one value to the other" [4-9]. The simplest demonstration of an antisymmetry operation is by color change. Figure 4-10 shows an identity operation and an antiidentity operation. Nothing changes, of course, in the former whereas merely the black-and-white coloring reverses in the latter. Antimirror symmetry along with mirror symmetry can be found in Figure 4-11, and further antimirror symmetries are presented in Figure 4-12. The Vasarely picture (Figure 4-12a) is a characteristic representative of geometrical art. There is more than geometrical correspondence in the Soviet poster from 1987 (Figure 4-12b). The text says "This is perestroika to some," implying dissatisfaction in the way reforms of that time were carried out, amounting to mere color changes rather than substantial ones.

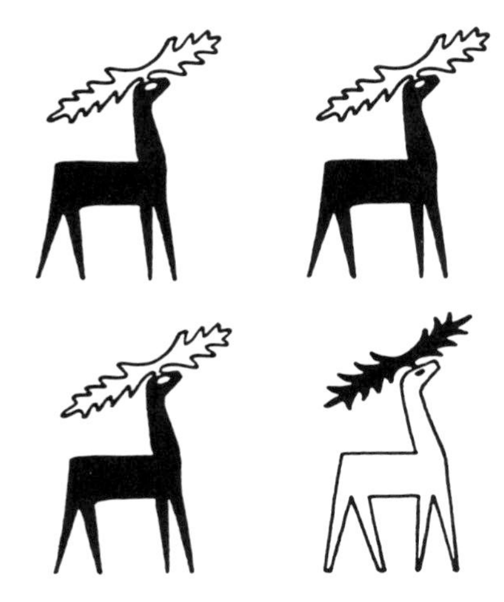

Figure 4-10. Identity operation (top) and antiidentity operation (bottom).

Symmetry elements other than a symmetry plane may also serve as antisymmetry elements. Thus, for example, twofold, fourfold, and sixfold antirotation axes appear in Figure 4-13, after Shubnikov [4-10]. The fourfold antirotation axis includes a twofold rotation axis, and the sixfold antirotation axis includes a threefold rotation axis. The antisymmetry elements have the same notation as the ordinary ones except that they are underlined. Antimirror rotation axes characterize the rosettes in the second row of Figure 4-13. The antirotation axes appear in combination with one or more symmetry planes perpendicular to the plane of the drawing in the third and fourth rows of Figure 4-13. Finally, the ordinary rotation axes are combined with one or more

Figure 4-11. Mirror symmetries and antimirror symmetries: 1–2 and 3–4 mirror symmetries 1–4 and 2–3, antimirror symmetries.

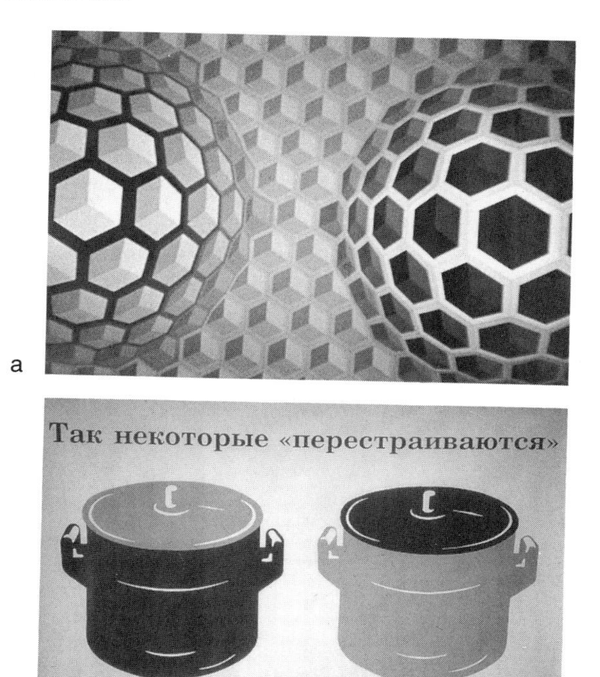

Figure 4-12. Illustrations of antimirror symmetry: (a) Picture by Victor Vasarely; used by permission; (b) Soviet (1987) poster on *perestroika*. Photograph by the authors.

antisymmetry planes in the three bottom rows of this figure. In fact, symmetry $1 \cdot \underline{m}$ here is the symmetry illustrated also in Figures 4-11 and 4-12.

The black-and-white variation is the simplest case of what is color symmetry. These considerations become very complicated quickly with increasing number of colors [4-10–4-13]. Our single example indicative of the complexity of color symmetry involves the Rubik's cube. In its monocolor version, the cube itself has many symmetry elements, among them fourfold axes going through the midpoints of opposite faces. In the unscrambled starting position of the Rubik's cube, each side has a different color. Thus, the original fourfold axis of the cube is no longer a symmetry element for the Rubik's cube. However, it is possible to specify this axis in such a way that it corresponds to the color changes of the Rubik's cube.

All the above examples applied to point groups. Such distinctions and further coloring, of course, may be introduced in space-group symmetries as

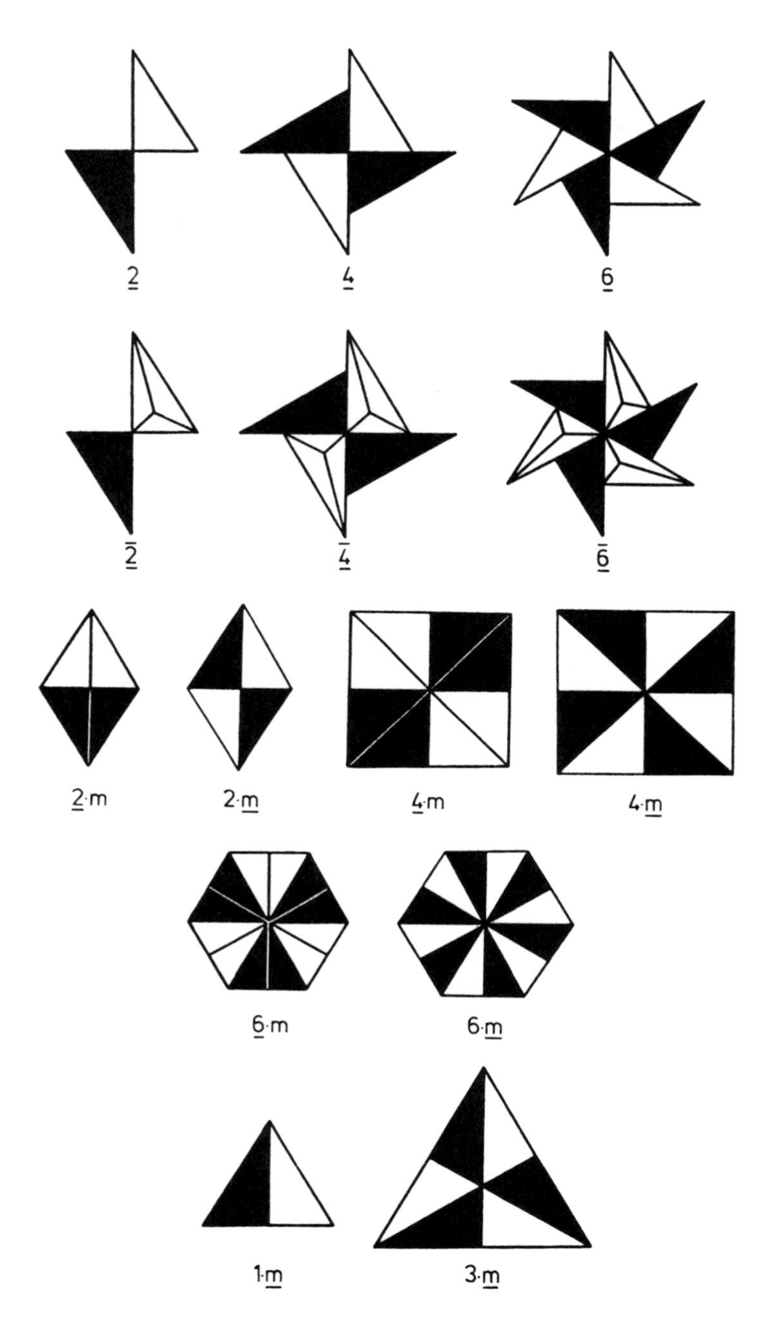

Figure 4-13. Antisymmetry operations: antirotation axes $\underline{2}$, $\underline{4}$, $\underline{6}$; antimirror rotation axes $\underline{\bar{2}}$, $\underline{\bar{4}}$, $\underline{\bar{6}}$; antirotation axes combined with ordinary mirror planes $\underline{2}\cdot m$, $\underline{4}\cdot m$, $\underline{6}\cdot m$; ordinary rotation axes combined with antimirror planes $1\cdot\underline{m}$, $2\cdot\underline{m}$, $3\cdot\underline{m}$, $4\cdot\underline{m}$, $6\cdot\underline{m}$. After Shubnikov [4-10]. Reproduced with permission from Nauka Publ. Co., Moscow.

well [4-10]. Antisymmetry also appears in space groups in Figures 8-31, 8-32, 8-40, and 9-46, in the discussion of space groups.

The color change is perhaps the simplest version of antisymmetry. The general definition of antisymmetry, at the beginning of this section, however, calls for a much broader interpretation and application. The relationship between matter and antimatter is a conspicuous example of antisymmetry. There is no limit to down-to-earth examples, as well as to abstract ones, especially if, again, symmetry is considered rather loosely.

We have already seen the contour of the oriental symbol Yin Yang representing twofold rotational symmetry in Figure 2-16a. The complete sign has also a black/white or red/blue color change as seen in Figure 4-14 and thus shows twofold antirotational symmetry. Besides color change, this symbol represents a whole array of opposites, such as night/day, hot/cold, male/female, young/old, etc.

The op art decoration of the car in Figure 4-15 involves a change in the motifs of the pattern in addition to color change. The change of the motifs appears as circle/square variation, and it may also be considered antisymmetric if the circle and the square are considered as each other's opposites.

Figure 4-16 shows the logo of a sporting goods store in Boston, Massachusetts. Geometrical correspondence is gone, yet we have no difficulty in recognizing the antimirror symmetry relationship. The antireflection plane relates a half-snowflake and a half-sun, symbolizing winter and summer. There are two Coke machines in the picture of Figure 4-17. There is no geometrical correspondence, but there is color reversal, and reversal of yet another, more important, property, the sugar content. This makes the two machines an example of antisymmetry with some abstraction.

Twofold rotational antisymmetry is shown by the ballet dancer couple in Figure 4-18, involving not only color change but gender change as well.

Figure 4-14. The flag of the Republic of Korea. Photograph by the authors.

Figure 4-15. Op art decoration of a car. Photograph by the authors.

Our final example shows two military jets and a seagull in Figure 4-19, symbolizing the contrast between war and peace.

The above examples of antisymmetry may have implied at least as much abstraction as any chemical application. The symmetric and antisymmetric behavior of orbitals describing electronic structure and vectors describing molecular vibrations may be perceived with greater ease after the preceding diversion. Before that, however, some more of group theory will be covered.

Figure 4-16. Logo of a sporting goods store in Boston, Massachusetts. Photograph by the authors.

Figure 4-17. Two Coke machines where color change and, even more importantly, reversal of sugar content make the antisymmetric relationship. Photograph by the authors.

Figure 4-18. Twofold rotational antisymmetry involving not only color change but gender change as well.

Figure 4-19. Military jets and sea gull, off Bodø, Norway, symbolizing the antisymmetric relationship between war and peace. Photograph by the authors (1982).

4.7 SHORTCUT TO DETERMINE A REPRESENTATION

It was quite easy to find the irreducible representation of R_z before, as the representation we worked out appeared to be an irreducible representation itself. In most cases, however, a reducible representation is found when the symmetry operations are applied to a certain basis. Now a simpler way will be shown (1) to describe the representation on a given basis without generating the matrices themselves and (2) to reduce them, if reducible, to irreducible representations.

The diimide molecule (**4-1**) is our example again, and the basis is the two N–H bond length changes (see Figure 4-7). It is easy to generate the matrices corresponding to each operation using such a simple basis; however, even this may not be necessary. As mentioned before, instead of the representations themselves, we can work with their characters. For this particular case, the characters of the representation have already been determined:

$$\Gamma_1 \quad 2 \quad 0 \quad 0 \quad 2$$

But how can we know the character of a matrix without writing down the whole matrix?

Looking back at the effect of the different symmetry operations on HNNH

(Figure 4-7), it is recalled, for example, that C_2 interchanges Δr_1 and Δr_2, so the diagonal elements of the matrix will all be 0. Consequently, these vectors do not contribute to the character.

This observation can be generalized so that those basis elements that are associated with an atom changing its position during the symmetry operation will have zero contribution to the character. The basis element that is unchanged by a given operation contributes +1 to the character. Finally, the basis element that is transformed into its negative contributes −1. The only complication arises with the rotational operations when the atom does not move during the symmetry operation but the basis element associated with it is rotated by a certain angle. Here the matrix of the rotation has to be constructed as shown in Section 4.2.

Returning to the diimide N–H bond length changes, let us see how the above simple rules work. The identity operation, E, leaves the molecule unchanged, so the two vectors, Δr_1 and Δr_2, will each contribute +1 to the character:

$$1 + 1 = 2$$

The effect of C_2 has already been looked at. Its character is 0. The effect of the inversion operation is the same as that of C_2, so the character will be

$$0 + 0 = 0$$

Finally, operation σ_h leaves the two bonds unchanged, so both of them contribute +1 to the character:

$$1 + 1 = 2$$

The result is the same as before:

$$\Gamma_1 \quad 2 \quad 0 \quad 0 \quad 2$$

Now, check the rules with a larger basis set, the Cartesian displacement coordinates of the atoms of HNNH (see Figure 4-8). Operation E leaves all the 12 vectors unchanged, so its character will be 12. C_2 brings each atom into a different position so their vectors will also be shifted. This means that all vectors will have zero contribution to the character. The same applies to the inversion operation. Finally, as already worked out before, the horizontal reflection leaves all the x and y vectors unchanged and brings the four z vectors into their negative selves. The result is

$$8 + (-4) = 4$$

The whole representation of the displacement vectors is:

$$\Gamma_2 \quad 12 \quad 0 \quad 0 \quad 4$$

Both representations that we have constructed here are reducible since there are no 2- and 12-dimensional representations in the C_{2h} character table (Table 4-7). The next question is how to reduce these representations.

4.8 REDUCING A REPRESENTATION

It was discussed before that the irreducible representations can be produced from the reducible representations by suitable similarity transformations. Another important point is that the character of a matrix is not changed by any similarity transformation. From this it follows that the sum of the characters of the irreducible representations is equal to the character of the original reducible representation from which they are obtained. We have seen that for each symmetry operation the matrices of the irreducible representations stand along the diagonal of the matrix of the reducible representation, and the character is just the sum of the diagonal elements. When reducing a representation, the simplest way is to look for the combination of the irreducible representations of that group—that is, the sum of their characters in each class of the character table—that will produce the characters of the reducible representation.

First, reduce the representation of the two N–H bond length changes of HNNH:

$$\Gamma_1 \quad 2 \quad 0 \quad 0 \quad 2$$

The C_{2h} character table shows that Γ_1 can be reduced to $A_g + B_u$:

C_{2h}	E	C_2	i	σ_h
A_g	1	1	1	1
B_g	1	−1	1	−1
A_u	1	1	−1	−1
B_u	1	−1	−1	1
$A_g + B_u$	2	0	0	2

It may be asked, of course, whether this is the only way of decomposing the Γ_1 representation. The answer is reassuring: *The decomposition of any reducible representation is unique*. If we find a solution just by inspection of the character table, it will be the only one. Often this is the fastest and simplest way to decompose a reducible representation.

A more general and more complicated way is to use a *reduction formula*:

$$a_i = (1/h)\sum_R \chi(R)\cdot\chi_i(R)$$

where a_i is the number of times the ith irreducible representation appears in the reducible representation, h is the order of the group, R is an operation of the group, $\chi(R)$ is the character of R in the reducible representation* and $\chi_i(R)$ is the character of R in the ith irreducible representation. The summation extends over all operations of the group.

The reduction formula can be simplified by grouping the equivalent operations into classes,

$$a_i = (1/h)\sum_Q N\cdot\chi(R)_Q\cdot\chi_i(R)_Q$$

where a_i is the number of times the ith irreducible representation appears in the reducible representation, h is the order of the group, Q is a class of the group, N is the number of operations in class Q, R is an operation of the group, $\chi(R)_Q$ is the character of an operation of class Q in the reducible representation, and $\chi_i(R)_Q$ is the character of an operation of class Q in the ith irreducible representation. The summation extends over all classes of the group.

The reduction formula can only be applied to finite point groups. For the infinite point groups, $D_{\infty h}$ and $C_{\infty v}$, the usual practice is to reduce the representations by inspection of the character table.

For illustration, let us find the irreducible representations of the two examples used before. First, on the basis of the two N–H distance changes of diimide (i.e., Γ_1):

C_{2h}	E	C_2	i	σ_h
A_g	1	1	1	1
B_g	1	−1	1	−1
A_u	1	1	−1	−1
B_u	1	−1	−1	1
Γ_1	2	0	0	2

The order of the group is 4. The number of times the irreducible representation A_g appears in the reducible representation is

*Here and hereafter, the short expression "character of R" stands for the character of the matrix corresponding to operation R, in accordance with our previous discussion.

$$a_{A_g} = (1/4)[1 \cdot 2 \cdot 1 + 1 \cdot 0 \cdot 1 + 1 \cdot 0 \cdot 1 + 1 \cdot 2 \cdot 1] = (1/4)(2 + 0 + 0 + 2) = 4/4 = 1$$

In the same way we can deduce the number of times the other irreducible representations appear in Γ_1:

$$a_{B_g} = (1/4)[1 \cdot 2 \cdot 1 + 1 \cdot 0 \cdot (-1) + 1 \cdot 0 \cdot 1 + 1 \cdot 2 \cdot (-1)] = (1/4)(2 + 0 + 0 - 2) = 0$$
$$a_{A_u} = (1/4)[1 \cdot 2 \cdot 1 + 1 \cdot 0 \cdot 1 + 1 \cdot 0 \cdot (-1) + 1 \cdot 2 \cdot (-1)] = (1/4)(2 + 0 + 0 - 2) = 0$$
$$a_{B_u} = (1/4)[1 \cdot 2 \cdot 1 + 1 \cdot 0 \cdot (-1) + 1 \cdot 0 \cdot (-1) + 1 \cdot 2 \cdot 1] = (1/4)(2 + 0 + 0 + 2) = 1$$

That is, $\Gamma_1 = A_g + B_u$, and the result is the same as before.

With the 12-dimensional reducible representation of the Cartesian displacement vectors of HNNH, the inspection method probably does not work. However, the reduction formula can be used. The reducible representation is

$$\Gamma_2 \quad 12 \quad 0 \quad 0 \quad 4$$

Applying the reduction formula, we obtain:

$$a_{A_g} = (1/4)[1 \cdot 12 \cdot 1 + 1 \cdot 0 \cdot 1 + 1 \cdot 0 \cdot 1 + 1 \cdot 4 \cdot 1] = (1/4)(12 + 4) = 4$$
$$a_{B_g} = (1/4)[1 \cdot 12 \cdot 1 + 1 \cdot 0 \cdot (-1) + 1 \cdot 0 \cdot 1 + 1 \cdot 4 \cdot (-1)] = (1/4)(12 - 4) = 2$$
$$a_{A_u} = (1/4)[1 \cdot 12 \cdot 1 + 1 \cdot 0 \cdot 1 + 1 \cdot 0 \cdot (-1) + 1 \cdot 4 \cdot (-1)] = (1/4)(12 - 4) = 2$$
$$a_{B_u} = (1/4)[1 \cdot 12 \cdot 1 + 1 \cdot 0 \cdot (-1) + 1 \cdot 0 \cdot (-1) + 1 \cdot 4 \cdot 1] = (1/4)(12 + 4) = 4$$

Thus,

$$\Gamma_2 = 4A_g + 2B_g + 2A_u + 4B_u$$

4.9 AUXILIARIES

A few additional things need to be mentioned before embarking on chemical applications of group theoretical methods. For detailed descriptions and proofs, we refer to Refs. [4-1]–[4-3].

4.9.1 Direct Product

Wave functions form bases for representations of the point group of the molecule [4-1]. Suppose that f_i and f_j are such functions; then the new set of functions $f_i f_j$, called the *direct product* of f_i and f_j, is also basis for a representation of the group. The characters of the direct product can be determined by the following rule: *The characters of the representation of a direct product are equal to the products of the characters of the representations of the original functions*. The direct product of two irreducible representations will be a new representation which is either an irreducible representation itself or can be reduced into irreducible representations. Tables 4-8 and 4-9 show some examples for direct products with the C_{2v} and C_{3v} point groups, respectively.

4.9.2 Integrals of Product Functions

Integrals of product functions often occur in the quantum-mechanical description of molecular properties, and it is helpful to know their symmetry behavior. Why? The reason is that an integral whose integrand is the product of two or more functions will vanish unless the integrand is invariant under all symmetry operations of the point group. There is only one irreducible representation whose characters are 1 for each symmetry operation of the point group, and this is the totally symmetric irreducible representation. Therefore, an *integral will be nonzero only if the integrand belongs to the totally symmetric irreducible representation of the molecular point group*.

The representation of a product function can be determined by forming the direct product of the original functions. The representation of a direct product will contain the totally symmetric representation only if the original functions whose product is formed belong to the *same* irreducible representation of the molecular point group. This follows directly from rules 2 and 3 in Section 4.5.

These rules can be extended to integrals of products of more than two functions. For a triple product the integral will be nonzero only if the representation of the product of any two functions is the same as, or contains, the representation of the third function. If the integral is

$$\int f_i \cdot f_j \cdot f_k \, d\tau$$

Table 4-8. Character Table and Some Direct Products for the C_{2v} Point Group

C_{2v}	E	C_2	σ_v	σ_v'	
A_1	1	1	1	1	
A_2	1	1	−1	−1	
B_1	1	−1	1	−1	
B_2	1	−1	−1	1	
$A_1 \cdot A_2$	1	1	−1	−1	$= A_2$
$A_2 \cdot B_1$	1	−1	−1	1	$= B_2$
$B_1 \cdot B_2$	1	1	−1	−1	$= A_2$

Table 4-9. Character Table and Direct Products for the C_{3v} Point Group

C_{3v}	E	$2C_3$	$3\sigma_v$	
A_1	1	1	1	
A_2	1	1	-1	
E	2	-1	0	
$A_2 \cdot A_2$	1	1	1	$= A_1$
$A_2 \cdot E$	2	-1	0	$= E$
$E \cdot E$	4	1	0	$= A_1 + A_2 + E$

then the above condition is expressed by

$$\Gamma_{fi} \cdot \Gamma_{fk} \subset \Gamma_{fj}$$

where Γ stands for the representation, and $\subset$ means "is or contains." Very often, f_j is a quantum-chemical operator, and then the expressions are

$$\int f_i \, \hat{\text{op}}. f_k \, d\tau$$

or with other notation,

$$\langle f_i | \hat{\text{op}}. | f_k \rangle$$

and

$$\Gamma_{fi} \cdot \Gamma_{fk} \subset \Gamma_{\hat{\text{op}}.}$$

This kind of condition appears in energy integrals and spectral selection rules and in the discussion of chemical reactions.

4.9.3 Projection Operator

The *projection operator* is one of the most useful concepts in the application of group theory to chemical problems [4-1, 4-2]. It is an operator which takes the non-symmetry-adapted basis of a representation and projects it along new directions in such a way that it belongs to a specific irreducible representation of the group. The projection operator is represented by $\hat{P}$ in the following form:

$$\hat{P}^i = (1/h) \sum_R \chi_i(R) \cdot \hat{R}$$

where h is the order of the group, i is an irreducible representation of the group, R is an operation of the group, $\chi_i(R)$ is the character of R in the ith irreducible representation, and $\hat{R}$ means the application of the symmetry operation R to our basis component. The summation extends over all operations of the group.

Consider now the construction of the A_1 symmetry group orbital of the hydrogen s atomic orbitals in ammonia as an example of the application of the projection operator. (The various kinds of orbitals will be discussed in detail in Chapter 6.) The projection operator for the A_1 irreducible representation in the C_{3v} point group is

$$\hat{P}^{A_1} = (1/6)\sum_R \chi_{A_1}(R)\cdot\hat{R}$$

Applying this operator to the s orbital of one of the hydrogens (H1) of ammonia, we obtain

$$\begin{aligned}\hat{P}^{A_1}s_1 &\approx 1\cdot E\cdot s_1 + 1\cdot C_3\cdot s_1 + 1\cdot C_3^2\cdot s_1 + 1\cdot\sigma\cdot s_1 + 1\cdot\sigma'\cdot s_1 + 1\cdot\sigma''\cdot s_1 \\ &= s_1 + s_2 + s_3 + s_1 + s_2 + s_3 \approx s_1 + s_2 + s_3\end{aligned}$$

The expression is an approximation here since the numerical factor of $\frac{1}{6}$ was omitted. The coefficient (the normalization factor) in the symmetry-adapted linear combinations can be determined at a later stage by normalization. In an actual calculation this is necessary, whereas here we are interested only in the symmetry aspects, which are well represented by the relative values. In fact, the normalization factors will be ignored throughout our discussions.

Application of the projection operator will also be demonstrated pictorially in forthcoming chapters. These representations will emphasize the results of summation of symmetry-sensitive properties while the absolute magnitudes will not be treated rigorously. Thus, for example, the directions of vectors will be summed in describing vibrations, and the signs of the angular components of the electronic wave functions will be summed in describing the electronic structure.

4.10 DYNAMIC PROPERTIES

Molecular properties can be of either static or dynamic nature. A static property remains unchanged by every symmetry operation carried out on the molecule. The geometry of the nuclear arrangement in the molecule is such a property: a symmetry operation transforms the nuclear arrangement into another which will be indistinguishable from the initial.* The mass and the energy of a molecule are also static properties.

*Unless, of course, identical atoms are distinguished by labels as, e.g., in Figures 4-2 and 4-3.

Figure 4-20. Symmetric (a) and antisymmetric (b) consequencies of the "mirror operation" for two movements. Drawing courtesy of György Doczi, Seattle, Washington.

Dynamic properties, on the other hand, may change under symmetry operations. Molecular motion itself is a most common dynamic property. In our previous discussions of molecular structure, the molecules were mostly assumed to be motionless, and only the symmetry of their nuclear arrangement was considered. However, real molecules are not motionless, and their chemical behavior is influenced by their motion to a great extent.

In order to appreciate the effects of symmetry operations on motion, an example from our macroscopic world is invoked here, following the idea of Orchin and Jaffe [4-14]. Suppose there exists a long wall of mirror, and one walks alongside this mirror (Figure 4-20a). Our mirror image will be walking with us with the same speed and in the same direction (its velocity will be the same as ours). If we walk now from a distance towards the mirror perpendicularly to it, our mirror image will have a different velocity from ours: the speed will be the same again, but the direction will be just the opposite. Both we and our mirror image will be walking toward the plane of the mirror, and if we do not stop in time, we shall collide in that plane (Figure 4-20b).

The consequences of the mirror operation were different for the two movements. One was symmetric, and the other was antisymmetric.

There are analogous phenomena for all kinds of molecular motion which may be symmetric and antisymmetric with respect to the various symmetry operations of the molecular point group. The two main kinds of motion in a molecule are nuclear and electronic. The nuclear motion may be translational, rotational, and vibrational (Chapter 5). The electronic motion is basically the changes in the electron density distribution (Chapter 6).

4.11 WHERE IS GROUP THEORY APPLIED?

It is primarily the description of the dynamic properties that is facilitated by group-theoretical methods. This is in fact an understatement. The dynamic properties cannot be fully discussed without group theory. On the other hand, this theory need not be used to determine the point-group symmetry of the nuclear arrangement of a molecule, as has been shown before (cf. Figure 3-5).

The first step in the symmetry determination of the dynamic properties is the selection of the appropriate basis. Appropriate here means the correct representation of the changes in the properties examined. In the investigation of molecular vibrations (Chapter 5), either Cartesian displacement vectors or internal coordinate vectors are used. In the description of the molecular electronic structure (Chapter 6), the angular components of the atomic orbitals are frequently used bases. Since the angular wave function changes its "sign" under certain symmetry operations, its behavior will be characteristic of the spatial symmetry of a particular orbital. Molecular orbitals can also be used as basis of representation. The simple scheme below shows some important areas in chemistry where group theory is indispensable, and the most convenient basis functions are also indicated:

Area	Basis functions
Construction of molecular orbitals	Atomic orbitals
Construction of hybrid orbitals	Position vectors pointing toward the ligands
Predicting the decrease of degeneracies of *d* orbitals under a ligand field	*d* Atomic orbitals
Predicting the allowedness of chemical reactions	Molecular orbitals
Determining the number and symmetries of molecular vibrations	Cartesian displacement vectors
Normal coordinate analysis (symmetry coordinates)	Internal coordinate displacements

Group theory is also used prior to calculations to determine whether a quantum-mechanical integral of the type $\int \psi_i \hat{op}. \psi_j \, d\tau$ is different from zero or not. This is important in such areas as selection rules for electronic transitions, chemical reactions, infrared and Raman spectroscopy, and other spectroscopies.

REFERENCES

[4-1] F. A. Cotton, *Chemical Applications of Group Theory*, 3rd ed., Wiley-Interscience, New York (1990).

[4-2] A. Nussbaum, *Applied Group Theory for Chemists, Physicists and Engineers*, Prentice-Hall, Englewood Cliffs, New Jersey (1971).

[4-3] L. H. Hall, *Group Theory and Symmetry in Chemistry*, McGraw-Hill, New York (1969).

[4-4] G. Burns, *Introduction to Group Theory with Applications*, Material Science Series (A. M. Alper and A. S. Nowich, eds.), Academic Press, New York (1977).

[4-5] A. Vincent, *Molecular Symmetry and Group Theory: A Programmed Introduction to Chemical Applications*, Wiley-Interscience, New York (1977).

[4-6] S. F. A. Kettle, *Symmetry and Structure*, John Wiley & Sons, Chichester, England (1985).

[4-7] B. E. Douglas and C. A. Hollingsworth, *Symmetry in Bonding and Spectra: An Introduction*, Academic Press, Orlando, Florida (1985).

[4-8] I. Hargittai and M. Hargittai, *Math. Intell.* **16**(2), 60 (1994).

[4-9] A. L. Mackay, *Acta Crystallogr.* **10**, 543 (1957).

[4-10] A. V. Shubnikov, *Simmetriya i antisimmetriya konechnikh figur*, Izd. Akad. Nauk SSSR, Moscow (1951).

[4-11] A. Loeb, *Color and Symmetry*. Wiley-Interscience, New York (1971).

[4-12] A. Loeb, in *Patterns of Symmetry* (M. Senechal and G. Fleck, eds.), University of Massachusetts Press, Amherst (1977).

[4-13] M. Senechal, *Acta Crystallogr., Sect. A* **39**, 505 (1983).

[4-14] M. Orchin and H. H. Jaffe, *Symmetry, Orbitals, and Spectra (S.O.S.)*, Wiley-Interscience, New York (1971).

Chapter 5

Molecular Vibrations

Vibration is a special kind of motion: the atoms of every molecule are constantly changing their relative positions at every temperature (even at absolute zero) without changing the position of the molecular center of mass. In terms of the molecular geometry, these vibrations amount to continuously changing bond lengths and bond angles. Symmetry considerations will be applied to the molecular vibrations in this chapter following primarily Refs. [5-1–5-3]. Our brief discussion is only an indication of yet another important application of symmetry considerations. The mentioned references and two other fundamental monographs [5-4, 5-5] on vibrational spectroscopy are suggested for further reading. Our primary concern will be to examine in simple terms the following question. What kind of information can be deduced about the internal motion of the molecule from the mere knowledge of its point-group symmetry?

5.1 NORMAL MODES

The seemingly random motion of molecular vibrations can always be decomposed into the sum of relatively simple components, called *normal modes of vibration*. Each of the normal modes is associated with a certain frequency. Thus, for a normal mode every atom of the molecule moves with the same frequency and in phase. Three characteristics of normal vibrations will be examined: their number, their symmetry, and their type.

5.1.1 Their Number

Since vibration is only one of the possible forms of motion, it has to be separated from the others, translation and rotation. Consider first a single atom. Its motion can be characterized by the three Cartesian coordinates of its instantaneous position as shown in Figure 5-1. In other words, the atom has three *degrees of motional freedom*. Consider next a diatomic molecule. It will have $2 \times 3 = 6$ degrees of freedom. We might think again that the three Cartesian coordinates of each atom describe the motion of the molecule in space. However, this is not quite so. Since the two atoms are not independent from each other, they must move together in space. This means that three degrees of freedom will account altogether for the *translation* of a diatomic molecule (see Figure 5-2)—or of any polyatomic molecule, for that matter. Two other degrees of freedom describe the *rotation* of the diatomic molecule around the center of mass (see Figure 5-3a). The rotation around the z axis (Figure 5-3b) need not be considered as it is the axis of the molecule, and the rotation around it does not change the position of the molecule.

Thus, of the six degrees of freedom, five have been accounted for. The sixth will describe the movement of the two atoms relative to each other without changing the center of mass. This is the *vibration* of the molecule.

The complete nuclear motion of an N-atomic molecule can be described with $3N$ parameters; that is, an N-atomic molecule has $3N$ degrees of freedom. The translation of a molecule can always be described by three parameters. The rotation of a diatomic or any linear molecule will be described by two parameters, and the rotation of a nonlinear molecule by three parameters. This means that there are always three translational and three (for linear molecules

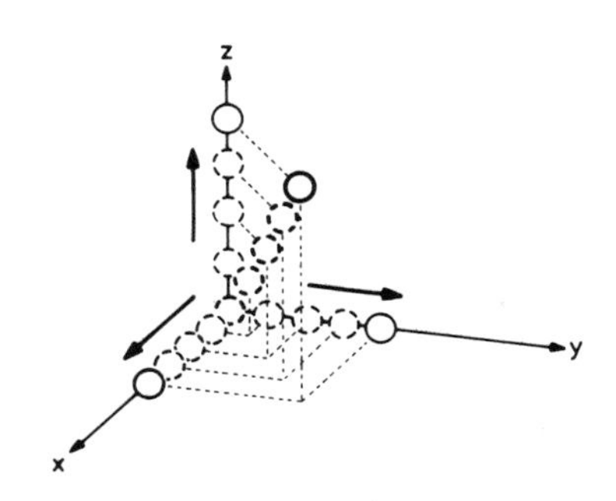

Figure 5-1. Three motional degrees of freedom of an atom.

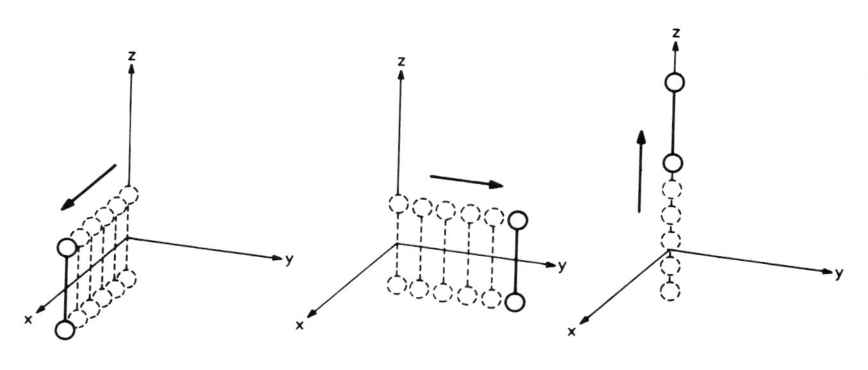

Figure 5-2. The three transitional degrees of freedom of a diatomic molecule.

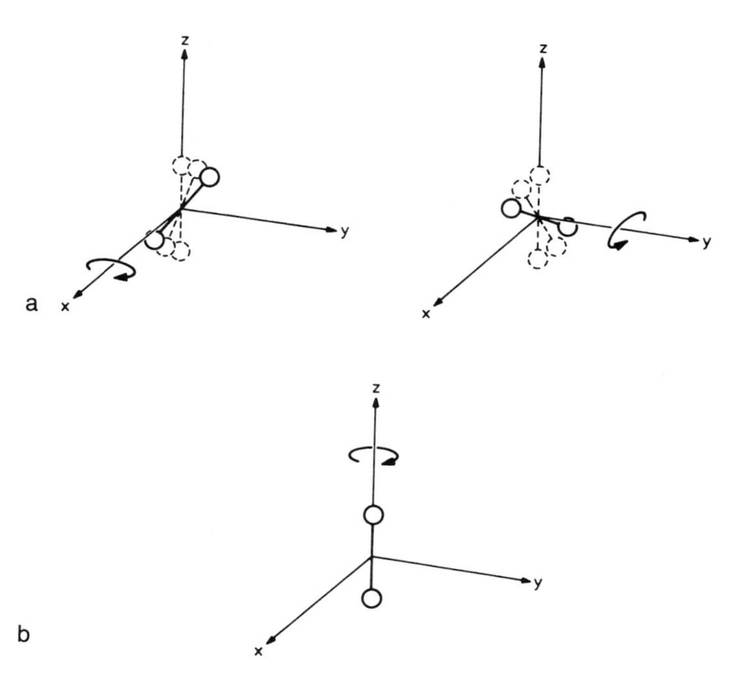

Figure 5-3. Rotation of a diatomic molecule. (a) Two rotational degrees of freedom describe the rotation of the molecule around the center of mass. (b) Rotation around the molecular axis does not change the position of the molecule.

two) rotational degrees of freedom. The remaining $3N - 6$ (for the linear case $3N - 5$) degrees of freedom account for the vibrational motion of the molecule. They give the number of normal vibrations.

The translational and rotational degrees of freedom, which do not change the relative positions of the atoms in the molecule, are often called *nongenuine modes*. The remaining $3N - 6$ (or $3N - 5$) degrees of freedom are called genuine vibrations or *genuine modes*.

5.1.2 Their Symmetry

The close relationship between symmetry and vibration is expressed by the following rule: *Each normal mode of vibration forms a basis for an irreducible representation of the point group of the molecule*.

Let us use the water molecule to illustrate the above statement. The normal modes of this molecule are shown in Figure 5-4. The point group is C_{2v}, and the character table is given in Table 5-1. It is seen that all operations bring ν_1 and ν_2 into themselves so their characters will be:

$$\begin{array}{ccccc} \Gamma_{\nu_1} & 1 & 1 & 1 & 1 \\ \Gamma_{\nu_2} & 1 & 1 & 1 & 1 \end{array}$$

The behavior of the third normal mode, ν_3, is different. While E and σ_v' leave it unchanged, both C_2 and σ_v bring it into its negative self: each atom moves in the opposite direction after the operation. This means that ν_3 is antisymmetric to these operations. The characters are:

$$\begin{array}{ccccc} \Gamma_{\nu_3} & 1 & -1 & -1 & 1 \end{array}$$

Looking at the C_{2v} character table, we can say that ν_1 and ν_2 belong to the totally symmetric irreducible representation A_1 and ν_3 belongs to B_2.

It was easy to determine the symmetry of the normal modes of the water molecule because we already knew their forms. Can the symmetry of the

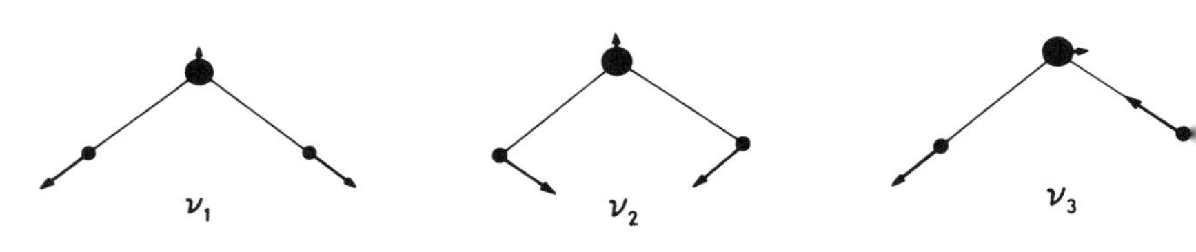

Figure 5-4. Normal modes of vibration for the water molecule. The lengths of the arrows indicate the relative displacements of the atoms.

Table 5-1. The C_{2v} Character Table

C_{2v}	E	C_2	$\sigma_v(xz)$	$\sigma_v'(yz)$		
A_1	1	1	1	1	z	x^2, y^2, z^2
A_2	1	1	−1	−1	R_z	xy
B_1	1	−1	1	−1	x, R_y	xz
B_2	1	−1	−1	1	y, R_x	yz

normal modes of a molecule be determined without any previous knowledge of the actual forms of the normal modes? The answer is fortunately yes. From the symmetry group of the molecule the symmetry species of the normal modes can be determined without any additional information.

First, an appropriate basis set has to be found. Considering that a molecule has $3N$ degrees of motional freedom, a system of $3N$ so-called *Cartesian displacement vectors* is a convenient choice. A set of such vectors is shown in Figure 5-5 for the water molecule. A separate Cartesian coordinate system is attached to each atom of the molecule, with the atoms at the origin. The orientation of the axes is the same in each system. Any displacement of the atoms can be expressed by a vector, and in turn this vector can be expressed as the vector sum of the Cartesian displacement vectors.

Next, the set of Cartesian displacement vectors is used as a basis for the representation of the point group. As discussed in Chapter 4, the vectors connected with atoms that change their position during an operation will not contribute to the character and thus can be ignored.

Continuing with the water molecule as an example, the basis of the Cartesian displacement vectors will consist of nine vectors (see Figure 5-5). Operation E brings all of them into themselves, and the character is 9. Operation C_2 changes the position of the two hydrogen atoms, so only the three coordinates of the oxygen atom have to be considered. The corresponding block of the matrix representation is

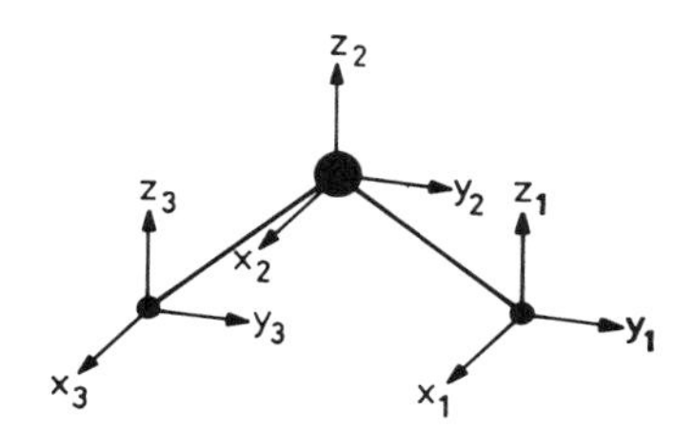

Figure 5-5. Cartesian displacement vectors as basis for representation of the water molecule.

$$C_2 = \begin{array}{c} \\ x_2' \\ y_2' \\ x_2' \end{array} \begin{array}{c} \begin{array}{ccc} x_2 & y_2 & z_2 \end{array} \\ \begin{bmatrix} -1 & 0 & 0 \\ 0 & -1 & 0 \\ 0 & 0 & 1 \end{bmatrix} \end{array}$$

The character is $(-1) + (-1) + 1 = -1$.

The next operation is σ_v. Again, only the oxygen coordinates have to be considered. Reflection through the xz plane leaves x_2 and z_2 unchanged and brings y_2 into $-y_2$. The character is $1 + 1 + (-1) = 1$.

Finally, operation σ_v' leaves all three atoms in their place, so all the nine coordinates have to be taken into account. Reflection through the yz plane leaves all y and z coordinates unchanged and takes all x coordinates into their negative selves. The character will be $(-1) + 1 + 1 + (-1) + 1 + 1 + (-1) + 1 + 1 = 3$.

The representation is

$$\Gamma_{\text{tot}} \quad 9 \quad -1 \quad 1 \quad 3$$

This is, of course, a reducible representation. Reduce it now with the reduction formula (see Chapter 4):

$$\begin{aligned} a_{A_1} &= (1/4)[1{\cdot}9{\cdot}1 + 1{\cdot}(-1){\cdot}1 + 1{\cdot}1{\cdot}1 + 1{\cdot}3{\cdot}1] \\ &= (1/4)(9 - 1 + 1 + 3) = 3 \\ a_{A_2} &= (1/4)[1{\cdot}9{\cdot}1 + 1{\cdot}(-1){\cdot}1 + 1{\cdot}1{\cdot}(-1) + 1{\cdot}3{\cdot}(-1)] \\ &= (1/4)(9 - 1 - 1 - 3) = 1 \\ a_{B_1} &= (1/4)[1{\cdot}9{\cdot}1 + 1{\cdot}(-1){\cdot}(-1) + 1{\cdot}1{\cdot}1 + 1{\cdot}3{\cdot}(-1)] \\ &= (1/4)(9 + 1 + 1 - 3) = 2 \\ a_{B_2} &= (1/4)\ [1{\cdot}9{\cdot}1 + 1{\cdot}(-1){\cdot}(-1) + 1{\cdot}1{\cdot}(-1) + 1{\cdot}3{\cdot}1] \\ &= (1/4)\ (9 + 1 - 1 + 3) = 3 \end{aligned}$$

The representation reduces to

$$\Gamma_{\text{tot}} = 3A_1 + A_2 + 2B_1 + 3B_2$$

These nine irreducible representations correspond to the nine motional degrees of freedom of the triatomic water molecule. To obtain the symmetry of the genuine vibrations, the irreducible representations of the translational and rotational motion have to be separated. This can be done using some considerations described in Chapter 4. The translational motion always belongs to those irreducible representations where the three coordinates, x, y, and z, belong.

Rotations belong to the irreducible representations of the point group indicated by R_x, R_y, and R_z in the third area of the character tables. In the C_{2v} point group,

$$\Gamma_{\text{tran}} = A_1 + B_1 + B_2$$

and

$$\Gamma_{\text{rot}} = A_2 + B_1 + B_2$$

Subtracting these from the representation of the total motion, we get

$$\begin{array}{rcl} \Gamma_{\text{tot}} & = & 3A_1 + A_2 + 2B_1 + 3B_2 \\ -(\Gamma_{\text{tran}} & = & A_1 \qquad\quad + \ B_1 + \ B_2) \\ -(\Gamma_{\text{rot}} & = & \qquad\ \ A_2 + \ B_1 + \ B_2) \\ \hline \Gamma_{\text{vib}} & = & 2A_1 \qquad\qquad\qquad + \ B_2 \end{array}$$

Thus, of the three normal modes of water, two will have A_1 and one will have B_2 symmetry. Let us stress again: this information could be derived purely from the molecular point-group symmetry.

5.1.3 Their Types

The normal modes can usually—though not always—be associated with a certain kind of motion. Those connected mainly with changes in bond lengths are the *stretching modes*. The ones connected mainly with changes of bond angles are the *deformation modes*. These may be mainly either in-plane or out-of-plane deformation modes. The simplest deformation mode is the *bending mode*.

Examine now the symmetries of these different types of vibration. For this purpose, a new type of basis set is used. Since we are interested in the changes of the geometrical parameters, these changes are an obvious choice for basis set. The geometrical parameters are also called *internal coordinates*, and the basis is the displacement of these internal coordinates.

Let us continue with the water molecule and determine the symmetry of its stretching modes. The molecule has two O–H bonds, so the basis will be the *changes* of these O–H bonds. The representation of this basis set is

$$\Gamma_{\text{str}} \quad 2 \quad 0 \quad 0 \quad 2$$

and with inspection of the C_{2v} character table we see that it reduces to $A_1 + B_2$. This means that the stretching of the O–H bonds contributes to the normal modes of A_1 and B_2 symmetry. (We shall later see that these are the symmetric and antisymmetric stretches, respectively.)

The third internal coordinate which can be considered in the water molecule is the bond angle, H–O–H. Its change will be the bending mode. All symmetry operations leave this basis unchanged, so the representation is

$$\Gamma_{\text{bend}} \quad 1 \quad 1 \quad 1 \quad 1$$

and it belongs to the totally symmetric representation, A_1. What can we conclude? B_2 appears only in the stretching mode, so the B_2 normal mode will be a pure stretching mode. The A_1 symmetry mode, however, appears in both the stretching and the bending mode. At this point we cannot say whether one of the A_1 normal modes will be purely stretch and the other purely bend or they will be a mixture. This depends on the energy of these vibrations. If they are energetically close, they can mix extensively. If they are separated by a large energy difference, they will not mix. In the case of H_2O, for example, the two A_1 symmetry modes are quite well separated, while in Cl_2O they are completely mixed.

Modes of different symmetry never mix, even if they are close in energy. (This is a general rule which will have its analogous version for the transitions among electronic states as will be seen later in Chapters 6 and 7.)

The above analysis of the types of normal modes brings us to the limit where simple symmetry considerations can take us. Nothing yet has been said about the pictorial manifestation of the various normal modes. Above we deduced, for example, that the B_2 normal mode of the water molecule is a pure stretch. The question may also be asked, how does it look? This question can be answered with the help of *symmetry coordinates*.

5.2 SYMMETRY COORDINATES

The symmetry coordinates are symmetry-adapted linear combinations of the internal coordinates. They always transform as one or another irreducible representation of the molecular point group.

Symmetry coordinates can be generated from the internal coordinates by the use of the projection operator introduced in Chapter 4. Both the symmetry coordinates and the normal modes of vibration belong to an irreducible representation of the point group of the molecule. A symmetry coordinate is always associated with one or another type of internal coordinate—that is, pure stretch, pure bend, etc.—whereas a normal mode can be a mixture of differen internal coordinate changes of the same symmetry. In some cases, as in H_2O the symmetry coordinates are good representations of the normal vibrations. I other cases, they are not. An example of such a case is Au_2Cl_6 [5-1], where th pure symmetry coordinate vibrations would be close in energy, so the rea normal vibrations are mixtures of the different vibrations of the same symme try type. The relationship between the symmetry coordinates and the normal vibrations can be established only by calculations called normal coordinate analysis [5-5, 5-6]. These calculations necessitate further data in addition to the knowledge of molecular symmetry and are not pursued here.

Return now to the symmetry coordinates of the water molecule. They can be generated using the projection operator. As has been mentioned before, here we are interested only in the symmetry aspects of the symmetry coordinates. Thus, the numerical factors are omitted, and normalization is not considered. First, let us work out the symmetry coordinate involving the stretching vibrations:

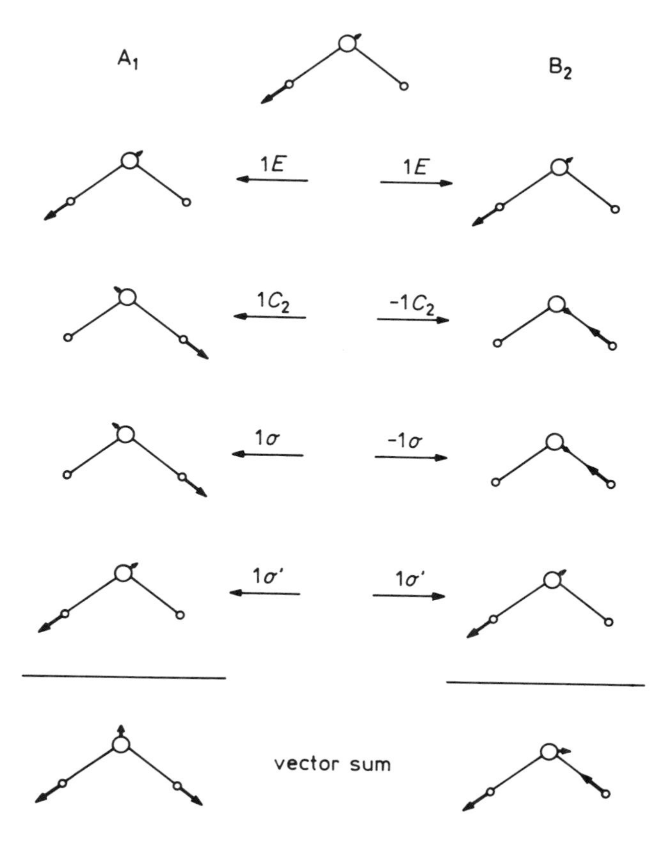

Figure 5-6. Generation of the symmetry coordinates representing bond stretching for H_2O.

$$\hat{P}^{A_1}\Delta r_1 \approx 1\cdot E\cdot\Delta r_1 + 1\cdot C_2\cdot\Delta r_1 + 1\cdot\sigma\cdot\Delta r_1 + 1\cdot\sigma'\cdot\Delta r_1$$
$$= \Delta r_1 + \Delta r_2 + \Delta r_2 + \Delta r_1 \approx \Delta r_1 + \Delta r_2$$
$$\hat{P}^{B_2}\Delta r_1 \approx 1\cdot E\cdot\Delta r_1 + (-1)\cdot C_2\cdot\Delta r_1 + (-1)\cdot\sigma\cdot\Delta r_1 + 1\cdot\sigma'\cdot\Delta r_1$$
$$= \Delta r_1 - \Delta r_2 - \Delta r_2 + \Delta r_1 \approx \Delta r_1 - \Delta r_2$$

The same procedure is presented pictorially in Figure 5-6. The bending mode of the water molecule stands alone (see the v_2 mode in Figure 5-4), so it will be a symmetry coordinate by itself.

Since the symmetry coordinates of water are good approximations of the normal vibrations, the pictorial representations are applicable to them as well. Indeed, the three normal modes of Figure 5-4 are the same as the symmetry coordinates we just derived. The A_1 symmetry stretching mode is called the symmetric stretch while the B_2 mode is the antisymmetric stretch.

5.3 SELECTION RULES

The vibrational wave function, as any wave function, must form a basis for an irreducible representation of the molecular point group [5-2].

The total vibrational wave function, ψ_v, can be written as the product of the wave functions $\psi_i(n_i)$, where ψ_i is the wave function of the ith normal vibration (i = 1 through m) in the nth state:

$$\psi_v = \psi_1(n_1)\cdot\psi_2(n_2)\cdot\psi_3(n_3)\cdots\psi_m(n_m)$$

In general, at any time, each of the normal modes may be in any state. There is, however, a situation when all the normal modes are in their ground states and only one of them gets excited into the first excited state. Such a transition is called a *fundamental transition*. The intensity of the fundamental transitions is much higher than the intensity of the other kinds of transitions.* Therefore, these are of particular interest.

The vibrational wave function of the ground state belongs to the totally symmetric irreducible representation of the point group of the molecule [5-2]. The wave function of the first excited state will belong to the irreducible representation to which the normal mode undergoing the particular transition belongs.

A fundamental transition will occur only if one of the following integrals has nonzero value:

*Were the vibrations strictly harmonic, only fundamental transitions would be observable.

$$\langle \psi_v^0 | x | \psi_v^i \rangle$$
$$\langle \psi_v^0 | y | \psi_v^i \rangle$$
$$\langle \psi_v^0 | z | \psi_v^i \rangle$$

Here, ψ_v^0 is the total vibrational wave function for the ground state, ψ_v^i is the total vibrational wave function for the first excited state referring to the ith normal mode, and x, y, and z are Cartesian coordinates.

The condition for an integral of product functions to have a nonzero value was given in Chapter 4. For the vibrational transitions this condition can be expressed in the following way:

$$\Gamma_{\psi_v^0} \cdot \Gamma_{\psi_v^i} \subset \Gamma_x \quad \text{or} \quad \Gamma_{\psi_v^0} \cdot \Gamma_{\psi_v^i} \subset \Gamma_y \quad \text{or} \quad \Gamma_{\psi_v^0} \cdot \Gamma_{\psi_v^i} \subset \Gamma_z$$

The considerations on the symmetries of the ground and excited states and the above conditions lead to the selection rule for infrared spectroscopy: *A fundamental vibration will be infrared active if the corresponding normal mode belongs to the same irreducible representation as one or more of the Cartesian coordinates.*

The selection rule for Raman spectroscopy can also be derived by similar reasoning. It says: *A fundamental vibration will be Raman active if the normal mode undergoing the vibration belongs to the same irreducible representation as one or more of the components of the polarizability tensor of the molecule.* These components are the quadratic functions of the Cartesian coordinates given in the fourth area of the character tables. The Cartesian coordinates themselves are given in the third area. Thus, the symmetry of the normal modes of a molecule is sufficient information to tell what transitions will be infrared active and what transitions will be Raman active. The normal modes of the water molecule belong to the A_1 and the B_2 irreducible representation of the C_{2v} point group. By using merely the C_{2v} character table, it can be deduced that all three vibrational modes will be active in both the infrared and Raman spectra.

Since a particular normal mode may belong to different symmetry species in different point groups, its behavior depends strongly on the molecular symmetry. Just to mention one example, the ν_1 symmetric stretching mode of an AX_3 molecule is not infrared active if the molecule is planar (D_{3h}). It is infrared active, however, if the molecule is pyramidal (C_{3v}). Vibrational spectroscopy is obviously one of the best experimental tools to determine the symmetry of molecules.

5.4 EXAMPLES

The utilization of symmetry rules in the description of molecular vibrations will be further illustrated by a few examples.

Diimide, HNNH. This molecule belongs to the C_{2h} point group (see Figure 4-7). The number of atoms is 4, so the number of normal vibrations is $(3 \times 4) - 6 = 6$.

Our first task is to generate the representation of the Cartesian displacement vectors of the four atoms of the molecule (see Figure 4-8a–c). As was shown in Chapter 4 (Section 4.7), the representation is

$$\Gamma_{\text{tot}} \quad 12 \quad 0 \quad 0 \quad 4$$

The reduction of this representation is also given in Chapter 4 (see p. 200). The result is

$$\Gamma_{\text{tot}} = 4A_g + 2B_g + 2A_u + 4B_u$$

These 12 irreducible representations account for the 12 degrees of motional freedom of HNNH. Subtracting the irreducible representations corresponding to the translation and rotation of the molecule (see C_{2h} character table, Table 5-2) leaves us the symmetry species of the normal modes of vibration:

$$\begin{aligned} \Gamma_{\text{tot}} &= 4A_g + 2B_g + 2A_u + 4B_u \\ -(\Gamma_{\text{tran}} &= \qquad\qquad\qquad\quad A_u + 2B_u) \\ -(\Gamma_{\text{rot}} &= A_g + 2B_g \qquad\qquad\qquad\quad) \\ \hline \Gamma_{\text{vib}} &= 3A_g \qquad\qquad + A_u + 2B_u \end{aligned}$$

Table 5-2. The C_{2h} Character Table and the Representations of the Internal Coordinates of Diimide

C_{2h}	E	C_2	i	σ_h		
A_g	1	1	1	1	R_z	x^2, y^2, z^2, xy
B_g	1	−1	1	−1	R_x, R_y	xz, yz
A_u	1	1	−1	−1	z	
B_u	1	−1	−1	1	x, y	
Γ_{NH}	2	0	0	2	$= A_g + B_u$	
Γ_{NN}	1	1	1	1	$= A_g$	
Γ_{NNH}	2	0	0	2	$= A_g + B_u$	
Γ_{HNNH}[a]	1	1	−1	−1	$= A_u$	

[a]Out-of-plane deformation mode.

Next we will see what kind of internal coordinate changes can account for each of these normal modes. There will have to be two N–H stretching modes and one N–N stretching mode. For deformation modes the two N–N–H angle bending modes are obvious choices, and they will be in-plane deformation modes. These constitute five normal vibrations so one is left to be accounted for. In deciding the nature of this normal mode, inspection of the character table may help. Of the above three different kinds of irreducible representations, A_g and B_u are symmetric with respect to σ_h so they must be vibrations within the molecular plane. The five vibrational modes suggested above then account for $3A_g + 2B_u$. The remaining A_u normal mode, however, is antisymmetric with respect to σ_h, so it must involve out-of-plane motion. Consequently, this normal mode will be an out-of-plane deformation mode.

We will work out next the representations of the internal coordinates. The representation of the two N–H distance changes has been given in Chapter 4 (Section 4.3). This and the other representations are all shown in Table 5-2, together with the C_{2h} character table. The Γ_{NH} representation has been reduced to $A_g + B_u$ in Chapter 4 (Section 4.8). The reduction of the Γ_{NNH} representation is the same. Both the N–N stretching and the out-of-plane deformation are already irreducible representations by themselves. Since A_g occurs three times, we cannot tell without calculation whether there will be three pure A_g modes, one N–H stretch, one N–N stretch, and one N–N–H bend, or each of the three A_g modes will be a mixture of these three vibrations. Similarly, there are two B_u symmetry normal vibrations, and they will be either pure N–H antisymmetric stretching and N–N–H bending modes or their mixtures. The only unambiguous assignment is that the A_u symmetry normal mode will be the out-of-plane deformation mode.

Let us generate the symmetry coordinates of HNNH by means of the projection operator (α is the N–N–H angle):

$$\begin{aligned}
\hat{P}^{A_g}\Delta r_1 &\approx 1\cdot E\cdot\Delta r_1 + 1\cdot C_2\cdot\Delta r_1 + 1\cdot i\cdot\Delta r_1 + 1\cdot\sigma_h\cdot\Delta r_1 \\
&= \Delta r_1 + \Delta r_2 + \Delta r_2 + \Delta r_1 \approx \Delta r_1 + \Delta r_2 \\
\hat{P}^{B_u}\Delta r_1 &\approx 1\cdot E\cdot\Delta r_1 + (-1)\cdot C_2\cdot\Delta r_1 + (-1)\cdot i\cdot\Delta r_1 + 1\cdot\sigma_h\cdot\Delta r_1 \\
&= \Delta r_1 - \Delta r_2 - \Delta r_2 + \Delta r_1 \approx \Delta r_1 - \Delta r_2 \\
\hat{P}^{A_g}\Delta\alpha_1 &\approx 1\cdot E\cdot\Delta\alpha_1 + 1\cdot C_2\cdot\Delta\alpha_1 + 1\cdot i\cdot\Delta\alpha_1 + 1\cdot\sigma_h\cdot\Delta\alpha_1 \\
&= \Delta\alpha_1 + \Delta\alpha_2 + \Delta\alpha_2 + \Delta\alpha_1 \approx \Delta\alpha_1 + \Delta\alpha_2 \\
\hat{P}^{B_u}\Delta\alpha_1 &\approx 1\cdot E\cdot\Delta\alpha_1 + (-1)\cdot C_2\cdot\Delta\alpha_1 + (-1)\cdot i\cdot\Delta\alpha_1 + 1\cdot\sigma_h\cdot\Delta\alpha_1 \\
&= \Delta\alpha_1 - \Delta\alpha_2 - \Delta\alpha_2 + \Delta\alpha_1 \approx \Delta\alpha_1 - \Delta\alpha_2
\end{aligned}$$

The same procedure is depicted in Figure 5-7. The forms of the symmetry coordinates of HNNH are shown in Figure 5-8. They might approximate well the normal modes of the molecule, and again, they might not.

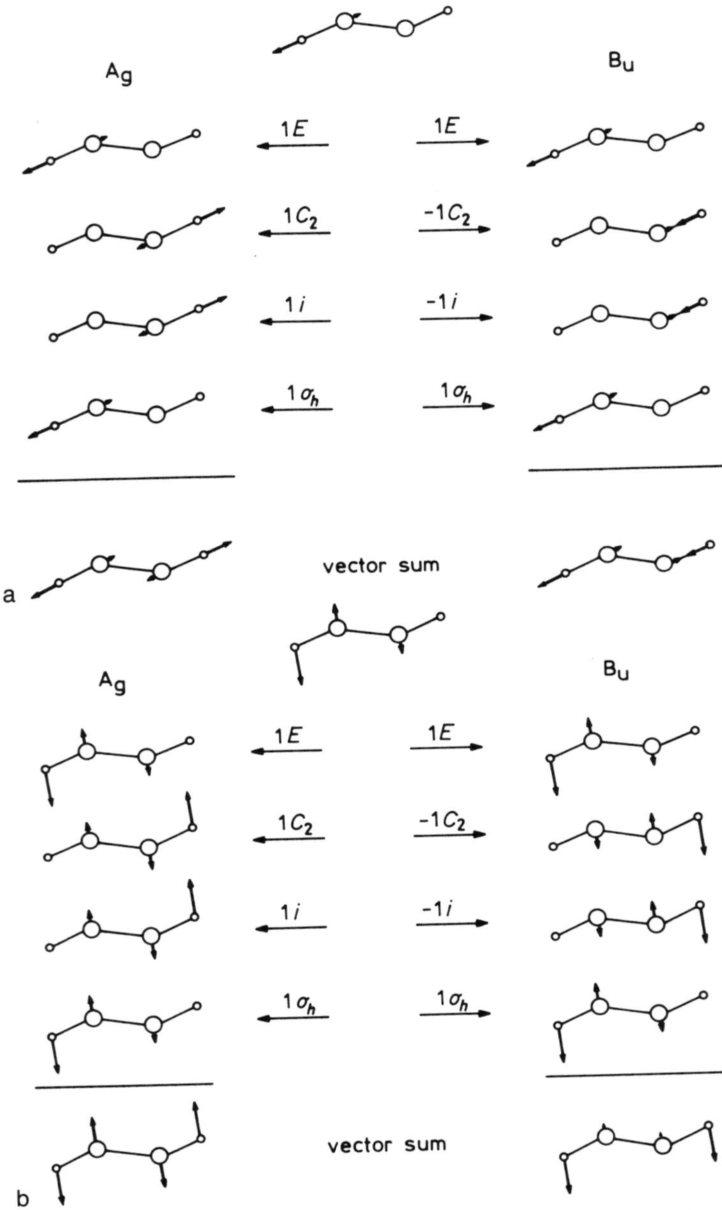

Figure 5-7. Generation of some symmetry coordinates of HNNH. (a) Symmetry coordinates corresponding to N–H bond stretches; (b) symmetry coordinates representing in-plane deformation.

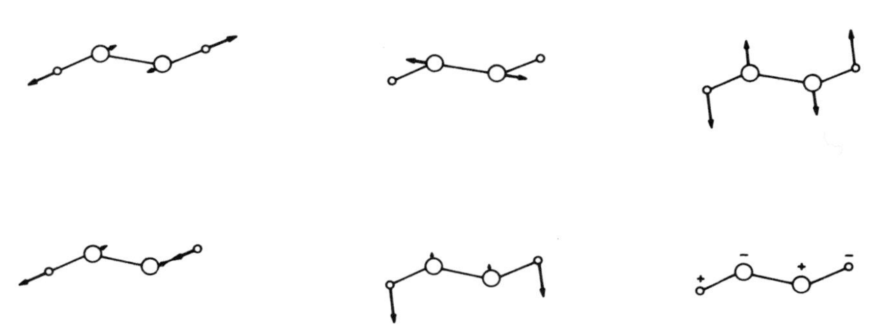

Figure 5-8. Symmetry coordinates for the HNNH molecule.

Finally, let us decide which normal modes will be infrared active and which ones will be Raman active. The Cartesian coordinates belong to the A_u and the B_u irreducible representation of the C_{2h} point group, while their binary products belong to A_g and B_g. Consequently, the selection rules are:

Infrared active: A_u, B_u
Raman active: A_g

This means that the A_g symmetry stretching modes and the A_g symmetry bending mode will be Raman active, while the B_u symmetry stretching and bending modes will be infrared active. Similarly, the A_u symmetry out-of-plane deformation mode will be infrared active.

Carbon Dioxide, CO_2. The molecule is linear and belongs to the $D_{\infty h}$ point group. The number of atoms is 3, so the number of normal vibrations is $(3 \times 3) - 5 = 4$. The set of Cartesian displacement vectors as basis for a representation is shown in Figure 5-9. The symmetry operations of the point group are also shown. The $D_{\infty h}$ character table is given in Table 5-3. Recall (Chapter 4) that the matrix of rotation by an angle Φ is

$$C^{\Phi} = \begin{bmatrix} \cos\Phi & \sin\Phi \\ -\sin\Phi & \cos\Phi \end{bmatrix}$$

The rotation by an arbitrary angle Φ will leave the three z coordinates unchanged and will mix the x and y coordinates according to the above expression. The following matrix represents the C^{Φ} rotation:*

* In the matrix, cos is abbreviated as c, and sin as s.

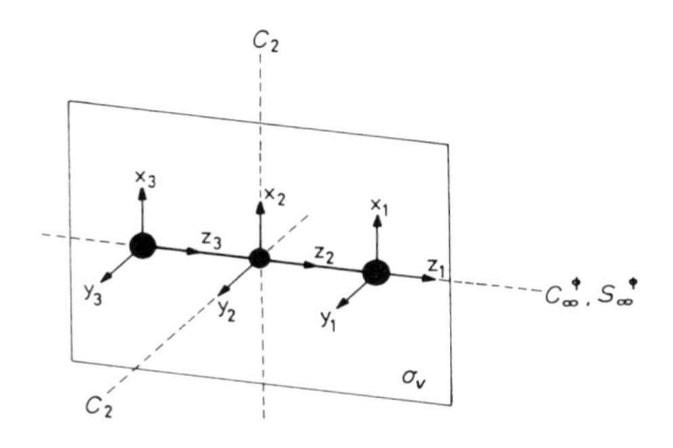

Figure 5-9. Cartesian displacement vectors of CO_2.

$$
\begin{array}{c|ccccccccc|}
 & x_1 & y_1 & z_1 & x_2 & y_2 & z_2 & x_3 & y_3 & z_3 \\
x_1' & c\Phi & s\Phi & 0 & 0 & 0 & 0 & 0 & 0 & 0 \\
y_1' & -s\Phi & c\Phi & 0 & 0 & 0 & 0 & 0 & 0 & 0 \\
z_1' & 0 & 0 & 1 & 0 & 0 & 0 & 0 & 0 & 0 \\
x_2' & 0 & 0 & 0 & c\Phi & s\Phi & 0 & 0 & 0 & 0 \\
y_2' & 0 & 0 & 0 & -s\Phi & c\Phi & 0 & 0 & 0 & 0 \\
z_2' & 0 & 0 & 0 & 0 & 0 & 1 & 0 & 0 & 0 \\
x_3' & 0 & 0 & 0 & 0 & 0 & 0 & c\Phi & s\Phi & 0 \\
y_3' & 0 & 0 & 0 & 0 & 0 & 0 & -s\Phi & c\Phi & 0 \\
z_3' & 0 & 0 & 0 & 0 & 0 & 0 & 0 & 0 & 1
\end{array}
$$

The character will be $3 + 6\cos\Phi$. The other relatively complicated operation is the mirror rotation by an arbitrary angle, S^Φ. This operation means a rotation around the z axis by angle Φ, followed by reflection through the xy plane. This reflection interchanges the positions of the two oxygen atoms so they need not be considered. The block matrix of the S^Φ operation will be:

$$
\begin{array}{c|ccc|}
 & x_2 & y_2 & z_2 \\
x_2' & \cos\Phi & \sin\Phi & 0 \\
y_2' & -\sin\Phi & \cos\Phi & 0 \\
z_2' & 0 & 0 & -1
\end{array}
$$

The character is $-1 + 2\cos\Phi$.

Table 5-3. The $D_{\infty h}$ Character Table

$D_{\infty h}$	E	$2C_\infty^\Phi$	...	$\infty\sigma_v$	i	$2S_\infty^\Phi$	...	∞C_2		
Σ_g^+	1	1	...	1	1	1	...	1		x^2, y^2, z^2
Σ_g^-	1	1	...	−1	1	1	...	−1	R_z	
Π_g	2	2cΦ[a]	...	0	2	−2cΦ	...	0	(R_x, R_y)	(xz, yz)
Δ_g	2	2c2Φ	...	0	2	2c2Φ	...	0		$(x^2 - y^2, xy)$
...	..	...	...	..	..	...	...	..		
Σ_u^+	1	1	...	1	−1	−1	...	−1	z	
Σ_u^-	1	1	...	−1	−1	−1	...	1		
Π_u	2	2cΦ	...	0	−2	2cΦ	...	0	(x, y)	
Δ_u	2	2c2Φ	...	0	−2	−2c2Φ	...	0		
...	..	...	...	..	..	...	...	..		

[a]c stands for cos.

Omitting the details of the determination of the remaining characters, the representation of the Cartesian displacement vectors is

$$\Gamma_{tot} \quad 9 \quad 3 + 6\cos\Phi \quad 3 \quad -3 \quad -1 + 2\cos\Phi \quad -1$$

Subtract the characters of the translational and rotational representations. Remember that CO_2 is linear and the rotation around the molecular axis need not be taken into account.

$$\begin{array}{lccccc}
\Gamma_{tot} = 9 & 3 + 6\cos\Phi & 3 & -3 & -1 + 2\cos\Phi & -1 \\
-(\Gamma_{tran} = 3 & 1 + 2\cos\Phi & 1 & -3 & -1 + 2\cos\Phi & -1) \\
-(\Gamma_{rot} = 2 & 2\cos\Phi & 0 & 2 & -\ 2\cos\Phi & 0) \\
\hline
\Gamma_{vib} = 4 & 2 + 2\cos\Phi & 2 & -2 & 2\cos\Phi & 0
\end{array}$$

The reduction formula cannot be applied to the infinite point groups (Chapter 4). Here inspection of the character table may help. Since $2\cos\Phi$ at S_∞^Φ appears with the Π_u irreducible representation, it is worth a try to subtract this one from Γ_{vib}:

$$\begin{array}{lccccc}
\Gamma_{vib} = 4 & 2 + 2\cos\Phi & 2 & -2 & 2\cos\Phi & 0 \\
-(\Gamma_{\Pi_u} = 2 & 2\cos\Phi & 0 & -2 & 2\cos\Phi & 0) \\
\hline
2 & 2 & 2 & 0 & 0 & 0
\end{array}$$

This representation can be resolved as the sum of Σ_g and Σ_u:

$$
\begin{array}{lcrrrrrr}
\Sigma_g & = & 1 & 1 & 1 & 1 & 1 & 1 \\
\Sigma_u & = & 1 & 1 & 1 & -1 & -1 & -1 \\
\hline
\Sigma_g + \Sigma_g & = & 2 & 2 & 2 & 0 & 0 & 0
\end{array}
$$

Thus, the normal modes of the CO_2 molecule will be

$$\Gamma_{\text{vib}} = \Sigma_g + \Sigma_u + \Pi_u$$

Since Π_u is a degenerate vibration, it counts as two, and so we indeed have the four necessary normal vibrations.

The obvious choice for the three internal coordinate changes is the stretching of the two C=O bonds and the bending of the O=C=O angle. Using these as bases for representations, we can build up the symmetry coordinates

$$\Gamma_{\text{str}} \quad 2 \quad 2 \quad 2 \quad 0 \quad 0 \quad 0$$

We have already seen before that this representation reduces as $\Sigma_g + \Sigma_u$. The Π_u normal mode will correspond to the bending vibration.

Since each of the three symmetry species, Σ_g, Σ_u, and Π_u appears onl once, the symmetry coordinates will be good representations of the norma modes. There is no possibility for mixing. Figure 5-10 shows the forms of th normal vibrations of the CO_2 molecule. The two bending modes are degener ate; they are of equal energy.

Finally, apply the vibrational selection rules to CO_2

Infrared active: Σ_u, Π_u
Raman active: Σ_g

Accordingly, the symmetric stretch C=O normal mode should appear i the Raman spectrum, while the antisymmetric stretch and the degenerat bending modes are expected to appear in the infrared spectrum.

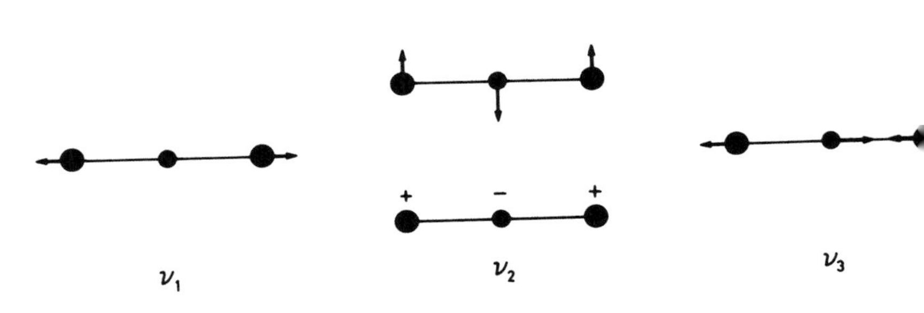

Figure 5-10. Normal modes of vibration of the CO_2 molecule.

REFERENCES

[5-1] D. C. Harris and M. D. Bertolucci, *Symmetry and Spectroscopy: An Introduction to Vibrational and Electronic Spectroscopy*, Oxford University Press, New York (1978).

[5-2] F. A. Cotton, *Chemical Applications of Group Theory*, 3rd ed., Wiley-Interscience, New York (1990).

[5-3] M. Orchin and H. H. Jaffe, *Symmetry, Orbitals, and Spectra (S.O.S)*, Wiley-Interscience, New York (1971).

[5-4] G. Herzberg, *Infrared and Raman Spectra*, Van Nostrand, Princeton, New Jersey (1959).

[5-5] E. B. Wilson, Jr., J. C. Decius, and P. C. Cross, *Molecular Vibrations*, McGraw-Hill, New York (1955).

[5-6] K. Nakamoto, *Infrared Spectra of Inorganic and Coordination Compounds*, 2nd ed., John Wiley & Sons, New York (1970).

Chapter 6

Electronic Structure of Atoms and Molecules

Everything that counts in chemistry is related to the electronic structure of atoms and molecules. The formation of molecules from atoms, their behavior, and their reactivity all depend on electronic structure. What is the role of symmetry in all this? In regard to various aspects of the electronic structure, symmetry can tell us a good deal; why certain bonds can form and others cannot, why certain electronic transitions are allowed and others are not, and why certain chemical reactions occur and others do not. Our discussion of these points is based primarily on some monographs listed in the references [6-1–6-7].

To describe the electronic structure, the electronic wave function $\psi(x, y, z, t)$ is used. As indicated, ψ depends, in general, on both space and time. Here, however, only the spatial dependence will be considered, $\psi(x, y, z)$. For detailed discussions of the nature of the electronic wave function, we refer to texts on the principles of quantum mechanics [6-1–6-3]. For a one-electron system the physical meaning of the electronic wave function is expressed by the product of ψ with its complex conjugate ψ^*. The product $\psi^* \cdot \psi \, d\tau$ gives the probability of finding an electron in the volume $d\tau = dx\,dy\,dz$ about the point (x, y, z).

A many-electron system is described by a similar but multivariable wave function:

$$\psi(x_1, y_1, z_1, \ldots, x_i, y_i, z_i, \ldots, x_n, y_n, z_n)$$

The product $\psi^* \cdot \psi \, d\tau$ gives the probability of finding the first electron in $d\tau_1$, about the point (x_1, y_1, z_1), and the ith electron in $d\tau_i$, about the point (x_i, y_i, z_i), all at the same time.

The symmetry properties of the electronic wave function and the energy of the system are two determining factors in chemical behavior. The relationship between the wave function characterizing the behavior of the electrons and the energy of the system—atoms and molecules—is expressed by the Schrödinger equation. In its general and time-independent form, it is usually written as follows:

$$\hat{H}\psi = E\psi \tag{6-1}$$

where $\hat{H}$ is the Hamiltonian operator, and E is the energy of the system.

The Hamiltonian operator is an energy operator, which includes both kinetic and potential energy terms for all particles of the system. In our discussion, only its symmetry behavior will be considered. With respect to the interchange of like particles (either nuclei or electrons), the *Hamiltonian must be unchanged* under a symmetry operation. A symmetry operation carries the system into an equivalent configuration, which is indistinguishable from the original. However, if nothing changes with the system, its energy must be the same before and after the symmetry operation. Thus, the Hamiltonian of a molecule is *invariant* to any symmetry operation of the point group of the molecule. This means that it belongs to the totally symmetric representation of the molecular point group.

A fundamental property of the wave function is that it can be used as basis for irreducible representations of the point group of a molecule [6-4]. This property establishes the connection between the symmetry of a molecule and its wave function. The preceding statement follows from Wigner's theorem which says that all eigenfunctions of a molecular system belong to one of the symmetry species of the group [6-8].

In the expression of the energy of a system the following type of integral appears:

$$\int \psi_i \hat{H} \psi_j \, d\tau$$

Depending on the problem, ψ_i and ψ_j may be atomic orbitals used to construct molecular orbitals, or they may represent two different electronic states of the same atom or molecule, etc. The energy, then, expresses the extent of interaction between the two wave functions ψ_i and ψ_j. As was shown in Chapter 4, an integral will have a nonzero value only if the integrand is invariant to the symmetry operations of the point group, i.e., belongs to the totally symmetric irreducible representation.

The above energy integral contains the $\hat{H}$ operator, which always belongs to the totally symmetric irreducible representation. Therefore, the symmetry of the whole integrand depends on the direct product of ψ_i and ψ_j. As was also shown in Chapter 4, the direct product of the representations of ψ_i and ψ_j belongs to, or contains, the totally symmetric irreducible representation only if ψ_i and ψ_j belong to the same irreducible representation. Consequently, *the energy integral will be nonzero only if ψ_i and ψ_j belong to the same irreducible representation of the molecular point group.*

6.1 ONE-ELECTRON WAVE FUNCTION

Before discussing many-electron systems, the hydrogen atom (a one-electron system) will be described. This is essentially the only atomic system for which an exact solution of the wave function is available. The spherical symmetry of the hydrogen atom makes it convenient to express the wave function in a polar coordinate system. Such a system is shown in Figure 6-1 with the proton at the origin. Ignoring the translational motion of the hydrogen atom, the Schrödinger equation can be simplified as follows [6-5]:

$$\hat{H}_e\psi = E\psi_e \tag{6-2}$$

where $\hat{H}_e$ depends only on the coordinates of the electron.

The electronic wave function can be represented as a product of a radial and an angular component:

$$\psi_e = R(r)\cdot A(\Theta, \Phi) \tag{6-3}$$

The radial wave function $R(r)$ depends on two quantum numbers, n and l. The *principal quantum number*, n, determines the electron *shell*. The numbers

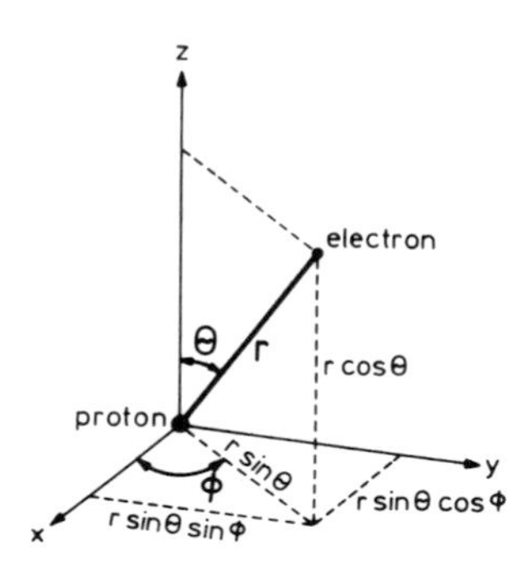

Figure 6-1. The relationship between Cartesian coordinates and spherical polar coordinates, illustrated for the hydrogen atom with the proton at the origin.

$n = 1, 2, 3, 4, \ldots$ correspond to the shells K, L, M, N, respectively. For the hydrogen atom, n completely determines the energy of the shell, which is inversely proportional to n^2. Since this energy is negative, E is smallest for the first (K) shell and increases with increasing n. The *azimuthal quantum number*, l, is associated with the total angular momentum of the electron and determines the shape of the *orbitals*. It may have integral values from 0 to $n - 1$. The $s, p, d, f, \ldots$ orbitals correspond to the azimuthal quantum numbers $l = 0, 1, 2, 3, \ldots$, respectively.

The angular wave function $A(\Theta, \Phi)$ depends also on two quantum numbers, l and m_l. The *magnetic quantum number*, m_l, is associated with the component of angular momentum along a specific axis in the atom. Since the hydrogen atom is spherically symmetrical, it is not possible to define a specific axis until the atom is placed in an external electric or magnetic field. This also means that the quantum number m_l has no effect on the energy and shape of the wave function of the hydrogen atom in the absence of such an external field. Generally, m_l may have values $-l, -l + 1, \ldots, 0, \ldots, l - 1, l$, altogether $2l + 1$ of them, and the orbitals are subdivided accordingly.

Usually, we refer to the energy of an orbital while what is really meant is the energy of an electron in that orbital. It was mentioned earlier that only the principal quantum number n influences the orbital energy in the hydrogen atom. This means that while 1s and 2s orbitals have different energies, the 2s and all three 2p orbitals have the same energy; i.e., these four $n = 2$ orbitals are degenerate in the hydrogen atom.

In many-electron atoms the value of l also influences the energy of the orbitals; thus, the 2s and 2p orbitals, or the 3s, 3p, and 3d orbitals, will no longer be degenerate. However, there are always three p orbitals and five d orbitals in each shell, and they differ only in the quantum number m_l and will be degenerate. As there are $2l + 1$ values of m_l for an orbital with quantum number l, the p orbitals ($l = 1$) will always be threefold degenerate while the d orbitals ($l = 2$) will always be fivefold degenerate.

Harris and Bertolucci [6-5] illustrated the relationship between symmetry and degeneracy of energy levels with a simple and attractive example. There are three parallelepipeds in Figure 6-2. Each of them has six stable resting positions. The potential energy of these positions depends on the height of the center of the mass above the supporting surface. This height, in turn, is determined by the choice of face on which the body rests. Three different positions are possible for the first parallelepiped (1) according to its three different kinds of faces. The potential energy of 1 will be largest when it stands on an *ab* face, since its center of mass is then at the highest possible position. There are only two energetically different positions for 2 since its center of mass is at the same height when it rests on face *bc* or on face *ac*. Parallelepiped 3 is indeed a cube, and all possible positions will be energetically equivalent.

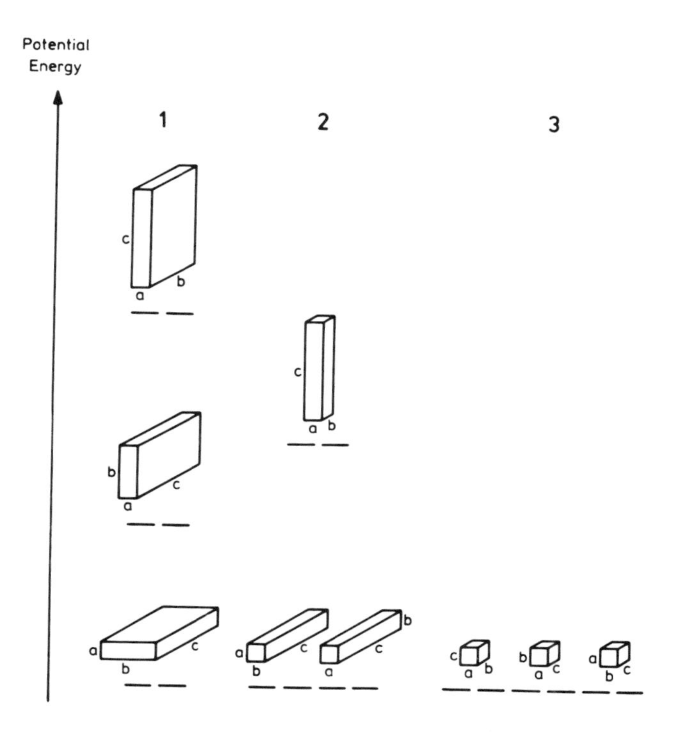

Figure 6-2. Illustration of the interrelation of symmetry and degeneracy, after Ref. [6-5]. Used with permission. See text for details.

Looking at the degeneracy of the most stable (lowest energy) position, it is twofold for 1, fourfold for 2, and sixfold for the cube. Thus, with increasing symmetry, the degree of degeneracy increases. The connection between symmetry and degeneracy is strikingly obvious. *The greater the degree of symmetry, the smaller will be the number of different energy levels and the greater will be the degeneracy of these levels.*

This correlation between symmetry and degeneracy of energy levels is fundamental to understanding the electronic structure of atoms and molecules. This relationship is valid not only when increasing symmetry renders the energy levels degenerate but also when energy levels are split as molecular symmetry decreases.

Let us now return to the wave function description of electronic structure. The separation of the wave function into two parts is convenient since these two parts relate to different properties. The radial part determines the energy of the system and is invariant to symmetry operations. The square of the radial function is related to probability. If we fix the angular variables, Θ and Φ, they define a direction from the nucleus. Then the square of the radial function is proportional to the probability of finding the electron in a volume element along this direction. In order to determine the probability of finding the electron anywhere in a spherical shell surrounding the nucleus at a distance r from the nucleus, integration over both angular variables must be performed. The result is the radial distribution function.

Consider now the angular part of the one-electron wave function. It says nothing about the energy of the system, but it can be altered by symmetry operations. Therefore, we shall be dealing with this function in greater detail. The function $A(\Theta, \Phi)$ may have different signs (+ and −) in different spatial regions. A change in sign indicates a drastic change in the wave function. These signs might be thought of as signs of the amplitudes of the wave function; they certainly have nothing to do with electric charges. The places where the wave function changes sign are called *nodes*. The number of nodes is $n - 1$, where n is the principal quantum number. Again, the squared function has physical significance; it is positive everywhere. The probability of finding an electron at a node is zero. However, as one proceeds in either direction from the nodes, the squared wave function has equal values relating to equal probabilities; to wit, the probability of finding the electron on the "positive" or on the "negative" side of the wave function is equal.

It usually helps to visualize and understand a problem in a pictorial way. However, since the wave function depends upon three variables, it can be represented only in four dimensions. To overcome this problem, symbolic representations are used to emphasize various properties of the wave function.

The angular wave function, $A(\Theta, \Phi)$, is shown for the H 1s and 2p_z orbitals in Figure 6-3a. The H 1s orbital is positive everywhere, but the 2p_z orbital has one node, through which it changes sign. The $A^2(\Theta, \Phi)$ function is shown for

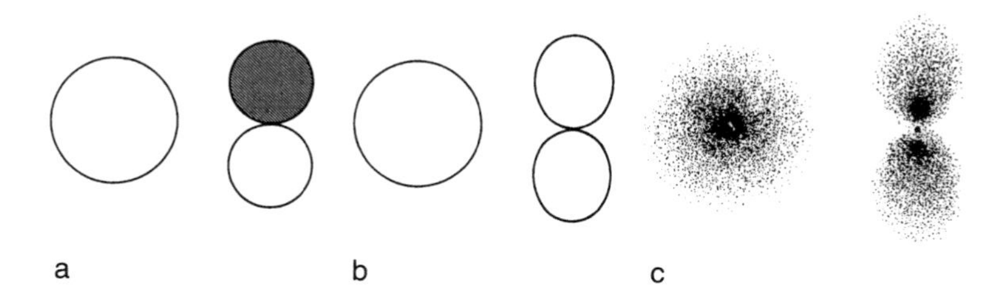

Figure 6-3. Representations of the hydrogen 1s and 2p_z orbitals: (a) Plot of the angular wave function, $A(\Theta, \Phi)$; (b) plot of the squared angular wave function, $A^2(\Theta, \Phi)$; (c) cross section of the squared total wave function, ψ^2, representing the electron density. Reprinted from Ref. [6-6] by permission of Thomas H. Lowry.

the same orbitals in Figure 6-3b. For both orbitals, the shape of this function is similar to the shape of the $A(\Theta, \Phi)$ function, but this function is positive everywhere. It represents the region in space where the electron can be found with a large probability (usually 90% or more). The boundary surface of this space is determined by the square of the angular function. The squared angular function does not say anything, however, about the variation of the probability density within this surface. That information is contained in the radial distribution function. A way to illustrate the latter is shown for the $1s$ and $2p_z$ orbitals in Figure 6-3c. A cross section of the electron density distribution is depicted. The varying amount of shading reflects the square of the radial function. Thus, this picture represents the squared total wave function, ψ^2. Rotating this picture around any axis for the $1s$ orbital and around the z axis for the $2p_z$ orbital would give the three-dimensional representation of the total wave function.

Whereas the square of the angular function has outstanding physical significance, the angular function itself contains valuable information regarding the symmetry properties of the wave function. These properties are lost in the squared angular function.

The well-known shapes of the one-electron orbitals are presented in Figure 6-4; these are, in fact, representations of the angular wave functions.

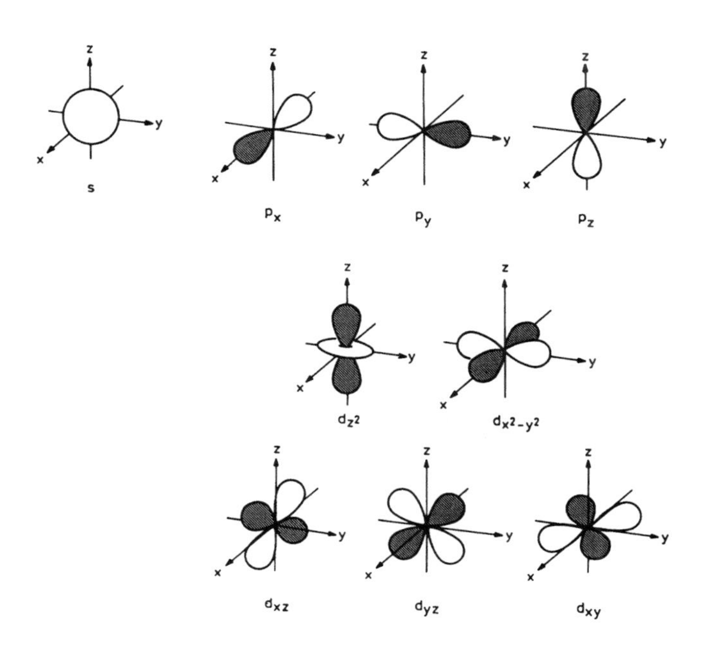

Figure 6-4. Shapes of one-electron orbitals. They are representations of the angular wave function, $A(\Theta, \Phi)$.

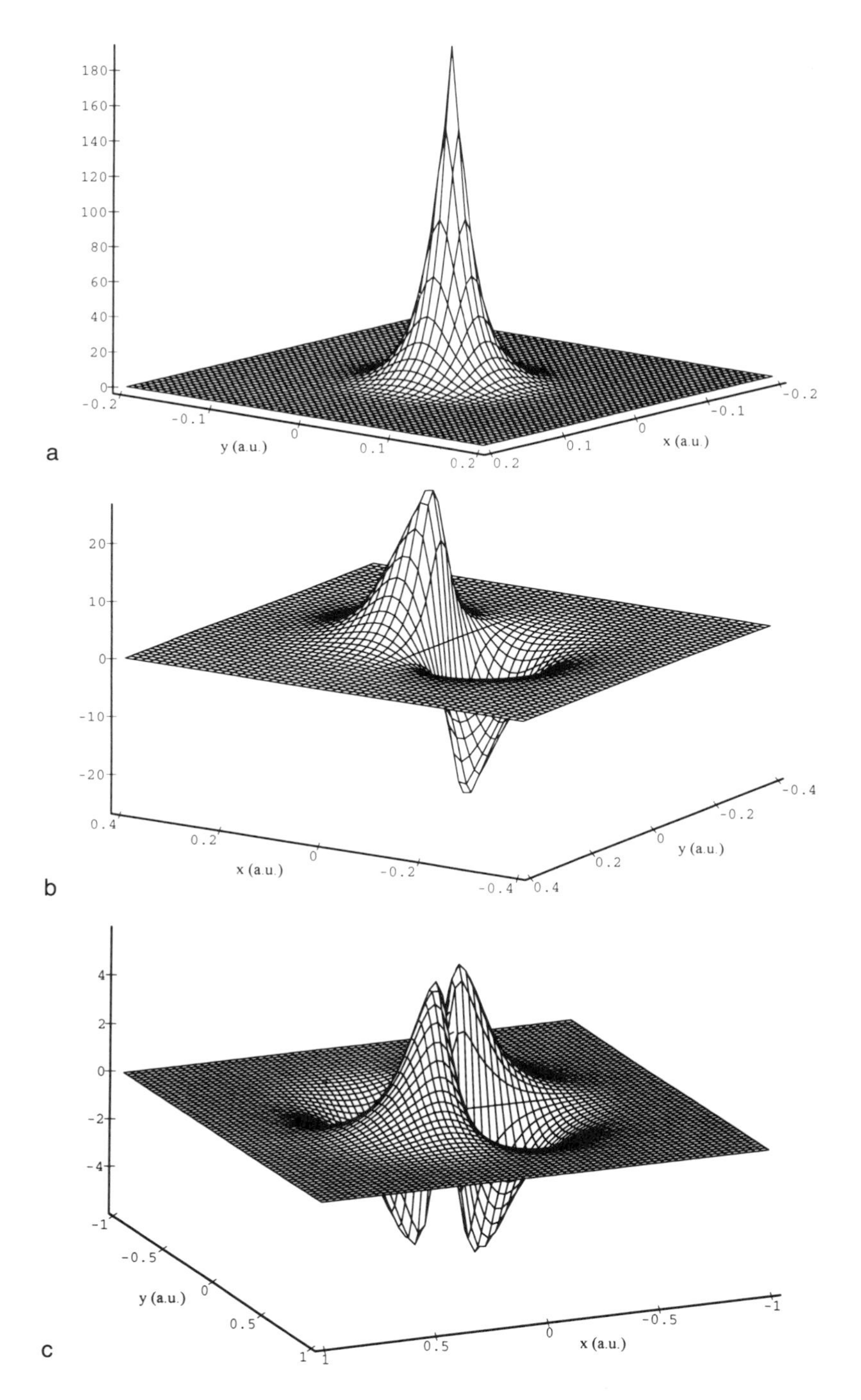

Figure 6-5. Three-dimensional computer drawings of the total wave function, ψ, of the iodine atom, calculated with a 3-21G basis set [6-9]. They show the values of ψ in a cross section. Courtesy of Dr. István Kolossváry. (a) 1s orbital; (b) 2p_x orbital; (c) 3d_{xy} orbital.

Such representations are used commonly for illustrations because they describe accurately the symmetry properties of the wave function. In order to give the total wave function, however, they must be multiplied by an appropriate radial function. Another representation, shown in Figure 6-5, is a three-dimensional computer drawing of the total function [6-9] including *both* the radial and the angular functions. These are *not* yet real "pictures" of the orbitals, since they represent a cross section of the wave function in one plane only. The vertical scale gives the value of ψ for each point in the xy plane. These diagrams show how the sign and magnitude of ψ vary in the xy plane, and they also help us visualize the electronic wave function as a wave. On the other hand, they do not illustrate its symmetry properties so well as do the simple diagrams in Figure 6-4.

As mentioned before, the symmetry properties of the one-electron wave function are shown by the simple plot of the angular wave function. But, what are the symmetry properties of an orbital and how can they be described? We can examine the behavior of an orbital under the different symmetry operations of a point group. This will be illustrated below via the inversion operation.

The s and d orbitals are transformed into themselves as the inversion operation is applied to them (Figure 6-6). Both the magnitude and the "sign" of the wave function will remain the same under the inversion operation. These orbitals are said to be *symmetric* with respect to inversion. The effect of the inversion operation on the p orbitals is demonstrated in Figure 6-7. Whereas the magnitude of the wave functions does not change, their "sign" changes upon inversion. These orbitals are said to be *antisymmetric* with respect to

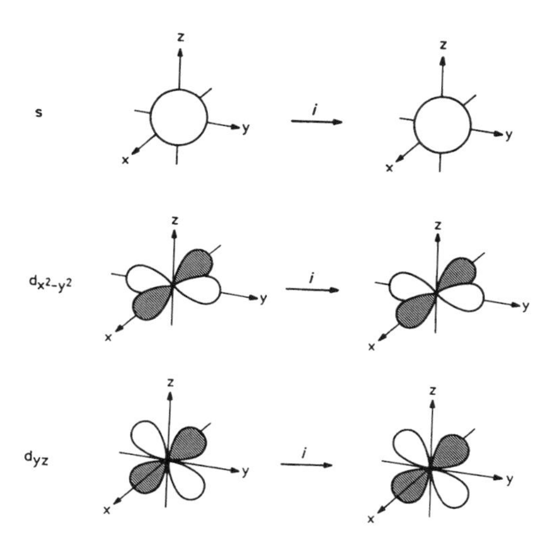

Figure 6.6. The effect of inversion on the s and d orbitals. They are symmetric to this operation.

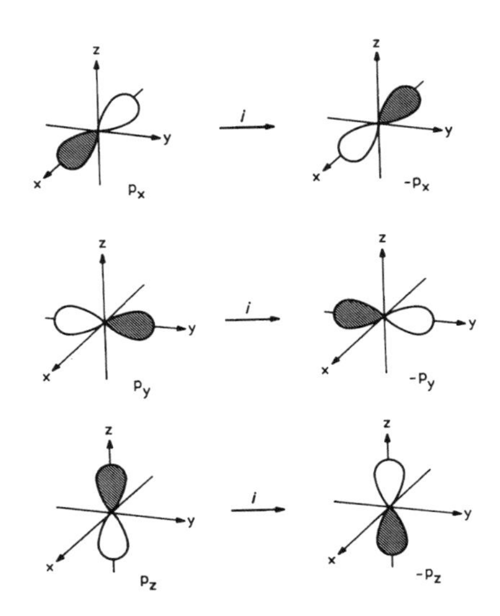

Figure 6-7. The effect of inversion on the *p* orbitals. They are antisymmetric to inversion, as the inversion operation changes their sign.

inversion. In the character tables, this is indicated by +1 for symmetric and −1 for antisymmetric behavior under each symmetry operation. As mentioned in Chapter 4, the atomic orbitals always belong to the same irreducible representations of the given point group as their subscripts ($x, y, z, xy, x^2 - y^2$, etc.).

6.2 MANY-ELECTRON ATOMS

There is interaction among all the electrons in a many-electron atom. Thus, the wave function for even one electron in a many-electron system will, in principle, be different from the wave function for the one electron in the hydrogen atom. Since the electrons are mutually indistinguishable, it is not possible to describe rigorously the properties of a single electron in such a system. There is no exact solution to this problem, and approximate methods must be adopted.

In the most commonly utilized approximation, the many-electron wave functions are written in terms of products of one-electron wave functions similar to the solutions obtained for the hydrogen atom. These one-electron functions used to construct the many-electron wave function are called *atomic orbitals*. They are also called "hydrogen-like" orbitals because they are one-electron orbitals and also because their shapes are similar to those of the hydrogen atom orbitals. Coulson referred to the atomic orbitals as "personal wave functions" [6-10] to emphasize that each electron is allocated to an individual orbital in this model.

At this point we can, again, appreciate the possibility of separating the total wave function into a radial and an angular wave function. The angular wave function does not depend on n and r, so it will be the same for every atom. This is why the "shapes" of atomic orbitals are always the same. Hence, symmetry operations can be applied to the orbitals of all atoms in the same way. The differences occur in the radial part of the wave function; the radial contribution depends on both n and r, and it determines the energy of the orbital, which is, of course, different for different atoms.

While the energy of a one-electron orbital depends only on n, in a many-electron atom the energy of the orbital is determined by both n and l. Thus, an electron in a $2p$ orbital has higher energy than an electron in a $2s$ orbital. The order of orbital energies in many-electron atoms is generally as follows:

$$1s < 2s < 2p < 3s < 3p < 4s \approx 3d < 4p < 5s < 4d < \ldots$$

There are some cases, however, when the order is changed somewhat. For example, the $3d$ orbital, sometimes lies below the $4s$ orbital. A diagram which illustrates the order of orbital energies is shown in Figure 6-8.

In addition to the three quantum numbers used to describe the one-electron wave function, the electron has also a fourth, the *spin quantum number*, m_s. It is related to the intrinsic angular momentum of the electron,

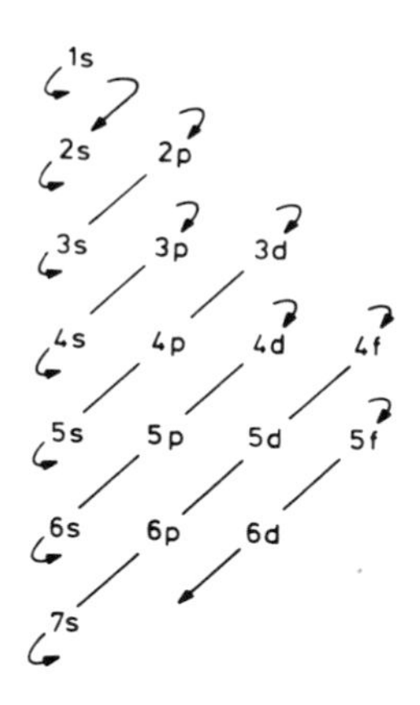

Figure 6-8. The sequence of orbital energies.

called *spin*. This quantum number may assume the values of $+\frac{1}{2}$ or $-\frac{1}{2}$. Usually, the sign of m_s is represented by arrows (↑ and ↓), or by the Greek letters α and β. Thus, the wave function of an orbital is expressed as

$$\psi_e = R(r) \cdot A(\Theta, \Phi) \cdot S(s) \tag{6-4}$$

rather than as in Eq. (6-3). However, the introduction of spin does not alter any of the properties discussed previously that relate to the shape and symmetry of the orbitals. The reason is that the spin function is independent of the spatial coordinates.

An important postulate in connection with the spin of the electron is called the *Pauli principle*. It states that if a system consists of identical particles with half-integral spins, then all acceptable wave functions must be antisymmetric with respect to the exchange of the coordinates of any two particles. In our case, the particles are electrons, and the Pauli principle is formulated accordingly: *No two electrons in an atom can have the same set of values for all four quantum numbers*.

The electronic configuration of an atom gives us the number of electrons that the atom has in its subshells. A *subshell* is a complete set of orbitals that have the same n and l. The building up of electronic configurations is governed by the Pauli principle and by *Hund's first rule*, according to which, for a given electronic configuration, *the state with the greatest number of unpaired spins has the lowest energy*.

There is a marked periodicity in the electronic configuration of the elements and this is the underlying idea of the periodic table (see Chapter 1). As the chemical properties of the atoms are determined by their electron configuration, atoms with similar electron configurations will have similar chemical properties.

6.3 MOLECULES

6.3.1 Constructing Molecular Orbitals

In the discussion of the electronic structure of atoms, the Schrödinger equation could be reduced to one involving only the electrons. This was achieved by separating the electronic energy of the atom from the nuclear kinetic energy, which is essentially determined by the translational motion of the atom.

Such a separation is exact for atoms. For molecules, only the translational motion of the whole system can be rigorously separated, while the kinetic energy includes all kinds of motion, vibration and rotation as well as translation. First, as in the case of atoms, the translational motion of the molecule is isolated. Then a two-step approximation can be introduced. The first is the separation of the rotation of the molecule as a whole, and thus the remaining equation describes only the internal motion of the system. The second step is the application of the Born–Oppenheimer approximation in order to separate the electronic and the nuclear motion. Since the relatively heavy nuclei move much more slowly than the electrons, the latter can be assumed to move about a fixed nuclear arrangement. Accordingly, not only the translation and rotation of the whole molecular system but also the internal motion of the nuclei is ignored. The molecular wave function is written as a product of the nuclear and electronic wave functions. The electronic wave function depends on the positions of both nuclei and electrons, but it is solved for the motion of the electrons only.

As was emphasized before (cf. Chapter 3), a molecule is not simply a collection of its constituent atoms. Rather, it is a system of atomic nuclei and a common electron distribution. Nevertheless, in describing the electronic structure of a molecule, the most convenient approach is to approximate the molecular electron distribution by the sum of atomic electron distributions. This approach is called the *linear combination of atomic orbitals* (LCAO) method. The orbitals produced by the LCAO procedure are called *molecular orbitals* (MOs). An important common property of the atomic and molecular orbitals is that both are one-electron wave functions. Combining a certain number of one-electron atomic orbitals yields the same number of one-electron molecular orbitals. Finally, the total molecular wave function is the sum of products of the one-electron molecular orbitals. Thus, the final scheme is as follows:

One-electron atomic orbitals (AOs)
↓ LCAO
One-electron molecular orbitals (MOs)
↓ Multiplication (and summation)
Total molecular wave function

Although both atomic orbitals and molecular orbitals are one-electron wave functions, the shape and symmetry of the molecular orbitals are different from those of the atomic orbitals of the isolated atom. The molecular orbitals extend over the entire molecule, and their spatial symmetry must conform to that of the molecular framework. Of course, the electron distribution is not uniform throughout the molecular orbital. In depicting these orbitals, usually only the portions with substantial electron density are emphasized.

When constructing molecular orbitals from atomic orbitals, there may be a large number of possible linear combinations of atomic orbitals. Many of these linear combinations, however, are unnecessary. Symmetry is instrumental as a criterion in choosing among them. The following statement is attributed to Michelangelo: "The sculpture is already there in the raw stone; the task of a good sculptor is merely to eliminate the unnecessary parts of the stone" (Figure 6-9). In the LCAO procedure, the knowledge of symmetry allows the unnecessary linear combinations to be eliminated. All those linear combinations must be eliminated that do not belong to any irreducible representation of the molecular point group. The reverse of this statement constitutes the fundamental principle of forming molecular orbitals: *Each possible molecular orbital must belong to an irreducible representation of the molecular point group*. Another equally important rule for the construction of molecular

Figure 6-9. One of Michelangelo's unfinished sculptures. It may be taken as an example of the sculpture existing already in the stone, the sculptor's task being merely to eliminate the unnecessary parts.

orbitals is that *only those atomic orbitals can form a molecular orbital which belong to the same irreducible representation of the molecular point group.* This rule follows from the general theorem (see p. 229) about the value of an energy integral. This theorem can be restated for the special case of MO construction as follows: *An energy integral will be nonzero only if the atomic orbitals used for the construction of molecular orbitals belong to the same irreducible representation of the molecular point group.*

The atomic orbitals in an isolated atom possess spherical symmetry. When they are used for MO construction, however, their symmetry must be considered in the symmetry group of the particular molecule. When two atomic orbitals of the same symmetry form a molecular orbital, the symmetry of the molecular orbital will be the same as that of the component atomic orbitals.

In addition to complying with the symmetry rules, successful MO construction requires certain energy conditions. In order for two orbitals to interact appreciably, their energies cannot be too different.

The so-called overlap integral S_{ij} is a useful guide in constructing molecular orbitals. It is symbolized as

$$S_{ij} = \int \psi_i \cdot \psi_j \, d\tau \tag{6-5}$$

where ψ_i and ψ_j are the two participating atomic orbitals. The physical meaning of S_{ij} is related to the measure of the volume in which there is electron density contributed by both atoms i and j. The knowledge of the sign and magnitude of S_{ij} is especially instructive; it can be arrived at via the following considerations.

Positive overlap results from the combination of adjacent lobes that have the same "sign." The electron density originating from both atoms will increase and concentrate in the region between the two nuclei. The resulting MO is a *bonding orbital*. Some typical bonding atomic orbital combinations are presented in Figure 6-10. Two kinds of molecular orbitals are shown in this figure. A σ orbital is concentrated primarily along the internuclear axis. On the other hand, a π orbital has a nodal plane going through this axis, and its electron density is highest on either side of this nodal plane. The σ orbitals are nondegenerate, while the π orbitals are always doubly degenerate.

Negative overlap results from the combination of adjacent lobes that have opposite "sign." In such an instance, there will be no common electron density in the region between the two nuclei; instead, electron density will concentrate in the outside regions. Such an MO is an *antibonding orbital* and is illustrated in Figure 6-11.

Zero overlap means that there is no net interaction between the two atomic orbitals. They have both positive and negative overlaps that cancel each other. Some examples are shown in Figure 6-12.

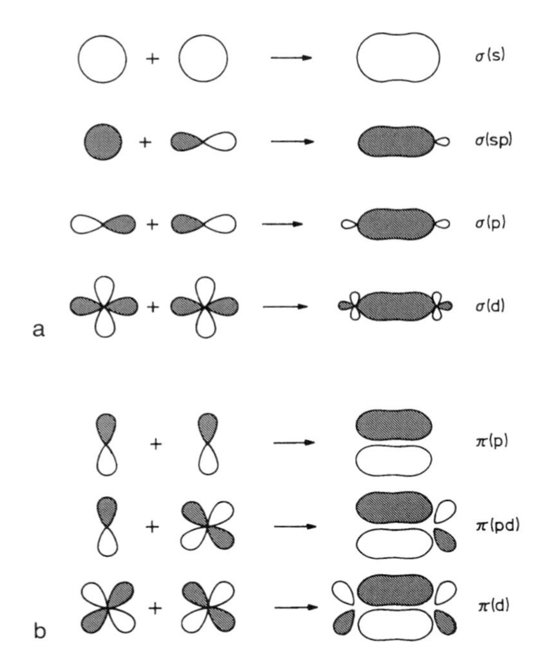

Figure 6-10. Illustration of positive overlap between atomic orbitals. The result is a bonding orbital. (a) σ orbitals; (b) π orbitals.

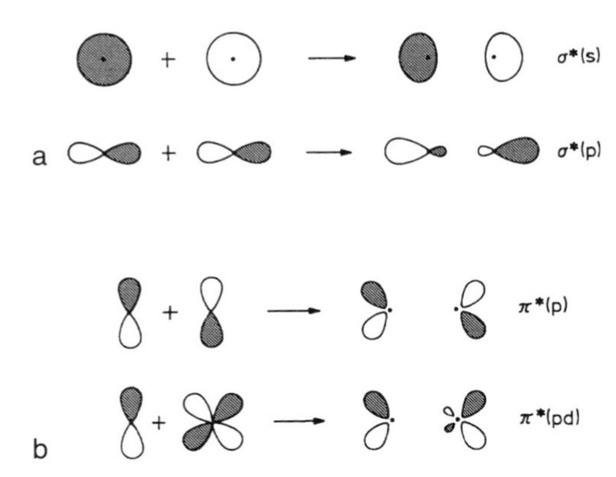

Figure 6-11. Formation of antibonding orbitals by the combination of different lobes of atomic orbitals. (a) σ antibonding orbitals; (b) π antibonding orbitals.

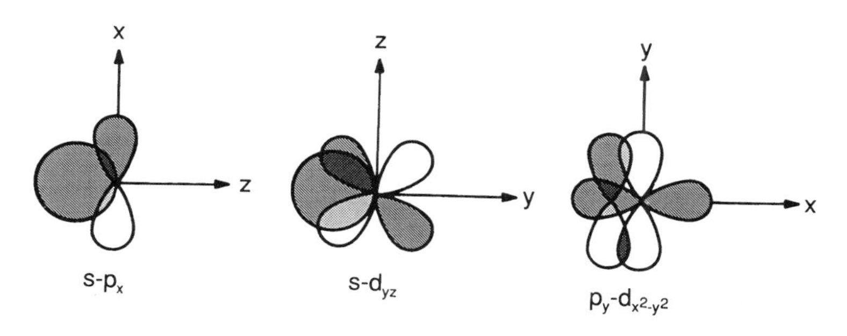

Figure 6-12. Zero overlap between atomic orbitals. There is no net interaction.

The energy changes in the formation of homonuclear and heteronuclear diatomic molecules are illustrated in Figure 6-13. The energy of the bonding MO is smaller (larger negative value) than the energy of the interacting atomic orbitals. On the other hand, the energy of the antibonding MO is larger than the energy of the interacting atomic orbitals. The largest energy changes occur when the two participating atomic orbitals have equal energies. As the energy difference between the participating atomic orbitals increases, the stabilization of the bonding MO decreases. Molecular orbitals are not formed when the participating atomic orbitals possess very different energies.

Thus, both symmetry and energy requirements must be fulfilled in order to form molecular orbitals. Energetically, the $2s$ and $2p$ atomic orbitals are sufficiently similar to form molecular orbitals with each other. For symmetry

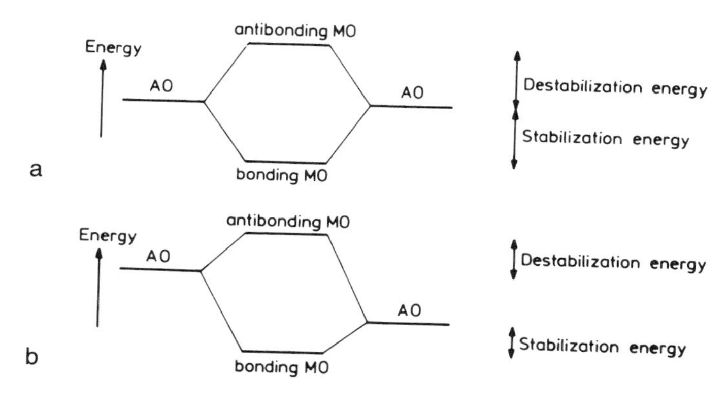

Figure 6-13. Energy changes during MO formation: (a) Homonuclear molecules; (b) heteronuclear molecules.

reasons, however, the p_x and p_y orbitals of one atom of a homonuclear diatomic molecule cannot combine with the 2*s* orbital of the other atom because they belong to different irreducible representations (see Figure 6-14). On the other hand, 3*d* orbitals of first-row transition metals are often prevented from forming molecular orbitals with the ligand orbitals for energy reasons, despite their matching symmetries. Quantum-chemical calculations on transition metal dihydrides [6-11] support this suggestion.

Knowledge of the symmetry of the MOs is important for practical reasons. The energy of the orbitals can be calculated by costly quantum-chemical calculations. The symmetry of the molecular orbitals, on the other hand, can be deduced from the molecular point group and with the use of character tables, a process that requires merely paper and pencil. Then, when all possible solutions that are not allowed by symmetry have been excluded, only the energies of the remaining orbitals must be calculated.

We are, of course, concerned with the symmetry aspects of the MOs and their construction. As was discussed before, the degeneracy of atomic orbitals is determined by m_l. Thus, all *p* orbitals are threefold degenerate, and all *d* orbitals are fivefold degenerate. The spherical symmetry of the atomic subshells, however, changes when the atoms enter the molecule, since the symmetry of molecules is nonspherical. The degeneracy of atomic orbitals will, accordingly, decrease; the extent of decrease will depend upon molecular symmetry.

Various methods (described in Chapter 4) can be used to determine the symmetry of atomic orbitals in the point group of a molecule, i.e., to determine the irreducible representations of the molecular point group to which the atomic orbitals belong. There are two possibilities depending on the position of the atoms in the molecule. For a central atom (such as O in H_2O or N in NH_3),

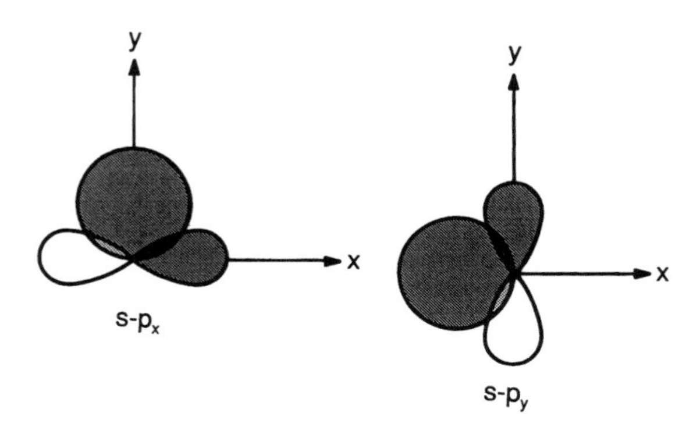

Figure 6-14. Combination of the 2*s* and $2p_x$ (or $2p_y$) atomic orbitals does not result in a molecular orbital.

the coordinate system can always be chosen in such a way that the central atom lies at the intersection of all symmetry elements of the group. Consequently, each atomic orbital of this central atom will transform as one or another irreducible representation of the symmetry group. The atomic orbitals will have the same symmetry properties as those basis functions in the third and fourth areas of the character table which are indicated in their subscripts. For all other atoms, so-called "group orbitals" or "symmetry-adapted linear combinations" (SALCs) must be formed from like orbitals. Several examples below will illustrate how this is done.

First, however, consider the symmetry properties of the central atom orbitals. Take the C_{4v} point group as an example. Its character table is presented in Table 6-1. The p_z and d_{z^2} atomic orbitals of the central atom belong to the totally symmetric irreducible representation A_1, the $d_{x^2-y^2}$ orbital belongs to B_1, and d_{xy} to B_2. The symmetry properties of the (p_x, p_y) and (d_{xz}, d_{yz}) orbitals present a good opportunity for illustrating two-dimensional representations. Taking the three p orbitals as basis functions, the symmetry operations of the C_{4v} point group are applied to them. This is shown in Figure 6-15. The matrix representations are given here:

$$E = \begin{bmatrix} 1 & 0 & 0 \\ 0 & 1 & 0 \\ 0 & 0 & 1 \end{bmatrix}$$

$$C_4 = \begin{bmatrix} 0 & 1 & 0 \\ -1 & 0 & 0 \\ 0 & 0 & 1 \end{bmatrix} \quad C_4^3 = \begin{bmatrix} 0 & -1 & 0 \\ 1 & 0 & 0 \\ 0 & 0 & 1 \end{bmatrix} \quad C_2 = \begin{bmatrix} -1 & 0 & 0 \\ 0 & -1 & 0 \\ 0 & 0 & 1 \end{bmatrix}$$

$$\sigma_v(xz) = \begin{bmatrix} 1 & 0 & 0 \\ 0 & -1 & 0 \\ 0 & 0 & 1 \end{bmatrix} \quad \sigma_v(yz) = \begin{bmatrix} -1 & 0 & 0 \\ 0 & 1 & 0 \\ 0 & 0 & 1 \end{bmatrix}$$

$$\sigma_d = \begin{bmatrix} 0 & 1 & 0 \\ 1 & 0 & 0 \\ 0 & 0 & 1 \end{bmatrix} \quad \sigma_d' = \begin{bmatrix} 0 & -1 & 0 \\ -1 & 0 & 0 \\ 0 & 0 & 1 \end{bmatrix}$$

Table 6-1. The C_{4v} Character Table

C_{4v}	E	$2C_4$	C_2	$2\sigma_v$	$2\sigma_d$		
A_1	1	1	1	1	1	z	$x^2 + y^2, z^2$
A_2	1	1	1	-1	-1	R_z	
B_1	1	-1	1	1	-1		$x^2 - y^2$
B_2	1	-1	1	-1	1		xy
E	2	0	-2	0	0	(x, y) (R_x, R_y)	(xz, yz)

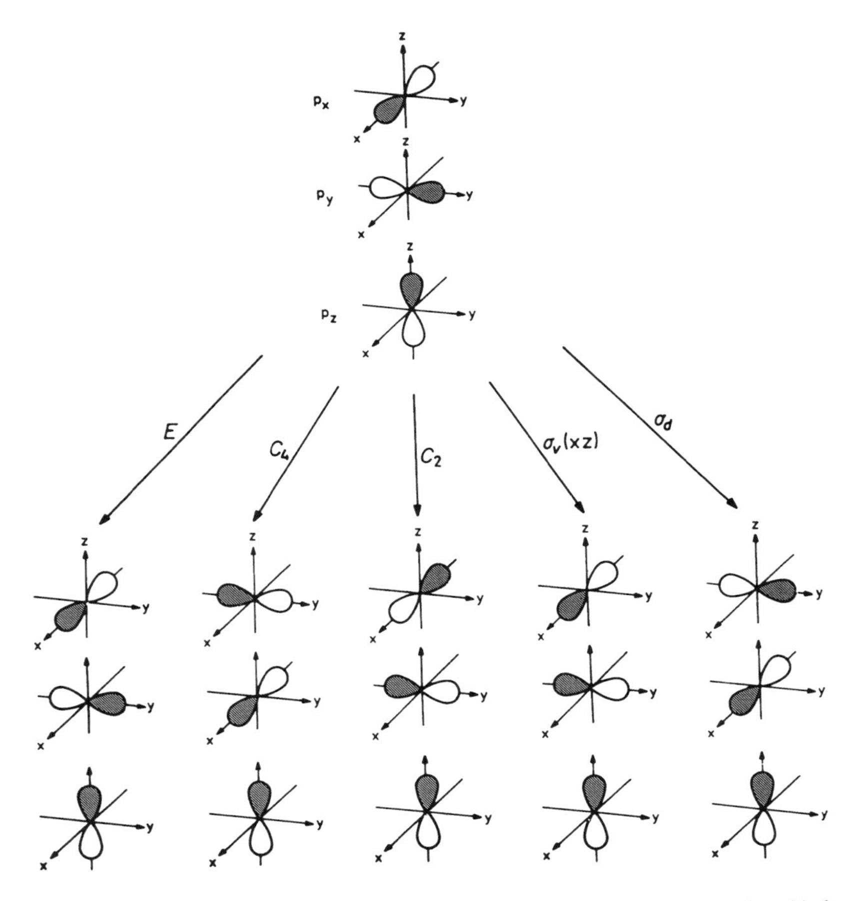

Figure 6-15. The symmetry operations of the C_{4v} point group applied to the $2p$ orbitals.

All these matrices can be simultaneously block-diagonalized into a 2×2 and a 1×1 matrix. The set of the 1×1 matrices corresponds to p_z and the set of the 2×2 matrices corresponds to p_x and p_y. The representations are:

	E	$2C_4$	C_2	$2\sigma_v$	$2\sigma_d$	
p_z	1	1	1	1	1	A_1
(p_x, p_y)	2	0	-2	0	0	E

Notice that the operations C_4 and σ_d transform p_x into p_y and vice versa. They cannot be separated from one another so they *together* belong to the two-dimensional representation E.

If two or more atomic orbitals are interrelated under a symmetry operation of the point group and, accordingly, they *together* belong to an irreducible representation, their energies will also be the same. In other words, these orbitals are *degenerate*. Such atomic orbitals are parenthesized in the character tables.

The direct connection between symmetry and degeneracy of the atomic orbitals is demonstrated here once again. The higher the symmetry of the molecule, the greater will be the interrelation of the orbitals upon symmetry operations. Consequently, their energies become less and less distinguishable. The following example shows how the degeneracy of p orbitals decreases with diminishing symmetry:

Free atom	Spherical symmetry	(p_x, p_y, p_z)	Threefold degenerate
O_h point group	T_{1u}	(p_x, p_y, p_z)	Threefold degenerate
C_{4v} point group	A_1	p_z	Nondegenerate
	E	$(p_x, p_y,)$	Twofold degenerate
C_{2v} point group	A_1	p_z	Nondegenerate
	B_1	p_x	Nondegenerate
	B_2	p_y	Nondegenerate

The degree of degeneracy of atomic orbitals always corresponds to the dimension of the irreducible representation to which these atomic orbitals belong. The same is true for molecular orbitals. Thus, knowing the symmetry of a molecule and looking at the character table, one can determine at once the maximum possible degeneracy of its molecular orbitals. The irreducible representation having the highest dimension will show this.

6.3.2 Electronic States

The orbitals and electronic configurations are useful descriptions. However, they are only models, and they employ approximations. The energy of an orbital has rigorous physical meaning for systems that contain only a single electron. In many-electron systems, the energy of the orbitals loses its physical meaning, and only the energies of the (ground and excited) states are real. It is these states that are described by the total electronic wave functions. Electronic transitions, in fact, represent changes in the state of an atom or a molecule and not necessarily in the electronic configurations.

We shall not be concerned with the atomic states. The systematic way of determining them is given, for example, in Refs. [6-3] and [6-5]. Molecular states and the determination of their symmetries, however, will be briefly introduced [6-4].

First, let us consider the customary notations. Assume that a hypothetical ground-state molecule of the C_{2v} point group has four electrons, two in an A_1 symmetry and two in a B_1 symmetry orbital. In shorthand notation this can be written as $a_1^2b_1^2$. An electron occupying an A_1 symmetry orbital is represented by a_1, the lower-case letter indicating that this is the symmetry of an *orbital* and not of an electronic state. If two electrons occupy this orbital, the notation is a_1^2. The symmetry of a state is represented by capital letters, just as are the irreducible representations.

The symmetry of the electronic states can be determined from the symmetry of the occupied orbitals. There are two different cases:

1. *States with fully occupied orbitals*. An electronic configuration in which all orbitals are completely filled possesses only one electronic state, and it will be totally symmetric. This can be seen for the case of nondegenerate orbitals. The wave function describing the electronic state can be written as the product of the one-electron orbitals. The symmetry of the product is given by the characters of the direct product representation. However, the product of any orbital with itself will always give the totally symmetric representation, no matter what characters it has, both $1 \cdot 1$ and $(-1) \cdot (-1)$ equal 1, i.e., in each class of the point group the characters of the product will be 1. The same is true for degenerate orbitals, although the procedure in this case is not as simple.
2. *States with partially occupied orbitals*. First of all, the completely filled orbitals are ignored for reasons described above. The symmetry of the state will be given by the direct product of the partially filled orbitals.

Let us consider some examples for the above hypothetical molecule. The supposed ground state and the configurations of two different singly excited states are represented in Figure 6-16.

The ground state $a_1^2b_1^2$ has only fully occupied orbitals, so its symmetry is A_1. The first excited state, $a_1^2b_1a_2$, has one fully occupied orbital, a_1^2, so this is not considered. The symmetry of this state will be given by the direct

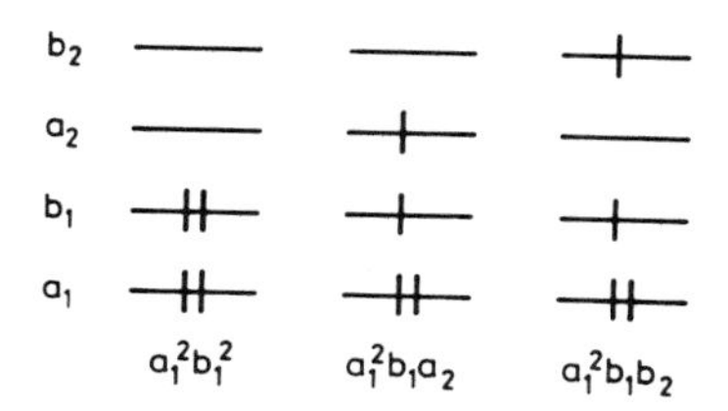

Figure 6-16. Different states of a molecule with C_{2v} symmetry.

product $B_1 \cdot A_2$. Table 6-2 lists the direct products under the C_{2v} character table. The symmetry of the state is B_2. The other excited state in our example has the configuration $a_1^2 b_1 b_2$. The direct product is given in Table 6-2; the state symmetry is A_2. Since we are concerned only with the spatial symmetry properties, the electron spin and its role in determining the electronic states have been neglected in the above description.

6.3.3 Examples of MO Construction

6.3.3.1 Homonuclear Diatomics

a. Hydrogen, H_2. There are two 1*s* hydrogen atomic orbitals available for bonding. The molecular point group is $D_{\infty h}$. This molecule does not have a central atom, so the symmetry operations of the point group are applied to both 1*s* orbitals, since they *together* form the basis for a representation of this point group. The 1*s* orbital of one hydrogen atom alone does not belong to any irreducible representation of the $D_{\infty h}$ point group. Several symmetry operations of this group transform one of the two 1*s* orbitals into the other rather than into itself (see Figure 6-17a). Thus, they must be treated together; in this way they

Table 6-2. C_{2v} Character Table and Some Direct Product Representations

C_{2v}	E	C_2	$\sigma_v\ (xz)$	$\sigma_v'\ (yz)$		
A_1	1	1	1	1	z	x^2, y^2, z^2
A_2	1	1	−1	−1	R_z	xy
B_1	1	−1	1	−1	x, R_y	xz
B_2	1	−1	−1	1	y, R_x	yz
$B_1 \cdot A_2$	1	−1	−1	1	B_2	
$B_1 \cdot B_2$	1	1	−1	−1	A_2	

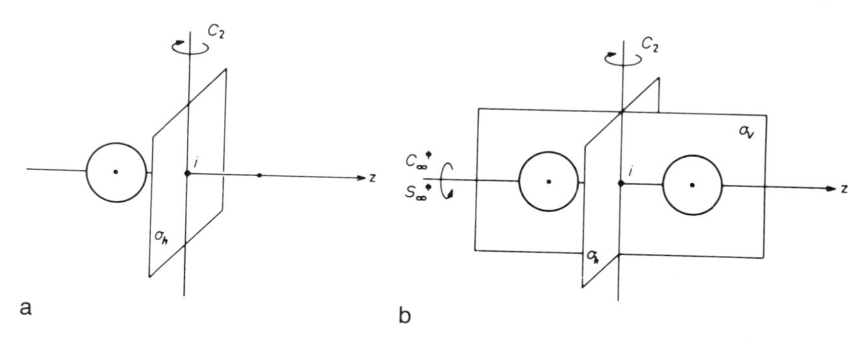

Figure 6-17. Some symmetry operations of the $D_{\infty h}$ point group applied to one $1s$ orbital in the hydrogen molecule (a) and the two $1s$ orbitals of the hydrogen molecule together (b).

form a basis for a representation. All symmetry operations are indicated in Figure 6-17b. The $D_{\infty h}$ character table is given in Table 5-3. The characters of this representation will be

$D_{\infty h}$	E	$2C_\infty^\Phi$	$\infty\sigma_v$	i	$2S_\infty^\Phi$	∞C_2
2H(1s)	2	2	2	0	0	0

This is a reducible representation of the $D_{\infty h}$ point group which reduces to $\sigma_g + \sigma_u$. Two molecular orbitals must be generated, one with σ_g and the other with σ_u symmetry. The two possible combinations are the bonding and antibonding orbitals which can be formed from the two $1s$ atomic orbitals.

+ → σ_u

+ → σ_g

The two electrons in the hydrogen molecule will occupy the lower energy bonding orbital, and none will go into the antibonding orbital.

σ_u antibonding

↑ ↓ σ_g bonding

Hence, the molecule is stable.

b. Other Homonuclear Diatomic Molecules. The principle utilized to construct molecular orbitals is the same as that for the hydrogen molecule. For helium, the MO picture is the same as for hydrogen except that here the additional two electrons occupy the antibonding σ_u orbital, and, therefore, the molecule is unstable.

In the series from lithium through neon, similar symmetry considerations apply, except that in these examples the second electron shell must be considered. The two $2s$ orbitals, as was found to be the case for the two $1s$ orbitals, form MOs that possess σ_g and σ_u symmetry. As regards the $2p$ orbitals, the two $2p_z$ orbitals lie along the molecular axis and belong to the same irreducible representation as the $2s$ orbitals. They also combine to give MOs that possess σ_g and σ_u symmetry.

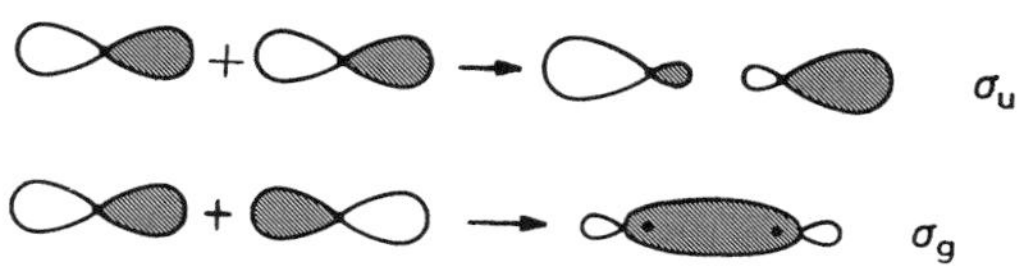

The $2s$ and $2p_z$ orbitals of the same atom belong to the same irreducible representation of the $D_{\infty h}$ point group. Their energies are also similar so they cannot be separated completely. Another way of making linear combinations is to first combine the $2s$ and $2p_z$ orbitals of the same atom

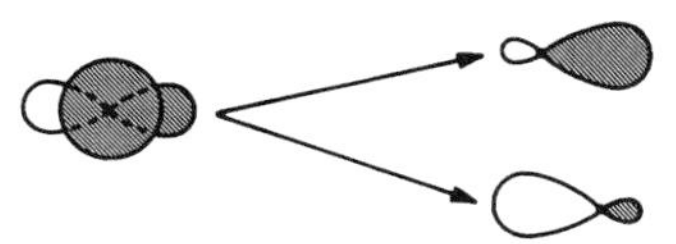

and then combine the resulting orbitals into MOs.

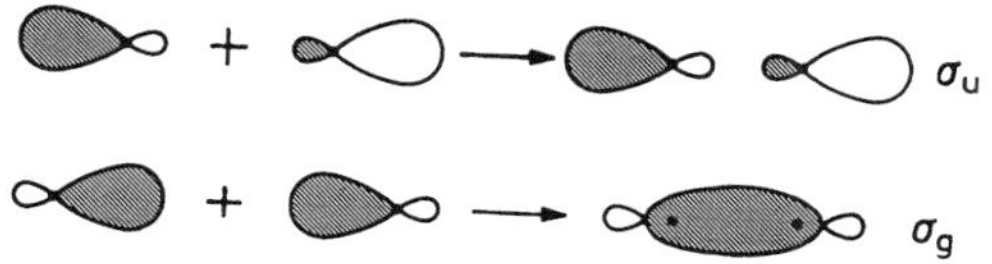

The result is essentially the same as before.

The $2p_x$ and $2p_y$ orbitals of the two atoms together form a representation that reduces to π_g and π_u. These correspond to two doubly degenerate π orbitals, one of which lies in the yz plane

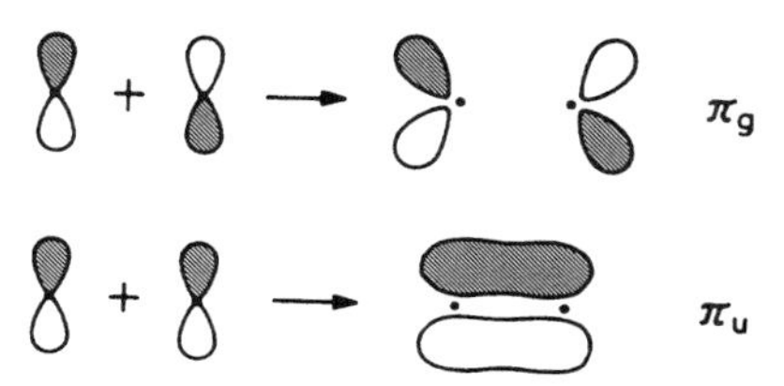

and the other in the xz plane. The relative energies of these orbitals are known from energy calculations. In most cases the order is as follows:

$$1\sigma_g < 1\sigma_u < 2\sigma_g < 2\sigma_u < 3\sigma_g < 1\pi_u < 1\pi_g < 3\sigma_u$$

while in some cases $1\pi_u < 3\sigma_g$.

6.3.3.2 Polyatomic Molecules

Before working out actual examples, let us recall what was said about the symmetry properties of atomic orbitals. If there is a central atom in the molecule, its atomic orbitals belong to some irreducible representation of the molecular point group. For the other atoms of these molecules, SALCs are formed from like orbitals. These new orbitals are then coupled with the atomic orbitals of the central atom to form MOs.

If the molecule does not have a central atom (e.g., C_6H_6), we begin with the second step, first forming different group orbitals and then combining them, if possible, into MOs. Examples will be given for both cases.

a. Water, H_2O. The molecular symmetry is C_{2v}. There are six atomic orbitals available for MO construction: two H 1*s*, one oxygen 2*s*, and three oxygen 2*p*. They can combine to produce six MOs. The molecule has a central atom, and its AOs will belong to some of the irreducible representations of the C_{2v} point group by themselves. Group orbitals must be formed from the H 1*s* orbitals. The symmetry operations applied to them are shown in Figure 6-18. The C_{2v} character table was given in Table 6-2. The reducible representation is:

C_{2v}	E	C_2	$\sigma_v(xz)$	$\sigma_v'(yz)$
2H(1*s*)	2	0	0	2

This representation reduces to $A_1 + B_2$. The projection operator (see Chapter 4) is used to form these SALCs. Since we are interested only in symmetry aspects, numerical factors and normalization are omitted.

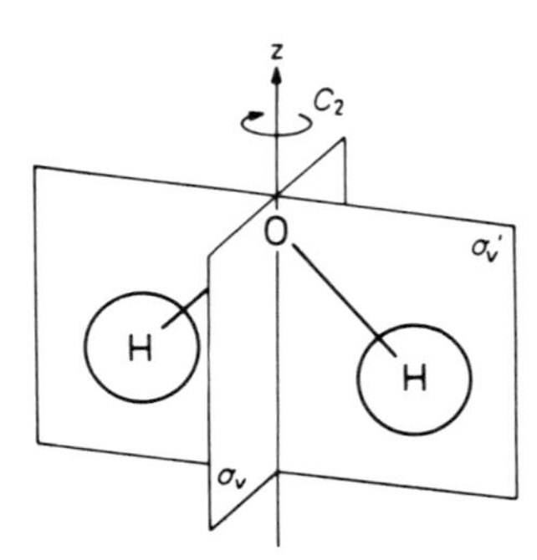

Figure 6-18. The C_{2v} symmetry operations applied to the two hydrogen 1*s* orbitals of water as basis functions.

$$\hat{P}^{A_1}s_1 \approx 1\cdot E\cdot s_1 + 1\cdot C_2\cdot s_1 + 1\cdot\sigma\cdot s_1 + 1\cdot\sigma'\cdot s_1 = s_1 + s_2 + s_2 + s_1$$
$$= 2s_1 + 2s_2 \approx s_1 + s_2$$
$$\hat{P}^{B_2}s_1 \approx 1\cdot E\cdot s_1 + (-1)\cdot C_2\cdot s_1 + (-1)\cdot\sigma\cdot s_1 + 1\cdot\sigma'\cdot s_1 = s_1 - s_2 - s_2 + s_1$$
$$= 2s_1 - 2s_2 \approx s_1 - s_2$$

Thus, the two hydrogen group orbitals (φ_1 and φ_2) will have the forms:

$$\varphi_2 = s_1 - s_2:$$

$$\varphi_1 = s_1 + s_2:$$

The available AOs are summarized according to their symmetry properties in Table 6-3. Since only orbitals of the same symmetry can overlap, two combinations are possible: one has A_1 symmetry and the other has B_2 symmetry. The remaining two orbitals of oxygen (one with A_1 and the other with B_1 symmetry) will be nonbonding in the water molecule.

If we choose the oxygen $2s$ orbital for bonding and leave the $2p_z$ orbital nonbonding (from the symmetry point of view, the opposite choice or a mixed orbital would do just as well; actually, if the two orbitals are close in energy, they mix), the MOs of the water molecule can be constructed as shown in Figure 6-19. These MOs are compared with the calculated contour diagrams of the water molecular orbitals in Figure 6-20 [6-12].

The construction of the molecular orbitals of the water molecule can also be represented by a qualitative MO diagram (see Figure 6-21). The relative energies of the orbitals are also indicated in Figure 6-21. What information can be deduced from such a diagram? First, there are two bonding orbitals occupied by four electrons; these correspond to the two O–H bonds of water. There are two nonbonding orbitals also occupied; these are the two lone pairs of oxygen. Finally, there are two antibonding orbitals that are empty, so there is a net energy gain in the formation of H_2O and the molecule is stable.

Table 6-3. The Atomic Orbitals of Water Grouped According to Their Symmetry Properties

	O orbitals	H group orbitals
A_1	$2s$, $2p_z$	φ_1
A_2		
B_1	$2p_x$	
B_2	$2p_y$	φ_2

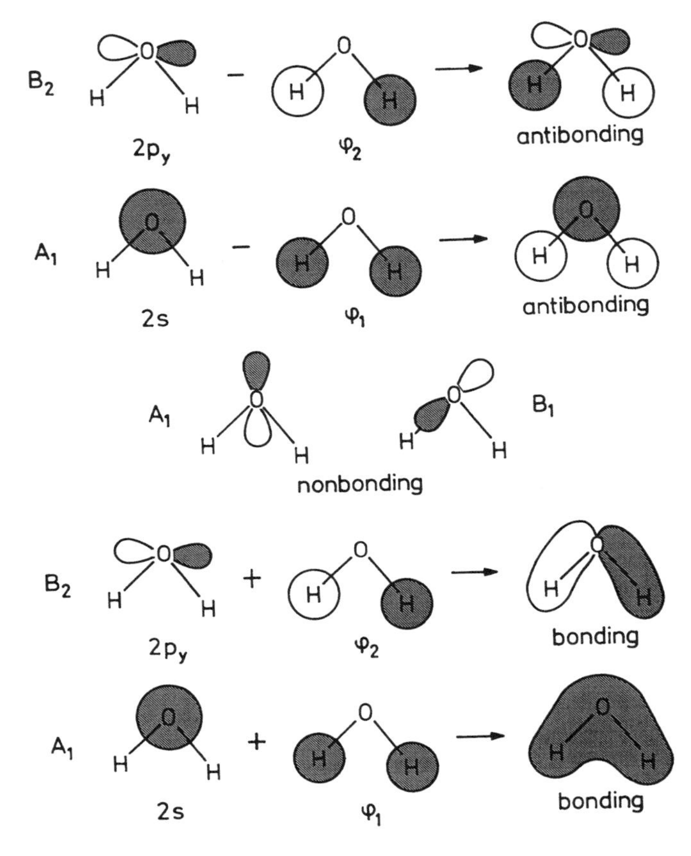

Figure 6-19. Construction of the molecular orbitals of water.

b. Ammonia, NH_3. This example is given primarily to illustrate t construction of degenerate molecular orbitals. The symmetry of the molecu is C_{3v}. There are seven atomic orbitals available for bonding: three H 1*s*, one 2*s*, and three *N* 2*p* AOs; hence, seven MOs must be formed. Since the nitrog atom is a central atom, the coordinate axes can be chosen so that its AOs lie all symmetry elements of the C_{3v} point group. The pertinent character table given in Table 6-4. The N 2*s* and $2p_z$ orbitals will have A_1 symmetry and the 2 and $2p_y$ orbitals together belong to the *E* irreducible representation. Gro orbitals must be formed from the three H 1*s* orbitals. The symmetry elements the C_{3v} point group applied to these orbitals are shown in Figure 6-22; th representation is given in Table 6-4.

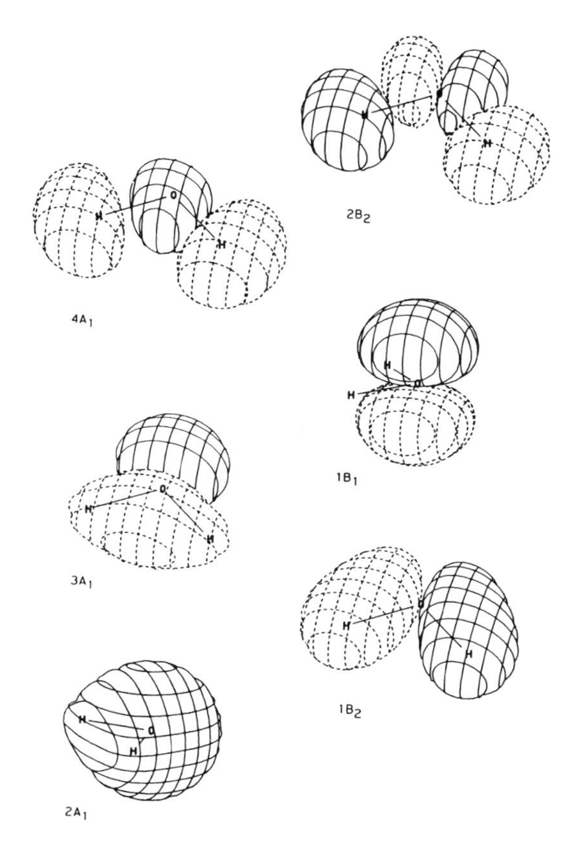

Figure 6-20. Contour diagrams of the molecular orbitals of water. Reproduced with permission from Ref. [6-12]. Copyright (1973) Academic Press.

This representation can now be reduced by using the reduction formula introduced in Chapter 4:

$$a_{A_1} = (\tfrac{1}{6})(1\cdot3\cdot1 + 2\cdot0\cdot1 + 3\cdot1\cdot1) = 1$$
$$a_{A_2} = (\tfrac{1}{6})(1\cdot3\cdot1 + 2\cdot0\cdot1 + 3\cdot1\cdot(-1)) = 0$$
$$a_E = (\tfrac{1}{6})(1\cdot3\cdot2 + 2\cdot0\cdot(-1) + 3\cdot1\cdot0) = 1$$

Thus, the representation reduces to $A_1 + E$. Next, let us use the projection operator to generate the form of these SALCs:

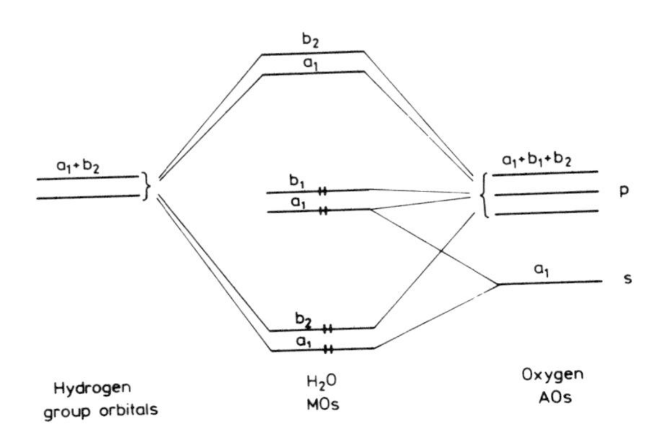

Figure 6-21. Qualitative MO diagram for water.

$$\hat{P}^{A_1}s_1 \approx 1\cdot E\cdot s_1 + 1\cdot C_3\cdot s_1 + 1\cdot C_3^2\cdot s_1 + 1\cdot\sigma\cdot s_1 + 1\cdot\sigma'\cdot s_1 + 1\cdot\sigma''\cdot s_1$$
$$= s_1 + s_2 + s_3 + s_1 + s_2 + s_3 = 2(s_1 + s_2 + s_3) \approx s_1 + s_2 + s_3$$

The same procedure is illustrated pictorially in Figure 6-23, after Ref. [6-5]

For the construction of the E symmetry group orbitals, a time-savin[g] simplification will be introduced [6-5]. First of all, it utilizes the fact that th[e] rotational subgroup C_n in itself contains all the information needed to construc[t] the SALCs in a molecule that possesses a principal axis C_n. The rotationa[l] subgroup of C_{3v} is C_3, and its character table is given in Table 6-5. If w[e] perform the three symmetry operations of the C_3 point group and check th[e] generation of the A_1 symmetry SALC of NH_3 (Figure 6-23), we see that th[e] application of these three operations suffices to define the form of this orbita[l]

The difficulty in applying the projection operator for this symmetry grou[p] arises from the fact that the C_3 character table contains imaginary characters fo[r]

Table 6-4. The C_{3v} Character Table and the Reducible Representation of the Hydrogen Group Orbitals of Ammonia

C_{3v}	E	$2C_3$	$3\sigma_v$		
A_1	1	1	1	z	$x^2 + y^2, z^2$
A_2	1	1	-1	R_z	
E	2	-1	0	(x, y) (R_x, R_y)	$(x^2 - y^2, xy)$ (xz, yz)
3 H(1s)	3	0	1		

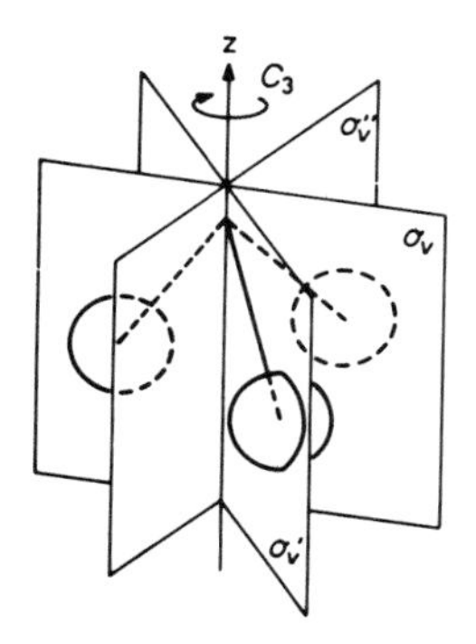

Figure 6-22. The C_{3v} symmetry operations applied to the three hydrogen atom 1*s* orbitals of ammonia as basis functions.

the *E* representation. They can be eliminated by following the procedure used in Ref. [6-5]. The character ε corresponds to $\exp(2\pi i/n)$, where n is the order of the rotation axis; in our case, 3. Using Euler's formula, $\exp(i\alpha) = \cos\alpha + i\sin\alpha$ (and the complex conjugate of ε will be $\varepsilon^* = \cos\alpha - i\sin\alpha$), the characters for the *E* representation will be:

$$\begin{Bmatrix} 1 & -\frac{1}{2} + i\sqrt{3}/2 & -\frac{1}{2} - i\sqrt{3}/2 \\ 1 & -\frac{1}{2} - i\sqrt{3}/2 & -\frac{1}{2} + i\sqrt{3}/2 \end{Bmatrix} \begin{matrix} \text{(a)} \\ \text{(b)} \end{matrix}$$

Using two different ways to obtain linear combinations of these characters will make it possible to eliminate the imaginary characters. One way may be summing Eqs. (a) and (b), resulting in

$$2 \quad -1 \quad -1$$

The other linear combination may be obtained by subtraction of Eq. (b) from Eq. (a) and dividing the result by $i\sqrt{3}$. This linear combination results in the "characters"

$$0 \quad 1 \quad -1$$

not satisfying all the relationships of irreducible representations, but it serves our purpose of showing the shape of the SALCs. Our "quasi character table" for the C_3 point group is now:

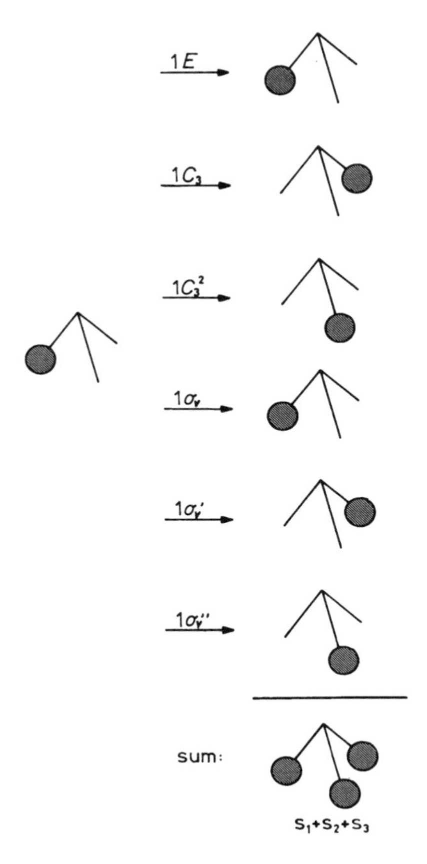

Figure 6-23. Generation of the A_1 symmetry orbital of the 3H group orbitals of ammonia.

$$
\begin{array}{ccccc}
A & & 1 & 1 & 1 \\
E & & \left\{\begin{array}{ccc} 2 & -1 & -1 \\ 0 & 1 & -1 \end{array}\right\} & &
\end{array}
$$

When the projection operator is applied to one of the l*s* orbitals of the hydrogen group orbitals with the two E representations, the two E symmetry doubly degenerate SALCs result:

$$\hat{P}^{E^1}s_1 \approx 2\cdot E\cdot s_1 + (-1)\cdot C_3\cdot s_1 + (-1)\cdot C_3^2\cdot s_1 = 2s_1 - s_2 - s_3$$
$$\hat{P}^{E^2}s_1 \approx 0\cdot E\cdot s_1 + 1\cdot C_3\cdot s_1 + (-1)\cdot C_3^2\cdot s_1 = s_2 - s_3$$

Figure 6-24 illustrates the same procedure pictorially.

Table 6-5. The C_3 Character Table

C_3	E	C_3	C_3^2		$\varepsilon = \exp(2\pi i/3)$
A_1	1	1	1	z, R_z	$x^2 + y^2, z^2$
E	$\begin{Bmatrix} 1 \\ 1 \end{Bmatrix}$	$\begin{matrix} \varepsilon \\ \varepsilon^* \end{matrix}$	$\begin{matrix} \varepsilon^* \\ \varepsilon \end{matrix}$	$(x, y)(R_x, R_y)$	$(x^2 - y^2, xy)(yz, xz)$

The next step is the MO construction. The orbitals used for this purpose are summarized in Table 6-6. An A_1 and a doubly degenerate E symmetry combination is possible here, and there will be a nonbonding orbital with A_1 symmetry left on nitrogen. Figure 6-25 illustrates the building of MOs. Again, the MOs can be compared with the calculated contour diagrams of the ammonia molecular orbitals in Figure 6-26. The qualitative MO diagram is given in Figure 6-27. The following conclusions can be drawn: (1) there are three bonding orbitals occupied by electrons; these correspond to the three N–H bonds; (2) there is a nonbonding orbital also occupied by electrons; this corresponds to the lone electron pair; and (3) the three antibonding orbitals are

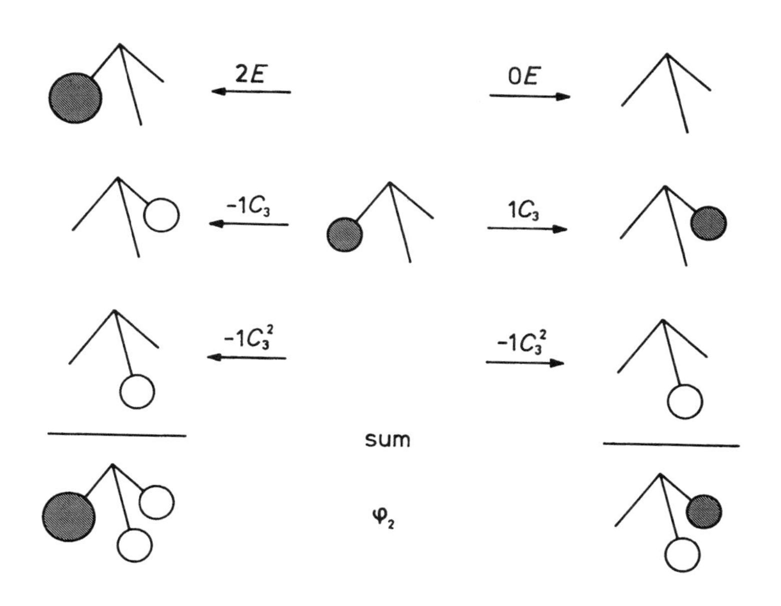

Figure 6-24. Projection of the two E symmetry group orbitals of the three H 1s orbitals in ammonia.

Table 6-6. The Atomic Orbitals of Ammonia Sorted According to Their Symmetry Properties

	N orbitals	H group orbitals
A_1	s, p_z	φ_1
E	(p_x, p_y)	φ_2

unoccupied, so the MO construction is energetically favorable, and the molecule is stable.

c. Benzene, C_6H_6. The molecular symmetry is D_{6h}. There are 30 AOs that can be used for MO construction: six H 1*s*, six C 2*s* and 18 C 2*p* orbitals. Since this molecule does not contain a central atom, each AO must be grouped into SALCs in such a way that they can transform according to the symmetry operations of the D_{6h} point group. It is straightforward to combine like orbitals for example, the hydrogen 1*s* orbitals, the carbon 2*s* orbitals, and so on. The following combinations will be used here:

$$\Phi_1(6\text{ H }1s),\ \Phi_2(6\text{ C }2s),\ \Phi_3(6\text{ C }2p_x,2p_y),\text{ and }\Phi_4(6\text{ C }2p_z)$$

The next step is to determine how these group orbitals transform in the D_{6h} point group. The D_{6h} character table is given in Table 6-7. Since most of the AOs in the suggested group orbitals are transformed into another AO by most of the symmetry operations, the representations will be quite simple, though still reducible:

Γ_{Φ_1}	6	0	0	0	2	0	0	0	0	6	0	2
Γ_{Φ_2}	6	0	0	0	2	0	0	0	0	6	0	2
Γ_{Φ_3}	12	0	0	0	0	0	0	0	0	12	0	0
Γ_{Φ_4}	6	0	0	0	−2	0	0	0	0	−6	0	2

These representations can be reduced by applying the reduction formula. First, Φ_1:

$$a_{A_{1g}} = (\tfrac{1}{24})(1\cdot6\cdot1 + 2\cdot0\cdot1 + 2\cdot0\cdot1 + 1\cdot0\cdot1 + 3\cdot2\cdot1 + 3\cdot0\cdot1 + 1\cdot0\cdot1 + 2\cdot0\cdot1 + 2\cdot0\cdot1 + 1\cdot6\cdot1 + 3\cdot0\cdot1 + 3\cdot2\cdot1) = (\tfrac{1}{24})(6 + 6 + 6 + 6) = \tfrac{24}{24} = 1$$

$$a_{A_{2g}} = (\tfrac{1}{24})(6 - 6 + 6 - 6) = 0$$

$$a_{B_{1g}} = (\tfrac{1}{24})(6 + 6 - 6 - 6) = 0$$

$$a_{B_{2g}} = (\tfrac{1}{24})(6 - 6 - 6 + 6) = 0$$

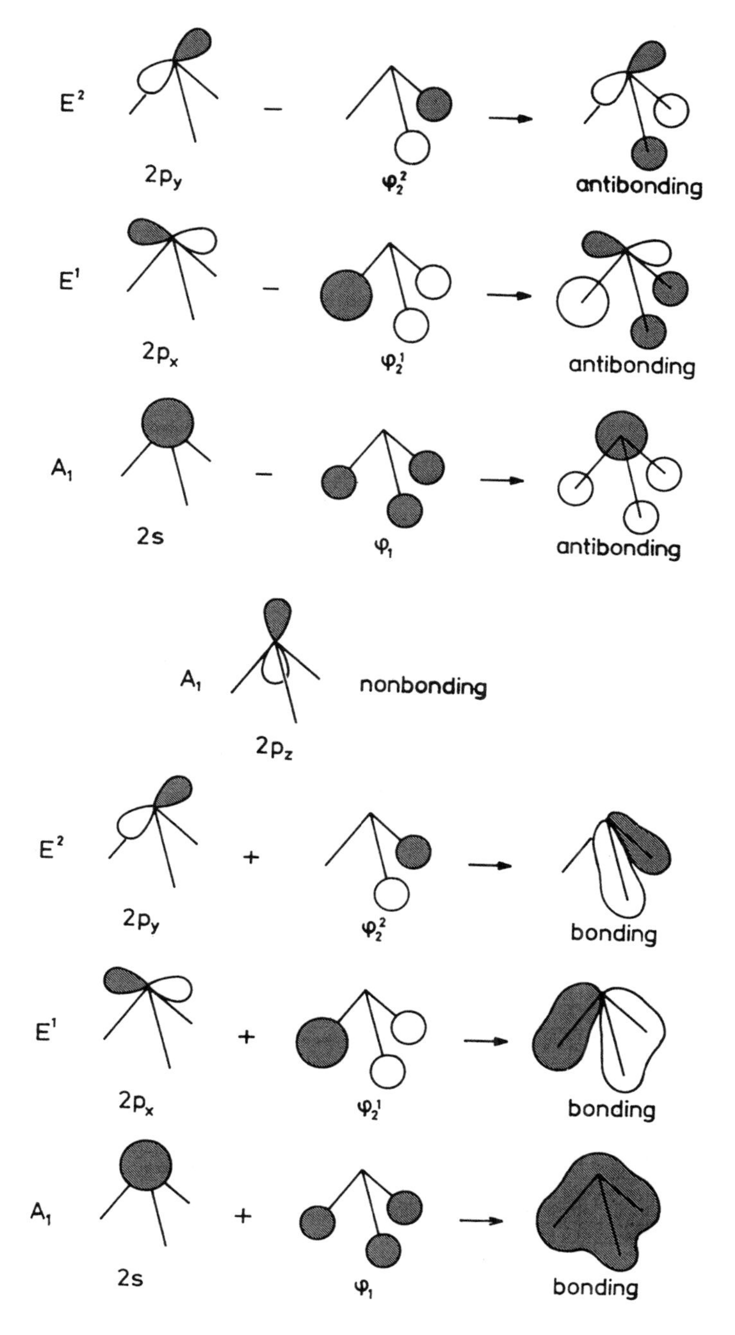

Figure 6-25. Construction of molecular orbitals for ammonia.

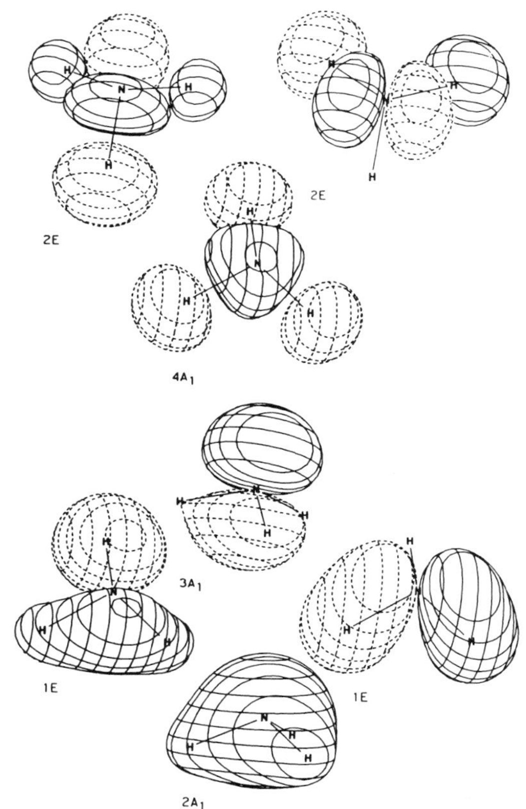

Figure 6-26. Contour diagrams of the MOs of ammonia. Reproduced with permission from Ref. [6-12]. Copyright (1973) Academic Press.

$$a_{E_{1g}} = (\tfrac{1}{24})(12 - 12) = 0$$
$$a_{E_{2g}} = (\tfrac{1}{24})(12 + 12) = 1$$
$$a_{A_{1u}} = (\tfrac{1}{24})(6 + 6 - 6 - 6) = 0$$
$$a_{A_{2u}} = (\tfrac{1}{24})(6 - 6 - 6 + 6) = 0$$
$$a_{B_{1u}} = (\tfrac{1}{24})(6 + 6 + 6 + 6) = 1$$
$$a_{B_{2u}} = (\tfrac{1}{24})(6 - 6 + 6 - 6) = 0$$
$$a_{E_{1u}} = (\tfrac{1}{24})(12 + 12) = 1$$
$$a_{E_{2u}} = (\tfrac{1}{24})(12 - 12) = 0$$

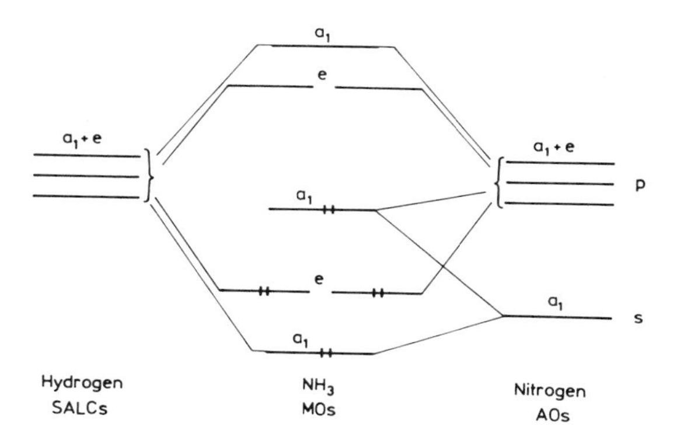

Figure 6-27. Qualitative MO diagram for ammonia.

Thus, the first representation reduces to the following irreducible representations:

$$\Gamma_{\Phi_1} = A_{1g} + E_{2g} + B_{1u} + E_{1u}$$

Without giving details of the other three reductions, the results are:

$$\Gamma_{\Phi_2} = A_{1g} + E_{2g} + B_{1u} + E_{1u}$$
$$\Gamma_{\Phi_3} = A_{1g} + A_{2g} + 2E_{2g} + B_{1u} + B_{2u} + 2E_{1u}$$
$$\Gamma_{\Phi_4} = B_{2g} + E_{1g} + A_{2u} + E_{2u}$$

Table 6-7. The D_{6h} Character Table

D_{6h}	E	$2C_6$	$2C_3$	C_2	$3C_2'$	$3C_2''$	i	$2S_3$	$2S_6$	σ_h	$3\sigma_d$	$3\sigma_v$		
A_{1g}	1	1	1	1	1	1	1	1	1	1	1	1		$x^2 + y^2, z^2$
A_{2g}	1	1	1	1	−1	−1	1	1	1	1	−1	−1	R_z	
B_{1g}	1	−1	1	−1	1	−1	1	−1	1	−1	1	−1		
B_{2g}	1	−1	1	−1	−1	1	1	−1	1	−1	−1	1		
E_{1g}	2	1	−1	−2	0	0	2	1	−1	−2	0	0	(R_x, R_y)	(xz, yz)
E_{2g}	2	−1	−1	2	0	0	2	−1	−1	2	0	0		$(x^2 - y^2, xy)$
A_{1u}	1	1	1	1	1	1	−1	−1	−1	−1	−1	−1		
A_{2u}	1	1	1	1	−1	−1	−1	−1	−1	−1	1	1	z	
B_{1u}	1	−1	1	−1	1	−1	−1	1	−1	1	−1	1		
B_{2u}	1	−1	1	−1	−1	1	−1	1	−1	1	1	−1		
E_{1u}	2	1	−1	−2	0	0	−2	−1	1	2	0	0	(x, y)	
E_{2u}	2	−1	−1	2	0	0	−2	1	1	−2	0	0		

Similarly to the case of ammonia, the rotational subgroup of D_{6h}, that is C_6, contains enough information to generate the SALCs of benzene. The C_6 character table is given in Table 6-8, and, again, contains imaginary characters. These can be handled in the same way as was done for ammonia, keeping in mind that the solution is right for the determination of the shape of the SALCs but the derived "quasi-characters" are not real characters.

These "quasi-characters" for the two E representations are:

$$E_1 \quad \begin{Bmatrix} 2 & 1 & -1 & -2 & -1 & 1 \\ 0 & 1 & 1 & 0 & -1 & -1 \end{Bmatrix}$$

$$E_2 \quad \begin{Bmatrix} 2 & -1 & -1 & 2 & -1 & -1 \\ 0 & 1 & -1 & 0 & 1 & -1 \end{Bmatrix}$$

Benzene consists of 30 MOs; only a few of these will be constructed and shown here. It may be a good exercise for the reader to construct the remaining MOs of benzene by following the procedure demonstrated here. The SALCs are sorted according to their symmetry properties in Table 6-9. Inspection of this table reveals that the first three group orbitals have common irreducible representations, so they can be mixed with each other. They consist of 24 AOs; thus, 24 MOs will be formed. Since each bonding MO has its antibonding counterpart, there will be 12 bonding and 12 antibonding molecular orbitals. The former will be the σ bonding orbitals of benzene, since there are six C–C and six C–H bonds. The fourth group orbital does not belong to any irreducible representation common to the other three, so it will not be mixed with them. This representation corresponds to the π orbitals of benzene by itself.

Let us now construct the A_{1g} and B_{1u} symmetry σ orbitals of benzene. The totally symmetric representation, A_{1g}, appears three times, once in each of Φ_1, Φ_2, and Φ_3. Two A_{1g} representations can be combined into an MO, and the third one can represent an MO by itself. These three SALCs can be generated by using the projection operator pictorially as shown in Figure 6-28. The forms of

Table 6-8. The C_6 Character Table

C_6	E	C_6	C_3	C_2	C_3^2	C_6^5		$\varepsilon = \exp(2\pi i/6)$
A	1	1	1	1	1	1	z, R_z	$x^2 + y^2, z^2$
B	1	-1	1	-1	1	-1		
E_1	1	ε	$-\varepsilon^*$	-1	$-\varepsilon$	ε^*	(x, y)	(xz, yz)
	1	ε^*	$-\varepsilon$	-1	$-\varepsilon^*$	ε	(R_x, R_y)	
E_2	1	$-\varepsilon^*$	$-\varepsilon$	1	$-\varepsilon^*$	$-\varepsilon$		$(x^2 - y^2, xy)$
	1	$-\varepsilon$	$-\varepsilon^*$	1	$-\varepsilon$	$-\varepsilon^*$		

Table 6-9. The Symmetry of the Different Group Orbitals of Benzene

	Φ_1 H group orbital	Φ_2 C 2s group orbital	Φ_3 C $2p_x$, $2p_y$ group orbital	Φ_4 C $2p_z$ group orbital
A_{1g}	+	+	+	
A_{2g}			+	
B_{1g}				
B_{2g}				+
E_{1g}				+
E_{2g}	+	+	++	
A_{1u}				
A_{2u}				+
B_{1u}	+	+	+	
B_{2u}			+	
E_{1u}	+	+	++	
E_{2u}				+

these group orbitals are such that $\Phi_2(A_{1g})$ can be taken as an MO by itself (C–C σ bond; cf. also the corresponding orbital, $2A_{1g}$, in the contour diagram in Figure 6-29a), and the other two group orbitals can be combined into molecular orbitals as shown in Figure 6-30. The contour diagram of the bonding MO is depicted by the $3A_{1g}$ orbital in Figure 6-29a.

The next MO will be of B_{1u} symmetry. This irreducible representation also appears in Φ_1, Φ_2, and Φ_3. Take this time the corresponding Φ_1 and Φ_2 group orbitals and combine them into molecular orbitals:

$$\begin{aligned}\hat{P}^{B_{1u}}s_1 &\approx 1\cdot E\cdot s_1 + (-1)\cdot C_6\cdot s_1 + 1\cdot C_3\cdot s_1 + (-1)\cdot C_2\cdot s_1 + 1\cdot C_3^2\cdot s_1 + (-1)\cdot C_6^5\cdot s_1 \\ &= s_1 - s_2 + s_3 - s_4 + s_5 - s_6\end{aligned}$$

or pictorially:

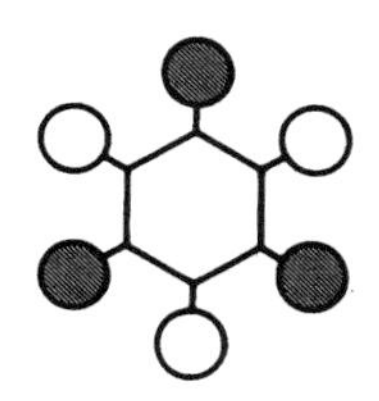

The B_{1u} symmetry SALC of Φ_2, i.e., the group orbital of the six C 2s AOs, will have a similar form:

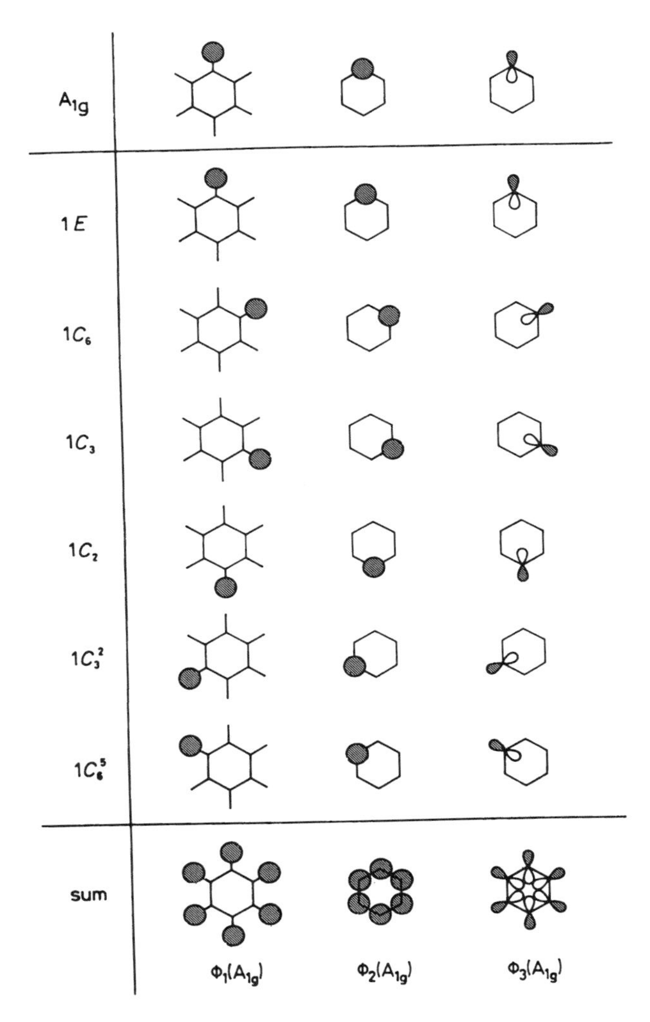

Figure 6-28. Generation of the A_{1g} symmetry group orbitals of benzene.

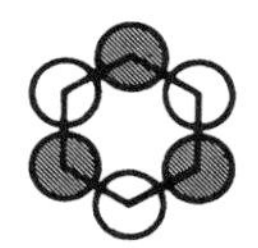

The combination of these Φ_1 and Φ_2 SALCs affords the bonding and antibonding combinations shown in Figure 6-31. The contour diagram corresponding to the bonding MO is the $2B_{1u}$ orbital in Figure 6-29a.

Since there is only one B_{2u} symmetry orbital among the SALCs, the one in Φ_3, it will be an MO by itself. Let us generate this MO:

$$\hat{P}^{B_{2u}}p_y(C_1) \approx 1 \cdot E \cdot p_{y_1} + (-1) \cdot C_6 \cdot p_{y_1} + 1 \cdot C_3 \cdot p_{y_1} + (-1) \cdot C_2 \cdot p_{y_1} + 1 \cdot C_3^2 \cdot p_{y_1} + (-1) \cdot C_6^5 \cdot p_{y_1} = p_{y1} - p_{y2} + p_{y3} - p_{y4} + p_{y5} - p_{y6}$$

This group orbital has the following shape:

Compare the above orbital with $1B_{2u}$ (Figure 6-29a).

The π orbitals of benzene will be the two doubly degenerate and the two non-degenerate combinations of the Φ_4 group orbital itself. All of these are shown below.

A_{2u} *symmetry orbital*: This corresponds to the totally symmetric representation in the rotational subgroup C_6; so, even without using the projection operator, its form can be given by:

$$\Phi_4(A_{2u}) = p_{z1} + p_{z2} + p_{z3} + p_{z4} + p_{z5} + p_{z6}$$

The corresponding orbital in Figure 6-29b will be the $1A_{2u}$ orbital.

B_{2g} *symmetry orbital*: Using the projection operator, we obtain:

$$\hat{P}^{B_{2g}}p_z(C_1) \approx 1 \cdot E \cdot p_{z_1} + (-1) \cdot C_6 \cdot p_{z_1} + 1 \cdot C_3 \cdot p_{z_1} + (-1) \cdot C_2 \cdot p_{z_1} + 1 \cdot C_3^2 \cdot p_{z_1} + (-1) \cdot C_6^5 \cdot p_{z_1} = p_{z1} - p_{z2} + p_{z3} - p_{z4} + p_{z5} - p_{z6}$$

This is the $1B_{2g}$ orbital of Figure 6-29b.

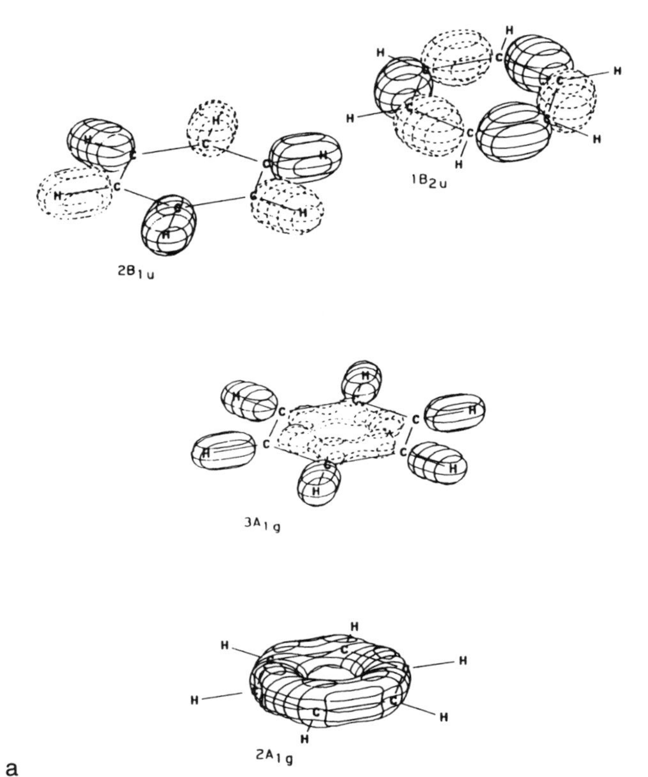

Figure 6-29. Contour diagrams of some molecular orbitals of benzene: (a) σ orbitals. *(Continued on next page)*

The two E_{1g} symmetry SALCs are constructed in Figure 6-32. Compare them to the contour diagram of the $1E_{1g}$ orbitals in Figure 6-29b.

Finally, the two E_{2u} symmetry orbitals are expressed as follows:

$$\hat{P}^{E^1_{2u}}p_z(C_1) \approx 2\cdot E\cdot p_{z1} + (-1)\cdot C_6\cdot p_{z1} + (-1)\cdot C_3\cdot p_{z1} + 2\cdot C_2\cdot p_{z1} + (-1)\cdot C_3^2\cdot p_z$$
$$+ (-1)\cdot C_6^5\cdot p_{z1} = 2p_{z1} - p_{z2} - p_{z3} + 2p_{z4} - p_{z5} - p_{z6}$$

$$\hat{P}^{E^2_{2u}}p_z(C_1) \approx 0\cdot E\cdot p_{z1} + 1\cdot C_6\cdot p_{z1} + (-1)\cdot C_3\cdot p_{z1} + 0\cdot C_2\cdot p_{z1} + 1\cdot C_3^2\cdot p_{z1}$$
$$+ (-1)\cdot C_6^5\cdot p_{z1} = p_{z2} - p_{z3} + p_{z5} - p_{z6}$$

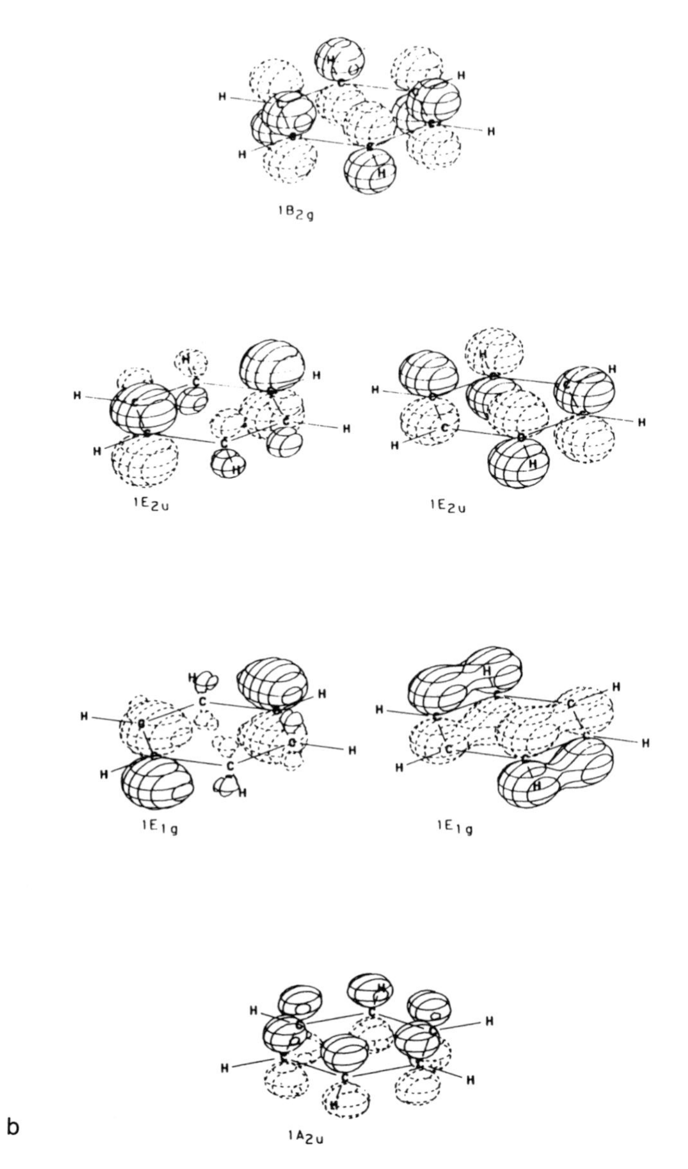

Figure 6-29. (*Continued*) (b) π orbitals. Reproduced with permission from Ref. [6-12]. Copyright (1973) Academic Press.

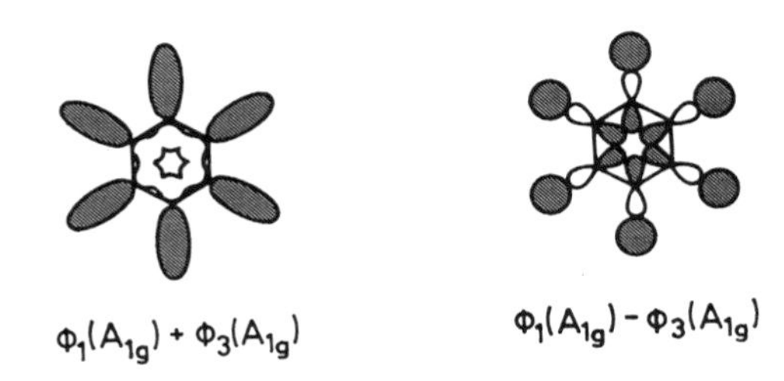

Figure 6-30. Bonding and antibonding combination of A_{1g} symmetry group orbitals of benzene.

Their forms are:

These SALCs correspond to the contour diagram of the $1E_{2u}$ orbitals (Figure 6-29b). Figure 6-33 shows the relative energies of the benzene π orbitals.

6.3.3.3 Short Summary of MO Construction

The steps of MO construction can now be summarized as follows:

1. Identify the symmetry of the molecule.
2. List all atomic orbitals that are intended to be used for MO construction.
3. See whether or not the molecule has a central atom. If it does, then look up in the character table the irreducible representations to which

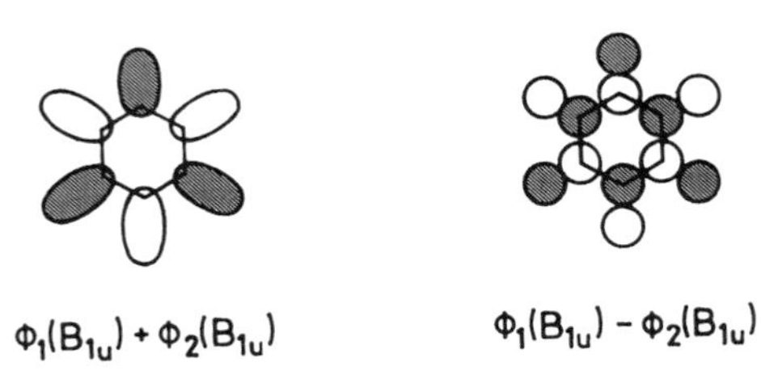

Figure 6-31. Bonding and antibonding combination of B_{1u} symmetry group orbitals of benzene

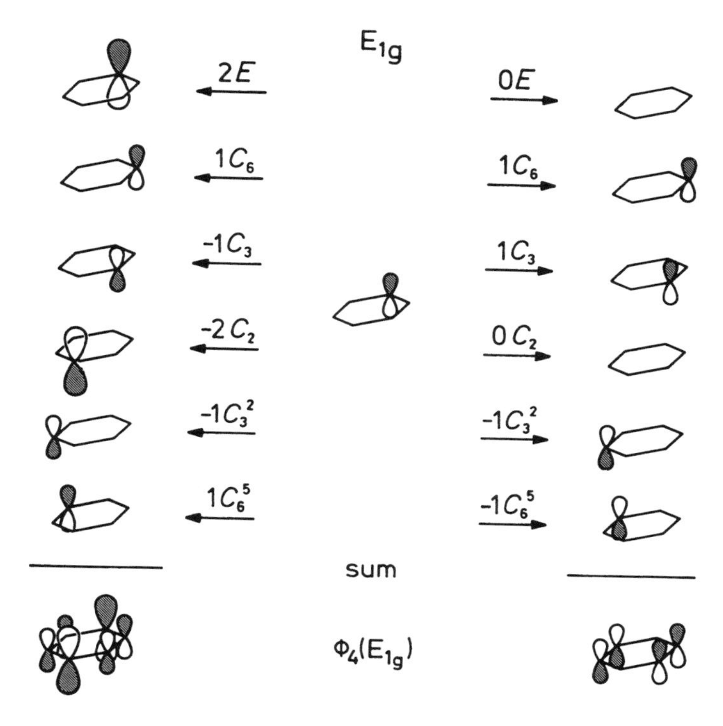

Figure 6-32. The two E_{1g} symmetry group orbitals formed from the carbon $2p_z$ orbitals in benzene.

its atomic orbitals belong. If there is no central atom in the molecule, proceed to the next step.

4. Construct group orbitals (SALCs) from the atomic orbitals of like atoms.
5. Use these orbitals as bases for representations of the point group.
6. Reduce these representations to their irreducible components.
7. Apply the projection operator to the AOs for each of these irreducible representations to obtain the forms of the SALCs.
8. These SALCs will either be MOs by themselves, or they can be combined with other SALCs or central atom orbitals of the same symmetry. Each of these combinations will give one bonding and one antibonding MO of the same symmetry.
9. Normalization has been ignored throughout our discussion. However, the SALCs must be properly normalized in all calculations [6-4].

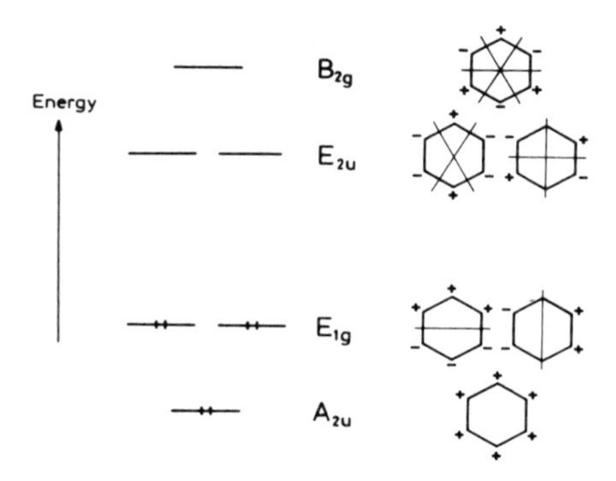

Figure 6-33. Relative energies of the benzene π orbitals.

This may be done at the end of the SALC construction, i.e., after step 7 in our list.

6.4 QUANTUM-CHEMICAL CALCULATIONS

The results of quantum chemical calculations usually yield the wave functions and the energies of a system. Numerous integrals must be evaluated even for the simplest molecules. Their number can be conveniently reduced, however, by applying the theorem according to which an energy integral, $\int \psi_i \hat{H} \psi_j \, d\tau$, is nonzero only if ψ_i and ψ_j belong to the *same* irreducible representation of the molecular point group.

Many chemical and physical properties of the molecule can be calculated, including the complete geometry, conformational properties, barrier to internal rotation, and relative stabilities of various isomers as well as different electronic states. The electron spectroscopic and vibrational spectroscopic constants and other parameters can also be determined. The present comments focus on just one of the many characteristics of the molecule, viz., its equilibrium geometry. State-of-the-art calculations of molecular geometry involving relatively light atoms are as reliable as the results of the best experiments (cf. Ref. [6-13]). While calculations provide information on the equilibrium geometry, the various experiments yield some effective geometries for the molecule, averaged over molecular vibrations. Depending on the magnitude of these vibrations and their structural influence, the equilibrium and average structures may differ to various extents. Examples of rather extreme effects were mentioned in Section 3.7.6. The results of calculations are less reliable for molecules involving heavier atoms, for example, transition metals. On the other hand, even less sophisticated calculations may be instructive if structural differences, rather than absolute values of the structural parameters, are sought. Important systematic errors usually cancel in the determination of structural differences in calculations as well as in experiments. The importance of small structural differences in understanding various effects in series of substances is increasingly becoming recognized [6-14].

Small structural changes are especially important in *molecular recognition*. It has been noted [6-15], for example, that "subtle changes of molecular structure may result in severe changes of inclusion behavior of a potential host molecule due to the complicated interplay of weak intermolecular forces that govern host–guest complex formation."

Quantum-chemical calculations have proved to be an especially important source of information on small structural differences. In this respect they have aided greatly the experimental determination of molecular geometry in that they can provide reliable constraints in the experimental analysis (see, e.g., the structural analysis of 2-nitrophenol [6-16]).

For direct comparison of parameters determined experimentally and computationally, however, the following caveat has been issued [6-17]: "For truly accurate comparison, experimental bond lengths (or, generally, geometries) should be compared with computed ones only following necessary corrections, bringing all information involved in the comparison to a common denominator."

The difference, however, is not merely practical; it is conceptual as well. R. D. Levine [6-18] distinguished between physical and chemical shapes. According to him, the physical shape corresponds to a hard space-filling model, whereas the chemical shape describes how molecular reactivity depends on the direction of approach and distance of the other reactant. In terms of geometrical representations, the chemical shape can be related to the average structures determined from the experiments and the physical shape to the motionless equilibrium structure.

Quantum-chemical calculations are, of course, the exclusive source of information for systems that are not amenable to experimental study. Such systems may include unstable or even unknown species and transition states. These calculations have proved to be complementary with experiments or can even be their alternatives. The situation is evolving, and a host of problems must still be resolved for individual systems (see, e.g., *Reviews in Computational Chemistry* [6-19]).

6.5 INFLUENCE OF ENVIRONMENTAL SYMMETRY

Symmetry has a major role in two widely used and successful theories of chemistry, viz., the crystal field and ligand field theories of coordination compounds. This topic has been thoroughly covered in textbooks and monographs on coordination chemistry. Therefore, it is mentioned here only in passing.

Bethe [6-20] showed that the degenerate electronic state of a cation is split by a crystal field into nonequivalent states. The change is determined entirely by the symmetry of the crystal lattice. Bethe's original work was concerned with ionic crystals, but his concept has more general applications. When an atom or an ion enters a ligand environment, the symmetry of the ligand arrangement will influence the electron density distribution of that atom or ion. The original spherical symmetry of the atomic orbitals will be lost, and the symmetry of the ligand environment will be adopted. As a consequence of the decrease of symmetry that usually results, the degree of degeneracy of the orbitals decreases.

The *s* electrons are already nondegenerate in the free atom, so their degeneracy does not change. They will always belong to the totally symmetric irreducible representation of the symmetry group. The *p* orbitals, however, are threefold degenerate, and the *d* orbitals are fivefold degenerate. To determine their splitting in a certain point group, we must use them, in principle, as bases for a representation of the group. In practice, we can find in the character table of the point group the irreducible representations to which the orbitals belong. An orbital always belongs to the same irreducible representation as do its subscripts. Some orbital splittings that accompany the decrease in environmental symmetry are shown in Table 6-10.

As environmental symmetry decreases, the orbitals will become split to an increasing extent. In the C_{2v} point group, for example, all atomic orbitals become split into nondegenerate levels. This is not surprising since the C_{2v} character table contains only one-dimensional irreducible representations. This result shows at once that there are no degenerate energy levels in this point group. This has been stressed in Chapter 4 in the discussion of irreducible representations.

Table 6-10. Splitting of Atomic Orbitals in Different Symmetry Environments

	s	p	d
O_h	a_{1g}	t_{1u}	$e_g + t_{2g}$
T_d	a_1	t_2	$e + t_2$
$D_{\infty h}$	σ_g	$\sigma_u + \pi_u$	$\sigma_g + \pi_g + \Delta_g$
D_{4d}	a_1	$b_2 + e_1$	$a_1 + e_2 + e_3$
D_{4h}	a_{1g}	$a_{2u} + e_u$	$a_{1g} + b_{1g} + b_{2g} + e_g$
C_{4v}	a_1	$a_1 + e$	$a_1 + b_1 + b_2 + e$
C_{2v}	a_1	$a_1 + b_1 + b_2$	$2a_1 + a_2 + b_1 + b_2$

The symmetry of the ligand environment gives an important but limited amount of information about orbital splitting. Both the octahedral and cubic ligand arrangements, for example, belong to the O_h point group, and we can tell that the *d* orbitals of the central atom will split into a doubly degenerate and a triply degenerate pair. Nothing is revealed, however, about the relative energies of these two sets of degenerate orbitals.

The problem of relative energies is dealt with by crystal field theory. This theory examines the repulsive interaction between the ligands and the central-atom orbitals. Consider first an octahedral molecule (Figure 6-34), and compare the positions of one e_g (e.g., $d_{x^2-y^2}$) and one t_{2g} (e.g., d_{yz}) orbital. The others need not be considered, as they are degenerate with, and thus have the same energy as, one of the e_g or t_{2g} orbitals. The lobes of the $d_{x^2-y^2}$ orbital point towards the ligands. The resulting electrostatic repulsion will destabilize this orbital, and its energy will increase accordingly. The d_{yz} orbital, on the other hand, points in directions between the ligands. This is an energetically more favorable position; hence, the energy of this orbital will decrease.

Examine now the cubic arrangement in Figure 6-35. It can be seen that the d_{yz} orbital is in a more unfavorable situation relative to the ligands than is the $d_{x^2-y^2}$ orbital, so their relative energies will be reversed (see Figure 6-36). Some other typical orbital splittings and the corresponding changes in the relative energies are shown in Figure 6-37.

Prediction of Structural Changes. Crystal field theory is frequently applied to account for, and even predict, structural and chemical changes. A well-known example is the variation of first-row transition metal ionic radii in an octahedral environment [6-21], as illustrated by curve A in Figure 6-38. The solid line connects the points for Ca, Mn, and Zn, i.e., atoms with a spherically symmetrical distribution of *d* electrons. Since the shielding of one *d* electron by another is imperfect, a contraction in the ionic radius is expected along this series. This in itself would account only for a steady decrease in the radii,

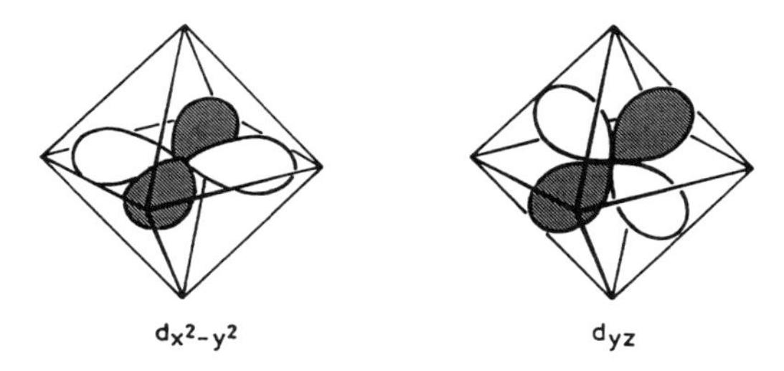

Figure 6-34. The orientation of the different symmetry *d* orbitals in an octahedral environment.

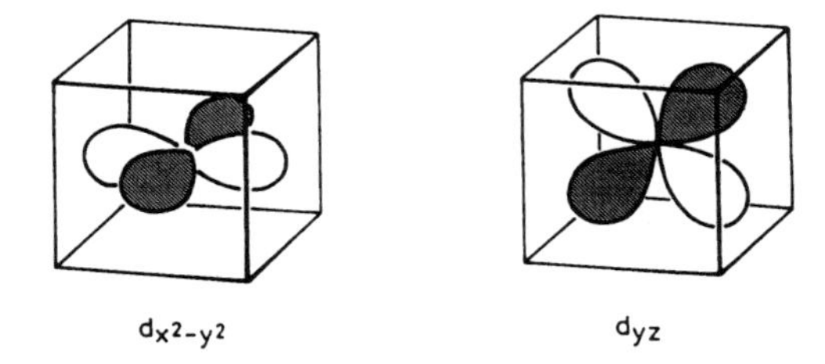

Figure 6-35. The orientation of the different symmetry d orbitals in a cubic environment.

whereas the ionic radii of all the other atoms are smaller than interpolation from the Ca–Mn–Zn curve would suggest. As is well known, the nonuniform distribution of d electrons around the nuclei is the origin of this phenomenon. In the octahedral environment the d orbitals split into orbitals with t_{2g} and e_g symmetry. The electrons, added gradually, occupy t_{2g} orbitals in Sc^{2+}, Ti^{2+}, and V^{2+} as well as in Fe^{2+}, Co^{2+}, and Ni^{2+}, if only high-spin configurations are considered. Since these orbitals are not oriented toward the ligands, the degree of shielding between the ligands and the positively charged atomic cores decreases along with the ionic radius. The fourth electron in Cr^{2+} as well as the ninth electron in Cu^{2+} occupy e_g symmetry orbitals. The degree of shielding thus somewhat increases, and, accordingly, there is a smaller relative decrease in the ionic radii.

The bond length variation among the first-row transition metal dihalides

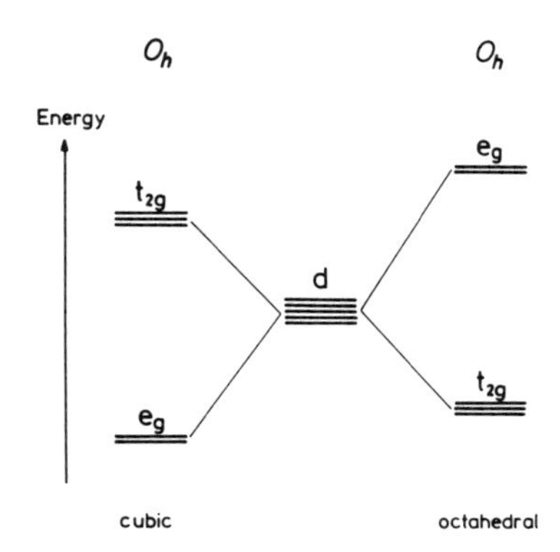

Figure 6-36. Relative energies of the d orbitals in an octahedral and a cubic ligand environment.

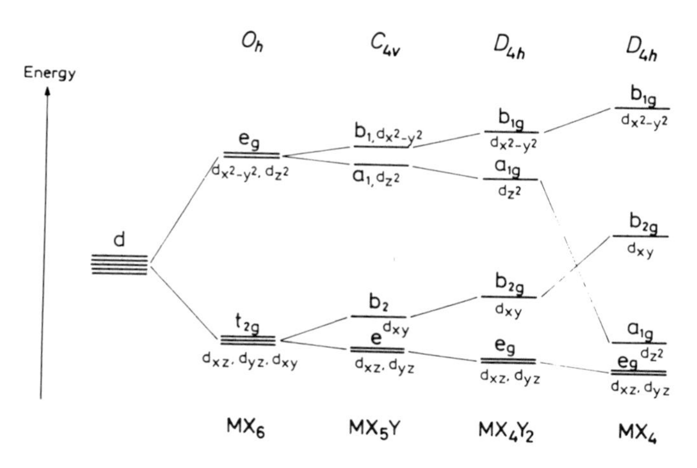

Figure 6-37. The *d* orbital splittings in different ligand environments.

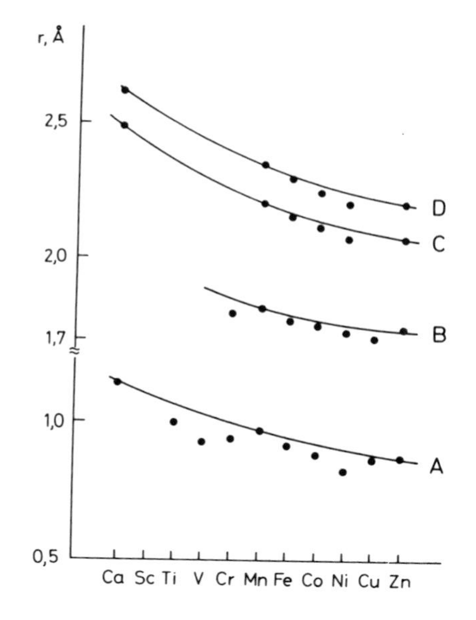

Figure 6-38. The variation of octahedral M^{2+} ionic radii according to Ref. [6-21] (curve A; ionic radii from Ref. [6-22]) and of the bond lengths of difluorides (curve B), dichlorides (curve C), and dibromides (curve D) in the first transition metal series.

has been interpreted in terms of similar symmetry arguments [6-23]. Curves B–D in Figure 6-38 represent the available experimental data on the bond lengths of gaseous first-row transition metal difluorides, dichlorides, and dibromides, respectively. The solid lines again connect the points for atoms that possess spherically symmetrical electron distributions.

The observed decrease in the bond lengths with increasing atomic number is even more pronounced than what was observed for the ionic radii. This difference between the slopes of the curves may originate from differences in coordination numbers. The coordination number is smaller in the dihalides than in the octahedral complexes. The electronic repulsion between the ligand lone pairs may counter the attraction by the central atom in the octahedral environment and may partially compensate for the imperfect shielding. On the other hand, this repulsion probably has a negligibly small influence on the metal–halogen bond length in the linear dihalides.

Even more interesting is the different extent of deviation from the smooth curves for the bond lengths and the crystal radii. Here, the explanation may be sought by considering molecular symmetry, i.e., the symmetry of the "ligand field." The splitting of *d* orbitals in these linear molecules will be different from that in an octahedral environment. Figure 6-39 depicts the two different orbital splittings. In the linear dihalides, only the d_{z^2} orbital is oriented toward the ligands. Since this is energetically the least favorable orbital, it will be occupied only by the fifth and the tenth electrons. Thus, the least shielding occurs with the fourth and ninth electrons. Accordingly, the largest deviations from the Ca–Mn–Zn line could be anticipated for the bond lengths of chromium and copper dihalides. According to curve B of Figure 6-38, the shortest bond occurs for copper difluoride, indeed. Also, the deviation of the chromium difluoride bond length from the continuation of the Mn–Zn line is conspicuously large, which is consistent with the prediction.

In a similar way, the variation of M^{3+} octahedral crystal radii can be

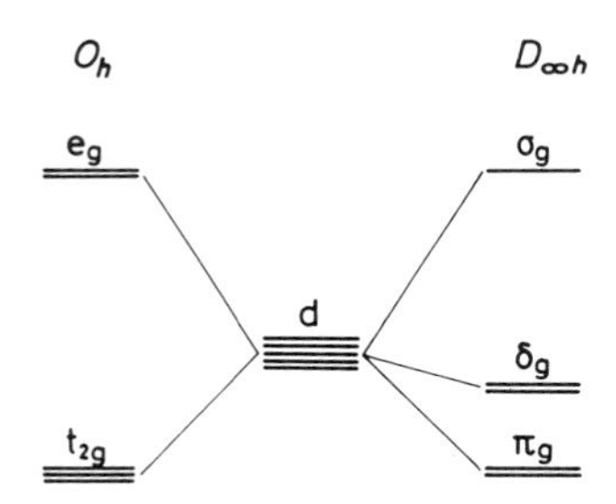

Figure 6-39. The *d* orbital splittings in an octahedral and a linear ligand environment.

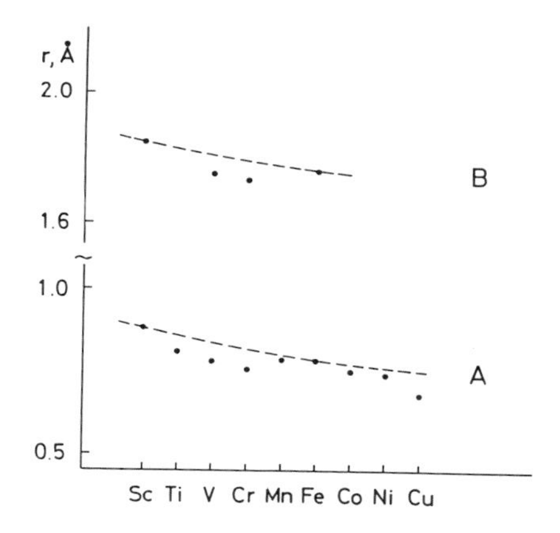

Figure 6-40. The variation of octahedral M^{3+} ionic radii [6-22] (curve A) and of the bond lengths of trifluorides (curve B) in the first transition metal series.

compared with the bond length variation of transition metal trihalides, as shown in Figure 6-40. In this case Sc^{3+} (d^0) and Fe^{3+}(d^5) have spherical *d* subshells, and there is no third such ion (d^{10}). Curve B in Figure 6-40 represents the experimental bond lengths for the planar gas-phase trifluorides, and the points for ScF_3 and FeF_3 are connected.

The *d* orbital splitting in a trigonal planar environment is shown in Figure 6-41 and is similar to that in the octahedral environment, although the degeneracy and the symmetry of the orbitals are different. There are two unfavorable orbitals in the trigonal planar situation just as in the octahedral situation. Consequently, the same qualitative trend can be expected in the bond length variation of MX_3 molecules as in the corresponding M^{3+} ions. The

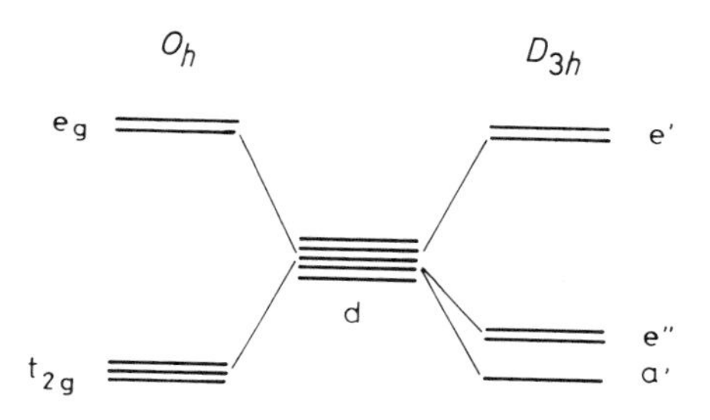

Figure 6-41. Relative energies of the *d* orbitals in an octahedral and a trigonal planar environment.

available experimental data fully support this reasoning. The difference between the ionic radii and the bond lengths of the trifluorides is nearly constant. Accordingly, we can predict CrF_3 to have the shortest bond in its vicinity. The overall shortest bond may be similarly predicted for CuF_3.

6.6 JAHN–TELLER EFFECT

"Somewhat paradoxically, symmetry is seen to play an important role in the understanding of the Jahn–Teller effect, the very nature of which is symmetry destruction" [6-24]. Only a brief discussion of this effect, also called the "first-order Jahn–Teller effect," will be given here. For more detail, we refer the reader to the literature [6-25–6-28]. According to the original formulation of the Jahn–Teller effect [6-29], a nonlinear symmetrical nuclear configuration in a degenerate electronic state is unstable and gets distorted, thereby removing the electronic degeneracy, until a nondegenerate ground state is achieved. This formulation indicates the strong relevance of this effect to orbital splitting and, generally, to the relationship of symmetry and electronic structure as well as molecular vibrations, discussed in previous sections. Owing to a coupling of the electronic and vibrational motions of the molecule, the ground-state orbital degeneracy is removed by distorting the highly symmetrical molecular structure to a lower symmetry structure.

Jahn–Teller distortion can only be expected if the energy integral

$$\left\langle \psi_0 \left| \frac{\partial E}{\partial q} \right| \psi_0 \right\rangle \tag{6-6}$$

has a nonzero value (ψ_0 is the ground-state electronic wave function of the high-symmetry nuclear configuration, and q is a normal mode of vibration). According to what has already been said about the value of an energy integral (Section 4.9.2), this can only happen if the direct product of ψ_0 with itself is, or contains, the irreducible representation of the q normal mode of vibration:

$$\Gamma_{\psi_0} \cdot \Gamma_{\psi_0} \subset \Gamma_q \tag{6-7}$$

Since ψ_0 is degenerate, its direct product with itself will always contain the totally symmetric irreducible representation and, at least, one other irreducible representation. For the integral to be nonzero, q must belong either to the totally symmetric irreducible representation or to one of the other irreducible representations contained in the direct product of ψ_0 with itself. A vibration belonging to the totally symmetric irreducible representation, however, does not decrease the symmetry of the molecule. Accordingly, in order to have a

Jahn–Teller type distortion, q must belong to one of the other irreducible representations.

Let us see a simple example, the H_3 molecule, which has the shape of an equilateral triangle. Its symmetry is D_{3h}, the electronic configuration is $a_1'^2e'$, and the symmetry of the ground electronic state is E'. Thus, the electronic state of the molecule is degenerate and is subject to Jahn–Teller distortion.

The symmetry of the normal mode of vibration that can take the molecule out of the degenerate electronic state will have to be such as to satisfy Eq. (6-7). The direct product of E' with itself (see Table 6-11) reduces to $A_1' + A_2' + E'$. The molecule has three normal modes of vibration $[(3 \times 3) - 6 = 3]$, and their symmetry species are $A_1' + E'$. A totally symmetric normal mode, A_1', does not reduce the molecular symmetry (this is the symmetric stretching mode), and thus the only possibility is a vibration of E' symmetry. This matches one of the irreducible representations of the direct product $E' \cdot E'$; therefore, this normal mode of vibration is capable of reducing the D_{3h} symmetry of the H_3 molecule. These types of vibrations are called Jahn–Teller active vibrations.

The two E' symmetry vibrations of the H_3 molecule are the angle bending and the asymmetric stretching modes (see Figure 6-42). They lead to the dissociation of the molecule into H_2 and H. Indeed, H_3 is so unstable that it cannot be observed as it would immediately dissociate into H_2 and H. This is one of the reasons why it has been so difficult to find experimental evidence of the Jahn–Teller effect. The structures that are predicted to be unstable are often not found, and the observed structures are so different from them that the connection is not obvious. Other reasons of the frequent difficulty encountered in observing the Jahn–Teller effect will be given later.

Obviously, only molecules with partially filled orbitals display Jahn–Teller distortion. As was shown in Section 6.3.2, the electronic ground state of molecules with completely filled orbitals is always totally symmetric and thus cannot be degenerate. In comparison with the unstable H_3 molecule, H_3^+ has

Table 6-11. The D_{3h} Character Table and the Reducible Representation $E' \cdot E'$

D_{3h}	E	$2C_3$	$3C_2$	σ_h	$2S_3$	$3\sigma_v$		
A_1'	1	1	1	1	1	1		$x^2 + y^2, z^2$
A_2'	1	1	−1	1	1	−1	R_z	
E'	2	−1	0	2	−1	0	(x, y)	$(x^2 - y^2, xy)$
A_1''	1	1	1	−1	−1	−1		
A_2''	1	1	−1	−1	−1	1	z	
E''	2	−1	0	−2	1	0	(R_x, R_y)	(xz, yz)
$E' \cdot E'$	4	1	0	4	1	0	$= A_1' + A_2' + E'$	

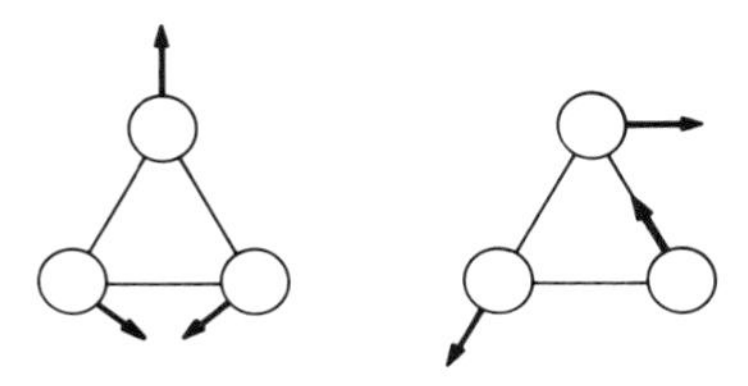

Figure 6-42. The two E' symmetry normal modes of vibration of the H_3 molecule leading to dissociation.

only two electrons in an a_1' symmetry orbital; therefore, its electronic ground state is totally symmetric, and the D_{3h} symmetry triangular structure of this ion is stable (see, e.g., Ref. [6-30]).

Transition metals have partially filled *d* or *f* orbitals, and therefore their compounds may be Jahn–Teller systems. Let us consider an example from among the much studied cupric compounds (cf. Ref. [6-21]). Suppose that the Cu^{2+} ion with its d^9 electronic configuration is surrounded by six ligands in an octahedral arrangement. We have already seen (Table 6-10 and Figure 6-36) that the *d* orbitals split into a triply (t_{2g}) and a doubly (e_g) degenerate level in an octahedral environment. For Cu^{2+} the only possible electronic configuration is $t_{2g}^6 e_g^3$.

Suppose now that of the two e_g orbitals, d_{z^2} is doubly occupied while $d_{x^2-y^2}$ is only singly occupied. Thus, the two ligands along the *z* axis are better screened from the electrostatic attraction of the central ion, and will move farther away from it, than the four ligands in the *xy* plane. The opposite happens if the unpaired electron occupies the d_{z^2} orbital. In both cases the octahedral arrangement undergoes tetragonal distortion along the *z* axis, in the former by elongation, and in the latter by compression. The original O_h symmetry reduces to D_{4h}. The symmetry-reducing vibrational mode here is of E_g symmetry and has the form shown in Figure 6-43. The splitting of *d* orbitals in both environments is given in Table 6-10 and is also shown here:

$$
\begin{array}{ccccc}
O_h & \rightarrow & & D_{4h} & \\
e_g & \rightarrow & a_{1g} & + & b_{1g} \\
(d_{x^2-y^2},\ d_{z^2}) & & (d_{z^2}) & & (d_{x^2-y^2}) \\
t_{2g} & \rightarrow & e_g & + & b_{2g} \\
(d_{xz},\ d_{yz},\ d_{xy}) & & (d_{xz},\ d_{xy}) & & (d_{xy})
\end{array}
$$

Figure 6-44 illustrates the tetragonal elongation and compression of an octahedron. For the Cu^{2+} ion the relative energies of the d_{z^2} and $d_{x^2-y^2}$ orbitals depend on the location of the unpaired electron.

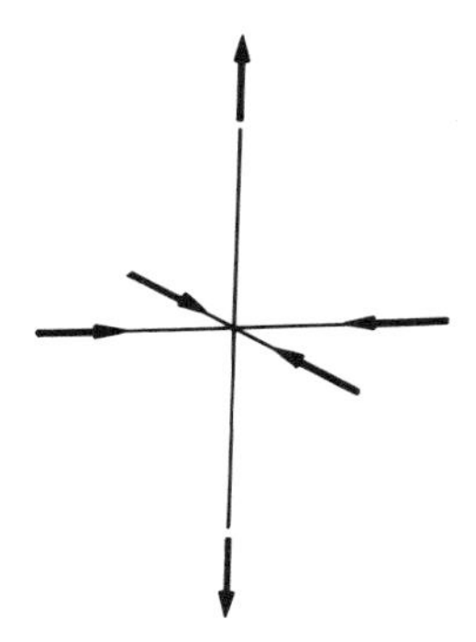

Figure 6-43. The symmetry-reducing vibrational mode of E_g symmetry for an octahedron.

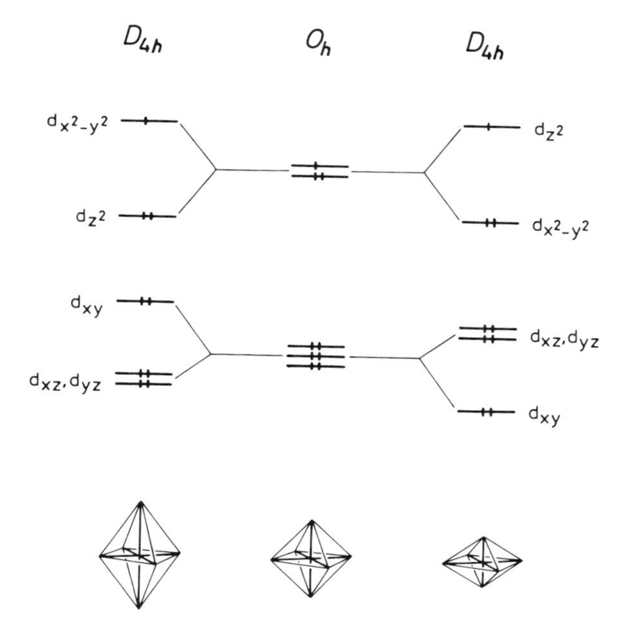

Figure 6-44. Tetragonal distortions of the regular octahedral arrangement around a d^9 ion.

Consider now a qualitative picture of the splitting of the t_{2g} orbitals. If the ligands are somewhat further away along the z axis, their interaction with the d_{xz} and d_{yz} orbitals will decrease, and so will their energy compared with that of the d_{xy} orbital. This is illustrated on the left-hand side of Figure 6-44. Tetragonal compression can be accounted for by similar reasoning (cf. right-hand side of Figure 6-44).

The splitting of the d orbitals in Figure 6-44 shows the validity of the "center of gravity rule." One of the e_g orbitals goes up in energy as much as the other goes down. From among the t_{2g} orbitals, the doubly degenerate pair goes up (or down) in energy half as much as the nondegenerate orbital goes down (or up). Thus, for the Cu(II) compounds the splitting of the fully occupied t_{2g} orbitals does not bring about a net energy change. The same is true for all other symmetrically occupied degenerate orbitals, such as t_{2g}^3, e_g^4, or e_g^2. On the other hand, the occupancy of the e_g orbitals of Cu^{2+} is unsymmetrical, since two electrons go down and only one goes up in energy, and here there is a net energy gain in the tetragonal distortion. This energy gain is the Jahn–Teller stabilization energy.

The above example referred to an octahedral configuration. Other highly symmetrical systems, for example, tetrahedral arrangements, can also display this effect. For general discussion, see, e.g., Refs. [6-25], [6-27], and [6-31].

The Jahn–Teller effect enhances the structural diversity of Cu(II) compounds [6-32]. Most of the octahedral complexes of Cu^{2+}, for example, show elongated tetragonally distorted geometry. Crystalline cupric fluoride and cupric chloride both have four shorter and two longer copper–halogen interatomic distances, 1.93 vs. 2.27 Å and 2.30 vs. 2.95 Å, respectively [6-32].

The square planar arrangement can be regarded as a limiting case of the elongated octahedral configuration. The four oxygen atoms are at 1.96 Å from the copper atom in a square configuration in crystalline cupric oxide, whereas the next nearest neighbors, two other oxygen atoms, are at 2.78 Å. The ratio of the two distances is much larger than that in a usual distorted octahedral configuration [6-32].

Tetragonal compression around the central Cu^{2+} ion is much rarer. K_2CuF_4 is an example with two shorter and four longer Cu–F distances, viz., 1.95 vs. 2.08 Å [6-32].

There are also numerous cases in which experimental investigation has failed to provide evidence for Jahn–Teller distortion. For example, several chelate compounds of Cu(II), as well as some compounds containing the $[Cu(NO_2)_6]^{4-}$ ion, show no detectable distortion from the regular octahedral structure (see Ref. [6-33] and references therein).

Bersuker [6-27, 6-28, 6-31] has shown the need for a more sophisticated approach to account for such phenomena. We attempt to convey at least the flavor of his ideas here. Jahn–Teller distortions are of a *dynamic* nature in systems under no external influence. This means that there may be many minimum-energy distorted structures in such systems. Whether an experiment will or will not detect such a dynamic Jahn–Teller effect depends on the relationship between the time scale of the physical measurement used for the investigation and the mean lifetime of the distorted configurations. If the time period of the measurement is longer than the mean lifetime of the distorted configurations, only an average structure, corresponding to the undistorted high-symmetry configuration, will be detected. Since different physical techniques have different time scales, one technique may detect a distortion which appears to be undetected by another.

The *static* Jahn–Teller effect can be observed only in the presence of an external influence. Bersuker [6-28, 6-31] stressed this point as the opposite statement is often found in the literature. According to the statement criticized, the effect is not to be expected in systems where low-symmetry perturbations remove electronic degeneracy. According to Bersuker, it is exactly the low-symmetry perturbations that make the Jahn–Teller distortions static and thus observable. Such a low-symmetry perturbation can be the substitution of one ligand by another. In this case one of the previously equivalent minimum-energy structures, or a new one, will become energetically more favorable than the others.

The so-called *cooperative* Jahn–Teller effect is another occurrence of the static distortions. Here, interaction, that is, cooperation between different crystal centers, makes the phenomenon observable. Without interaction, the nuclear motion around each center would be independent and of a dynamic character.

Lattice vibrations tend to destroy the correlation among Jahn–Teller centers. Thus, with increasing temperature, these centers may become independent of each other at a certain point, and their static Jahn–Teller effects convert to dynamic ones. At this point, the crystal as a whole becomes more symmetric. This temperature-dependent static ⇔ dynamic transition is called a *Jahn–Teller phase transition*. Below the temperature of the phase transition, the cooperative Jahn–Teller effect governs the situation, providing static distortion; the overall structure of the crystal is of a lower symmetry. Above this temperature, cooperation breaks down, the Jahn–Teller distortion becomes dynamic, and the crystal itself becomes more symmetric.

The temperature of the Jahn–Teller phase transition is very high for CuF_2, $CuCl_2$, and K_2CuF_4 among the examples mentioned above [6-31]. Therefore, at room temperature their crystal structures display distortions. Other compounds have symmetric crystal structures at room temperatures as their Jahn–Teller phase transition occurs at lower temperatures. Cupric chelate compounds and $[Cu(NO_2)_6]^{4-}$ compounds, such as $K_2PbCu(NO_2)_6$ and $Tl_2PbCu(NO_2)_6$, can be mentioned as examples [6-33]. Further cooling, however, may make even these structures distorted [6-33].

REFERENCES

[6-1] M. W. Hanna, *Quantum Mechanics in Chemistry*, 2nd ed. W. A. Benjamin, Inc., New York (1969).
[6-2] J. N. Murrel, S. F. A. Kettle, and J. M. Tedder, *Valence Theory*, 2nd ed. John Wiley & Sons, New York (1970).
[6-3] D. V. George, *Principles of Quantum Chemistry*, Pergamon Press, New York (1972).
[6-4] F. A. Cotton, *Chemical Applications of Group Theory*, 3rd ed., Wiley-Interscience, New York (1990).
[6-5] D. C. Harris and M. D. Bertolucci, *Symmetry and Spectroscopy: An Introduction to Vibrational and Electronic Spectroscopy*, Oxford University Press, New York (1978).
[6-6] T. H. Lowry and K. S. Richardson, *Mechanism and Theory in Organic Chemistry*, 3rd ed., Harper & Row, New York (1987).
[6-7] M. Orchin and H. H. Jaffe, *Symmetry Orbitals, and Spectra (S.O.S.)*, Wiley-Interscience, New York (1971).
[6-8] E. P. Wigner, *Group Theory and Its Application to the Quantum Mechanics of Atomic Spectra*, Academic Press, New York (1959).
[6-9] MAPLE V, Release 2, Waterloo Maple Software, University of Waterloo, Ontario, Canada.
[6-10] C. A. Coulson, *The Shape and Structure of Molecules*, Clarendon Press, Oxford (1973).
[6-11] J. Demuynck and H. F. Schaefer III, *J. Chem. Phys.* **72**, 311 (1980).
[6-12] W. L. Jorgensen and L. Salem, *The Organic Chemist's Book of Orbitals*, Academic Press, New York (1973).
[6-13] J. E. Boggs, in *Accurate Molecular Structures: Their Determination and Importance* (A. Domenicano and I. Hargittai, eds.), p. 322, Oxford University Press, Oxford (1992).
[6-14] I. Hargittai and M. Hargittai, in *Molecular Structure and Energetics*, Vol. 2 (J. F. Liebman and A. Greenberg, eds.), Chapter 1, VCH Publishers, New York (1987).
[6-15] R. Hilgenfeld and W. Saenger, *Angew. Chem. Suppl.* **1982**, 1690.
[6-16] K. B. Borisenko, C. W. Bock, and I. Hargittai, *J. Phys. Chem.* **98**, 1442 (1994).
[6-17] M. Hargittai and I. Hargittai, *Int. J. Quant. Chem.* **44**, 1057 (1992).
[6-18] R. D. Levine, *J. Phys. Chem.* **94**, 8872 (1990).
[6-19] K. B. Lipkowitz and D. B. Boyd (eds.), *Reviews in Computational Chemistry*, Vols. 1–3, VCH Publishers, New York [1990 (Vol. 1), 1991 (Vol 2.), 1992 (Vol 3)].
[6-20] H. Bethe, *Ann. Phys.* **3**, 133 (1929).
[6-21] F. A. Cotton, G. Wilkinson, and P. L. Gaus, *Basic Inorganic Chemistry*, 2nd ed., John Wiley & Sons, New York (1987).
[6-22] R. D. Shannon, *Acta Crystallogr., Sect. A* **32**, 751 (1976).
[6-23] M. Hargittai, *Inorg. Chim. Acta* **53**, L111 (1981); **180**, 5 (1991).
[6-24] A. Ceulemans, D. Beyens, and L. G. Vanquickenborne, *J. Am. Chem. Soc.* **106**, 5824 (1984).
[6-25] R. Englman, *The Jahn–Teller Effect in Molecules and Crystals*, John Wiley & Sons, New York (1972).
[6-26] R. G. Pearson, *Symmetry Rules for Chemical Reactions: Orbital Topology and Elementary Processes*, Wiley-Interscience, New York (1976).
[6-27] I. B. Bersuker, *The Jahn-Teller Effect and Vibronic Interactions in Modern Chemistry*, Plenum Press, New York (1984).
[6-28] I. B. Bersuker and V. Z. Polinger, *Vibronic Interactions in Molecules and Crystals*, Springer-Verlag, Berlin (1989).
[6-29] H. A. Jahn and E. Teller, *Proc. R. Soc. London, Ser. A* **161**, 220 (1937).

[6-30] M. J. Gaillard, D. S. Gemmell, G. Goldring, I. Levine, W. J. Pietsch, J. C. Poizat, A. J. Ratkowski, J. Remillieux, Z. Vager, and B. J. Zabransky, *Phys. Rev. A* **17**, 1797 (1978); J. S. Wright and G. A. DiLabio, *J. Phys. Chem.* **96**, 10793 (1992).
[6-31] I. B. Bersuker, *Coord. Chem. Rev.* **14**, 357 (1957).
[6-32] A. F. Wells, *Structural Inorganic Chemistry*, 4th ed., Clarendon Press, Oxford (1975).
[6-33] J. E. Huheey, *Inorganic Chemistry: Principles of Structure and Reactivity*, 3rd ed., Harper & Row, New York (1983).

Chapter 7

Chemical Reactions

The chemical reaction is the "most chemical" event. Our encounter with the role of symmetry in chemistry would certainly be one-sided without looking at chemical reactions. In fact, this is perhaps the most flourishing, booming area today of all chemistry-related applications of the symmetry concept. For this very reason, we shall present only a short survey and refer to the vast recent literature (see, e.g., Refs. [7-1]–[7-12]). Our discussion fully relies on these papers and monographs.

The first application of symmetry considerations to chemical reactions can be attributed to Wigner and Witmer [7-13]. The Wigner–Witmer rules are concerned with the conservation of spin and orbital angular momentum in the reaction of diatomic molecules. Although symmetry is not explicitly mentioned, it is present implicitly in the principle of conservation of orbital angular momentum. The real breakthrough in recognizing the role that symmetry plays in determining the course of chemical reactions has occurred only recently, mainly through the activities of Woodward and Hoffmann, Fukui, Bader, Pearson, and others.

The main idea in their work is that symmetry phenomena may play as important a role in chemical reactions as they do in the construction of molecular orbitals or in molecular spectroscopy. It is even possible to make certain symmetry-based "selection rules" for the "allowedness" and "forbiddenness" of a chemical reaction, just as is done for spectroscopic transitions.

Before describing these rules, however, we would like to mention some limitations. Symmetry rules can usually be applied to comparatively simple reactions, the so-called *concerted reactions*. In a concerted reaction all relevant changes occur simultaneously; the transformation of reactants into products happens in one step with no intermediates.

At first sight, it would seem logical that symmetry rules can be applied only to symmetrical molecules. However, even nonsymmetric reactants can be "simplified" to related symmetrical parent molecules. As Woodward and Hoffmann put it, they can be "reduced to their highest inherent symmetry" [7-3]. This is, in fact, a necessary criterion if symmetry principles are to be applied.

What does this mean? For example, propylene, $H_2C{=}CHCH_3$, must be treated as its "parent molecule," ethylene. The reason is that it is the double bond of propylene which changes during the reaction, and it nearly possesses the symmetry of ethylene. Salem calls this feature "pseudosymmetry" [7-7].

The statement "a chemical reaction is 'symmetry allowed' or 'symmetry forbidden' " should not be taken literally. When a reaction is symmetry allowed, it means that it has a low activation energy. This makes it possible for the given reaction to occur, though it does not mean that it always will. There are other factors which can impose a substantial activation barrier. Such factors may be steric repulsions, difficulties in approach, and unfavorable relative energies of orbitals. Similarly, "symmetry forbidden" means that the reaction, as a concerted one, would have a high activation barrier. However, various factors may make the reaction still possible; for example, it may happen as a stepwise reaction through intermediates. In this case, of course, it is no longer a concerted reaction.

Most of the symmetry rules explaining and predicting chemical reactions deal with changes in the electronic structure. However, a chemical reaction is more than just that. Breakage of bonds and formation of new ones are also accompanied by nuclear rearrangements and changes in the vibrational behavior of the molecule. (Molecular translation and rotation as a whole can be ignored.)

As has been shown previously, both the vibrational motion and the electronic structure of the molecules strongly depend on symmetry. This dependence can be fully utilized when discussing chemical reactions.

Describing the structures of both reactant and product molecules with the help of symmetry would not add anything new to our previous discussion. What is new and important is that certain symmetry rules can be applied to the transition state in between the reactants and products. This is indeed the topic of the present chapter.

7.1 POTENTIAL ENERGY SURFACE

The potential energy surface is the cornerstone of all theoretical studies of reaction mechanisms [7-7]. The topography of a potential energy surface contains all possible information about a chemical reaction. However, how this potential energy surface can be depicted is another matter.

The total energy of a molecule consists of the potential energy and the kinetic energy of both the nuclei and the electrons. The coulombic energy of the nuclei and the electronic energy together represent the whole potential energy under whose influence the nuclei carry out their vibrations. Since the energies of the (ground and various excited) electronic states are different, each state has its own potential energy surface. We are usually interested in the lowest energy potential surface, which corresponds to the ground state of the molecule. An N-atomic molecule has $3N - 6$ internal degrees of freedom (a linear molecule has $3N - 5$). The potential energy for such a molecule can be represented by a $(3N - 6)$-dimensional hypersurface in a $(3N - 5)$-dimensional space. Clearly, the actual representation of this surface is impossible in our limited dimensions.

There are ways, however, to plot parts of the potential energy hypersurface. For example, the energy is plotted with respect to the change of two coordinates during a reaction in Figures 7-1a and b. Such drawings help to visualize the real potential energy surface. It is like a rough topographic map with mountains of different heights, long valleys of different depths, mountain paths, and holes. Since energy increases along the vertical coordinate, the mountains correspond to energy barriers, and the holes and valleys to different energy minima.

Studying reaction mechanisms means essentially finding the most economical way to go from one valley to another. Two adjacent valleys are connected by a mountain path: this is the road that the reactant molecules must follow if they want to reach the valley on the other side, which will correspond to the product(s). The top of the pass is called the *saddle point* or *col*. The name

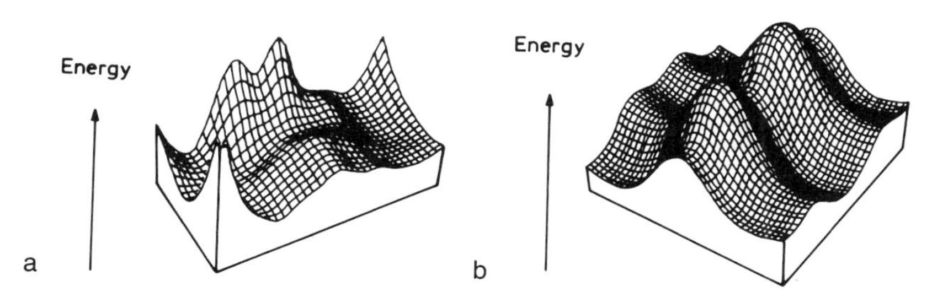

Figure 7-1. Three-dimensional potential energy surfaces. (a) Energy hypersurface for FSSF↔SSF_2 isomerization (detail). Reproduced with permission from Solouki and Bock [7-14]. Copyright (1977) American Chemical Society. (b) Rotation-inversion surface of $^{-}CH_2OH$ (detail). Reproduced with permission from Bernardi *et al.* [7-15]. Copyright (1975) American Chemical Society.

saddle point refers to the saddle on a horse. Starting from the center of the saddle, it is going up in the direction of the head as well as the tail, and it is going down in the direction of both sides. The configuration of nuclei at the saddle point is sometimes called a *transition state*, sometimes a *transition structure*, in other cases an *activated complex*, and yet in other cases a *supermolecule*. Transition state is the most commonly used term, although it is somewhat ambiguous (see Section 7.1.1).

7.1.1 Transition State, Transition Structure

The region of the potential energy surface indicating the transition state is illustrated in Figure 7-2, while a modern sculpture reminiscent of a potential energy surface at and around the saddle point is shown in Figure 7-3.

The term *transition state* is sometimes used interchangeably with the term *transition structure*, although in a strict sense the two are not identical. *Transition state* is the quasi-thermodynamic state of the reacting system as defined by Eyring [7-17]. The *transition structure*, on the other hand, is the molecular structure at the saddle point. As was shown by Houk *et al.* [7-18], when a reaction has a large activation barrier and a slowly varying entropy in the region of the potential energy maximum, the transition-state geometry and

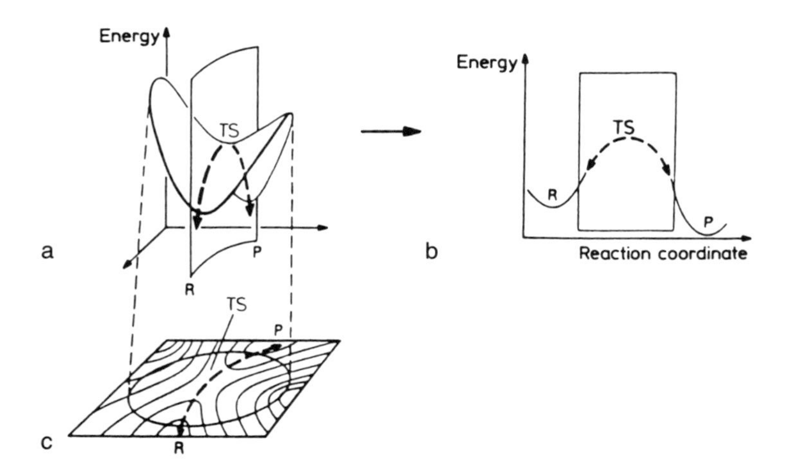

Figure 7-2. Potential energy surface by Williams [7-16] in the region of the transition structure (TS) in different representations: (a) Three-dimensional representation of the saddle-shaped potential energy surface; (b) two-dimensional potential energy curve produced by a vertical cut through the surface in (a) along the reaction path (indicated by bold dashed line) from reactants (R) to products (P); (c) energy contours produced by horizontal cuts through the potential energy surface in (a). Adapted with permission from Ref. [7-16].

Figure 7-3. Saddle-shaped sculpture in Madrid, Spain. Photograph by the authors.

the transition structure are about the same. This is illustrated in Figure 7-4a. However, when the barrier of the reaction is low and the entropy varies rapidly in the region of the potential energy minimum, the transition-state geometry differs from the transition structure (Figure 7-4b).

As Williams [7-16] stated:

> The transition state is of strategic importance within the field of chemical reactivity. Owing to its location in the region of the highest energy point on the most accessible route between reactants and products it commands both the direction and the rate of chemical change. Questions of selectivity ("Which way is it to the observed product?") and efficiency ("How easy is it to get there?") may be answered by a knowledge of the structure and properties of the transition state.

The development of transition-state theory is due to Eyring and Polanyi [7-17], while the term *transition state* was first used by Evans and Polanyi [7-19]. Since then, it has been obvious that the properties of the region between the reactants and the products need to be known in order to understand reaction mechanisms. However, the lifetime of the transition state is usually less than 10^{-12} s, and, therefore, for a long time this state could only be studied by theoretical methods. Only recently have experimental techniques become available that make the study of elementary reactions possible in real time. Direct measurements of the transition state have been carried out using

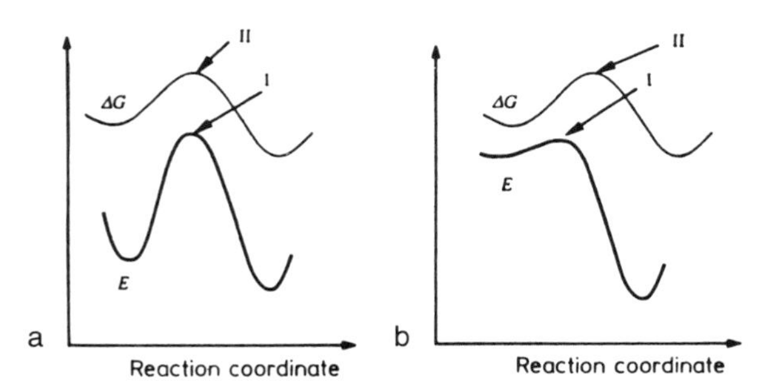

Figure 7-4. Variation of ΔG and E along a reaction path, after Houk *et al.* [7-18]. I, Transition structure; II, transition state. (a) The transition structure and the transition state coincide; (b) the transition structure and the transition state differ. Adapted with permission.

different sophisticated spectroscopic techniques (see, e.g., Refs. [7-20]–[7-22]). An example is the laser experiments that make it possible to record snapshots of chemical reactions in the femtosecond (10^{-15} s) time scale, thus providing direct real-time observations of the transition state [7-22].

At the same time, with the ever increasing capabilities of computational techniques, it has become possible to calculate the details of transition-state geometries and energetics with great precision [7-18]. Due to the increasing reliability of quantum-chemical calculations on the one hand and to the possibility of real-time experimental observation of transition-state geometries on the other, the investigation of the structure and dynamics of elementary chemical reactions has become one of the most exciting areas of modern chemical research [7-16, 7-18, 7-20].

7.1.2 Reaction Coordinate

How does symmetry come into the picture? It happens through the movement of the nuclei along the potential energy surface. As discussed in detail in Chapter 5, all possible internuclear motions of a molecule can be resolved into sets of special motions corresponding to the normal modes of the molecule. These normal modes already have a symmetry label since they belong to one of the irreducible representations of the molecular point group. The changing nuclear positions during the course of a reaction are collectively described by the term *reaction coordinate*. In simple cases, we may assume that the chemical reaction is dominated by one of the normal modes of vibration, and thus this vibrational mode is the reaction coordinate. By selecting this coordinate, we may cut a slice through the potential energy hypersurface along this particular motion. This was done by Williams [7-16] in

Figure 7-2 by cutting a slice of (a) in order to produce (b). Figure 7-2b shows the reaction path along the reaction coordinate. Points R and P are minima, corresponding to the initial (reactants) and final (product) stages of the reaction, while TS is the saddle point corresponding to the transition structure and the energy barrier.

This diagram in Figure 7-2b has several important features. First of all, it represents only a slice of the potential energy hypersurface. It is the variation along one coordinate, and it is supposed that all the other possible motions of the nuclei, that is, all the other normal vibrations, are at their optimum value, so their energy is at minimum. Therefore, this reaction path can be taken as a *minimum energy path*. All other possible nuclear motions will be orthogonal to the reaction coordinate and will not contribute to it. In other words, if we would try to leave the reaction path sideways, that is, along some other vibrational mode, the energy would invariably increase.

Figure 7-2a illustrates this point. The bold line shows the reaction path. It goes through a maximum point, which is the reaction barrier. The surface, however, rises on both sides of the reaction coordinate. Thus, with respect to the energy of the other vibrational coordinates, the reaction follows a minimum energy path indeed.

7.1.3 Symmetry Rules for the Reaction Coordinate

Symmetry rules to predict reaction mechanisms through the analysis of the reaction coordinate were first applied by Bader [7-23] (see, also, Ref. [7-24]) and were further developed by Pearson [7-6].

The energy variation along the reaction path can be characterized in the following way. The energy of all vibrational modes, except the reaction coordinate, is minimal all along the path; i.e.,

$$\frac{\partial E}{\partial Q_i} = 0 \quad \text{and} \quad \frac{\partial^2 E}{\partial Q_i^2} > 0 \tag{7-1}$$

where Q_i is any coordinate ($3N - 7$ for nonlinear molecules), except Q_r, the reaction coordinate. With respect to symmetry, these vibrations are unrestricted. (Of course, every normal mode must belong to one or another irreducible representation of the molecular point group.)

The energy variation of the reaction coordinate is different. At every point, except at the maximum and minimum values, it is nonzero:

$$\frac{\partial E}{\partial Q_r} \neq 0 \tag{7-2}$$

This is simply the slope of the curve on the potential energy diagram. At the minimum points (R and P in Figure 7-2a):

$$\frac{\partial E}{\partial Q_r} = 0 \quad \text{and} \quad \frac{\partial^2 E}{\partial Q_r^2} > 0 \tag{7-3}$$

At the saddle point (TS in Figure 7-2a):

$$\frac{\partial E}{\partial Q_r} = 0 \quad \text{and} \quad \frac{\partial^2 E}{\partial Q_r^2} < 0 \tag{7-4}$$

In order to predict reaction mechanisms and to estimate energy barriers, the energy can be expressed in terms of the reaction coordinate using second-order perturbation theory in such a way that the expression contains symmetry-dependent terms (see Refs. [7-6] and [7-23] for details).

The expression of energy contains two different types of energy integrals,

$$\left\langle \psi_0 \middle| \frac{\partial E}{\partial Q_r} \middle| \psi_0 \right\rangle \quad \text{and} \quad \left\langle \psi_0 \middle| \frac{\partial E}{\partial Q_r} \middle| \psi_i \right\rangle \tag{7-5}$$

where ψ_0 and ψ_i are the wave functions of the ground state and an excited state, respectively. In the actual calculations, these wave functions are approximated by molecular orbitals, but their relationship remains the same.

Examine now the two energy integrals separately, bearing in mind what was said about the conditions necessary for an integral to have nonzero value (Chapter 4). The first integral contains only the ground-state wave function. It appears in the first-order perturbation energy term that expresses the effect of changing the nuclear positions on the original electron distribution. This integral will have a nonzero value only if

$$\Gamma_{\psi_0} \cdot \Gamma_{\psi_0} \subset \Gamma_{Q_r} \tag{7-6}$$

that is, if the direct product of the representation of ψ_0 with itself (a function with the *same* symmetry) contains the representation of Q_r.

Concerning ψ_0 there are two possibilities: it can be degenerate or non-degenerate. If ψ_0 is degenerate, the molecule will be unstable (this is the case of the first-order Jahn–Teller effect; see Section 6.6) and it will undergo a distortion that reduces the molecular symmetry and destroys the degeneracy of ψ_0. Consider now the case when ψ_0 is nondegenerate. We know that the direct product of two nondegenerate functions with the same symmetry always belongs to the totally symmetric irreducible representation. Therefore, Q_r must also belong to the totally symmetric irreducible representation so that the integral will have a nonzero value. We can conclude that, except at a maximum or at a minimum, *the reaction coordinate belongs to the totally symmetric irreducible representation* of the molecular point group.

The reaction coordinate is just one particular normal mode in the simplest case. It must always be, however, a symmetric mode, and this is so even if more complicated nuclear motion is considered for the reaction coordinate. Such a motion can always be written as a sum of normal modes. Of these modes, however, only those which are totally symmetric will contribute to the reaction coordinate. The nonsymmetric modes may contribute only at the extremes of the potential energy function.

The second integral in Eq. (7-5) appears in the second-order perturbation energy term, and it expresses the mixing in of the first excited state into the ground state during the reaction:

$$\left\langle \psi_0 \left| \frac{\partial E}{\partial Q_r} \right| \psi_i \right\rangle \tag{7-7}$$

This integral will be nonzero only if the direct product of the representations of the wave functions ψ_0 and ψ_i contains the representation to which the reaction coordinate belongs,

$$\Gamma_{\psi_0} \cdot \Gamma_{\psi_i} \subset \Gamma_{Q_r} \tag{7-8}$$

This expression contains important information regarding the symmetry of the excited states. Only those excited states can participate in the reaction whose symmetry matches the symmetry of both the ground state and the reaction coordinate. We already know that Q_r belongs to the totally symmetric irreducible representation except at maxima and minima. This implies that only those excited states can participate in the reaction whose symmetry is the same as that of the ground state. This information is instrumental in the construction of correlation diagrams, as will be seen later.

The reaction coordinate can possess any symmetry at maxima and minima provided that the condition of Eq. (7-8) is fulfilled. This also means that at the maximum point the symmetry of the excited state may differ from that of the ground state. However, any minute distortion will remove the system from the saddle point. The reaction coordinate must then become again totally symmetric. How can this happen? Obviously, the answer is by changing the point group of the system. By reducing the symmetry, nonsymmetric vibrational modes may become symmetric, and the reaction coordinate may become totally symmetric. This reasoning may even help in predicting how the symmetry will be reduced; we just have to find the point group in which the reaction coordinate becomes totally symmetric.

Two examples will illustrate how these rules work. One involves the reduction of symmetry which occurs when a linear molecule becomes bent [7-6]. The other example involves transforming a planar molecule into a pyramidal one.

For a linear AX_2 molecule of $D_{\infty h}$ symmetry, the normal mode that

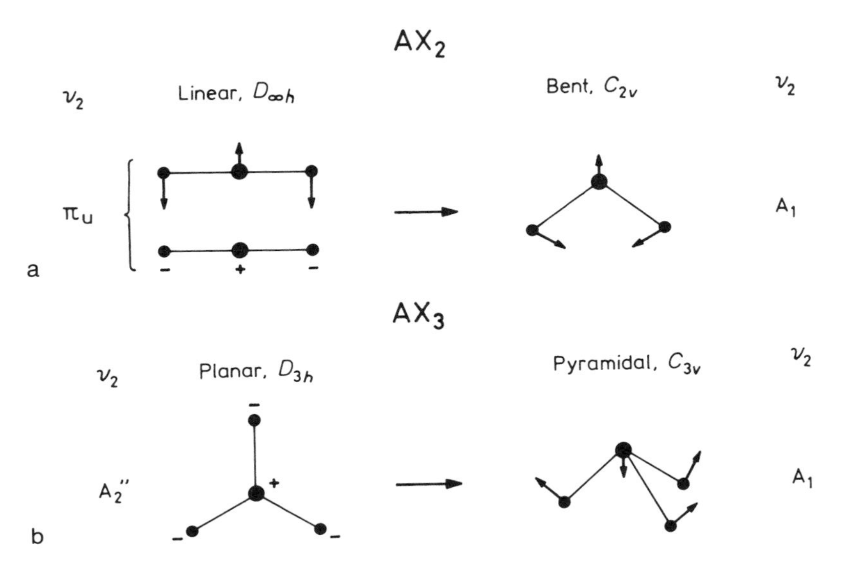

Figure 7-5. The effect of the reduction of symmetry on the reaction coordinate: (a) Bending of a linear AX_2 molecule $[\nu_2(\pi_u) \rightarrow \nu_2(A_1)]$; (b) puckering of a planar AX_3 molecule $[\nu_2(A_2'') \rightarrow \nu_2(A_1)]$

reduces the symmetry to C_{2v} is the π_u bending mode (Figure 7-5a). In the C_{2v} point group, this normal mode becomes totally symmetric. (The other component of the π_u mode becomes the rotation of the molecule.)

For an AX_3 planar molecule the symmetry is D_{3h}. The puckering mode (Figure 7-5b) of A_2'' symmetry reduces the symmetry to C_{3v}. In the C_{3v} point group, the symmetry of this vibration is A_1.

Concerning the energy integral in Eq. (7-7), Bader [7-23] called attention to an interesting phenomenon. If the excited state ψ_i lies very close to the ground state ψ_0, a distortion occurs that will push the two states apart. This phenomenon is similar to the Jahn-Teller effect and is called the *second-order Jahn–Teller effect*. The symmetry of the distortion is predicted by Eq. (7-7).

It is stressed that the physical bases for the first-order and the second-order Jahn–Teller effects are quite different. The first-order Jahn–Teller effect operates between states that are of equal energy and are degenerate, and the effect destroys degeneracy by lowering the symmetry. The second-order Jahn–Teller effect, on the other hand, appears between states that are only close in energy and are not degenerate. The effect here pushes the states further apart. The two states, ψ_0 and ψ_i, must belong to the same irreducible representation in the new point group as before and can continue to interact, which is not the case with the first-order Jahn–Teller effect.

7.2 ELECTRONIC STRUCTURE

7.2.1 Changes during a Chemical Reaction

A chemical reaction is a consequence of interactions between molecules. The electronic aspects of these interactions can be discussed in much the same way as the interactions of atomic electron distributions in forming a molecule. The difference is that while molecular orbitals (MOs) are constructed from the atomic orbitals (AOs) of the constituent atoms, in describing a chemical reaction the MOs of the product(s) are constructed from the MOs of the reactants. Before a reaction takes place (i.e., while the reacting molecules are still far apart), their electron distribution is unperturbed. When they approach each other, their orbitals begin to overlap, and distortion of the original electron distribution takes place. There are two requirements for a constructive interaction between molecules: symmetry matching and energy matching. These two factors can be treated in different ways. The approaches of Fukui [7-1, 7-2], and of Woodward and Hoffmann [7-3, 7-4] differ somewhat. Since these are the two most successful methods in this field, we shall concentrate on them. First, the basis of each method will be presented briefly, followed by a few classical examples, each of which will be treated in some detail.

7.2.2 Frontier Orbitals: HOMO and LUMO

A successful chemical reaction requires both energy and symmetry matching between the MOs of the reactants. The requirements are essentially the same as in the case of constructing MOs from AOs. Only orbitals of the same symmetry and comparable energy can overlap successfully. The strongest interactions occur between those orbitals which are close to each other in energy. However, the interaction between filled MOs is destabilizing since the energy of one orbital increases by about as much—actually a little more—as that of the other decreases (see Figure 7-6a). The most important interactions occur between the filled orbitals of one molecule and the vacant orbitals of the other. Moreover, since the interaction is strongest for energetically similar orbitals, the most significant interactions can be expected between the highest occupied molecular orbital (HOMO) of one molecule and the lowest unoccupied molecular orbital (LUMO) of the other (Figure 7-6b). The labels HOMO and LUMO were incorporated by Fukui into a descriptive collective name: *frontier orbitals*. The first article on this topic appeared in 1952 [7-25], and the idea has been extended to a host of different reactions in the succeeding years (see, e.g., Refs. [7-1] and [7-2]).

Fukui [7-1] recognized the importance of the symmetry properties of HOMOs and LUMOs perhaps for the first time in connection with the Diels–

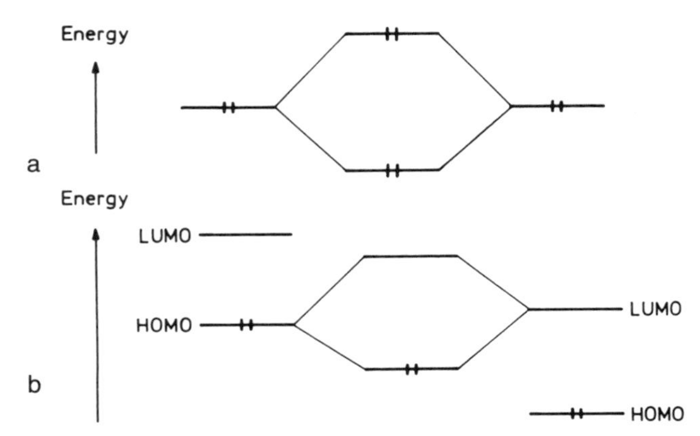

Figure 7-6. (a) Interaction of two filled orbitals. The interaction is destabilizing, and so the reaction will not occur. (b) Interaction of the highest occupied MO (HOMO) of one molecule with the lowest unoccupied MO (LUMO) of another molecule.

Alder reaction. According to his Nobel lecture [7-12], however, it was only after the appearance of the papers by Woodward and Hoffmann in 1965 that he "became fully aware that not only the density distribution but also the nodal property"—that is, symmetry—"of the particular orbitals have significance in . . . chemical reactions."

The concept of frontier orbitals simplifies the MO description of chemical reactions enormously, since only these MOs of the reactant molecules need to be considered. Several examples of this approach will be given in Section 7.3.

7.2.3 Conservation of Orbital Symmetry

The first papers by Woodward and Hoffmann outlining and utilizing the idea of conservation of orbital symmetry appeared in 1965 [7-26–7-28]. Salem [7-7] called the discovery of orbital symmetry conservation a revolution in chemistry:

> It was a major breakthrough in the field of chemical reactions in which notions preexisting in other fields (orbital correlations by Mulliken, and nodal properties of orbitals in conjugated systems, by Coulson and Longuet-Higgins) were applied with great conceptual brilliance to a far-reaching problem. Chemical reactions were suddenly adorned with novel significance.

The idea and the principles of drawing correlation diagrams follows directly from the atomic correlation diagrams of Hund [7-29] and of Mulliken

[7-30]. They are very useful for predicting the "allowedness" of a given concerted reaction. In constructing correlation diagrams, both the energy and the symmetry aspects of the problem must be considered. On one side of the diagram the approximate energy levels of the reactants are drawn, while on the other side those of the product(s) are indicated. A particular geometry of approach must be assumed. Furthermore, the symmetry properties of the molecular orbitals must be considered in the framework of the point group of the supermolecule. In contrast to the frontier orbital method, it is not necessarily the HOMOs and LUMOs that are considered. Instead, attention is focused upon those molecular orbitals which are associated with bonds that are broken or formed during the chemical reaction. We know that each acceptable molecular orbital must belong to one irreducible representation of the point group of the system. At least for nondegenerate point groups, this MO must be either symmetric or antisymmetric with respect to any symmetry element that may be present. (The character under any operation is either 1 or -1.)

Among all possible symmetry elements, those must be considered which are maintained throughout the approach and which bisect bonds that are either formed or broken during the reaction. There must always be at least one such symmetry element. The next step is to connect levels of like symmetry without violating the so-called *noncrossing rule*. According to this rule, two orbitals of the same symmetry cannot intersect [7-31]. Thus, the correlation diagram is completed. These diagrams yield valuable information about the transition state of the chemical reaction. The method will be illustrated with examples in Section 7.3.

7.2.4 Analysis in Maximum Symmetry

In the analysis in maximum symmetry approach two points are considered when predicting whether or not a chemical reaction can occur. One such point involves the allowedness of an electron transfer from one orbital to another. The other involves consideration of the reaction-decisive normal vibration. In both cases symmetry arguments are used. This approach, developed by Halevi [7-10, 7-32, 7-33], is thorough and rigorous. It is similar in part to the Bader/Pearson method and in part to the Woodward–Hoffmann method. It incorporates several features of each of these methods. First, the transformation of *both* the molecular orbitals (electronic structure) and the displacement coordinates (vibration) are examined in the context of the full symmetry group of the reacting system. All ways of breaking the symmetry of the system are explored, and no symmetry elements which are retained along the pathway are ignored. The correlation diagrams are called "correspondence diagrams" in this approach to distinguish them from the Woodward–Hoffmann diagrams.

Halevi's method to determine whether a chemical reaction is allowed or forbidden considers both the electronic and vibrational changes in the molecule. Of course, its high degree of rigor may render its application more complicated as compared with the methods which focus only upon changes in the electronic structure. The approaches introduced by Fukui and Woodward and Hoffmann, mentioned previously, seem to have received more widespread acceptance and utilization.

7.3 EXAMPLES

7.3.1 Cycloaddition

7.3.1.1 Ethylene Dimerization

The interaction of two ethylene molecules will be considered in two geometrical arrangements. The two molecules adopt a mutually parallel approach in one arrangement and a mutually perpendicular approach in the other. Applications of various methods will be considered briefly.

a. Parallel Approach, HOMO–LUMO. According to the frontier orbital method, only the HOMOs and the LUMOs of the two ethylene molecules need to be considered. A further simplification is introduced in the pictorial description of the interactions. Although the molecular orbitals of the reactants are used to construct the MOs of the products, the former are usually drawn schematically as the atomic orbitals from which they are built. The reason is that the form of the atomic orbitals is better defined and better understood than is the form of the molecular orbitals, unless one resorts to actual molecular orbital calculations.

The MOs of ethylene can be constructed according to the principles given in the preceding chapter. The HOMO of ethylene is the bonding MO, and the LUMO is the antibonding MO composed of the two p_z orbitals of carbon. These MOs are of B_{1u} and B_{2g} symmetry, respectively, in the D_{2h} point group. Figure 7-7 shows them both in a simplified way along with the corresponding contour diagrams.

Consider first the frontier orbital interactions between two ethylene molecules that approach one another in parallel planes ("face to face"). Their HOMOs and LUMOs are indicated in Figure 7-8 on the left and right, respectively. Also shown is the behavior of these orbitals with respect to the symmetry plane bisecting the two breaking π bonds. Since the HOMOs are symmetric and the LUMOs are antisymmetric with respect to this operation, there is a symmetry mismatch between the HOMO of one molecule and the LUMO of the other. The symmetry-allowed combination is between the two filled HOMOs. Since the interaction of two filled molecular orbitals of the same energy is destabilizing, the reaction will not occur thermally.

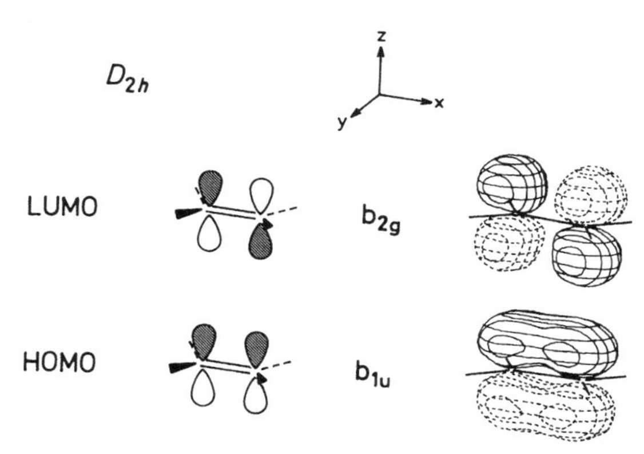

Figure 7-7. The HOMO and LUMO of ethylene. The contour diagrams are reproduced by permission from Ref. [7-34]. Copyright (1973) Academic Press.

b. Parallel Approach, Correlation Diagram. Now consider the ethylene dimerization using the Woodward–Hoffmann approach. There is again the important condition mentioned before which must be fulfilled: for the whole reacting system, at least one symmetry element must persist throughout the entire process. Let us consider the reaction in this respect. Each separated

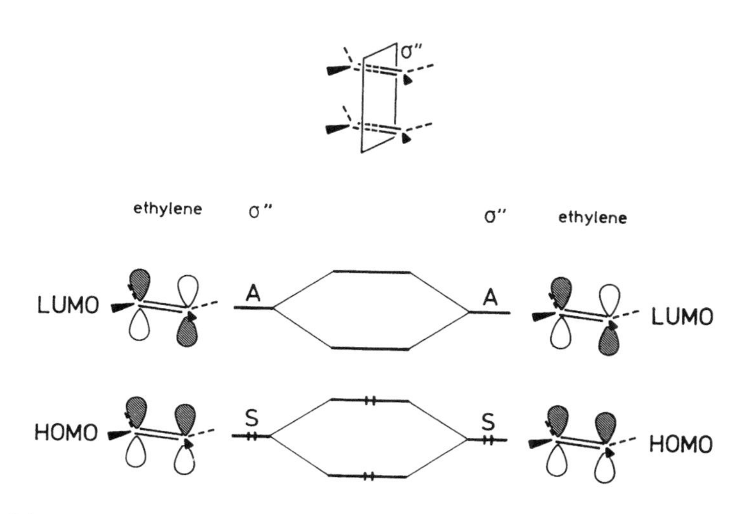

Figure 7-8. Frontier orbital interactions in the face-to-face approach of two ethylene molecules. S indicates symmetric and A indicates antisymmetric behavior with respect to the σ'' symmetry plane.

ethylene molecule has D_{2h} symmetry. When two of these molecules approach one another with their molecular planes parallel as shown in Figure 7-9, the whole system retains this symmetry. Finally, the product cyclobutane is of D_{4h} symmetry. Since D_{2h} is a subgroup of D_{4h}, the symmetry elements of D_{2h} persist.

One of the symmetry elements in the D_{2h} point group is the symmetry plane σ' (Figure 7-9). All of the MOs considered in this reaction, that is, those associated with the broken π bonds of the two ethylene molecules and the two new σ bonds of cyclobutane, lie in the plane of this symmetry element. All of them will be symmetric to reflection in this plane. There will be no change in their behavior with respect to this symmetry operation during the reaction. This brings us back to a very important point in the construction of correlation diagrams: the symmetry element chosen to follow the reaction must bisect bonds broken or made during the process. Adding extra symmetry elements, like σ' above, will not change the result. It is not wrong to include them; it is just not necessary. Considering, however, only such symmetry elements could lead to the erroneous conclusion that every reaction is symmetry allowed.

As was found to be the case when constructing MOs from AOs, the symmetry of the reacting system as a whole must be considered rather than just the symmetry of the individual molecules alone. Figure 7-10 illustrates this point with respect to one of the reflection planes. The σ plane transforms the MO drawn as the two p_z orbitals of the two carbon atoms of one ethylene molecule into the molecular orbital of the other ethylene molecule. Thus, each MO of the reacting system has a contribution from each p_z orbital.

The possible combinations of the π and π^* orbitals of the two ethylene

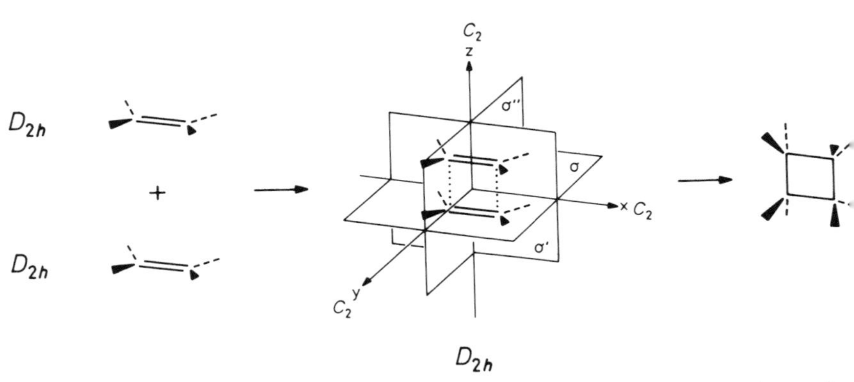

Figure 7-9. The symmetry of reactants, transition structure, and product in the face-to-face dimerization of ethylene.

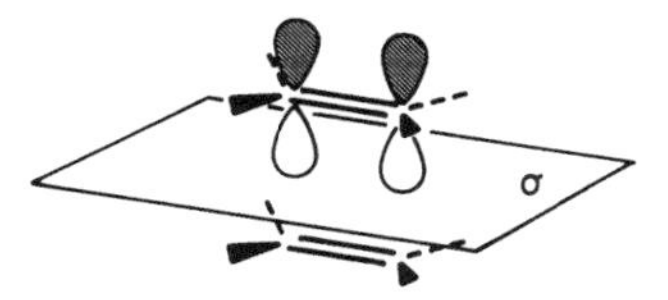

Figure 7-10. The π MO of one ethylene molecule alone does not belong to any irreducible representation of the point group of the system of two ethylene molecules.

molecules are presented on the left-hand side of Figure 7-11 in order of increasing energy. Consideration of these MOs shows that $\pi_1 + \pi_2$ and $\pi_1^* + \pi_2^*$ are in proper phase to form a bonding MO (that is, closing the ring). The right-hand side of Figure 7-11 illustrates this, together with the formation of the antibonding orbitals of cyclobutane.

The construction of the correlation diagram is shown in Figure 7-12. The two crucial symmetry elements are indicated in the upper part of the figure. The molecular orbitals of the reactants are shown in order of increasing energy at the left side of the diagram, and their behavior with respect to these symmetry elements is indicated; the corresponding product MOs are shown at the right in this same figure.

Since σ and σ'' are maintained throughout the reaction, there must be a continuous correlation of orbitals of the same symmetry type. Therefore, orbitals of like symmetry correlate with one another, and they can be connected. This, the fundamental idea of the Woodward–Hoffmann method, is shown in the central part of the diagram.

Inspection of this correlation diagram immediately reveals that there is a problem. One of the bonding orbitals at the left correlates with an antibonding orbital on the product side. Consequently, if orbital symmetry is to be conserved, two ground-state ethylene molecules cannot combine via face-to-face approach to give a ground-state cyclobutane, or vice versa. This concerted reaction is *symmetry forbidden**.

c. State Correlation. The correlation diagram in Figure 7-12 refers to molecular orbitals. The molecular orbitals and the corresponding electronic configurations are, however, only substitutes for the real wave functions which describe the actual electronic states. It is the electronic states that have definite energy and not the electronic configurations (cf. Chapter 6). Since electronic transitions occur physically between electronic states, the correlation of these

*Note that considerations of either one of the two crucial symmetry elements, σ and σ'', alone would give the same result.

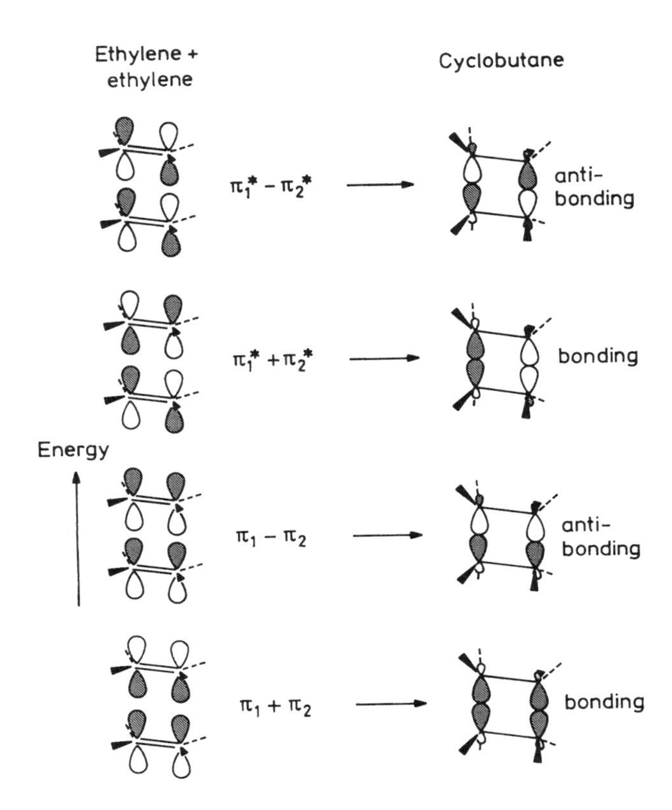

Figure 7-11. Molecular orbitals of the ethylene–ethylene system and the construction of molecular orbitals of cyclobutane. (The energy scale refers to the reactant orbitals only.)

states is of interest. It was Longuet-Higgins and Abrahamson [7-35] who drew attention to the importance of state-correlation diagrams.

The rules for the state correlation diagrams are the same as for the orbital correlation diagrams; only states that possess the same symmetry can be connected. In order to determine the symmetries of the states, first the symmetries of the MOs must be determined. These are given for the face-to-face dimerization of ethylene in Table 7-1. The D_{2h} character table (Table 7-2) shows that the two crucial symmetry elements are the symmetry planes $\sigma(xy)$ and $\sigma''(yz)$. The MOs are all symmetric with respect to the third plane, $\sigma'(xz)$ (*vide supra*). The corresponding three symmetry operations will unambiguously determine the symmetry of the MOs. Another possibility is to take the simplest subgroup of D_{2h} which already contains the two crucial symmetry operations, that is, the C_{2v} point group (cf. Ref. [7-36]). In these two ap-

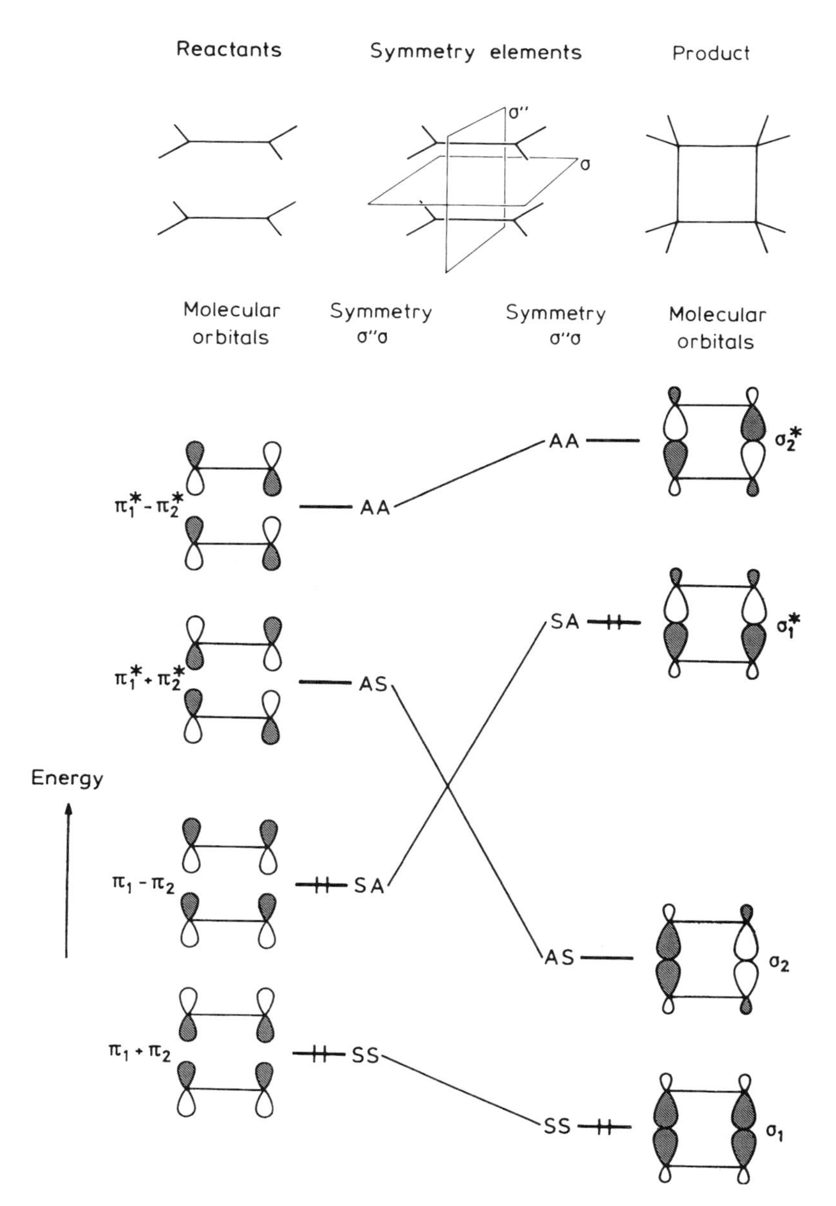

Figure 7-12. Construction of the correlation diagram for ethylene dimerization with parallel approach. S denotes symmetric and A denotes antisymmetric behavior with respect to the indicated symmetry planes. Adaptation of Figure 10.19 from Ref. [7-9]. Reprinted by permission of Thomas H. Lowry.

Table 7-1. The Symmetry of Molecular Orbitals in the Face-to-Face Dimerization of Ethylene[a]

Ethylene + ethylene					Cyclobutane				
Character under:				Orbital occupation			Character under:		
σ′ (xz)	σ (xy)	σ″ (yz)	D_{2h}			D_{2h}	σ′ (xz)	σ (xy)	σ″ (yz)
1	−1	−1	b_{2g}	——	——	b_{2g}	1	−1	−1
1	1	−1	b_{3u}	——	——	b_{1u}	1	−1	1
1	−1	1	b_{1u}	—++—	—++—	b_{3u}	1	1	−1
1	1	1	a_g	—++—	—++—	a_g	1	1	1

[a]The orientation of the coordinate axes is given in Figure 7-9.

proaches, only the designation of the orbitals and states is different; the outcome, i.e., the state correlation diagram, is the same.

In determining the symmetries of the states (see Chapter 6), we must remember that states with completely filled orbitals are always totally symmetric. In other cases, the symmetry of the state is determined by the direct product of the incompletely filled orbitals.

The ground-state configuration of the two-ethylene system is $a_g^2 b_{1u}^2$ (see Table 7-1). This state is totally symmetric, A_g. The excitation of an electron from the HOMO to the LUMO will give the electron configuration $a_g^2 b_{1u} b_{3u}$. The direct product is

$$b_{1u} \cdot b_{3u} = b_{2g}$$

This yields a state of B_{2g} symmetry. The electronic configuration of the product is $a_g^2 b_{3u}^2$, again with A_g symmetry. This electron configuration corresponds to a

Table 7-2. The D_{2h} Character Table

D_{2h}	E	$C_2(z)$	$C_2(y)$	$C_2(x)$	i	$\sigma(xy)$	$\sigma(xz)$	$\sigma(yz)$		
A_g	1	1	1	1	1	1	1	1		x^2, y^2, z^2
B_{1g}	1	1	−1	−1	1	1	−1	−1	R_z	xy
B_{2g}	1	−1	1	−1	1	−1	1	−1	R_y	xz
B_{3g}	1	−1	−1	1	1	−1	−1	1	R_x	yz
A_u	1	1	1	1	−1	−1	−1	−1		
B_{1u}	1	1	−1	−1	−1	−1	1	1	z	
B_{2u}	1	−1	1	−1	−1	1	−1	1	y	
B_{3u}	1	−1	−1	1	−1	1	1	−1	x	

doubly excited state of the reactants. Finally, the state correlation diagram can be drawn (Figure 7-13).

An obvious connection between states that possess the same electronic configuration would be the one indicated by dashed lines in Figure 7-13. This does not occur, however, because states of the same symmetry cannot cross. This is again a manifestation of the noncrossing rule, which applies to electronic states as well as to orbitals. Instead of crossing, when two states are coming too close to each other, they will turn away, and so the two ground states, both of A_g symmetry, and also two A_g symmetry excited states will each mutually correlate.

The solid line connecting the two ground states in Figure 7-13 indicates that there is a substantial energy barrier for the ground-state-to-ground-state process; this reaction is said to be "thermally forbidden."

Consider now one electron in the reactant system excited photochemically to the B_{2g} state. Since this state correlates directly with the B_{2g} state of the product, this reaction does not have any energy barrier and may occur directly. It is said that the reaction is "photochemically allowed." Indeed, it is an experimental fact that olefin dimerization occurs smoothly under irradiation.

This observation can be generalized. *If a concerted reaction is thermally forbidden, it is photochemically allowed and vice versa; if it is thermally allowed, then it is photochemically forbidden.*

Although the state correlation diagram is physically more meaningful than the orbital correlation diagram, the latter is usually used because of its simplicity. This is similar to the kind of approximation made when the electronic wave function is replaced by the products of one-electron wave functions in MO theory. The physical basis for the rule that only orbitals of the

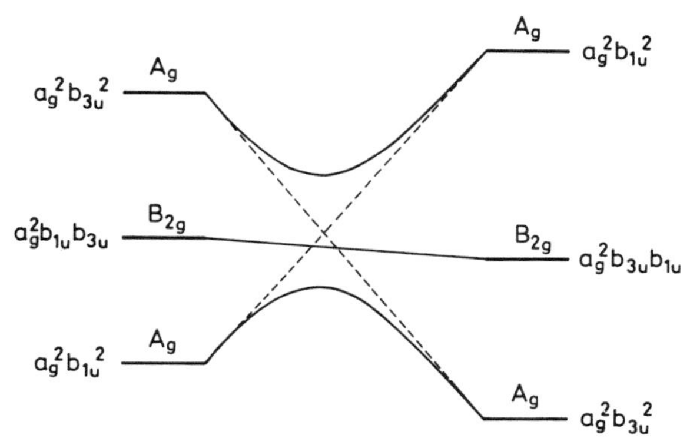

Figure 7-13. State correlation diagram for the ethylene dimerization.

same symmetry can correlate is that only in this case can constructive overlap occur. This again has its analogy in the construction of molecular orbitals. The physical basis for the noncrossing rule is electron repulsion. It is important that this rule applies to orbitals—or states—of the same symmetry only. Orbitals of different symmetry cannot interact anyway, so their correlation lines are allowed to cross.

d. Parallel Approach, Orbital Correspondence Analysis. It is worthwhile to see what additional information can be learned from orbital correspondence analysis [7-7, 7-10, 7-32]. The correspondence diagram of the ethylene dimerization reaction is drawn after Halevi [7-32] in Figure 7-14. It is essentially the same as the correlation diagram in Figure 7-12 with the following difference: Here the maximum symmetry of the system, D_{2h}, is taken into consideration, and the irreducible representation of each MO in this point group is shown. The solid lines of the diagram connect molecular orbitals of the same symmetry. This is the same as the correlation diagram derived from consideration of the crucial symmetries. In addition, we can see that the required transition toward producing a stable ground-state cyclobutane would be from an MO of b_{1u} symmetry to another MO of b_{3u} symmetry. The symmetry of the necessary vibration is given by the direct product of these MOs:

$$b_{1u} \cdot b_{3u} = b_{2g}$$

The B_{2g} symmetry motion of a rectangle of D_{2h} symmetry would be an in-plane vibration that shortens one of the diagonals and lengthens the other:

This result suggests a stepwise mechanism. The first step is the formation of a transoid tetramethylene biradical. Then, this intermediate rotates, thereby permitting closure of the cyclobutane ring in a second step. Recent high-quality *ab initio* calculations [7-37] support this mechanism. The reverse of ethylene dimerization, the pyrolysis of cyclobutane, is experimentally observed [7-38]. Both quantum-chemical calculations [7-39] and thermochemical considerations [7-40] suggest that the pyrolysis proceeds through a 1,4-biradical intermediate. This shows the value of the additional information yielded by the orbital correspondence approach.

e. Orthogonal Approach. Let us consider ethylene dimerization in yet another approach. Assume that the orientation of the two molecules is orthogonal

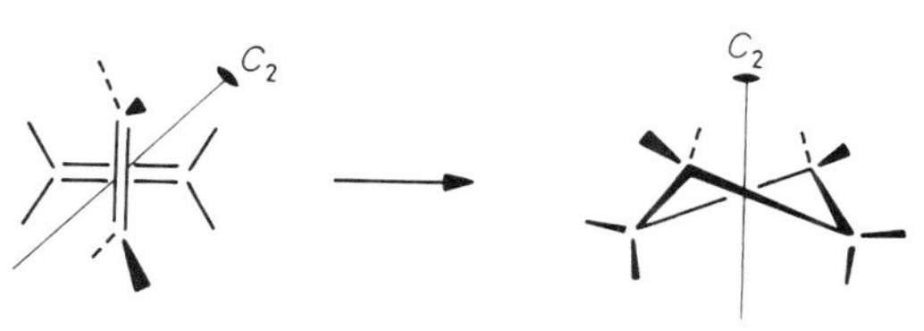

There is one symmetry element that is maintained in this arrangement, i.e., the C_2 rotation. Considering the behavior of the reactant π MOs and the product σ MOs under the C_2 operation, the correlation diagram shown in Figure 7-15 can be drawn. It shows that both bonding MOs of the reactant side correlate with a

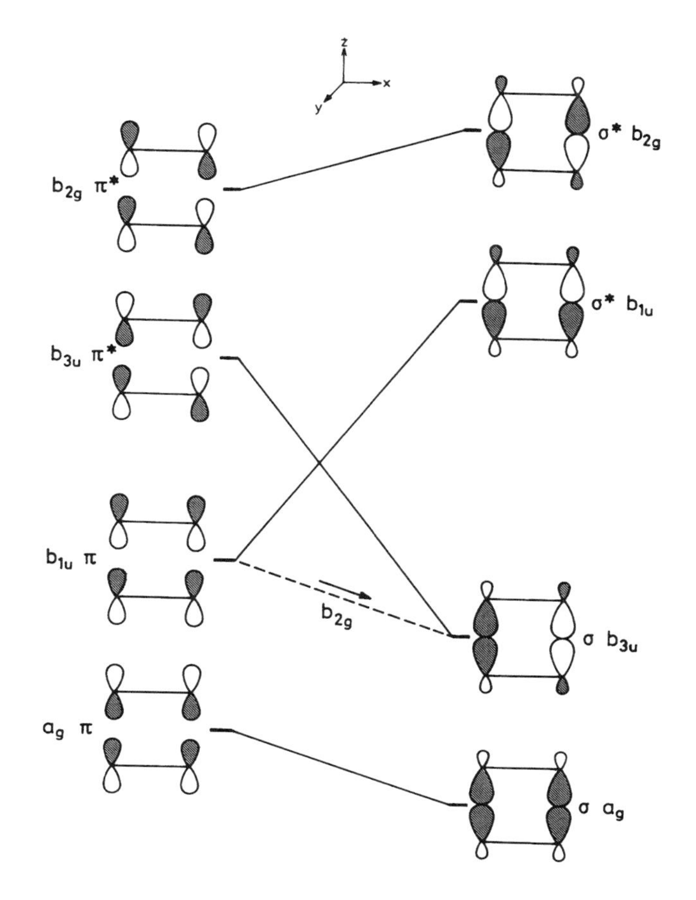

Figure 7-14. Correspondence diagram for the face-to-face dimerization of ethylene. After Ref. [7-32]; reproduced with permission.

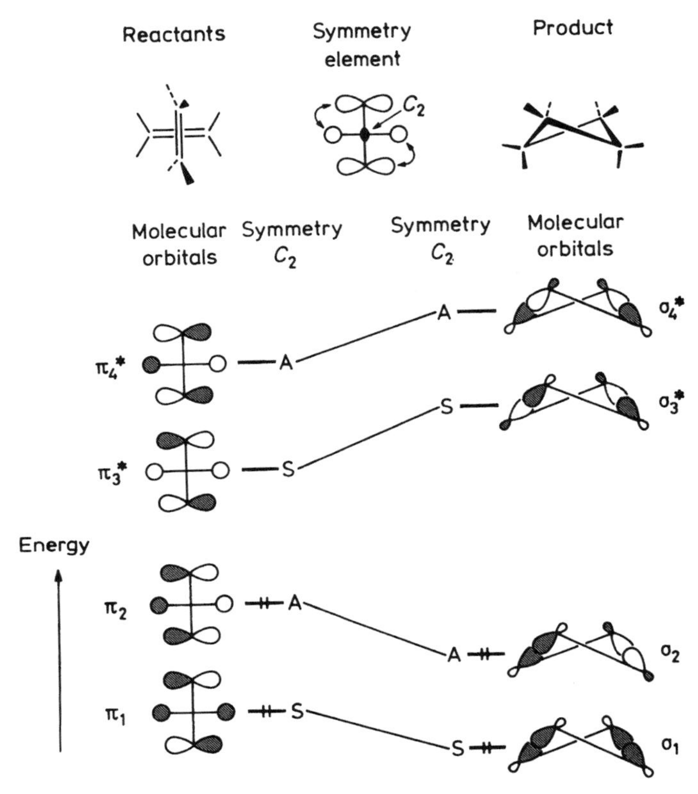

Figure 7-15. Correlation diagram for the orthogonal orientation of two ethylene molecules in the dimerization reaction. Adaptation of Figure 10.22 from Ref. [7-9]. Reprinted by permission of Thomas H. Lowry.

bonding MO on the product side. There is a net energy gain in the reaction, and the process is "thermally allowed."

One of the ethylene molecules enters the above reaction *antarafacially*; this means that the two new bonds are formed on opposite sides of this molecule:

The other ethylene molecule enters the reaction *suprafacially*; this means that the two new bonds are formed on the same side of this second molecule:

Thus, in the orthogonal approach the two molecules enter the reaction differently: one of them antarafacially and the other suprafacially. On the other hand, in the parallel approach of two ethylenes, both molecules enter the reaction suprafacially:

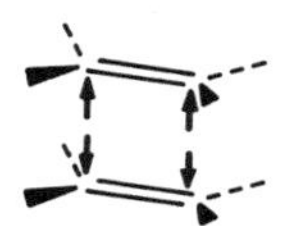

The following abbreviations are often used in the literature: $_{\pi}2_s + {}_{\pi}2_s$ means that both ethylene molecules are approaching in a suprafacial manner, while $_{\pi}2_s + {}_{\pi}2_a$ indicates that the same molecules are reacting in a process which is suprafacial for one component and antarafacial for the other. The number $_{\pi}2$ indicates that two π electrons are contributed by each ethylene molecule.

Just for the sake of completeness, it is worthwhile mentioning that, according to the orbital correspondence analysis, this $_{\pi}2_s + {}_{\pi}2_a$ cycloaddition of ethylene is also thermally forbidden [7-10, 7-32]. Recent quantum-chemical calculations [7-37] reported a transition structure for this thermally allowed concerted reaction, but due to steric repulsions between some of the hydrogens, this transition structure is very high in energy. Indeed, this reaction is not observed experimentally.

7.3.1.2 Diels–Alder Reaction

a. HOMO–LUMO Interaction. Another famous example that demonstrates the applicability of symmetry rules in determining the course of chemical reactions is the Diels–Alder reaction. It was discussed in Fukui's seminal paper [7-1] on the frontier orbital method. Figure 7-16 illustrates the HOMOs and LUMOs of ethylene (dienophile) and butadiene (diene). The only symmetry element common to both the diene and the dienophile is the reflection plane that passes through the central 2,3-bond of the diene and the double bond of the dienophile. The symmetry behavior of the MOs with respect to this symmetry element is also shown.

There are two favorable interactions here. One is between the HOMO of ethylene and the LUMO of butadiene, and the other is between the HOMO of butadiene and the LUMO of ethylene. These two interactions occur simul-

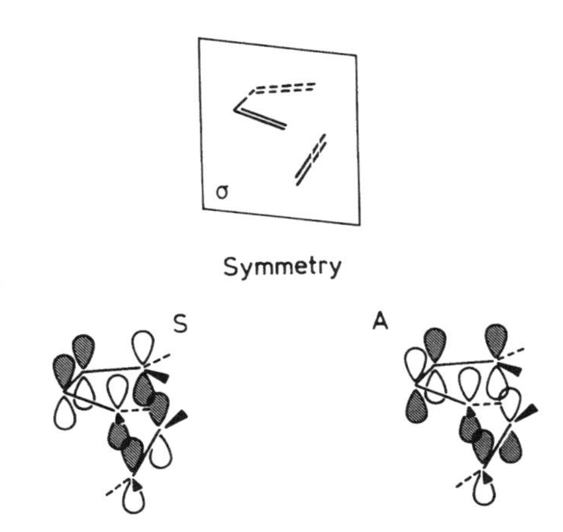

Figure 7-16. HOMO–LUMO interaction in the Diels–Alder reaction.

taneously. There is, however, a difference in the role of these two interactions because of their different symmetry behavior. The HOMO of ethylene and the LUMO of butadiene are symmetric with respect to the symmetry element that is maintained throughout the reaction. There is no nodal plane at this symmetry element, so the electrons can be delocalized over the whole new bond. Thus, both carbon atoms of ethylene are bound synchronously to both terminal atoms of butadiene.

The situation is different with the other HOMO–LUMO interaction. These orbitals are antisymmetric with respect to the symmetry element, anc the two ends of the new linkage are separated by a nodal plane. Therefore, twc separated chemical bonds will form, each connecting an ethylene carbon aton with a terminal butadiene carbon atom. From this consideration, it follows tha the first symmetric interaction is the dominant one. Also, the symmetric pai (HOMO of ethylene and LUMO of butadiene) are closer in energy and thu give a stronger interaction.

b. Orbital Correlation Diagram. The ethylene and butadiene mole cules must approach each other in the manner indicated at the top of Figur 7-17 in order to participate in a concerted reaction. There is only one persistin symmetry element in this arrangement, viz., the σ plane which bisects the 2,3 bond of the diene and the double bond of the dienophile. The orbitals affecte by the reaction are the π orbitals of the reactants which will be broken; two ne σ bonds and one new π bond are formed in the product. The π orbitals an their antibonding pairs for the reactants are shown on the left-hand side c Figure 7-17. The new σ and π orbitals, both bonding and antibonding, of th

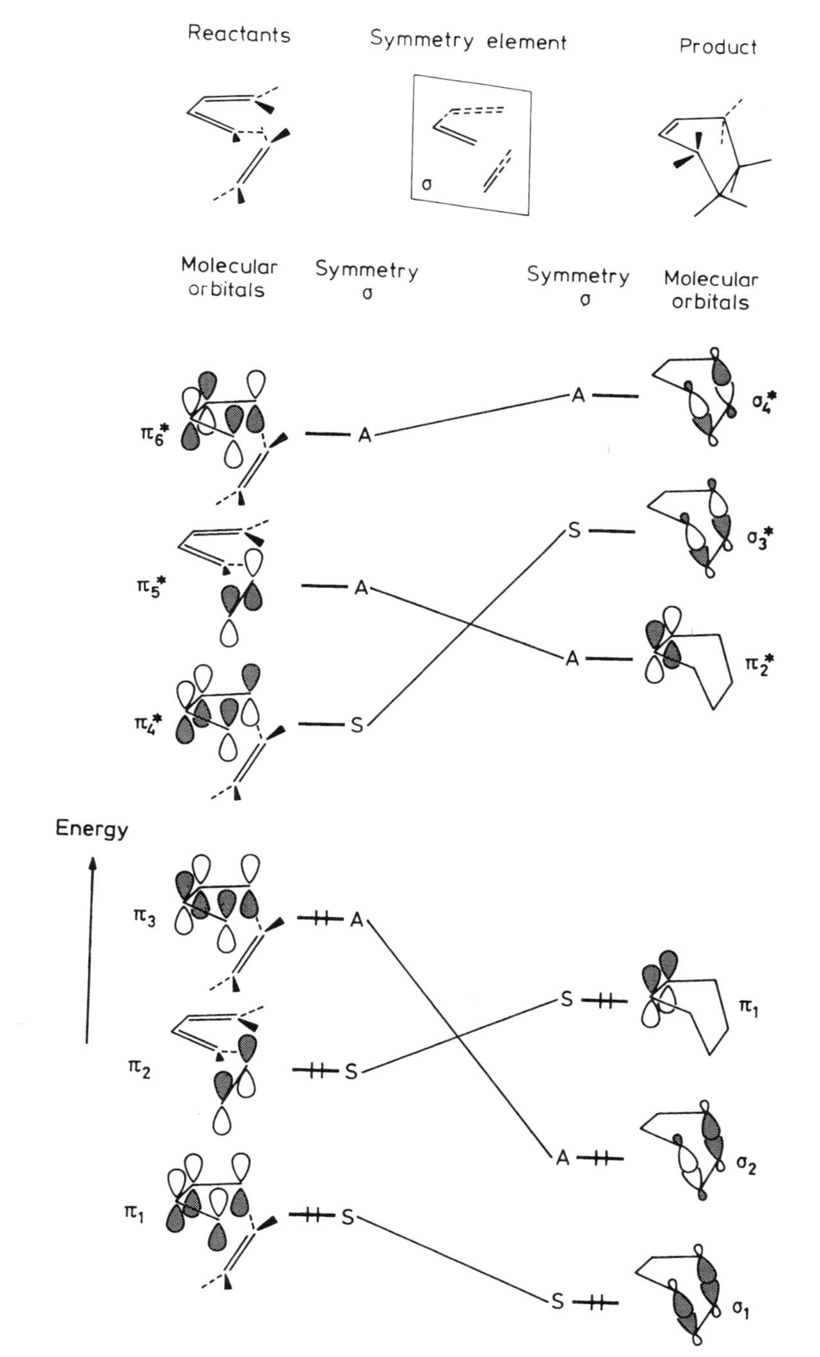

Figure 7-17. Orbital correlation diagram for the ethylene–butadiene cycloaddition. Adaptation of Figure 10.20 from Ref. [7-9]. Reprinted by permission of Thomas H. Lowry.

product cyclohexene are on the right-hand side of this figure. These are the orbitals that are affected by the reaction. The behavior of these orbitals with respect to the vertical symmetry plane is also indicated. The correlation diagram shows that all the filled bonding orbitals of the reactants correlate with filled ground-state bonding orbitals of the product. The reaction, therefore, is symmetry allowed. The predictions that arise by application of the correlation method and by application of the HOMO–LUMO treatment are identical.

The ethylene–butadiene cycloaddition is a good example to illustrate that symmetry allowedness does not necessarily mean that the reaction occurs easily. This reaction has a comparatively high activation energy, 144 kJ/mol [7-7]. A large number of quantum-chemical calculations has been devoted to this reaction with conflicting results (for recent references, see Ref. [7-18]). It seems, however, that the concerted nature of the prototype Diels–Alder reaction is well established. The reason for the relatively high activation energy is that substantial distortion must occur in the reactants before frontier orbital interactions can stabilize the product.

7.3.2 Intramolecular Cyclization

a. Orbital Correlation for the Butadiene/Cyclobutene Interconversion. The electrocyclic interconversion between an open-chain conjugated polyene and a cyclic olefin is another example for the application of the symmetry rules. The simplest case is the interconversion of butadiene and cyclobutene:

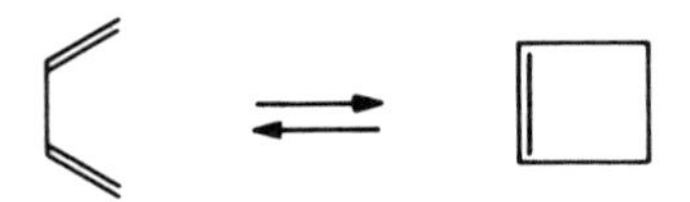

This process can occur in principle in two ways. In one, the two ends of the open chain turn in the opposite direction into the transition state. This is called a *disrotatory* reaction.

The other possibility is a *conrotatory* process, in which the two ends of the open chain turn in the same direction.

The ring opening of substituted cyclobutenes proceeds at relatively low temperatures and always in conrotatory fashion [7-3, 7-41], as illustrated by the isomerization of *cis*- and *trans*-3,4-dimethylcyclobutene [7-41]:

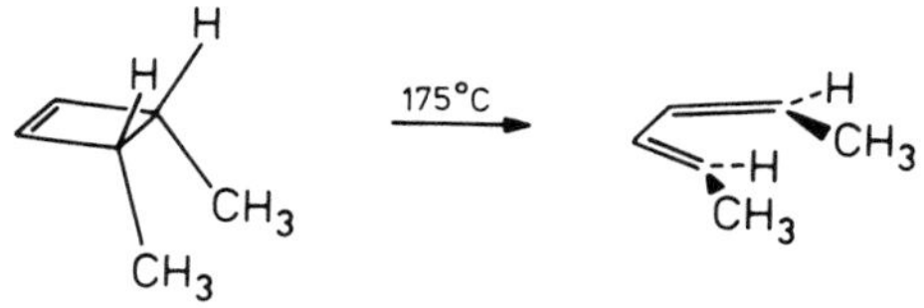

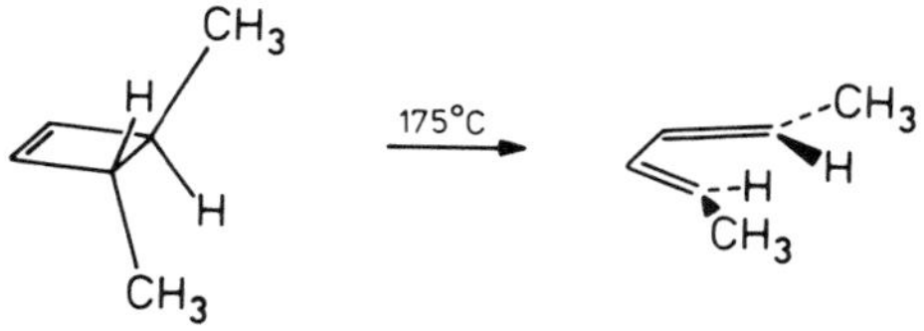

This stereospecificity is well accounted for by the correlation diagrams constructed for the unsubstituted butadiene/cyclobutene isomerization in Figures 7-18 and 7-19. Since two double bonds in butadiene are broken and a new double bond and a single bond are formed during the cyclization, two bonding and two antibonding orbitals must be considered on both sides. The persisting symmetry element is a plane of symmetry in the disrotatory process. The correlation diagram (Figure 7-18) shows a bonding electron pair moving to an antibonding level in the product, and, thus, the right-hand side corresponds to an excited-state configuration. The disrotatory ring opening is thus a thermally forbidden process.

Figure 7-19 shows the same reaction with conrotatory ring closure. Here, the symmetry element maintained throughout the reaction is the C_2 rotation axis. After connecting orbitals of like symmetry, it is seen that all ground-state reactant orbitals correlate with ground-state product orbitals, so the process is thermally allowed.

b. Symmetry of the Reaction Coordinate—Cyclobutene Ring Opening. It is interesting to consider the butadiene–cyclobutene reaction from a somewhat different viewpoint, viz., to determine whether the symmetry of the reaction coordinate does indeed predict the proper reaction. Let us look at the reaction from the opposite direction, i.e., the cyclobutene ring opening process. From the symmetry point of view, this change of direction is irrelevant.

The symmetry group of both cyclobutene and butadiene is C_{2v}, but the transition state is of C_2 symmetry in the conrotatory and C_s in the disrotatory mode. Pearson [7-6] suggested that this reaction might be visualized in the

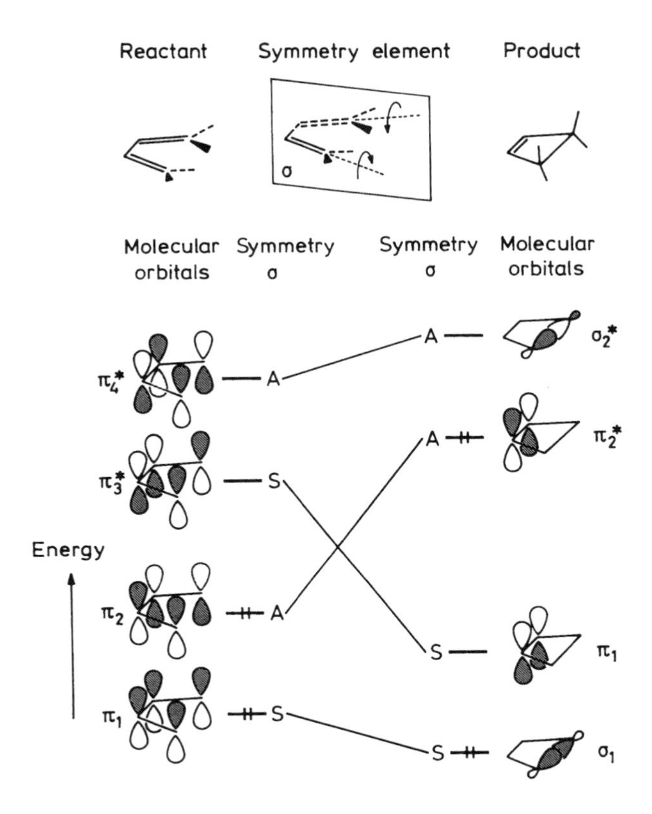

Figure 7-18. Correlation diagram for the disrotatory closure of butadiene. Adaptation of Figure 10.14 from Ref. [7-9]. Reprinted by permission of Thomas H. Lowry.

following way. In the cyclobutene–butadiene transition, two bonds of cyclobutene are destroyed, to wit the ring-closing σ bond and the opposite π bond. Hence, four orbitals are involved in the change, the filled and empty σ and σ* orbitals and the filled and empty π and π* orbitals. These orbitals are indicated in Figure 7-20. Their symmetry is also given for the three point groups involved.

Figure 7-21 demonstrates the nuclear movements involved in the conrotatory and disrotatory ring openings. These movements define the reaction coordinate, and they belong to the A_2 and the B_1 representation of the C_{2v} point group, respectively.

The two bonds of cyclobutene can be broken either by removing electrons from a bonding orbital or by putting electrons into an antibonding orbital.

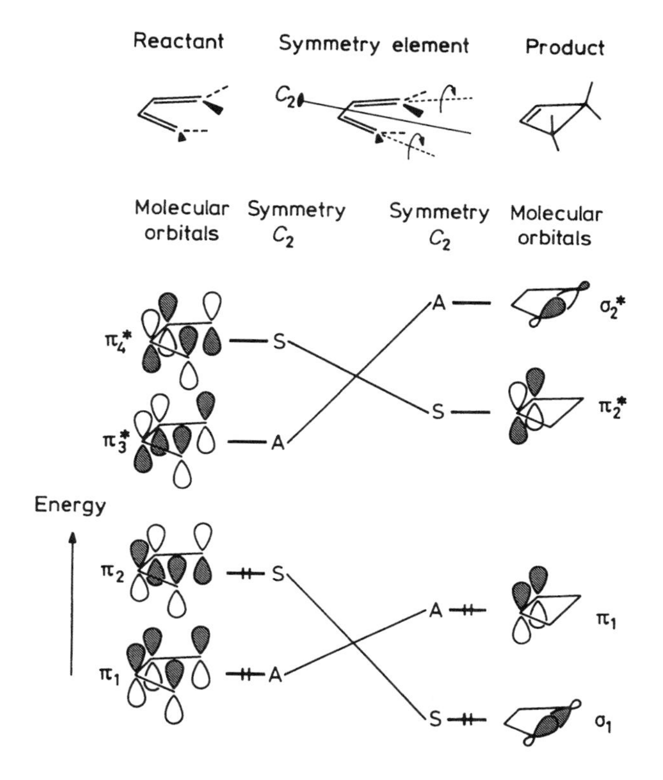

Figure 7-19. Correlation diagram for the conrotatory ring closure in the butadiene–cyclobutene isomerization. Adaptation of Figure 10.12 from Ref. [7-9]. Reprinted by permission of Thomas H. Lowry.

Consider the $\sigma \rightarrow \pi^*$ and the $\pi \rightarrow \sigma^*$ transitions. According to Pearson [7-6], the direct product of the two representations must contain the reaction coordinate:

$$\sigma \rightarrow \pi^*: \ a_1 \cdot a_2 = a_2$$
$$\pi \rightarrow \sigma^*: \ b_1 \cdot b_2 = a_2$$

A_2 is the irreducible representation of the conrotatory ring opening motion, so this type of ring opening seems possible. We can test the rules further. During a conrotatory process, the symmetry of the system decreases to C_2. The symmetry of the relevant orbitals also changes in this point group (see Figure

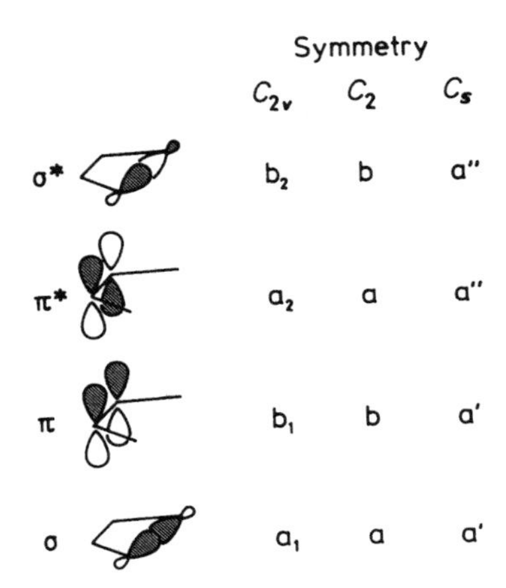

Figure 7-20. The molecular orbitals participating in the cyclobutene ring opening.

7-20). Both a_1 and a_2 become a, and both b_1 and b_2 become b. Therefore, thes orbitals are able to mix. Also, the symmetry of the reaction coordinat becomes A. This is consistent with the rule saying that the reaction coordinate except at maxima and minima, must belong to the totally symmetric represen tation of the point group.

The next step is to test whether the disrotatory ring opening is possible The $\sigma \rightarrow \pi^*$ and $\pi \rightarrow \sigma^*$ transitions obviously cannot be used here, since the correspond to the conrotatory ring opening of A_2 symmetry. Let us consider th $\sigma \rightarrow \sigma^*$ and the $\pi \rightarrow \pi^*$ transitions:

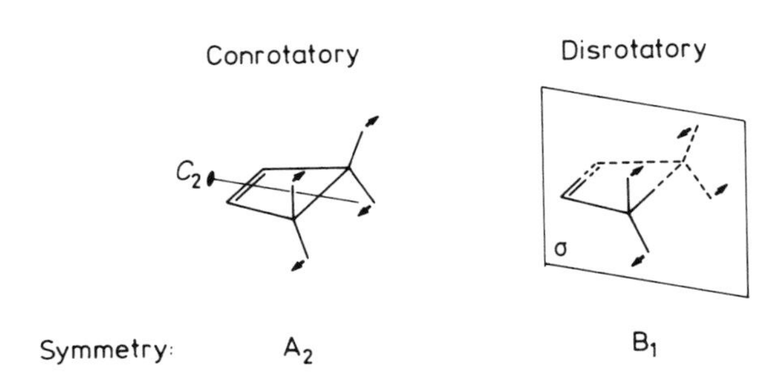

Figure 7-21. The symmetry of the reaction coordinate in the conrotatory and disrotatory r opening of cyclobutene.

$$\sigma \rightarrow \sigma^*: a_1 \cdot b_2 = b_2$$
$$\pi \rightarrow \pi^*: b_1 \cdot a_2 = b_2$$

Both direct products contain the B_2 irreducible representation. It corresponds to an in-phase asymmetric distortion of the molecule, which cannot lead to ring opening. The symmetry of the disrotatory reaction coordinate is B_1 (Figure 7-21). Moreover, if we consider the symmetry of the orbitals in the C_s symmetry point group of the disrotatory transition, it appears that σ and σ^* as well as π and π^* belong to different irreducible representations. Hence, their mixing would not be possible anyway. The prediction from this method is the same as the prediction from the orbital correlation diagrams. While examination of the reaction coordinate gives more insight into what is actually happening during a chemical reaction, it is somewhat more complicated than using orbital correlation diagrams.

An *ab initio* calculation for the cyclobutene-to-butadiene ring opening [7-42] led to the following observation. In the conrotatory process, first the C–C single bond lengthens followed by twisting of the methylene groups. The C–C bond lengthening is a symmetric stretching mode of A_1 symmetry, and the methylene twist is an A_2 symmetry process which was earlier supposed to be the reaction coordinate. This apparent controversy was resolved by Pearson [7-6], who emphasized the special role of the totally symmetric reaction coordinate. The effect of the C–C stretch is shown in Figure 7-22. The energies of the σ and σ^* orbitals increase and decrease, respectively, as a consequence of the bond lengthening. The A_1 symmetry vibration mode does not change the molecular symmetry. The crucial $\sigma \rightarrow \pi^*$ and $\pi \rightarrow \sigma^*$ transitions, which are symmetry related to the A_2 twisting mode, occur more easily. Apparently, the large energy difference between these orbitals is the determining factor in the actual process. The transition structure for this electrocyclic reaction has been

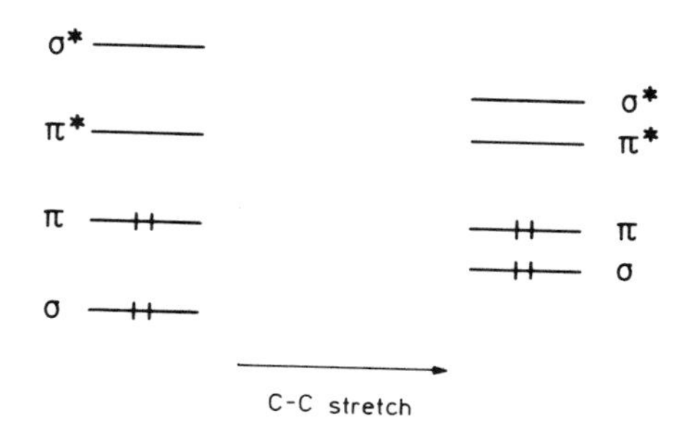

Figure 7-22. The effect of C–C stretching on the energies of the critical orbitals in the cyclobutene ring opening.

studied extensively in recent years by means of high-level quantum-chemical calculations [7-18]. Their results fully support the conclusions of the above reasoning.

7.3.3 Generalized Woodward–Hoffmann Rules

The selection rules for chemical reactions derived by using symmetry arguments show a definite pattern. Woodward and Hoffmann generalized the selection rules [7-3] on the basis of orbital symmetry considerations applied to a large number of systems. Two important observations are summarized here, and the reader may refer to the literature [7-3, 7-9] for further details.

a. Cycloaddition. The reaction between two molecules is thermally allowed if the total number of electrons in the system is $4n + 2$ (n is an integer), and both components are either suprafacial or antarafacial. If one component is suprafacial and the other is antarafacial, the reaction will be thermally allowed if the total number of electrons is $4n$.

b. Electrocyclic Reactions. The rules are similar to those given above. A disrotatory process is thermally allowed if the total number of electrons is $4n + 2$, and a conrotatory process is allowed thermally if the number of delocalized electrons is $4n$. For a photochemical reaction, both sets of rules are reversed.

7.4 HÜCKEL–MÖBIUS CONCEPT

There are a number of other methods used to predict and interpret chemical reactions without relying upon symmetry arguments. It is worthwhile to compare at least some of them with symmetry-based approaches.

The so-called "aromaticity rules" are chosen for comparison, as they provide a beautiful correspondence with the symmetry-based Woodward–Hoffmann rules. A detailed analysis [7-43] showed the equivalence of the generalized Woodward–Hoffmann selection rules and the aromaticity-based selection rules for pericyclic reactions. Zimmermann [7-44] and Dewar [7-45] have made especially important contributions in this field.

The word "aromaticity" usually implies that a given molecule is stable, compared to the corresponding open-chain hydrocarbon. For a detailed account on aromaticity, see, e.g., Ref. [7-46]. The aromaticity rules are based on the Hückel–Möbius concept. A cyclic polyene is called a Hückel system if its constituent p orbitals overlap everywhere in phase, i.e., the p orbitals all have the same sign above and below the nodal plane (Figure 7-23). According to

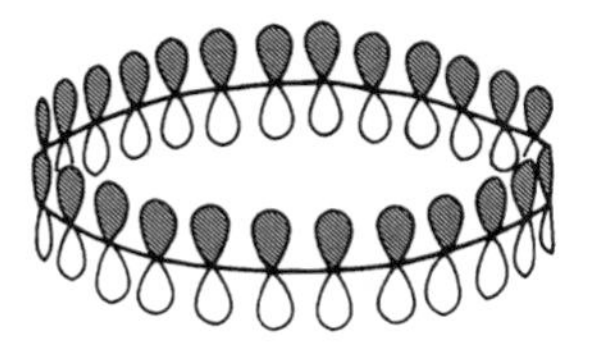

Figure 7-23. Illustration of a Hückel ring. Reproduced by permission from Ref. [7-49].

Hückel's rule [7-47], if such a system has $4n + 2$ electrons, the molecule will be aromatic and stable. On the other hand, a Hückel ring with $4n$ electrons will be antiaromatic.

If the Hückel ring is twisted once, as shown in Figure 7-24a, the situation is reversed [7-48]. Therefore, Dewar [7-45] referred to this twisted ring as an "anti-Hückel system." It is also called a "Möbius system" [7-44, 7-49], an appropriate name indeed. A Möbius strip is a continuous, one-sided surface which is formed by twisting the strip by 180° around its own axis and then joining its two ends. There is a phase inversion at the point where the two ends meet, as seen in Figures 7-24a and b. Figure 7-24c and d depict yet other Möbius strips.

According to Zimmermann [7-44] and Dewar [7-45], the allowedness of a concerted pericyclic reaction can be predicted in the following way. A cyclic array of orbitals belongs to the Hückel system if it has zero or an even number of phase inversions. For such a system, a transition state with $4n + 2$ electrons will be thermally allowed due to aromaticity, while the transition state with $4n$ electrons will be thermally forbidden due to antiaromaticity.

A cyclic array of orbitals is a Möbius system if it has an odd number of phase inversions. For a Möbius system, a transition state with $4n$ electrons will be aromatic and thermally allowed, while that with $4n + 2$ electrons will be antiaromatic and thermally forbidden. For a concerted photochemical reaction, the rules are exactly the opposite to those for the corresponding thermal process.

Each of these rules has its counterpart among the Woodward–Hoffmann selection rules. There was a marked difference between the suprafacial and antarafacial arrangements in the application of the Woodward–Hoffmann treatment of cycloadditions. The disrotatory and conrotatory processes in electrocyclic reactions presented similar differences. The suprafacial arrangement in both of the reacting molecules in the cycloaddition as well as the disrotatory ring closure in Figure 7-25 correspond to the Hückel system. On the other hand, the suprafacial–antarafacial arrangement as well as the conrotatory cyclization have a phase inversion (Figure 7-26), and they can be

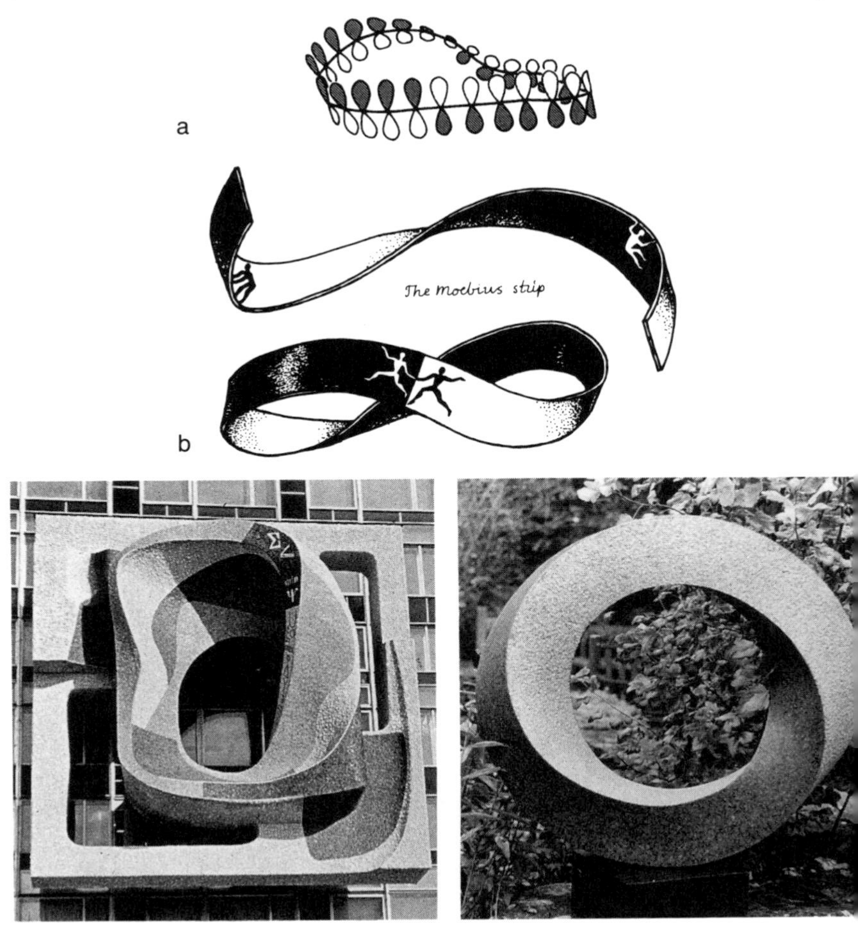

Figure 7-24. (a) Illustration of a Möbius ring. Reproduced by permission from Ref. [7-4 (b) Möbius strip. Drawing by György Doczi, Seattle, Washington. (c) Möbius strip on the faca of a Moscow scientific institute. Photograph by the authors. (d) Möbius strip sculptu *Dependent Beings* by John Robinson. Photograph courtesy of Professor Alan L. Mackay, Londo

regarded as Möbius systems. All the selection rules mentioned above a summarized in Table 7-3; their mutual correspondence is evident.

Both the Woodward–Hoffmann approach and the Hückel–Möbius co cept are useful for predicting the course of concerted reactions. They both ha their limitations as well. The application of the Hückel–Möbius concept probably preferable for systems with low symmetry. On the other hand, th

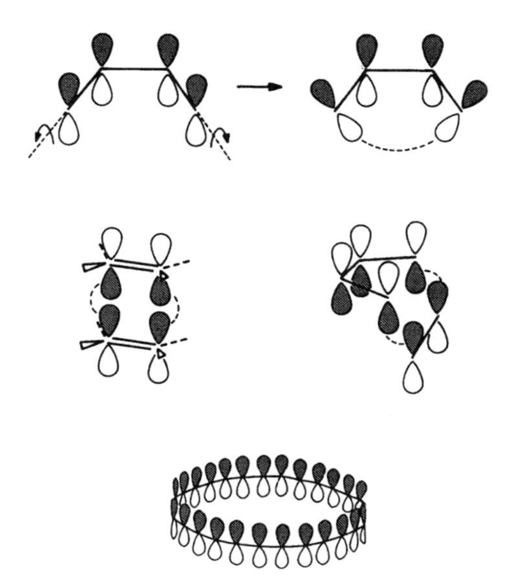

Figure 7-25. Comparison of the disrotatory ring closure and the ${}_{\pi}2_s + {}_{\pi}2_s$ reaction with the Hückel ring.

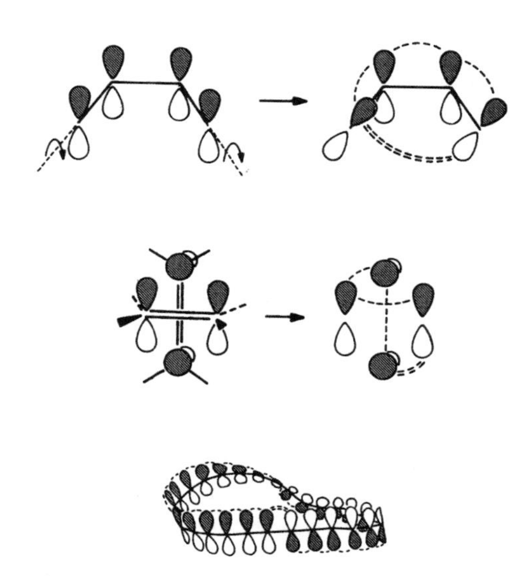

Figure 7-26. Comparison of the conrotatory ring closure and the ${}_{\pi}2_s + {}_{\pi}2_a$ reaction with the Möbius ring.

Table 7-3. Selection Rules for Chemical Reactions from Different Approaches

Approach[a]	Reaction	Thermally allowed	Thermally forbidden
1	s + s a + a	$4n + 2$	$4n$
	s + a	$4n$	$4n + 2$
2	Disrotatory	$4n + 2$	$4n$
	Conrotatory	$4n$	$4n + 2$
3	Hückel system: sign inversion even or 0	$4n + 2$	$4n$
	Möbius system: sign inversion odd	$4n$	$4n + 2$

[a]1, Woodward–Hoffmann cycloaddition; 2, Woodward–Hoffmann electrocyclic reaction; 3 Hückel–Möbius concept.

concept can only be applied when there is a cyclic array of orbitals. The conservation of orbital symmetry approach does not have this limitation.

7.5 ISOLOBAL ANALOGY

So far, our discussion of reactions has been restricted to organic molecules. However, all the main ideas are applicable to inorganic systems as well. Thus, for example, the formation of inorganic donor–acceptor complexes may be conveniently described by the HOMO–LUMO concept. A case in point is the formation of the aluminum trichloride–ammonia complex (cf. Figure 3-22). This complex can be considered to result from interaction between the LUMO of the acceptor ($AlCl_3$) and the HOMO of the donor (NH_3).

The potential of a unified treatment of organic and inorganic systems has been expressed eloquently in Roald Hoffmann's Nobel lecture [7-11], entitled "Building Bridges between Inorganic and Organic Chemistry."

The main idea is to examine the similarities between the structures of relatively complicated inorganic complexes and relatively simple and well understood organic molecules. Then the structure and possible reactions of the former can be understood and even predicted by applying the considerations that work so well for the latter. Two important points were stressed by Hoffmann:

1. "It is the resemblance of the frontier orbitals of inorganic and organic moieties that will provide the bridge that we seek between the subfields of our science."

2. Many aspects of the electronic structure of the molecules discussed and compared are heavily simplified, but "the time now, here, is for building conceptual frameworks, and so similarity and unity take temporary precedence over difference and diversity."

One of the fastest growing areas of inorganic chemistry is transition metal organometallic chemistry. In a general way, the structure of transition metal organometallic complexes can be thought of as containing a transition metal–ligand fragment, such as $M(CO)_5$, $M(PF_3)_5$, M(allyl), and MCp (Cp = cyclopentadienyl) or, in general, ML_n. All these fragments may be derived from an octahedral arrangement:

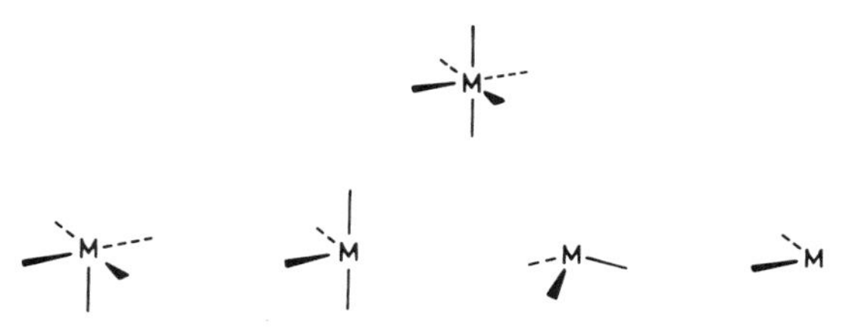

In describing the bonding in these fragments, first the six octahedral hybrid orbitals on the metal atom are constructed. Hybridization is not discussed here, but symmetry considerations are used in constructing hybrid orbitals just as in constructing molecular orbitals [7-36]. In an octahedral complex, the six hybrid orbitals point toward the ligands, and together they can be used as a basis for a representation of the point group. The O_h character table and the representation of the six hybrid orbitals are given in Table 7-4. The representation reduces to

$$\Gamma_h = A_{1g} + E_g + T_{1u}$$

Inspection of the O_h character table shows that the only possible combination from the available nd, $(n + 1)s$, and $(n + 1)p$ orbitals of the metal is:

$$\begin{array}{ccc} s, & p_x, p_y, p_z, & d_{x^2-y^2}, d_{z^2} \\ a_{1g} & t_{1u} & e_g \end{array}$$

These six orbitals will participate in the hybrid, and the remaining t_{2g} symmetry orbitals (d_{xz}, d_{yz}, and d_{xy}) of the metal will be nonbonding.

Six ligands approach the six hybrid orbitals of the metal in forming an octahedral complex. These ligands are supposed to be donors, or, in other words, Lewis bases with even numbers of electrons. Six bonding σ orbitals and six antibonding σ* orbitals are formed, with the ligand electron pairs occupy-

Table 7-4. The O_h Character Table and the Representation of the Hybrid Orbitals of the Transition Metal in an ML_6 Complex

O_h	E	$8C_3$	$6C_2$	$6C_4$	$3C_2$ $(=C_4^2)$	i	$6S_4$	$8S_6$	$3\sigma_h$	$6\sigma_d$		
A_{1g}	1	1	1	1	1	1	1	1	1	1		$x^2 + y^2 + z^2$
A_{2g}	1	1	−1	−1	1	1	−1	1	1	−1		
E_g	2	−1	0	0	2	2	0	−1	2	0		$(2z^2 - x^2 - y^2, x^2 - y^2)$
T_{1g}	3	0	−1	1	−1	3	1	0	−1	−1	(R_x, R_y, R_z)	
T_{2g}	3	0	1	−1	−1	3	−1	0	−1	1		(xz, yz, xy)
A_{1u}	1	1	1	1	1	−1	−1	−1	−1	−1		
A_{2u}	1	1	−1	−1	1	−1	1	−1	−1	1		
E_u	2	−1	0	0	2	−2	0	1	−2	0		
T_{1u}	3	0	−1	1	−1	−3	−1	0	1	1	(x, y, z)	
T_{2u}	3	0	1	−1	−1	−3	1	0	1	−1		
Γ_h	6	0	0	2	2	0	0	0	4	2		

ing the bonding orbitals as seen in Figure 7-27. As a consequence of the strong interaction, all six hybrid orbitals of the metal are removed from the frontier orbital region, and only the unchanged metal t_{2g} orbitals remain there.

We can also deduce the changes that will occur in the five-, four-, and three-ligand fragments as compared to the ideal six-ligand case with the help of Figure 7-27. The situation is illustrated in Figure 7-28. With five ligands, only five of the six metal hybrid orbitals will interact; the sixth orbital, the one pointing toward where no ligand is, will be unchanged. Consequently, this orbital will remain in the frontier orbital region, together with the t_{2g} orbitals. With four ligands, two of the six hybrid orbitals remain unchanged, and with three ligands, three. Always, those metal hybrid orbitals which point toward the missing ligands in the octahedral site remain unchanged.

Now we shall seek analogies between transition metal complexes and simple, well-studied organic molecules or fragments. In principle, any hydrocarbon can be constructed from methyl groups (CH_3), methylenes (CH_2), methynes (CH), and quaternary carbon atoms. They can be imagined as being derived from the methane molecule itself, which has a tetrahedral structure:

The essence of the "isolobal analogy" concept is to establish similarities between these simple organic fragments and the transition metal ligand

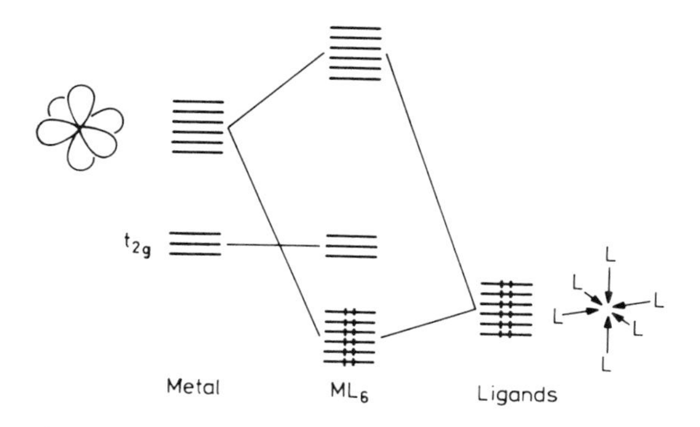

Figure 7-27. Molecular orbital construction in an ideal octahedral complex formation. Reproduced by permission from Hoffmann [7-11]. © The Nobel Foundation 1982.

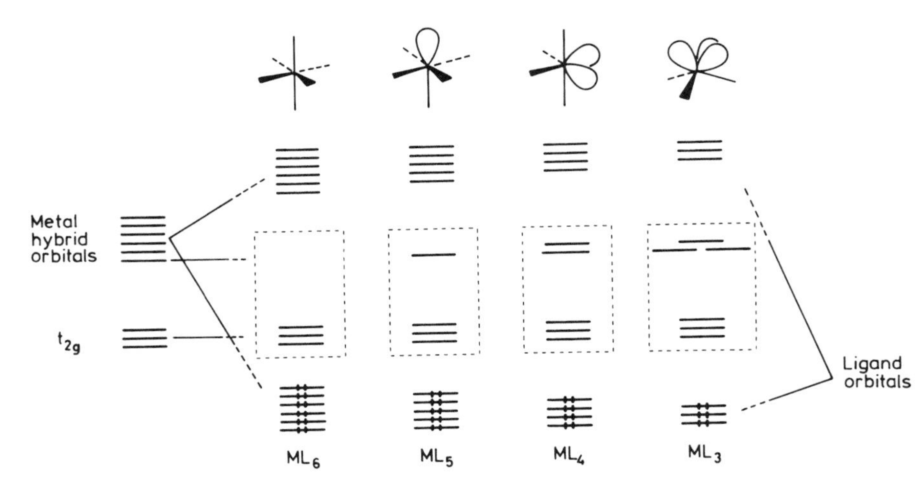

Figure 7-28. Molecular orbitals in different ML_n transition metal–ligand fragments. Adapted with permission from Ref. [7-11]. © The Nobel Foundation 1982.

fragments and then to build up the organometallic compounds. As defined by Hoffmann, "two fragments are called *isolobal*, if the number, symmetry properties, approximate energy and shape of the frontier orbitals and the number of electrons in them are similar—not identical, but similar" [7-11]. However, the molecules involved are not and need not be either isoelectronic or isostructural.

The first analogy considered here is a d^7-metal–ligand fragment, for example, $Mn(CO)_5$, and the methyl radical, CH_3:

a1 a1

t2g

d^7-ML_5 CH_3

Though the two fragments belong to different point groups, C_{4v} and C_{3v}, respectively, the orbitals that contain the unpaired electron belong to the totally symmetric representation in both cases. Since the three occupied t_{2g} orbitals of the ML_5 fragment are comparatively low-lying, the frontier orbital pictures of the two fragments should be similar. If this is so, then they are expected to show some similarity in their chemical behavior, notably in reactions. Indeed, both of them dimerize, and even the organic and inorganic fragments can codimerize, giving $(CO)_5MnCH_3$:

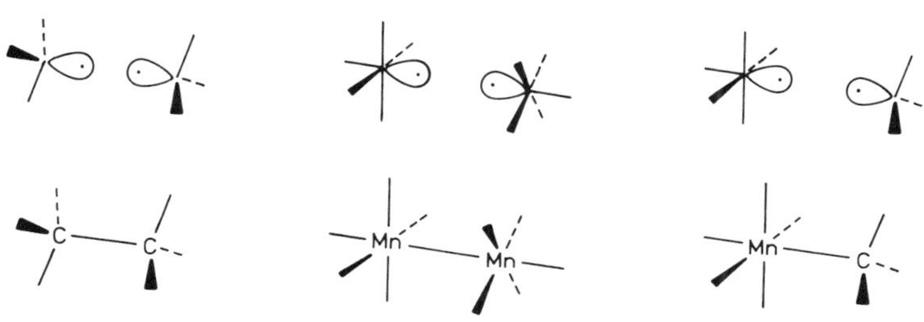

Following this analogy, the four-ligand d^8-ML_4 fragments [e.g., $Fe(CO)_4$] are expected to be comparable with the methylene radical, CH_2:

a_1 | b_2 | t_{2g} | d^8-ML_4

b_2 | a_1 | CH_2

Both fragments belong to the C_{2v} point group, and the representation of the two hybrid orbitals with the unpaired electrons is:

C_{2v}	E	C_2	σ	σ'
Γ	2	0	0	2

This reduces to $a_1 + b_2$. Although the energy ordering differs in the two fragments, this is not important, since both will participate in bonding when they interact with another ligand, and their original ordering will thereby change anyway.

Consider the possible dimerization process: two methylene radicals give ethylene, which is a known reaction. Similarly, the mixed product, $(CO)_4FeCH_2$, or at least its derivatives, can be prepared. The $Fe_2(CO)_8$ dimer, however, is unstable and has only been observed in a matrix [7-50]. This illustrates that the

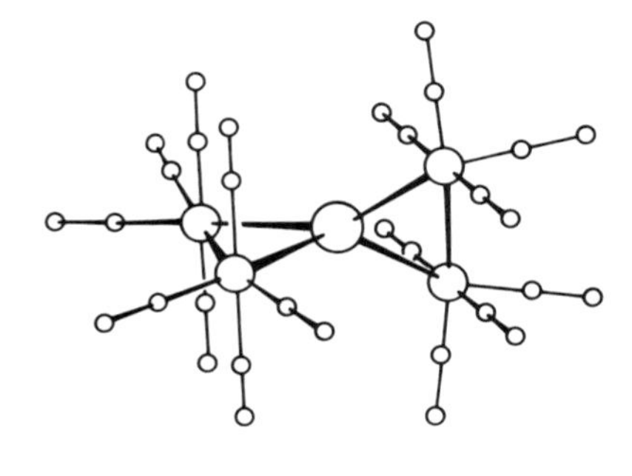

Figure 7-29. (a) The molecular geometry of $Sn[Fe_2(CO)_8]_2$. Reproduced by permission from Hoffmann [7-11]. © The Nobel Foundation 1982. (b) The organic analog, spiropentane.

isolobal analogy suggests only the possible consequences of similarity in the electronic structure of two fragments. It says nothing, however, about the thermodynamic and kinetic stability of any of the possible reaction products.

Although $Fe_2(CO)_8$ is unstable, it can be stabilized by complexation. The molecule in Figure 7-29a consists of two $Fe_2(CO)_8$ units connected through a tin atom [7-51]. Using the inorganic/organic analogy, this molecule can be compared to spiropentane (Figure 7-29b).

An example of a d^9-ML_3 fragment is $Co(CO)_3$. This is isolobal to a methyne radical, CH:

a_1 e

e a_1

t_{2g}

d^9-ML_3 CH

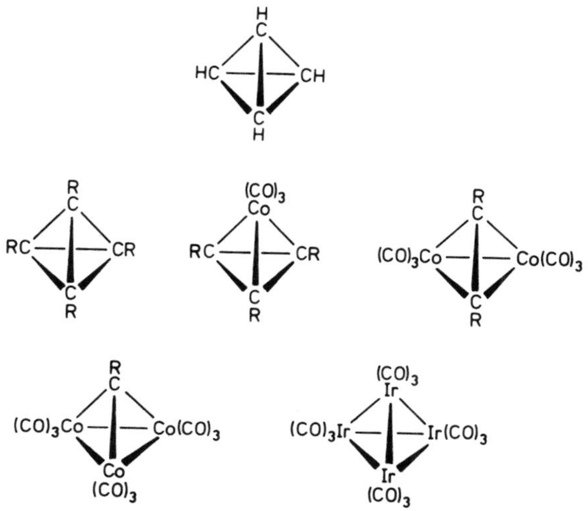

Figure 7-30. Molecular geometries from tetrahedrane to its inorganic analog. Adapted with permission from Ref. [7-11]. © The Nobel Foundation 1982.

Both have C_{3v} symmetry. The representation of the three hybrid orbitals with an unpaired electron is:

C_{3v}	E	$2C_2$	$3\sigma_v$
Γ	3	0	1

It reduces to $a_1 + e$. Again, the ordering is different, but the similarity between their electronic structures is obvious. A series of molecules and their similarities are illustrated in Figure 7-30. The first molecule is tetrahedrane, and the last one is a cluster with metal–metal bonds that can be considered as being the inorganic analog of tetrahedrane.

Only a few examples have been given to illustrate the isolobal analogy.

Table 7-5. Isolobal Analogies

Organic fragment	Transition metal coordination number				
	9	8	7	6	5
CH_3	d^1–ML_8	d^3–ML_7	d^5–ML_6	d^7–ML_5	d^9–ML_4
CH_2	d^2–ML_7	d^4–ML_6	d^6–ML_5	d^8–ML_4	d^{10}–ML_3
CH	d^3–ML_6	d^5–ML_5	d^7–ML_4	d^9–ML_3	

Hoffmann and his co-workers have extended this concept to other metal–ligand fragment compositions with various *d* orbital participations. Some of these analogies are summarized in Table 7-5. Hoffmann's Nobel lecture [7-11] contained several of them, and many more can be found in the references given therein and in later works.

REFERENCES

[7-1] K. Fukui, in *Molecular Orbitals in Chemistry, Physics and Biology* (P. O. Löwdin and B. Pullmann, eds.), Academic Press, New York (1964).

[7-2] K. Fukui, *Theory of Orientation and Stereoselection*, Springer-Verlag, Berlin (1975); K. Fukui, *Top. Curr. Chem.* **15**, 1 (1970).

[7-3] R. B. Woodward and R. Hoffmann, *The Conservation of Orbital Symmetry*, Verlag Chemie, Weinheim (1970).

[7-4] R. B. Woodward and R. Hoffmann, *Angew. Chem. Int. Ed. Engl.* **8**, 781 (1969).

[7-5] H. E. Simmons and J. F. Bunnett (eds.), *Orbital Symmetry Papers*, American Chemical Society, Washington, D.C. (1974).

[7-6] R. G. Pearson, *Symmetry Rules for Chemical Reactions: Orbital Topology and Elementary Processes*, Wiley-Interscience, New York (1976); R. G. Pearson, in *Special Issues on Symmetry*, Computers & Mathematics with Applications **12B**, 229 (1986).

[7-7] L. Salem, *Electrons in Chemical Reactions: First Principles*, Wiley-Interscience, New York (1982).

[7-8] A. P. Marchand and R. E. Lehr (eds.), *Pericyclic Reactions*, Vols. 1 and 2, Academic Press, New York (1977).

[7-9] T. H. Lowry and K. S. Richardson, *Mechanism and Theory in Organic Chemistry*, 3rd ed., Harper & Row, New York (1987).

[7-10] E. A. Halevi, *Orbital Symmetry and Reaction Mechanisms: The OCAMS View*, Springer-Verlag, Berlin (1992).

[7-11] R. Hoffmann, *Angew. Chem. Int. Ed. Engl.* **21**, 711 (1982).

[7-12] K. Fukui, *Science* **218**, 747 (1982).

[7-13] E. Wigner and E. E. Witmer, *Z. Phys.* **51**, 859 (1928).

[7-14] B. Solouki and H. Bock, *Inorg. Chem.* **16**, 665 (1977).

[7-15] F. Bernardi, I. G. Csizmadia, A. Mangini, H. B. Schlegel, M.-H. Whangbo, and S. Wolfe, *J. Am. Chem. Soc.* **97**, 2209 (1975).

[7-16] I. H. Williams, *Chem. Soc. Rev.* **1993**, 277.

[7-17] H. Eyring and M. Polanyi, *Z. Phys. Chem. B* **12**, 279 (1931); H. Eyring, *J. Chem. Phys.* **3**, 107 (1935); H. Eyring, *Chem. Rev.* **17**, 65 (1935).

[7-18] K. N. Houk, Y. Li, and J. D. Evanseck, *Angew. Chem. Int. Ed. Engl.* **31**, 682 (1992).

[7-19] M. G. Evans and M. Polanyi, *Trans. Faraday Soc.* **31**, 875 (1935).

[7-20] *Structure and Dynamics of Reactive Transition States*, *Faraday Discuss. Chem. Soc.* **91** (1991).

[7-21] E. R. Lovejoy, S. K. Kim, and C. B. Moore, *Science* **256**, 1541 (1992).

[7-22] A. H. Zewail, *Science* **242**, 1645 (1988).

[7-23] R. F. W. Bader, *Can. J. Chem.* **40**, 1164 (1962).

[7-24] R. F. W. Bader, P. L. A. Popelier, and T. A. Keith, *Angew. Chem. Int. Ed. Engl.* **33**, 620 (1994).

[7-25] K. Fukui, T. Yonezawa, and H. Shingu, *J. Chem. Phys.* **20**, 722 (1952).

[7-26] R. B. Woodward and R. Hoffmann, *J. Am. Chem. Soc.* **87**, 395 (1965).
[7-27] R. Hoffmann and R. B. Woodward, *J. Am. Chem. Soc.* **87**, 2046 (1965).
[7-28] R. B. Woodward and R. Hoffmann, *J. Am. Chem. Soc.* **87**, 2511 (1965).
[7-29] F. Hund, *Z. Phys.* **40**, 742 (1927); **42**, 93 (1927); **51**, 759 (1928).
[7-30] R. S. Mulliken, *Phys. Rev.* **32**, 186 (1928).
[7-31] J. von Neumann and E. Wigner, *Phys. Z.* **30**, 467 (1929); E. Teller, *J. Phys. Chem.* **41**, 109 (1937).
[7-32] F. A. Halevi, *Helv. Chim. Acta* **58**, 2136 (1975).
[7-33] J. Katriel and E. A. Halevi, *Theor. Chim. Acta* **40**, 1 (1975).
[7-34] W. L. Jorgensen and L. Salem, *The Organic Chemist's Book of Orbitals*, Academic Press, New York (1973).
[7-35] H. C. Longuet-Higgins and E. W. Abrahamson, *J. Am. Chem. Soc.* **87**, 2045 (1965).
[7-36] F. A. Cotton, *Chemical Applications of Group Theory*, 3rd ed., Wiley-Interscience, New York (1990).
[7-37] F. Bernardi, M. Olivucci, J. J. W. McDouall, and M. A. Robb, *J. Am. Chem. Soc.* **109**, 544 (1987); F. Bernardi, A. Bottoni, M. A. Robb, H. B. Schlegel, and G. Tonachini, *J. Am. Chem. Soc.* **107**, 2260 (1985).
[7-38] R. W. Carr, Jr., and W. D. Walters, *J. Phys. Chem.* **67**, 1370 (1963).
[7-39] R. Hoffmann, S. Swaminathan, B. G. Odell, and R. Gleiter, *J. Am. Chem. Soc.* **92**, 7091 (1970).
[7-40] H. E. O'Neal and S. W. Benson, *J. Phys. Chem.* **72**, 1866 (1968).
[7-41] R. E. K. Winter, *Tetrahedron Lett.* **1965**, 1207.
[7-42] K. Hsu, R. J. Buenker, and S. D. Peyerimhoff, *J. Am. Chem. Soc.* **93**, 2117 (1971).
[7-43] A. C. Day, *J. Am. Chem. Soc.* **97**, 2431 (1975).
[7-44] H. E. Zimmermann, *J. Am. Chem. Soc.* **88**, 1564, 1566 (1966); *Acc. Chem. Res.* **4**, 272 (1971).
[7-45] M. J. S. Dewar, *Tetrahedron Suppl.* **8**, 75 (1966); M. J. S. Dewar, *Angew. Chem. Int. Ed. Engl.* **10**, 761 (1971); M. J. S. Dewar, *The Molecular Orbital Theory of Organic Chemistry*, McGraw-Hill, New York (1969).
[7-46] V. I. Minkin, M. N. Glukhovtsev, and B. Ya. Simkin, *Aromaticity and Antiaromaticity: Electronic and Structural Aspects*, John Wiley & Sons, New York (1994).
[7-47] E. Hückel, *Z. Phys.* **70**, 204 (1931); **76**, 628 (1932); **83**, 632 (1933).
[7-48] E. Heilbronner, *Tetrahedron Lett.* **1964**, 1923.
[7-49] K.-w. Shen, *J. Chem. Educ.* **50**, 238 (1973).
[7-50] M. Poliakoff and J. J. Turner, *J. Chem. Soc. A* **1971**, 2403.
[7-51] J. D. Cotton, S. A. R. Knox, I. Paul, and F. G. A. Stone, *J. Chem. Soc. A* **1967**, 264.

Chapter 8

Space-Group Symmetries

8.1 EXPANDING TO INFINITY

Up to this point, structures of mostly finite objects have been discussed. Thus, point groups were applicable. A simplified compilation of various symmetries was presented in Figure 2-41 and Table 2-2. The point-group symmetries are characterized by the lack of periodicity in any direction. Periodicity may be introduced by translational symmetry. If periodicity is present, space groups are applicable for the symmetry description. There is a slight inconsistency here in the terminology. Even a three-dimensional object may have point-group symmetry. On the other hand, the so-called dimensionality of the space group is not determined by the dimensionality of the object. Rather, it is determined by its periodicity. The following groups are space-group symmetries, where the superscript refers to the dimensionality of the object and the subscript to the periodicity.

$$\begin{array}{lll} G_1^1 & & \\ G_1^2 & G_2^2 & \\ G_1^3 & G_2^3 & G_3^3 \end{array}$$

Objects or patterns which are periodic in one, two, and three directions will have one-, two-, and three-dimensional space groups, respectively. The dimensionality of the object/pattern is merely a necessary but not a satisfactory condition for the "dimensionality" of their space groups. We shall first describe a planar pattern, after Budden [8-1], in order to get the flavor of space-group symmetry. Also, some new symmetry elements will be introduced. Later in this chapter, the simplest one-dimensional and two-dimensional space groups will be presented. A separate chapter, the next one, will be devoted to the obviously most important three-dimensional space groups which characterize crystal structures.

A pattern expanding to infinity always contains a basic unit, a motif, which is then repeated infinitely throughout the pattern. Figure 8-1a presents a planar decoration. The pattern shown is only part of the whole as the latter expands to infinity! The pattern is obviously highly symmetrical. In Figure 8-1b, the system of mutually perpendicular symmetry planes are indicated by solid lines. Some of the fourfold and twofold rotation axes are also indicated in

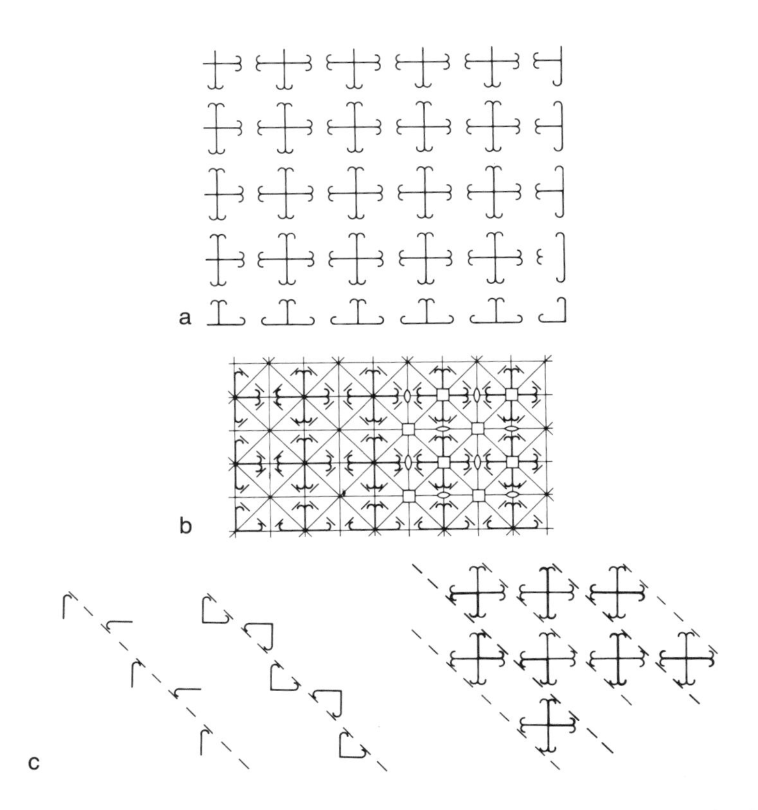

Figure 8-1. (a) Part of a planar decoration with two-dimensional space group, after Budden [8-1]. (b) Symmetry elements of the pattern shown in (a). (c) Some of the glide reflection planes and their effects in the pattern shown in (a).

this figure. A new symmetry element in our discussion is the glide reflection, which is shown by a dashed line. Some of these glide reflections are indicated separately in Figure 8-1c. A glide-reflection plane is a combination of translation and reflection. It is a symmetry element that can be present in space groups only. The glide-reflection plane involves an infinite sequence of consecutive translations and reflections. Whereas in a simple canon, there is only repetition of the tune at certain intervals in time, as shown in Figure 8-2a, Figure 8-2b shows a canon in which the repetition is combined with reflection. Two further patterns with glide-mirror symmetry are given in Figure 8-3. They are also thought to extend to infinity, at least in our imagination.

Simple *translation* is the most obvious symmetry element of the space groups. It brings the pattern into congruence with itself over and over again. The shortest displacement through which this translation brings the pattern into coincidence with itself is the elementary translation or elementary period. Sometimes it is also called the identity period. The presence of translation is seen well in the pattern in Figure 8-1. The symmetry analysis of the whole pattern was called by Budden [8-1] the analytical approach. The reverse procedure is the synthetic approach, in which the infinite and often complicated pattern is built up from the basic motif. Thus, the pattern of Figure 8-1a may be built up from a single crochet. There are several ways to proceed. Thus, for example, the crochet may be subjected to simple translation, then reflec-

Figure 8-2. Canons illustrating repetition (a) and repetition combined with reflection (glide-mirror symmetry) (b).

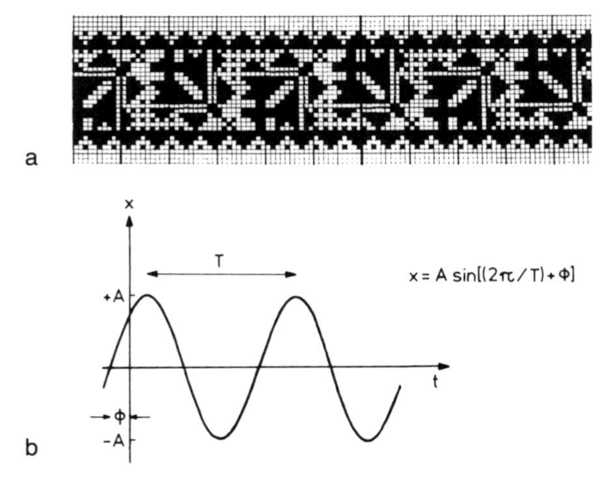

Figure 8-3. Illustrations of glide mirrors: (a) Pillow edge from Buzsák, Hungary; (b) function describing simple harmonic motion; reflection occurs following translation along the t axis by half a period, $T/2$.

tion, and then transverse reflection; these steps are illustrated in Figure 8-4a. The horizontal array obtained in this way is a one-dimensional pattern. It can be extended to a two-dimensional pattern by simple translation as in Figure 8-4a or by glide reflection as is shown in Figure 8-4b. Eventually, the complete two-dimensional pattern of Figure 8-1 can be reconstructed. In this synthetic approach, instead of the single crochet, any other motif combined from it could be selected for the start. If the crosslike motif were chosen, which contains eight of these crochets, then only translations in two directions would be needed to build up the final pattern. To learn most about the structure of a pattern, it is advantageous to select the smallest possible motif for the start.

The one-dimensional space groups are the simplest of the space groups. They have periodicity only in one direction. They may refer to one-dimensional, two-dimensional, or three-dimensional objects, cf. G_1^1, G_1^2, and G_1^3 of Table 2-2, respectively. The "infinite" carbon chains of the carbide molecules

$$\ldots =\!C\!=\!C\!=\!C\!=\!C\!=\!C\!= \ldots$$
$$\ldots -C\equiv C-C\equiv C-C\equiv \ldots$$

present one-dimensional patterns. The elementary translation or identity period is the length of the carbon–carbon double bond in the uniformly bonded

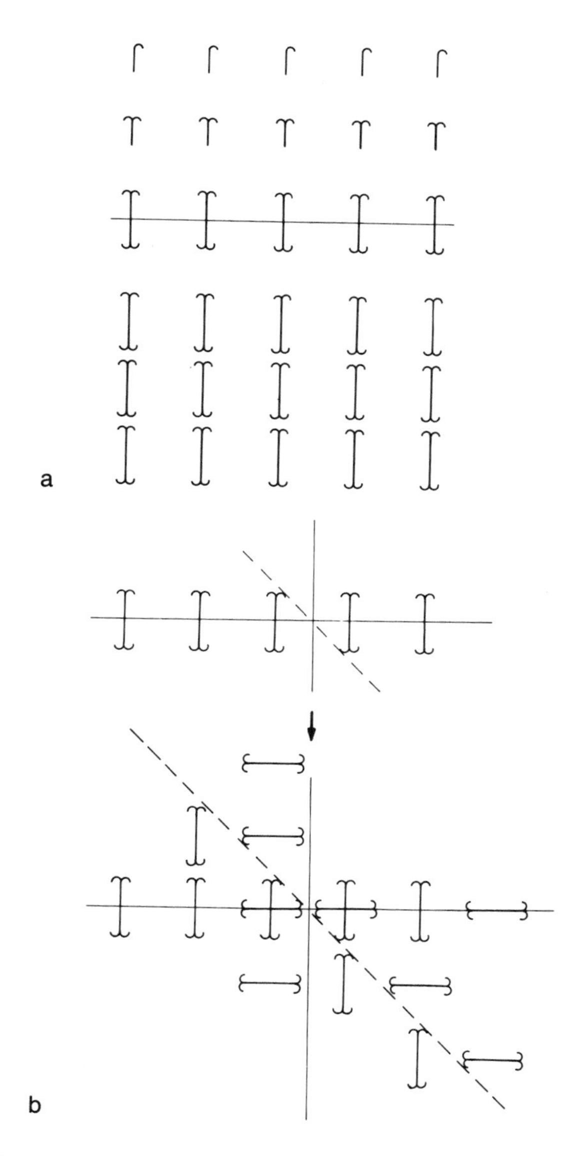

Figure 8-4. Pattern generation: (a) Starting with a single crochet, then applying horizontal translation/reflection/transverse reflection/vertical translation; (b) applying glide reflection.

chain while it is the sum of the lengths of the two different bonds in the chain consisting of alternating bonds. As the chain molecule extends along the axis of the carbon–carbon bonds, this axis can be called the translation axis. The carbon–carbon axis is a singular axis, and it is not polar as the two directions along the chain are equivalent. Earlier, we have seen the binary array . . . ABAB . . . in a crystal. The unequal spacings between the atom A and the two adjacent atoms B produced a polar axis (cf. Section 2.6 on polarity).

8.2 ONE-SIDED BANDS

Figure 8-5 presents two band decorations; one of them has a nonpolar axis while the other has a polar axis. An important feature of these patterns is that they have a polar singular plane, which is the plane of the drawing. This plane is left unchanged during the translation. Such two-dimensional patterns with periodicity in one direction are called one-sided bands [8-2].

There are altogether seven symmetry classes of one-sided bands. They are illustrated in Figure 8-6 for a suitable motif, a black triangle. A brief characterization of the seven classes is given here, together with their notation:

1. (a). The only symmetry element is the translation axis. The translation period is the distance between two identical points of the consecutive black triangles.
2. $(a)\cdot\tilde{a}$. Here the symmetry element is a glide-reflection plane ($\tilde{a}$). The black triangle comes into coincidence with itself after translation through half of the translation period ($a/2$) and reflection in the plane perpendicular to the plane of the drawing.

Figure 8-5. Byzantine mosaics from Ravenna, Italy, with one-dimensional space-group symmetry and polar (a) and nonpolar axes (b). Photographs by the authors.

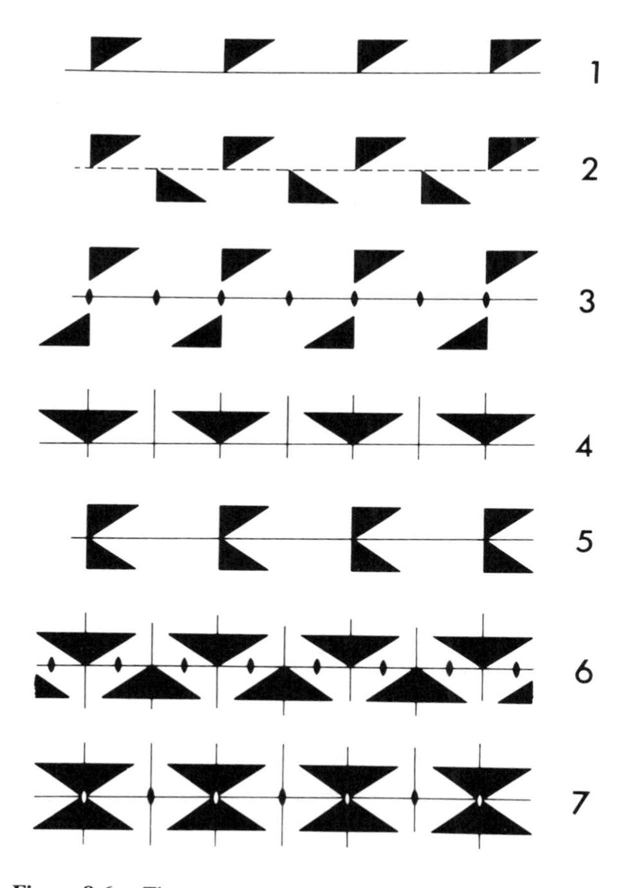

Figure 8-6. The seven symmetry classes of one-sided bands.

3. (a):2. There is a translation and a twofold rotation axis in this class. The twofold rotation axis is perpendicular to the plane of the one-sided band.
4. (a):m. The translation is achieved by transverse symmetry planes in this pattern.
5. $(a)\cdot m$. Here the translation axis is combined with a longitudinal symmetry plane.
6. $(a)\cdot\tilde{a}$:m. Combination of a glide-reflection plane with transverse symmetry planes characterizes this class. These elements generate new ones such as twofold rotation. Consequently, there are alternative descriptions of this symmetry class. One of them is by combining twofold rotation with glide reflection—the corresponding nota-

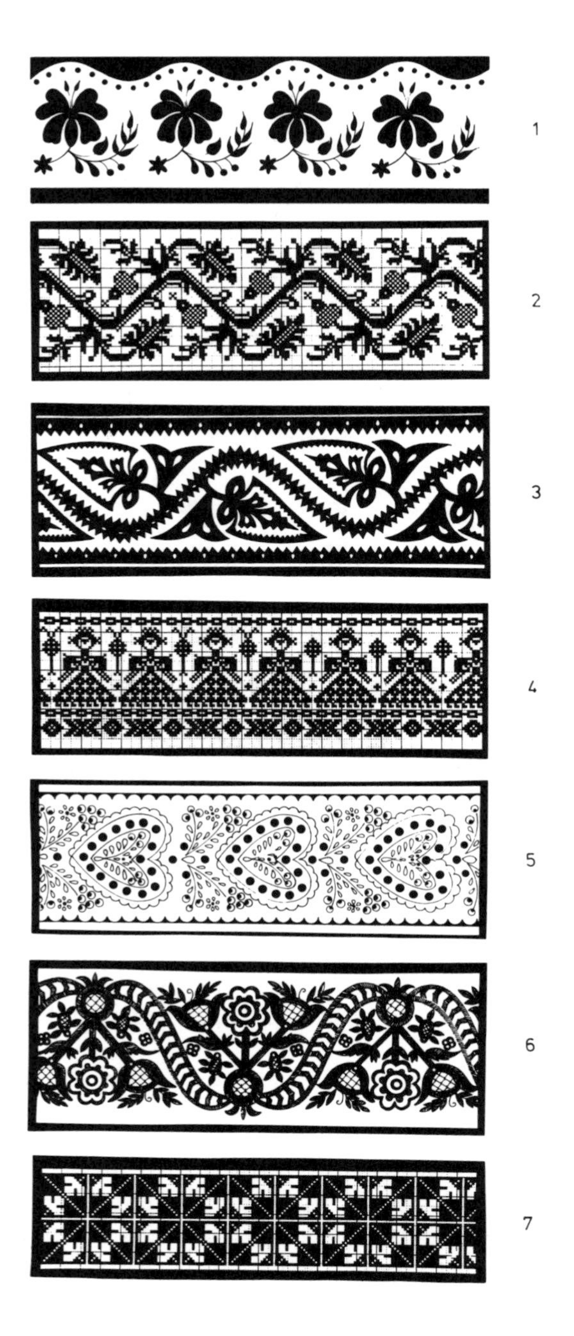
1
2
3
4
5
6
7

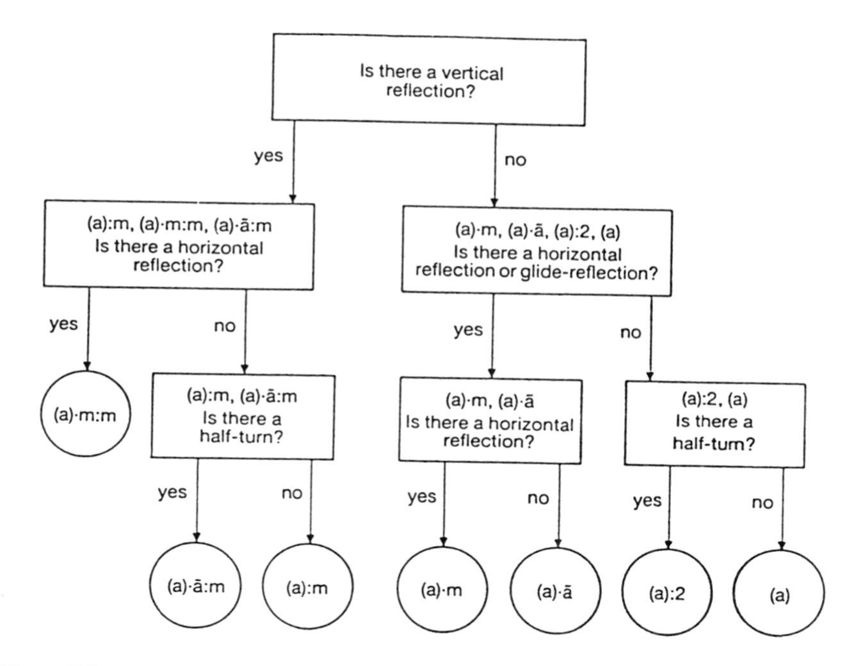

Figure 8-8. Scheme for establishing the symmetry of a one-sided band, after Crowe [8-5].

tion is $(a){:}2{\cdot}\tilde{a}$. Another is by combining twofold rotation with transverse reflection, for which the notation is $(a){:}2{:}m$.

7. $(a){\cdot}m{:}m$. This pattern has the highest symmetry, achieved by a combination of transverse and longitudinal symmetry planes. In this description the twofold axes perpendicular to the plane of the drawing are generated by the other symmetry elements. An alternative description is $(a){:}2{\cdot}m$.

←

Figure 8-7. Illustration of the seven symmetry classes of one-sided bands by Hungarian needlework [8-3]. The numbering corresponds to that of Figure 8-6. A brief description of the origin of the needlework is given here: (1) Edge decoration of table cover from Kalocsa, southern Hungary; (2) pillow-end decoration from Tolna county, southwest Hungary; (3) decoration patched onto a long embroidered felt coat of Hungarian shepherds in Bihar county, eastern Hungary; (4) embroidered edge decoration of bed sheet from the 18th century (note the deviations from the described symmetry in the lower stripes of the pattern); (5) decoration of shirtfront from Karád, southwest Hungary; (6) pillow decoration pattern from Torocko (Rimetea), Transylvania, Romania; (7) grape-leaf pattern from the territory east of the river Tisza.

The seven one-dimensional symmetry classes for the one-sided bands are illustrated by patterns of Hungarian needlework in Figure 8-7. This kind of needlework is a real "one-sided band." Figure 8-8 presents a scheme to facilitate establishing the symmetry class of one-sided bands [8-4, 8-5].

8.3 TWO-SIDED BANDS

If the singular plane of a band is not polar, the band is two-sided. The one-sided bands are a special case of the two-sided bands. Figure 8-9a shows a one-sided band generated by translation of a leaf motif. Figure 8-9b depicts a two-sided band characterized by a glide-reflection plane. There is a translation by half of the translation period and then a reflection in the plane of the drawing. The leaf patterns are paralleled by patterns of the triangle in Figure 8-9. A new symmetry element is illustrated in Figure 8-9c, the twofold (or second-order) *screw axis*, 2_1. The corresponding transformation is a translation by half the translation period and a 180° rotation around the translation axis. Bands have altogether 31 symmetry classes, including the seven one-sided band classes.

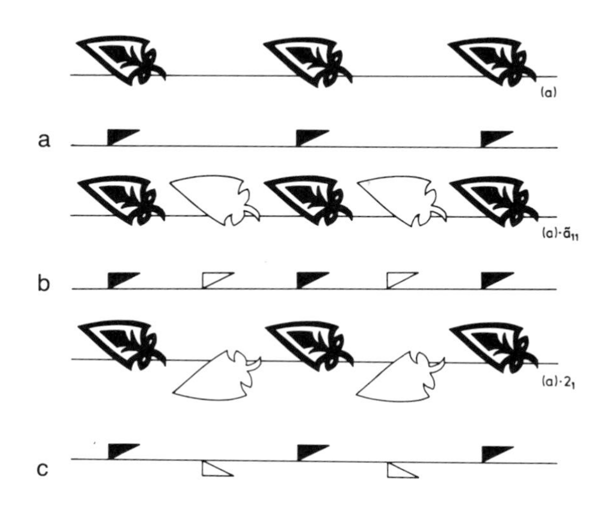

Figure 8-9. (a) One-sided bands generated by simple translation of the leaf motif and black triangle motif. The plane of the drawing is a polar singular plane. (b) Two-sided bands generated from the one-sided bands by introducing a glide-reflection plane. The singular plane in the plane of the drawing is no longer polar. The glide-reflection plane coinciding with the plane of the drawing is labeled $\tilde{a}_{11}$ [8-2]. Note that the two sides of the leaves are of different color (black and white). (c) Two-sided bands generated from the one-sided bands by introducing a screw axis of the second order, 2_1.

Table 8-1. Examples of Notations of Band Symmetries

Noncoordinate notation	Coordinate (international) notation
(a)	$p1$
$(a)\cdot\tilde{a}$	$p1a1$
$(a):2$	$p112$
$(a):m$	$pm11$
$(a)\cdot m$	$p1m1$
$(a)\cdot 2:m \equiv (a):2\cdot\tilde{a} \equiv (a):\tilde{a}\cdot m$	$pma2$
$(a)\cdot m:m \equiv (a):2\cdot m$	$pmm2$
$(a)\cdot 2_1$	$p2_111$
$(a)\cdot\tilde{a}_{11}$	$p11a$

Table 8-1 gives two different notations for the seven one-sided band classes and the two two-sided ones shown in Figure 8-9 as illustrations.

The so-called coordinate, or international, notation refers to the mutual orientation of the coordinate axes and symmetry elements [8-2]. The notation always starts with the letter p, referring to the translation group. Axis a is directed along the band, axis b lies in the plane of the drawing, and axis c is perpendicular to this plane. The first, second, and third positions after the letter p indicate the mutual orientation of the symmetry elements with respect to the coordinate axes. If no rotation axis or normal of a symmetry plane coincides with a coordinate axis, the number 1 is placed in the corresponding position in the notation. The coincidence of a rotation axis, 2 or 2_1, or the normal of a symmetry plane, m or $\tilde{a}$, with one of the coordinate axes is indicated by placing m or a, respectively, in the corresponding position in the notation.

8.4 RODS, SPIRALS, AND SIMILARITY SYMMETRY

The "infinite" carbide molecule is, of course, of finite width. It is indeed a three-dimensional construction with periodicity in one direction only. Thus, it has one-dimensional space-group symmetry (G_1^3). It is like an infinitely long rod. For a rod, the axis is a singular axis, and there is no singular plane. All kinds of symmetry axes may coincide with the axis of the rod, such as a translation axis, a simple rotation axis, or a screw-rotation axis. Of course, these symmetry elements, except the simple rotating axis, may characterize the rod only if it expands to infinity. As regards symmetry, a tube, a screw, or various rays are as much rods as are the stems of plants, vectors, or spiral stairways. Considering them to be infinite is a necessary assumption in describing their symmetries by space groups.

Real objects are not infinite. For symmetry considerations, it may be convenient to look only at some portions of the whole, where the ends are not yet in sight, and extend them in thought to infinity. A portion of an iron chain and a chain of beryllium dichloride in the crystal are shown in Figure 8-10. Translation from unit to unit is accompanied by a 90° rotation around the translation axis. A portion of a spiral stairway displaying screw-axis symmetry is shown in Figure 8-11a. The imaginary impossible stairway of Figure 8-11b indeed seems to go on forever.

A screw axis brings the infinite rod into coincidence with itself after a translation through a distance t accompanied by a rotation through an angle α. The screw axis is of the order $n = 360°/\alpha$. It is a special case when n is an integer. The iron chain and the beryllium dichloride chain have a fourfold (or fourth-order) screw axis, 4_2. Their overall symmetry is $(a)\cdot m\cdot 4_2{:}m$. For the screw axis of the second order, the direction of the rotation is immaterial. Other screw axes may be either left-handed or right-handed. The pair of left-handed and right-handed helices of Figure 2-50 is an example.

Figure 8-10. Examples of rods with 4_2 screw axis: (a) Iron chain in a park near the Royal Palace in Madrid; photograph by the authors; (b) beryllium dichloride chain in the crystal.

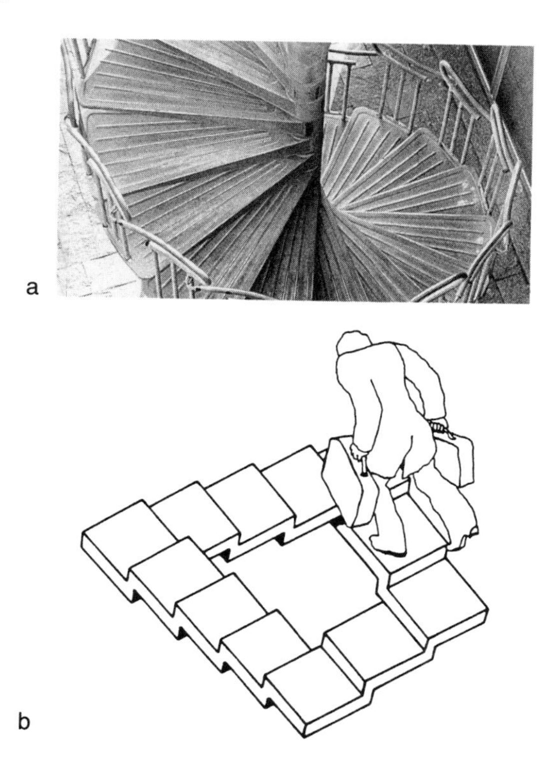

Figure 8-11. (a) A portion of a spiral stairway displaying screw-axis symmetry. Photograph by the authors. (b) An impossible stairway with proper space-group symmetry (as we can walk around this stairway *ad infinitum*). The idea for this drawing originated from a movie poster advertising *Glück im Hinterhaus*.

The scattered leaf arrangement around the stems of many plants is a beautiful occurrence of screw-axis symmetry in nature. The stem of *Plantago media* shown in Figure 8-12 certainly does not extend to infinity. It has been suggested, however, that for plants the plant/seed/plant/seed . . . infinite sequence, at least in time, provides enough justification to apply space groups in their symmetry description. Let us consider now the relative positions of the leaves around the stem of *Plantago media*. Starting from leaf "0," leaf "8" will be in eclipsed orientation to it. In order to reach leaf "8" from leaf "0," the stem has to be circled three times. The ratio of the two numbers, viz., 3/8, tells us that a new leaf occurs at each three-eighths of the circumference of the stem. The ratio 3/8 is characteristic in phyllotaxis, as are 1/2, 1/3, 2/5, and even

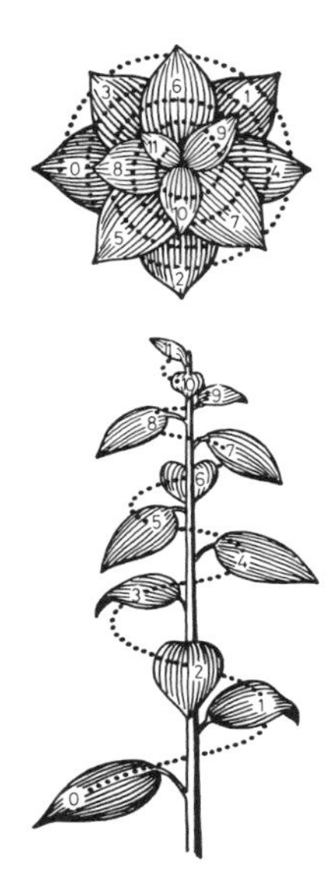

Figure 8-12. The scattered leaf arrangement (phyllotaxis) of *Plantago media*.

5/13. Very little is known about the origin of phyllotaxis. What has been note
a long time ago is that the numbers occurring in these characteristic ratios, viz.

$$1, 1, 2, 3, 5, 8, 13, \ldots$$

are members of the so-called Fibonacci series, in which each succeedin
number is the sum of the previous two. Fibonacci numbers can be observe
also in the numbers of the spirals of the scales of pine cones as viewed fro
below, displaying 13 left-bound and 8 right-bound spirals of scales as in Figur
8-13a. Left-bound and right-bound spirals in strictly Fibonacci numbers ar
found in other plants as well. The plate of seeds of the sunflower can b
considered as if it were a compressed scattered arrangement around th
stem. Figure 8-13b shows an example. It is most striking that the continuatio
of the ratios of the characteristic leaf arrangements eventually leads to a
extremely important irrational number, 0.381966 . . ., expressing the *golde*
mean!

Returning to screw axes, Figure 8-14 shows an infinite anion with a 1(
screw axis [8-6]. An important application of one-dimensional space grou
is for polymeric molecules in chemistry [8-7]. Figure 8-15 illustrates th
structure and symmetry elements of an extended polyethylene molecule. Th

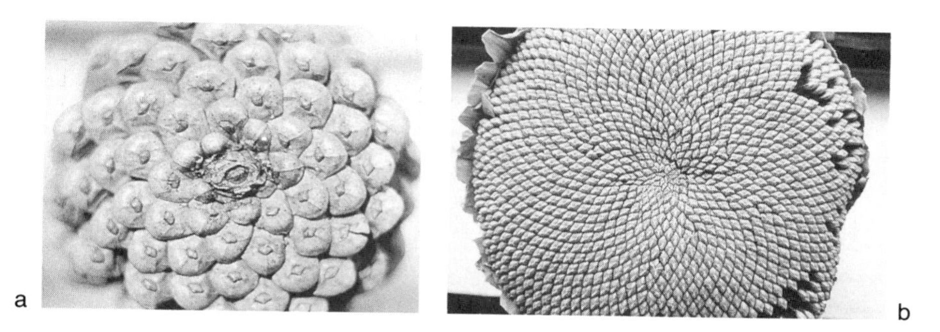

Figure 8-13. (a) Pine cone from below. (b) Sunflower seed plate.

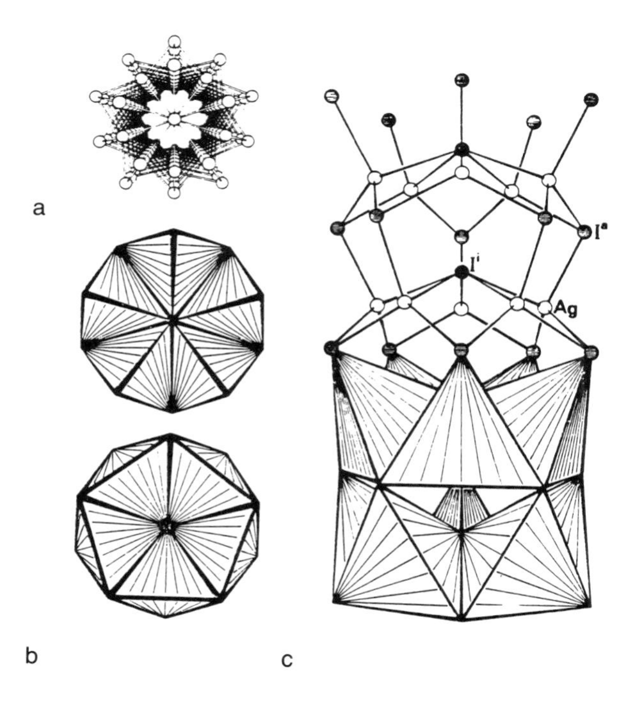

Figure 8-14. The polymeric anion (Ag_5I_6) with a 10_5 screw axis: (a) View along the screw axis; the sequence of the atoms from inside to outside is I–Ag–I; (b) views of the five condensed AgI_4 tetrahedra from the bottom and from the top. Reproduced from Ref. [8-6] with permission; (c) view perpendicular to the molecular axis .

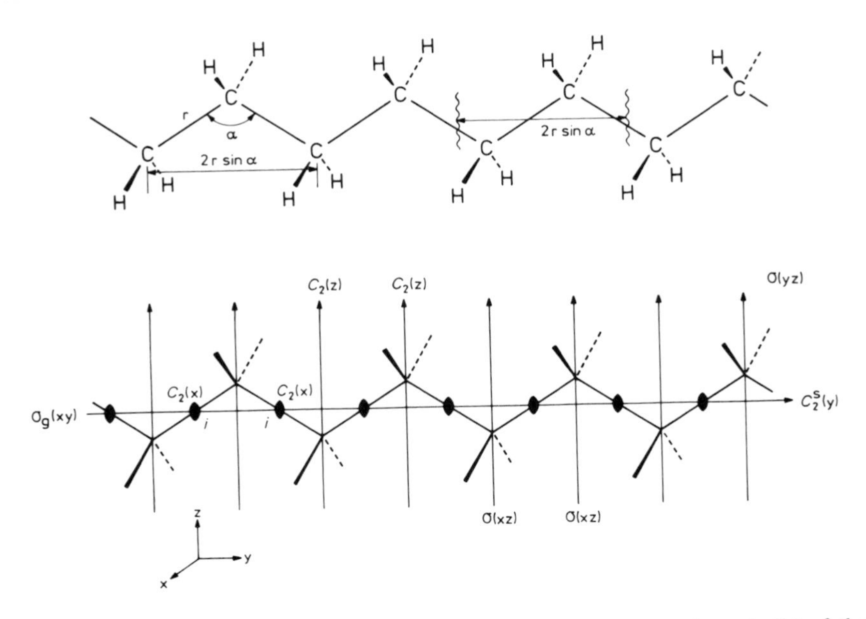

Figure 8-15. The structure and translation period (a) and symmetry elements (b) of the polyethylene chain molecule.

translation, or identity, period is shown; this is the distance between two carbo atoms separated by a third one. However, any portion with this length may b selected as the identity period along the polymeric chain. The translationa symmetry of polyethylene is characterized by this identity period. In addition there is a host of other symmetries as shown in Figure 8-15.

Biological macromolecules are often distinguished by their helical struc tures, to which one-dimensional space-group symmetries are applicable Figure 8-16a shows the polypeptide chain of the α helix, while Figure 8-16 depicts a polypeptide molecule in solution. The repeating units are the same i the two systems, viz., planar CCONHC skeletons. The linear rodlike structur of the α helix is accomplished by the hydrogen bonds, whereas these hydroge bonds are disrupted in solution [8-8]. A double helix, the structure of deoxyribonucleic acid molecule, is shown in Figure 8-17 in two representa tions. The double helix is held together by the hydrogen bonds of the base pai in between the two helices (see, e.g., Ref. [8-9]).

Whereas *helical symmetry* is characterized by a constant amount translation accompanied by a constant amount of rotation, in *spiral symmetr* the amounts of translation and rotation change gradually and regularly. A spir

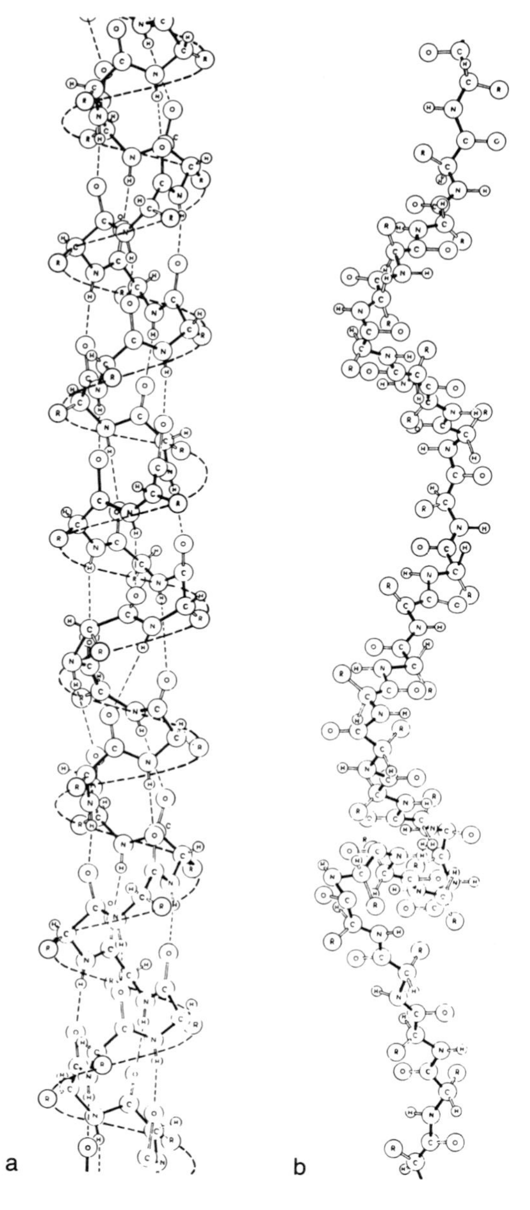

Figure 8-16. (a) Linear, rodlike, helical structure of the α helix. (b) Random chain of the polypeptide molecule as the hydrogen bonds of the α helix are disrupted in solution. After Ref. [8-8]. Copyright (1957) Scientific American. Used with permission.

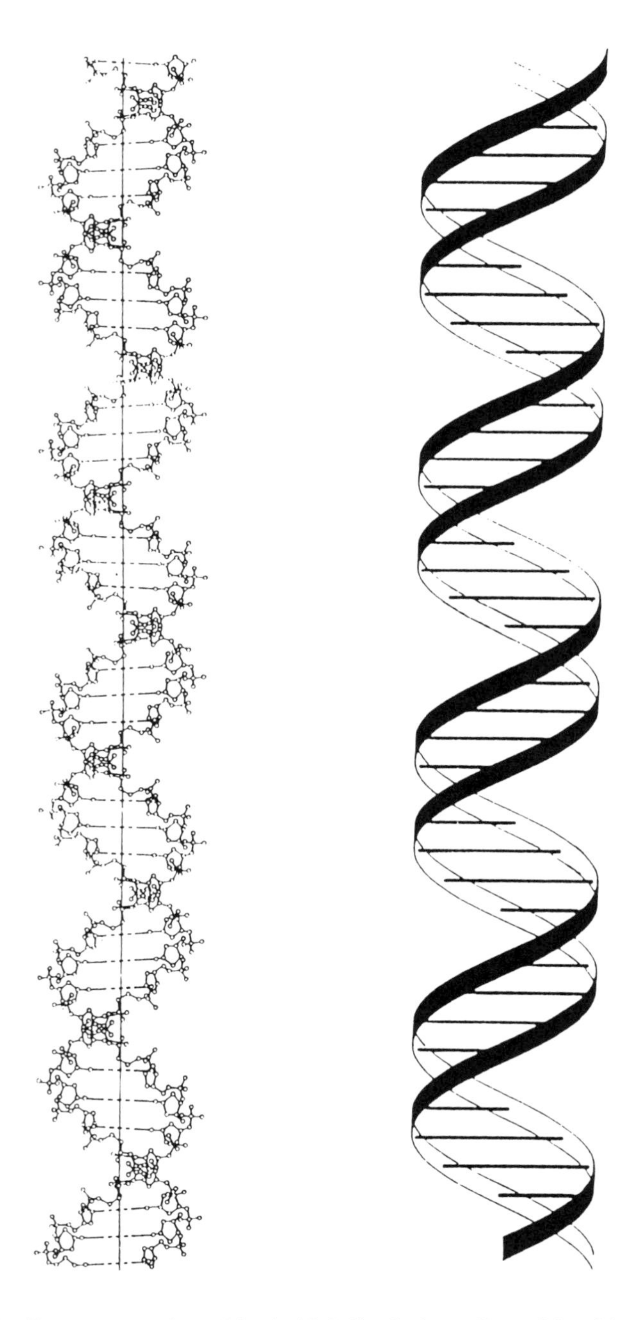

Figure 8-17. Two representations of the double helix of a deoxyribonucleic acid molecule, after Ref. [8-9].

may form along a rod or in a plane, and the scattered leaf arrangement and the sunflower seed plate may serve as their respective examples. An artistic double spiral is seen in Figure 8-18 as detail of a sculpture from the garden of a research institute where the structures of biological macromolecules are investigated.

Interesting chemical examples of spirals occur in systems with chemical oscillations. Oscillating reactions are often called Belousov–Zhabotinsky reactions. B. P. Belousov communicated his first observation in an obscure Russian medical publication [8-10] in the fifties, and it was followed by A. M. Zhabotinsky's first systematic studies [8-11] in the sixties. Although the chemical community was somewhat slow in catching up and many viewed the first reports on oscillating reactions with skepticism, research on nonlinear chemical phenomena has greatly expanded by now along with research on nonlinear phenomena in other fields. Recently, Körös [8-12] gave an overview of oscillations, waves, and spirals in chemical systems. Figure 8-19a illustrates the development of spiral structure in a Belousov–Zhabotinsky reaction. Incidentally, the two spirals shown make a heterochiral pair, paralleled by the one on a tombstone in Figure 8-19b.

Figure 8-18. Artistic double spiral. Detail of a sculpture in the garden of the Weizmann Institute, Rehovot, Israel. Photograph by the authors.

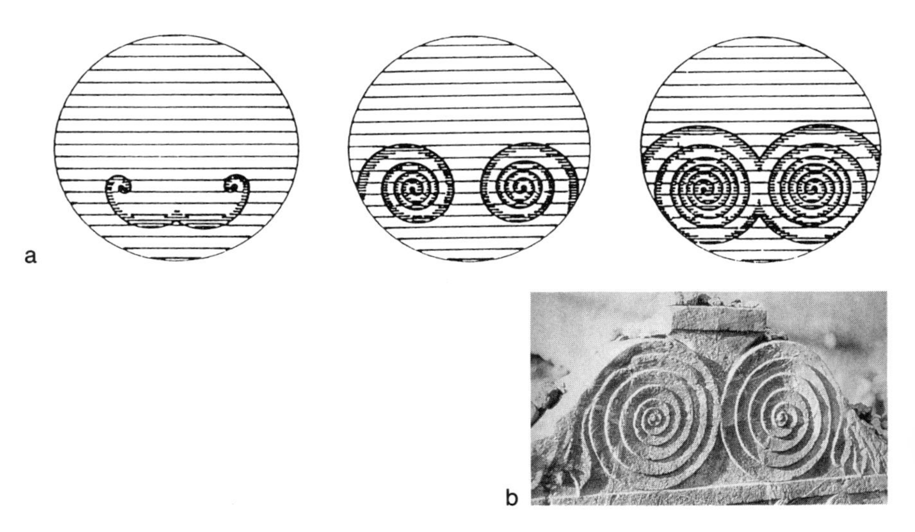

Figure 8-19. (a) Evolution of a spiral ring pattern in a reacting Belousov–Zhabotinsky system, after Körös [8-12]. (b) Tombstone in the Jewish cemetery, Prague. Photograph by the authors.

Spirals abound in Nature. Some examples are shown in Figure 8-20.

It has been suggested that the spiral shape of the seashell served as an example for developing the first screws in ancient times [8-14].* Decorations in a spiral shape have been known from the chalcolitic period. The link between the spiral decorations and the screw was found recently. It is a bronze needle with a special twisted shape. It is supposed to have been produced in the Bronze Age. The thread of this bronze needle is shown in Figure 8-21a. Sensitive analytical techniques were used to detect small amounts of impurities which gave clues as to the origin of the material used for producing the needle. A screw as a modern sculpture is seen in Figure 8-21b.

A gradual and regular change in size may appear by itself, that is, without being part of a spiral. A regular change in size characterizes, for example, homologous series, such as the alkanes, C_nH_{2n+2},

$$\ldots C_4H_{10}, C_5H_{12}, C_6H_{14}, C_7H_{16}, \ldots$$

with the increment of a methylene group, CH_2. Examples from outside chemistry are shown in Figure 8-22, where, again, it is up to our imagination to extend the series to infinity. All the spirals above and the phenomenon of phyllotaxis as well as the homologous series and the series of railway wheels

*We thank Professor G. Horányi, Budapest, for this reference.

Figure 8-20. Examples of spirals in Nature. (a) Galaxy [8-13]. The original photograph was taken with a near-infrared filter by Debra Meloy Elmegreen at the 48-inch Schmidt telescope on Mount Palomar in California. Image enhancement was made by Bruce Elmegreen, Debra Elmegreen, and Philip Seiden at the T. J. Watson Research Center in Yorktown Heights, New York. Photograph courtesy of Dr. B. Elmegreen, Yorktown Heights, New York. (b) Nautilus. Photograph courtesy of Lloyd Kahn, Bolinas, California. (c) Fern from the Big Island, Hawaii. Photograph by the authors. (d) Horn. Photograph courtesy of Fred Lipschultz, Storrs, Connecticut. (e) Plant from Pécs, Hungary. Photograph by the authors.

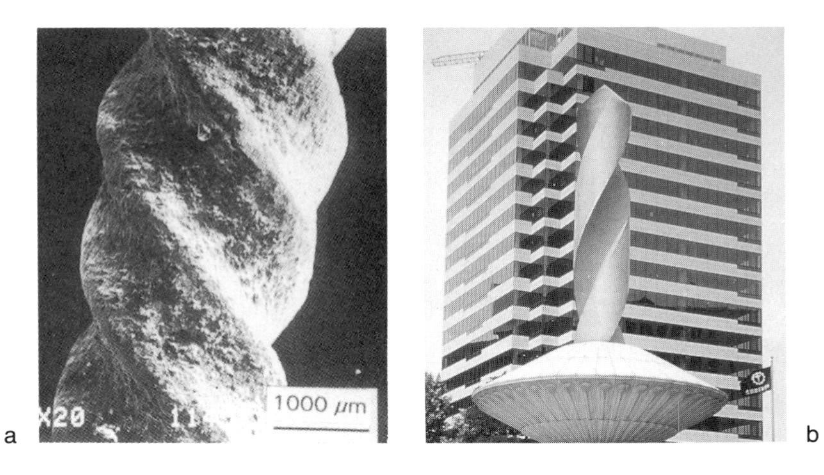

Figure 8-21. (a) Needle from the Bronze Age, after Ronen and Rozenak [8-14]. Photograph courtesy of Professor Ronen. (b) Screw as sculpture in Seoul. Photograph by the authors.

Figure 8-22. Examples of similarity symmetry: (a) Railway wheels, Technical Museum, Budapest; (b) mountain goats, Budapest Zoo. Photographs by the authors.

and the family of mountain goats in Figure 8-22 can be considered as examples of *similarity symmetry* (see, e.g., Ref. [8-15]).

8.5 TWO-DIMENSIONAL SPACE GROUPS

There are altogether 17 symmetry classes of one-sided planar networks. Figure 8-23 illustrates them in a way analogous to that in which the seven symmetry classes of the one-sided bands were illustrated (Figures 8-6 and 8-7). The most important symmetry elements and the coordinate notations of the symmetry classes are also given. The first letter (*p* or *c*) in this notation refers to translation. The next three positions carry information on the presence of various symmetry elements, where *m* denotes a symmetry plane, *g* a glide-reflection plane, 2, 3, 4, or 6 a rotation axis. The number 1, or a blank, indicates the absence of a symmetry element. The representations of the symmetry classes in Figures 8-6 and 8-23 were inspired by the illustrations inside the covers of Buerger's *Elementary Crystallography* [8-16]. Along with the purely geometrical configurations, Figure 8-23 presents 17 Hungarian needlework patterns. A brief description of their sources is given in the legend [8-17]. A scheme for establishing the symmetry class of one-sided two-dimensional space groups is given in Figure 8-24 [8-4, 8-5].

The lattice of the planar networks with two-dimensional space groups is defined by two noncollinear translations. Such a lattice is shown in Figure 8-25a. Given a particular lattice, the question is, which pair of translations should be selected to describe it? An infinite number of choices exists for each translation because a line joining any two lattice points is a translation of the lattice. Figure 8-25b shows a plane lattice and some of the possible choices for translation pairs to describe it. A primitive cell is defined by choices of translation pairs such as t_1 and t_2 or t_3 and t_4. Only one lattice point is associated with each primitive cell. This is understood if each lattice point in Figure 8-25 is considered to belong to four adjacent cells, or only one-fourth of each point to belong to any one cell. As each cell contains four corners, all this adds up to one whole point. Alternatively, by displacing any one primitive cell, each primitive cell will contain only one lattice point. On the other hand, a multiple cell contains one or more lattice points in addition to the one shared at the corners. The translation pair t_5 and t_6, for instance, defines a double cell. A cell is called a unit cell if the entire lattice can be derived from it by translations. Thus, a unit cell may be either primitive or multiple. The unit cell is chosen usually to represent best the symmetry of the lattice. The translations selected as the edges of the plane unit cell are *a* and *b*, and for a space lattice, *a*, *b*, and *c*. The latter are called the crystallographic axes. The angles between the edges of the three-dimensional unit cell are α, β, and γ, but only γ is needed for the plane lattice.

Figure 8-23. The 17 symmetry classes of one-sided planar networks, with the most important symmetry elements and the notations of the classes indicated. Along with the geometrical configurations, Hungarian needlework patterns are presented for illustration. A brief description of the origin of these patterns if given here [8-17]: *p*1 and *p*4, Patterns of indigo-dyed decorations on textiles for clothing, Sellye, Baranya county, 1899; *p*2, indigo-dyed decoration with palmette motif for curtains, currently a very popular pattern; *p*3, *p*6, *p6mm*, *p*3*m*1, and *p*31*m*, decorations with characteristic bird motifs from peasant vests, northern Hungary. (*Continued on next page*)

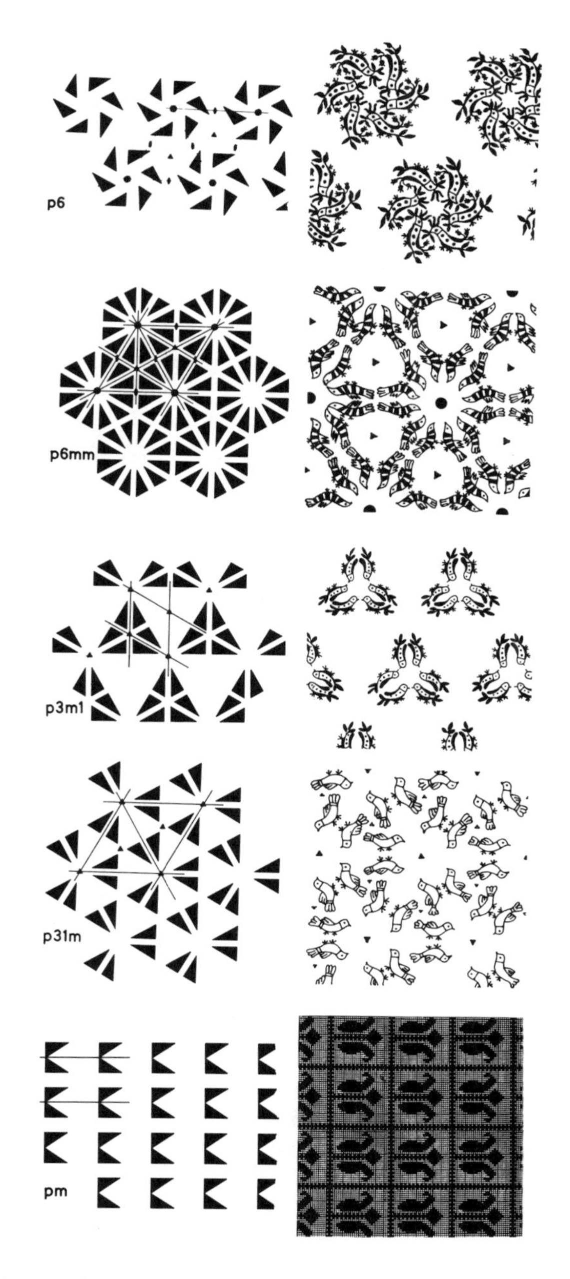

Figure 8-23. (*Continued*) *pm*, decoration with tulip motif for tablecloth, cross-stitched needlework, from the turn of the century. (*Continued on next page*)

Figure 8-23. (*Continued*) *pmm*2, bed sheet border decoration with pomegranate motif, northwest Hungary, 19th century; *p4mm*, pillow-slip decoration with stars, cross-stitched needlework, Transylvania, 19th century; *cm*, pillow-slip decoration with peacock tail motif, cross-stitched needlework, much used throughout Hungary around the turn of the century; *cmm*2, bed-sheet border decoration with cockscomb motif, cross-stitched needlework, Somogy county, 19th century. (*Continued on next page*)

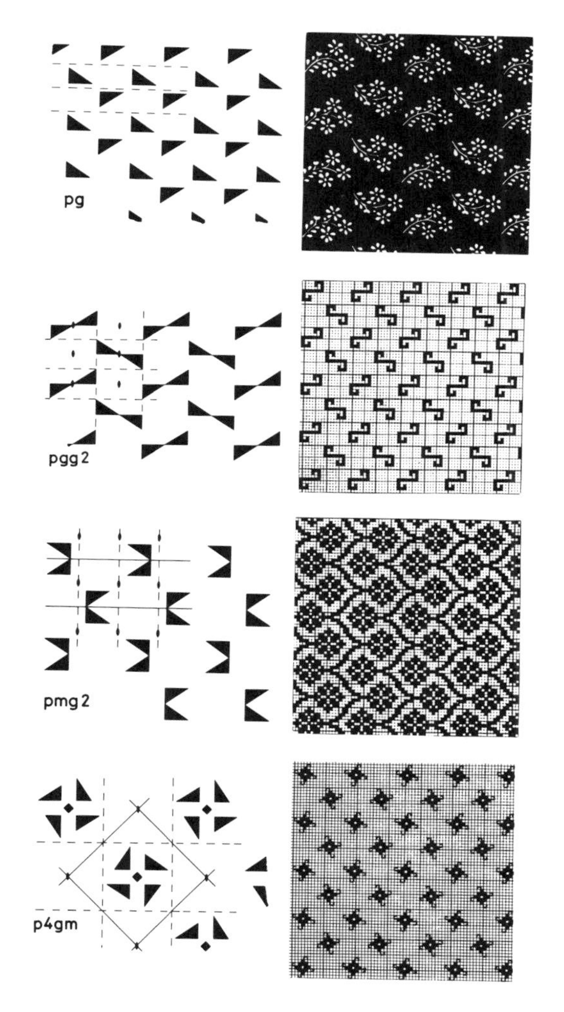

Figure 8-23. (*Continued*) *pg*, from a pattern book of indigo-dyed decorations. Pápa, Veszprém county, 1856; *pgg*2, children's bag decoration, Transylvania, turn of the century; *pmg*2, pillow-slip decoration with scrolling stem motif, much used throughout Hungary around the turn of the century; *p*4*gm*, blouse-arm embroidery, Bács-Kiskun county, 19th century.

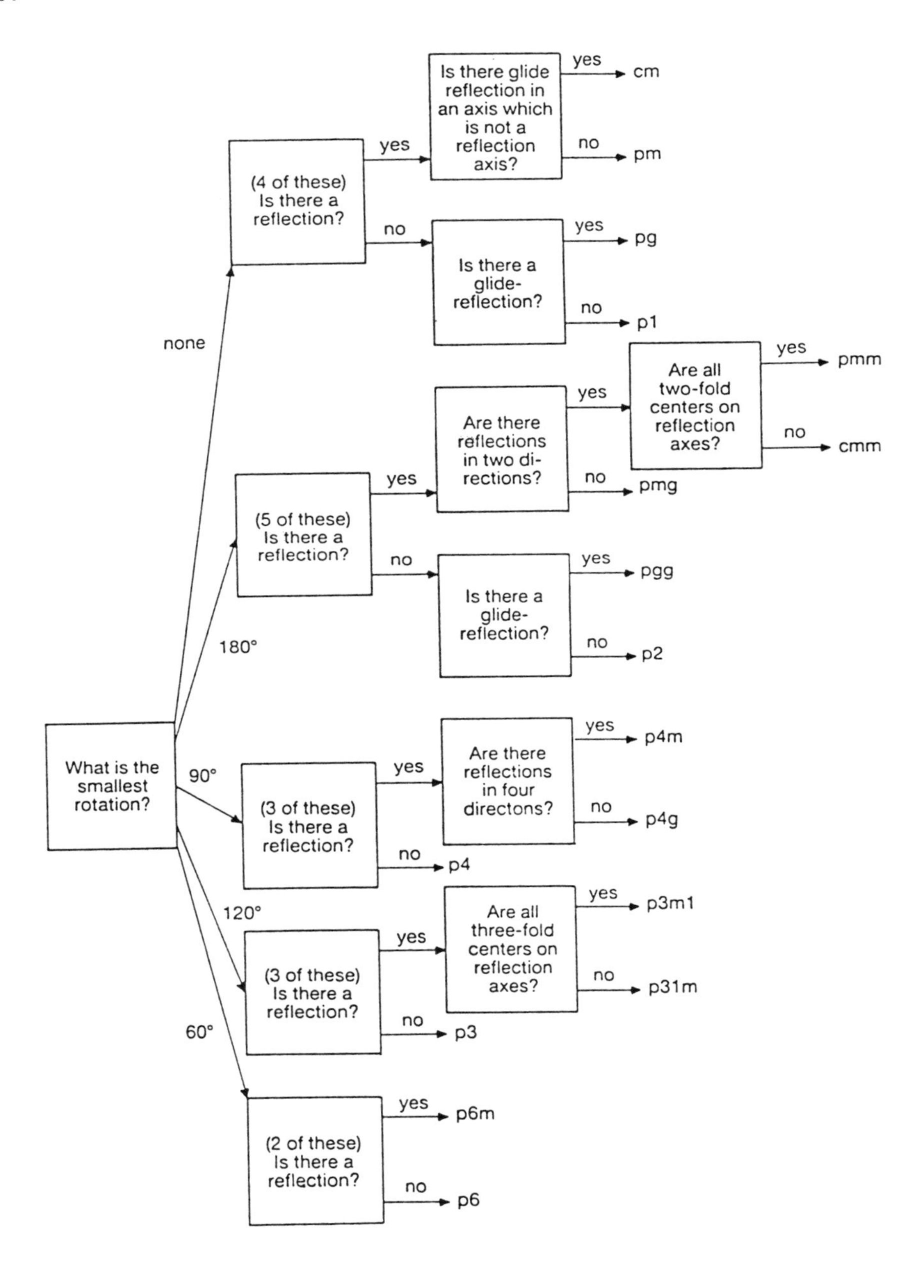

Figure 8-24. Scheme for establishing the symmetry of planar networks, after Crowe [8-5].

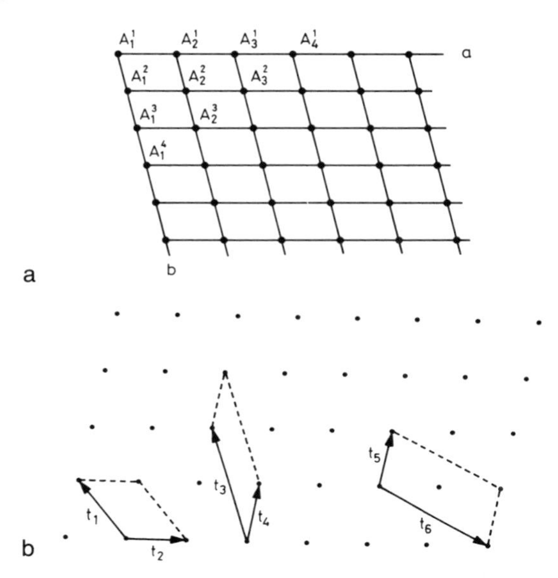

Figure 8-25. (a) Plane lattice defined by two noncollinear translations. (b) Illustration of primitive and unit cells on a plane lattice, after Azaroff [8-18]. Copyright (1960) McGraw-Hill, Inc. Used with permission.

Figure 8-26 shows three planar networks based on the same plane lattice. Two and only two lines intersect in each point of all three networks. Accordingly, the parallelograms of all three networks have the same area. All of them are unit cells, in fact, primitive cells. Each of these parallelograms is determined by two sides a and b and the angle γ between them. These are called the cell parameters.

The general plane lattice (a) shown in Figure 8-27 is called a parallelogram lattice. The other four plane lattices of Figure 8-27 are special cases of the general lattice. The rectangular lattice (b) has a primitive cell with unequal sides. The so-called diamond lattice (c) has a unit cell with equal sides. A

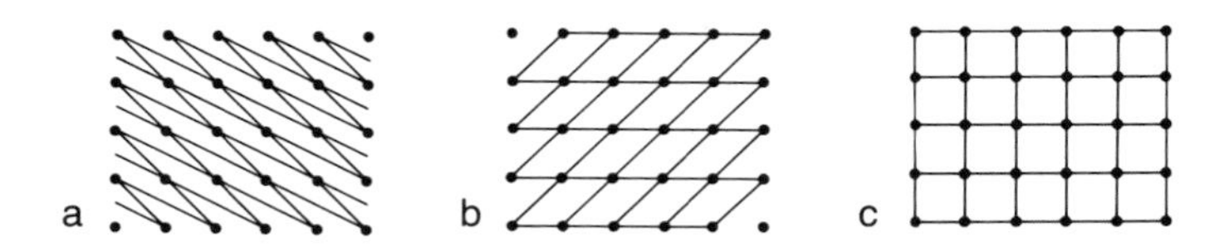

Figure 8-26. Different networks based on the same plane lattice.

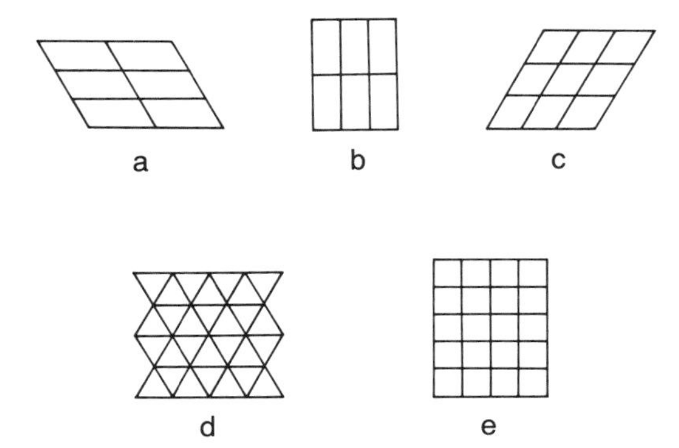

Figure 8-27. The five unique plane lattices (for identification of a through e, see text).

special case of the diamond lattice is that in which the angle between the equal sides of the unit cell is 120°, and this lattice (d) is then called rhombic, or triangular since the short cell diagonal divides the unit cell into two equilateral triangles. This lattice may also be considered as having hexagonal symmetry. Finally, there is the square lattice (e).

The five unique plane lattices were described above under the assumption that the lattice points themselves have the highest possible symmetry. In this case these five unique lattices will have the symmetries listed in Table 8-2.

When the point-group symmetries are combined with the plane lattices, 17 two-dimensional space groups can be produced. Severe limitations are imposed on the possible point groups that may be combined with lattices to produce space groups. Some symmetry elements, such as the fivefold rotation

Table 8-2. Symmetries of the Five Unique Plane Lattices of Figure 8-27

Lattice	Space group	
	Noncoordinate notation	Coordinate (international) notation
(a) Parallelogram lattice	(b/a):2	$p2$
(b) Rectangular lattice	$(b{:}a){:}2\cdot m$	$pmm2$
(c) Diamond Lattice	$(a/a){:}2\cdot m$	$cmm2$
(d) Hexagonal or triangular lattice	$(a/a){:}6\cdot m$	$p6mm$
(e) Square lattice	$(a{:}a){:}4\cdot m$	$p4mm$

axis, are not compatible with translational symmetry. This will be examined in detail in Section 9.3.

8.5.1 Some Simple Networks

The simplest two-dimensional space group is represented in four variations in Figure 8-28. This space group does not impose any restrictions on the parameters a, b, and γ. The equal motifs repeated by the translations may be completely separated from one another, they may consist of disconnected parts, they may intersect each other, and, finally, they may fill the entire plane without any gaps. Of course, such variations are possible for any of the more complicated two-dimensional space groups as well.

Especially intriguing are those variations which cover the whole available surface without gaps. Of the regular polygons, this is possible only with the equilateral triangle, the square, and the regular hexagon. For the latter, characteristic examples are shown in Figure 8-29, including some in which the hexagons have only approximately regular shapes.

Planar motifs of irregular shape can be used in infinite numbers to construct planar patterns completely covering the whole available surface.

M. C. Escher is especially famous for his periodic drawings which fill the plane [8-21]. Their symmetry aspects have been discussed in detail by the Dutch crystallographer Caroline MacGillavry [8-22]. The pattern in Figure 8-30 is from her book. It has $p1$ symmetry. The unit cell is the combination of a fish and a boat.

Canadian crystallographer François Brisse has designed a series of two-dimensional space-group drawings related to Canada [8-23]. The series was

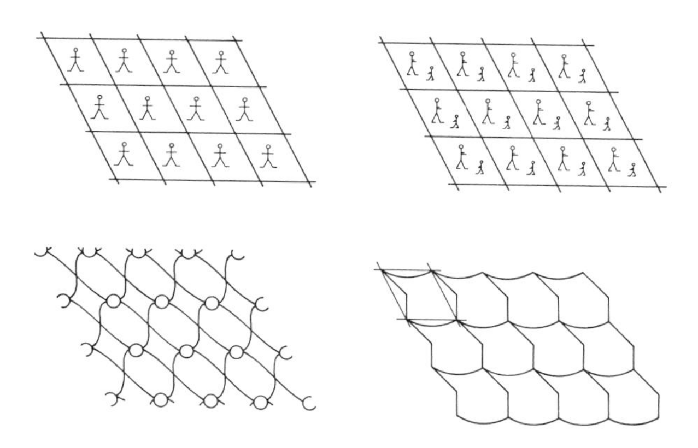

Figure 8-28. The simplest two-dimensional space group in four variations.

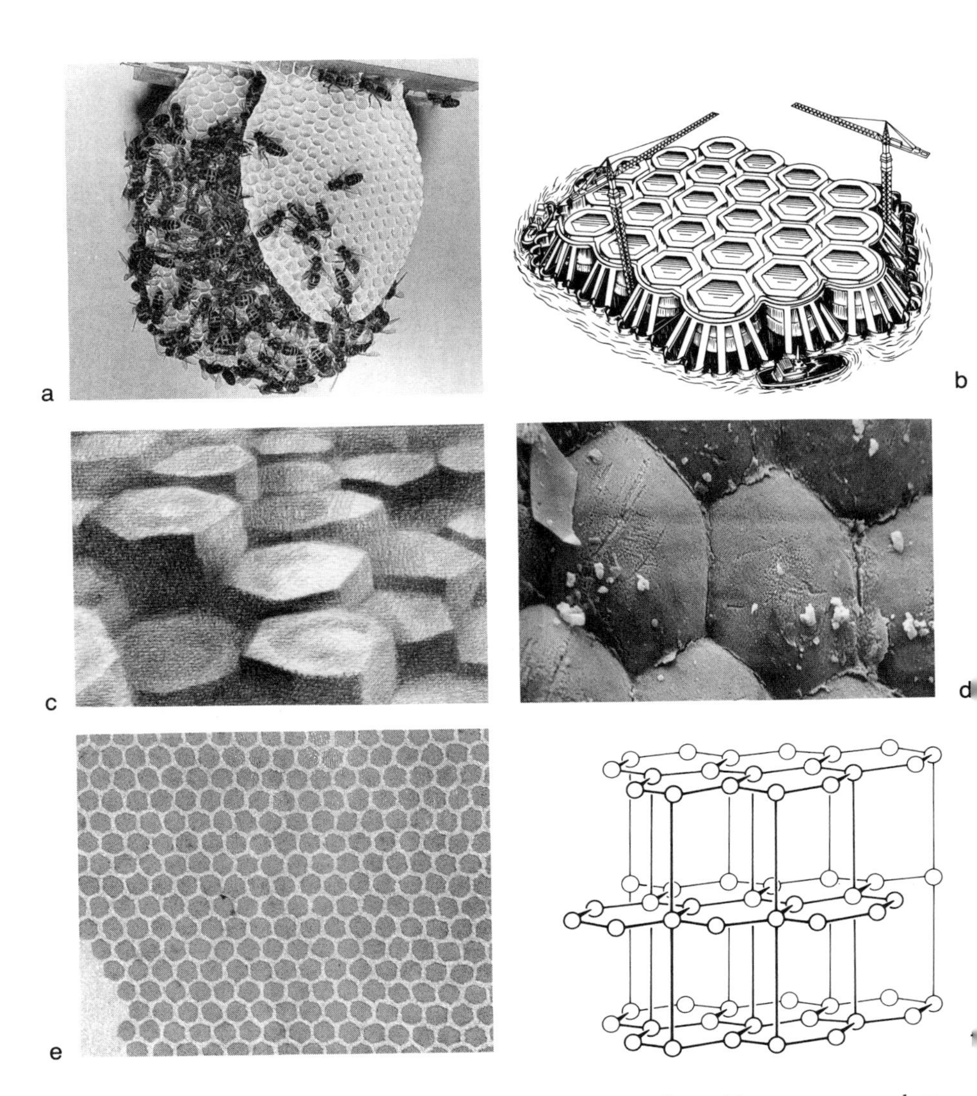

Figure 8-29. Networks of regular hexagons covering the surface without gaps or overlaps. (a) Honeycomb. Photograph courtesy of Professor Pál Zoltán Örösi, Budapest, 1982. (b) Oil platform under construction in the North Sea. Drawn after a lithograph in the 1979 report of the Statoil Company, Norway. (c) Columnar basalt joints. Drawing by Ferenc Lantos, after Ref. [8-19]. (d) Moth compound eye (magnified approx. × 2000). Courtesy of Dr. J. Morral, The University of Connecticut, 1984. (e) Filament material [8-20]. Photograph courtesy of Eric Gregory, Waterbury, Connecticut. (f) Structure of graphite layer.

Figure 8-30. (a) Escher's periodic drawing of fish and boats with space group $p1$ from MacGillavry's book [8-22]. Reproduced with permission from the International Union of Crystallography. (b) The unit cell consisting of a fish and a boat.

dedicated to the XIIth Congress of the International Union of Crystallography (IUCr), held in Ottawa in 1981. Drawings have been prepared to represent the Canadian provinces and territories. One of them is shown in Figure 8-31. The polar bear is a symbol for Canada's Northwest Territories. Its stylized representation is the asymmetric unit to which first a twofold rotation is applied, and then the translations. For unit cell it is convenient to choose two polar bears

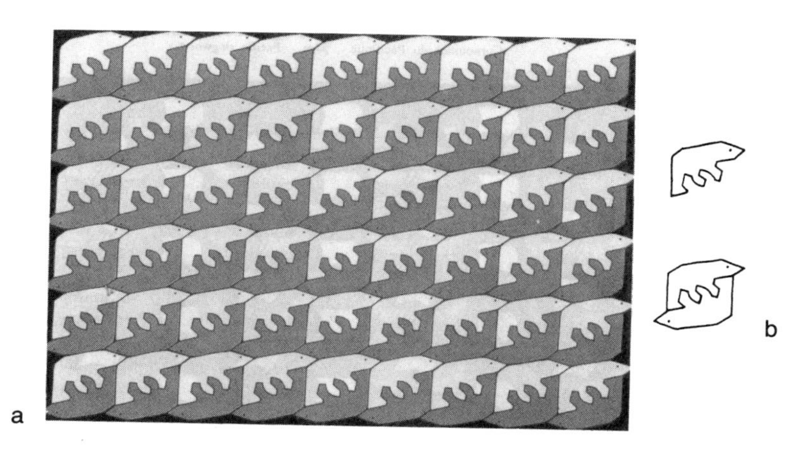

Figure 8-31. (a) Brisse's periodic drawing *Northwest Territories*. Reproduced with permission from *La symétrie bidimensionelle et le Canada* [8-23]. (b) The primitive cell and a unit cell displaying twofold rotation.

related by twofold rotation. White and blue polar bears alternate in the original drawing, but the colors are disregarded in the present discussion.

The symbol of the XIIth IUCr Congress was a unit of four stylized maple leaves related by fourfold rotation. It is shown in Figure 8-32a. The maple leaf is Canada's symbol and is shown in a more natural appearance on a stamp. The two-dimensional drawing in Figure 8-32b is created by repetition from the above unit. Again, the original alternating white/red coloring is disregarded in

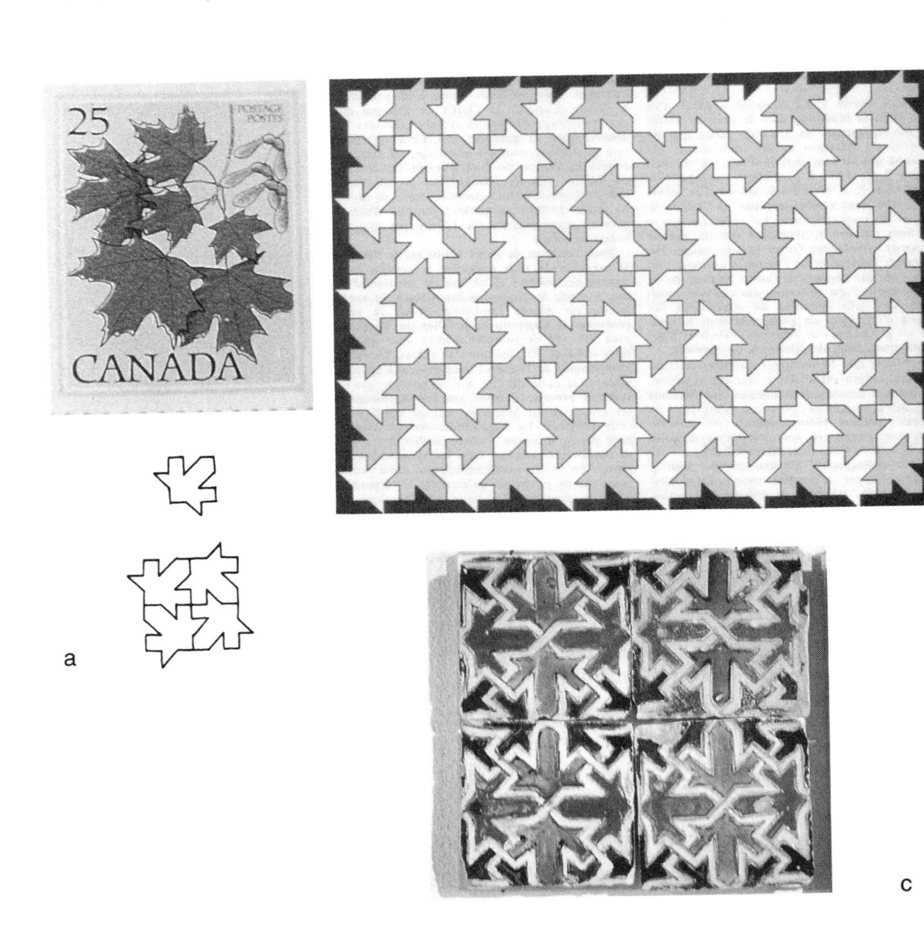

Figure 8-32. (a) The maple leaf and its stylized version. The unit cell displaying fourfold rotation was the symbol of the XIIth Congress of the International Union of Crystallography, Ottawa, 1981. (b) Brisse's periodic drawing *Canada*. Reproduced with permission from *La symétrie bidimensionelle et le Canada* [8-23]. (c) Portuguese tile decoration. Photograph by the authors.

our discussion. The two-dimensional space group of the pattern is then *p4gm* [8-23]. The pattern shown in Figure 8-32b has already been used by Pólya [8-24] among his representations of the 17 two-dimensional space groups. It may also be found among typical decorations, both as a two-dimensional pattern [8-25] and in its one-dimensional variation as a band ornament [8-26]. A Portuguese tile decoration is shown as an example in Figure 8-32c.

The repetition of the flies, butterflies, falcons, and bats in the Escher drawing in Figure 8-33 is accomplished by mirror planes. The two-dimensional space group is *pmm*, and the mirror planes are indicated separately as the borders of the primitive cell. The next two periodic drawings were created by Khudu Mamedov, an Azerbaijani crystallographer, and appeared in a remarkable little book *Decorations Remember* [8-27]. The space group of the drawing *Unity* of Figure 8-34 is *p*1, with the basic motif consisting of an old and a young man. The repetition of the uniform shapes truly satisfies the requirement of the two-dimensional space group. A closer look, however, reveals distinct individuality of facial expressions, especially for the old men. The other drawing by Mamedov in Figure 8-35 is entitled *Sea-Gulls*. The basic motif is a single sea gull. However, the sea gulls turn their heads in alternating directions, and the unit cell consists of four sea gulls.

A comprehensive and in-depth treatise of tilings and patterns has been published by Grünbaum and Shephard [8-28].

Figure 8-33. (a) Escher's periodic drawing of flies, butterflies, falcons and bats, from MacGillavry's book [8-22]. Reproduced with permission from the International Union of Crystallography. (b) The primitive cell framed by the square whose sides are parts of the mirror planes in the periodic drawing.

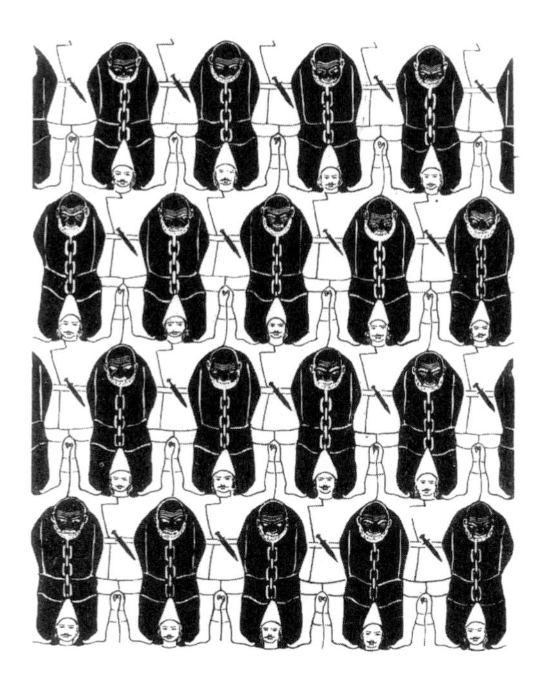

Figure 8-34. Mamedov's periodic drawing *Unity* [8-27].

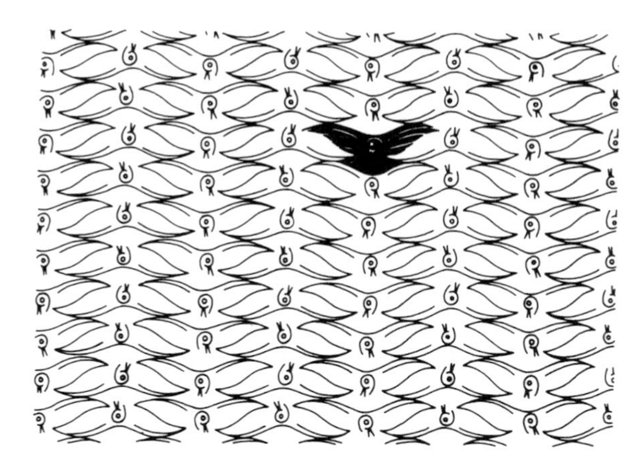

Figure 8-35. Mamedov's periodic drawing *Sea-gulls* [8-27].

8.5.2 Side Effects of Decorations

Shubnikov and Koptsik [8-2] have analyzed the influence of the various space groups of bands and networks on people's perception of movement. A one-sided band decoration without a polar axis induces no feeling of movement. The vertical symmetry planes of the fence and wall in Figure 8-36 act as if preventing motion. On the other hand, the bands with polar axes in Figure 8-37 act as if inducing left-bound (top) and right-bound (middle) movement.

a

b

Figure 8-36. Fence on the Liberty Bridge in Budapest (a) and The Great Wall, off Beijing (b) with vertical symmetry planes. Photographs by the authors.

Figure 8-37. Band decorations with polar axis.

Simple geometrical patterns can achieve similar effects, as seen, for example, in Figure 8-5a (inducing motion) and 8-5b (preventing it).

A symmetry plane always conveys the impression of preventing motion perpendicular to it. Symmetry planes are supposed to induce calmness and thus may be recommended for decorating the walls of halls for serious meetings. Examples are shown in Figure 8-38. On the other hand, the walls of dancing halls should probably be decorated with patterns of rotational symmetry only (Figure 8-39). The pattern of the Escher-like airline advertisement in Figure 8-40 conveys a strong feeling of motion, left-bound, induced by the white birds, and right-bound, induced by the dark ones. Thus, translational symmetry is combined with antisymmetry.

8.5.3 Moirés

The so-called Moiré pattern is created by superimposing infinite planar patterns. The resulting pattern is a new two-dimensional network. The simplest case is illustrated in Figure 8-41. Two identical systems of lines on transparent paper are superimposed.

The starting and resulting systems have a period of λ and d, respectively, and they are superimposed at an angle Θ. These parameters have the following relationship:

$$\lambda = 2d \sin(\Theta/2)$$

This expression is well known as the Bragg law in X-ray diffraction of crystals, where λ is the wavelength of the X rays, d is the distance between atomic layers in the crystal, and $\Theta/2$ is the angle at which the X rays hit the atomic layer.

The patterns produced in Figure 8-41 have twofold rotation axes perpen-

Figure 8-38. Wall decorations with symmetry planes from the Alhambra, Granada, Spain. Photographs by the authors.

Figure 8-39. Wall decorations with no symmetry planes from the Alhambra, Granada, Spain. Photographs by the authors.

Figure 8-40. Airline advertisement. There is a color change associated with the reversal of the direction of motion. Thus, translational symmetry is combined with antisymmetry. From a highway near O'Hare Airport, Chicago. Photograph by the authors.

dicular to their plane. Thus, the superposition at angle $180 + \Theta$ will produce the same result as that at angle Θ.

Figure 8-42 shows the interference of two identical infinite systems of small circles at a series of angles.

Moiré patterns occur in the most diverse phenomena. Two examples are mentioned here. It has been observed that only certain drop sizes cause supernumerary bows in a rain shower. The interference pattern produced by the rain wave front folding over on itself can be simulated with Moiré patterns.

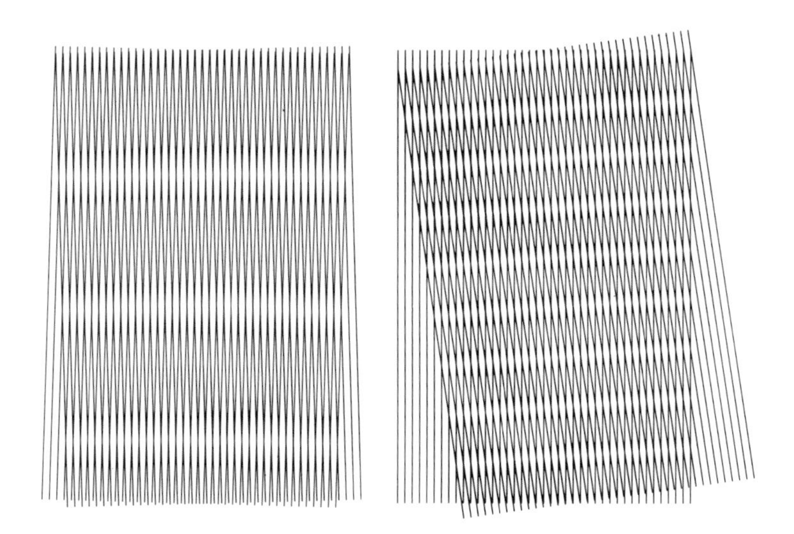

Figure 8-41. Moiré patterns from the superposition of two line systems at increasing angles.

Figure 8-42. Moiré patterns from the superposition of two circle systems at increasing angles.

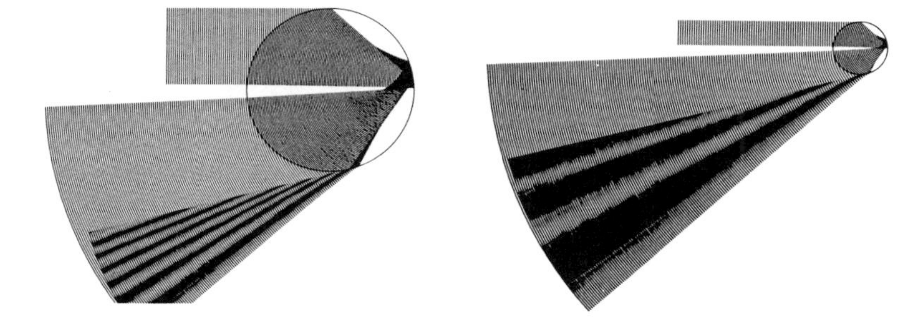

Figure 8-43. Moiré patterns simulate the formation of supernumerary bows in rain showers. The larger drops produce closely spaced bows at the top, while at the bottom the smaller drops produce more widely spaced bows. After Fraser [8-29], reproduced with permission.

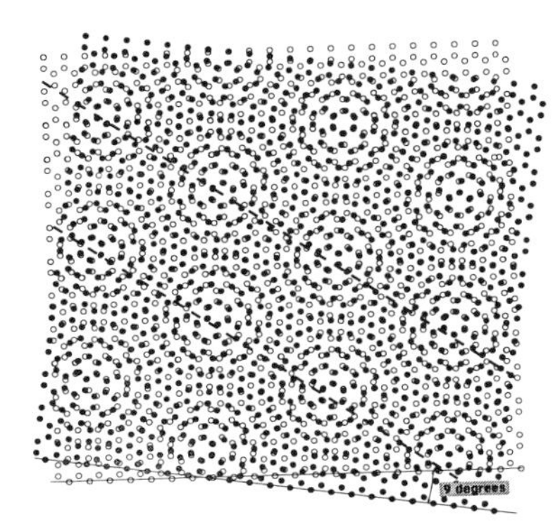

Figure 8-44. A model of the Moiré pattern produced by two misoriented graphite sheets. Illustration courtesy of Professor K. Sattler, University of Hawaii.

Figure 8-45. Werner Witschi and his Moiré sculptures. Photographs by the authors.

This is illustrated in Figure 8-43, after Fraser [8-29]. The spacing in the bows clearly depends on the drop size.

The other example is related to fullerene tubular structures [8-30]. Such structures have been generated by vapor condensation of carbon on atomically flat graphite surfaces. Due to a misorientation of the top layer relative to the second layer, a Moiré pattern is created whose lattice parameter is determined by the angle of misorientation. The structural model of the superpattern produced by two misoriented sheets is illustrated by Figure 8-44.

An analysis of Moiré patterns has been given by Hans Giger [8-31]. Moirés are often used in artistic expression [8-32]. The work of the Swiss sculptor Werner Witschi is illustrated by Figure 8-45.

REFERENCES

[8-1] F. J. Budden, *The Fascination of Groups*, Cambridge University Press, Cambridge (1972).

[8-2] A. V. Shubnikov and V. A. Koptsik, *Symmetry in Science and Art*, Plenum Press, New York (1974). [Russian original: *Simmetriya v nauke i iskusstve*, Nauka, Moscow (1972).]

[8-3] I. Hargittai and G. Lengyel, *J. Chem. Educ.* **61**, 1033 (1984).

[8-4] I. Hargittai, in *Symmetrie in Geistes- und Naturwissenschaft* (R. Wille, ed.), Springer-Verlag, Berlin (1988).

[8-5] D. W. Crowe, in *The Geometric Vein. The Coxeter Fertschrift* (C. Davis, B. Grünbaum, and F. A. Sherk, eds.), Springer-Verlag, New York (1982).

[8-6] K. Peters, W. Ott, and H. G. v. Schnering, *Angew. Chem. Int. Ed. Engl.* **21**, 697 (1982).

[8-7] H. Tadokoro, *Structure of Crystalline Polymers*, Wiley-Interscience, New York (1979).

[8-8] P. Doty, in *The Molecular Basis of Life* (R. H. Haynes and P. C. Hanewalt, eds.), W. H. Freeman and Co., San Francisco (1968).

[8-9] B. K. Vainshtein, V. M. Fridkin, and V. L. Indenbom, *Sovremennaya Kristallografiya*, Vol. 2, *Struktura Kristallov*, Nauka, Moscow (1979).

[8-10] B. P. Belousov, in *Sbornik Referatov po Radiatsionnoi Medicine*, pp. 145–147, Medgiz, Moscow (1958).

[8-11] A M. Zhabotinsky, *Biofizika* **9**, 306 (1964).

[8-12] E. Körös, in *Spiral Symmetry* (I. Hargittai and C. A. Pickover, eds.), p. 221, World Scientific, Singapore (1992).

[8-13] B. G. Elmegreen, in *Spiral Symmetry* (I. Hargittai and C. A. Pickover, eds.), p. 95, World Scientific, Singapore (1992); B. G. Elmegreen, D. M. Elmegreen, and P. E. Seiden, *Astrophys. J.* **343**, 602 (1989).

[8-14] Y. Ronen and P. Rozenak, *J. Mater. Sci.* **28**, 5576 (1993).

[8-15] A. V. Shubnikov, *Sov. Phys. Crystallogr.* **5**, 469 (1961); reprinted in *Crystal Symmetries* (I. Hargittai and B. K. Vainshtein, eds.), p. 365, Pergamon Press, Oxford (1988).

[8-16] M. J. Buerger, *Elementary Crystallography, An Introduction to the Fundamental Geometrical Features of Crystals*, 4th printing, John Wiley & Sons, New York (1967).

[8-17] I. Hargittai and G. Lengyel, *J. Chem. Educ.* **62**, 35 (1985).

[8-18] L. V. Azaroff, *Introduction to Solids*, McGraw-Hill, New York (1960).

[8-19] A. Holmes, *Principles of Physical Geology*, The Ronald Press Co., New York (1965).

[8-20] E. Gregory, *Proc. IEEE* **77**, 1110 (1989).

[8-21] D. Schattschneider, *Visions of Symmetry, Notebooks, Periodic Drawings, and Related Work of M. C. Escher*, W. H. Freeman and Co., New York (1990).
[8-22] C. H. MacGillavry, *Symmetry Aspects of M. C. Escher's Periodic Drawings*, Bohn, Scheltema and Holkema, Utrecht (1976).
[8-23] F. Brisse, *Can. Mineral.* **19**, 217 (1981).
[8-24] G. Pólya, *Z. Kristallogr.* **60**, 278 (1924).
[8-25] I. El-Said and A. Parman, *Geometric Concepts in Islamic Art*, World of Islam Festival Publ. Co., London (1976).
[8-26] P. D'Avennes (ed.), *Arabic Art Color*, Dover, New York (1978).
[8-27] Kh. S. Mamedov, I. R. Amiraslanov, G. N. Nadzhafov, and A. A. Muzhaliev, *Decorations Remember*, Azerneshr, Baku (1981) [in Azerbaijani].
[8-28] B. Grünbaum and G. C. Shephard, *Tilings and Patterns*, W. H. Freeman and Co., New York (1987).
[8-29] A. H. Fraser, *J. Opt. Soc. Am.* **73**, 1626 (1983).
[8-30] M. Ge and K. Sattler, *Science* **260**, 515 (1993); J. Xhie, K. Sattler, M. Ge, and N. Venkateswaran, *Phys. Rev. B* **47**, 15835 (1993).
[8-31] H. Giger, in *Symmetry, Unifying Human Understanding* (I. Hargittai, ed.), p. 329, Pergamon Press, New York (1986).
[8-32] W. Witschi, in *Symmetry, Unifying Human Understanding* (I. Hargittai, ed.), p. 363, Pergamon Press, New York (1986).

Chapter 9

Symmetries in Crystals

> . . . But I must speak again about crystals, shapes, colors. There are crystals as huge as the colonnade of a cathedral, soft as mould, prickly as thorns; pure, azure, green, like nothing else in the world, fiery black; mathematically exact, complete, like constructions by crazy, capricious scientists, or reminiscent of the liver, the heart . . . There are crystal grottos, monstrous bubbles of mineral mass, there is fermentation, fusion, growth of minerals, architecture and engineering art . . . Even in human life there is a hidden force towards crystallization. Egypt crystallizes in pyramids and obelisks, Greece in columns; the middle ages in vials; London in grinny cubes . . . Like secret mathematical flashes of lightning the countless laws of construction penetrate the matter. To equal nature it is necessary to be mathematically and geometrically exact. Number and phantasy, law and abundance—these are the living, creative strengths of nature; not to sit under a green tree but to create crystals and to form ideas, that is what it means to be at one with nature!

These are the words of Karel Čapek, the Czech writer, after his visit to the mineral collection of the British Museum [9-1].* He added a drawing (Figure 9-1) to his words to express his humility in front of these miracles of nature.

*The English version cited in the text was kindly provided by Professor Alan L. Mackay.

Figure 9-1. Čapek's drawing after his visit to the mineral collection of the British Museum [9-1]. Reproduced with permission.

Figure 9-2 displays a few stamps with crystals from different countries and several pictures of crystals.

The word crystal comes from the Greek *krystallos*, meaning clear ice. The name originated from the mistaken belief that the beautiful transparent quartz stones found in the Alps were formed from water at extremely low temperatures. By the 18th century the name crystal was applied to other solids that were also bounded by many flat faces and had generally beautiful symmetrical shapes. Crystals have also been considered to be mystical. A sad

a

Figure 9-2. Crystals: (a) Stamps with crystals. (*Continued on next page*)

Figure 9-2. (*Continued*) (b) electron microscope pictures; (c) photographs.

angel looks hopelessly at the huge rhombohedric crystal in Albrecht Dürer's *Melancholia* (Figure 9-3). The polyhedron in the picture is a truncated rhombohedron, and there has been considerable discussion as to whether Dürer meant a particular mineral by it and, if so, which one [9-2, 9-3]. It was concluded that this polyhedron "is simply an exercise in accurate draughtsmanship and that the art historians have made rather heavy weather of its explanation . . . The integral proportions show that no particular mineral was intended" [9-3]. Dürer's drawings have been carefully analyzed by Schröder [9-4], who "has satisfactorily settled the matter with a technological rather than a mystical explanation" [9-3].

Space-group symmetries have played an outstanding role in Escher's graphic art. So what he wrote about crystals is also of interest [9-5]:

> Long before there were men on this globe, all the crystals grew within the earth's crust. Then came a day when, for the very first

Figure 9-3. Albrecht Dürer's *Melancholia*.

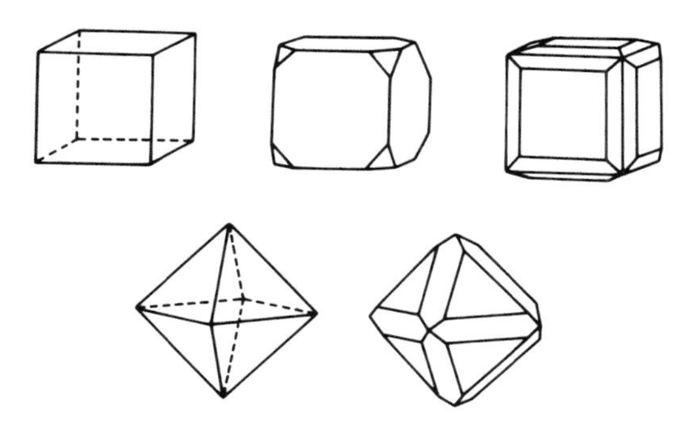

Figure 9-4. Different shapes of sodium chloride crystals as a consequence of the influence of impurities.

> time, a human being perceived one of these glittering fragments of regularity; or maybe he struck against it with his stone ax; it broke away and fell at his feet; then he picked it up and gazed at it lying there in his open hand. And he marveled.
>
> There is something breathtaking about the basic laws of crystals. They are in no sense a discovery of the human mind; they just "are"—they exist quite independently of us. The most that man can do is to become aware, in a moment of clarity, that they are there, and take cognizance of them.

The symmetry of the shapes of crystals is their most easily recognizable feature. The Russian crystallographer E. S. Fedorov remarked that "the crystals glitter with their symmetry." Obviously, this outer symmetry is a consequence of the inner structure. However, with the same inner structure, crystals may grow into different forms. Besides, under natural conditions, crystals seldom grow into their well-known regular forms. Under different conditions, in the presence of different impurities, for example, different forms may grow. Figure 9-4 shows the influence of impurities upon the form of sodium chloride crystals.

9.1 BASIC LAWS

It was recognized already in the earliest stages in the history of crystallography* that the most important characteristic of the outer symmetry of the

*There is a beautiful book on the history of crystallography: see Ref. [9-6].

crystals is not really the form itself but rather two phenomena expressed by two rules. One is the constancy of the angles made by the crystal faces. The other is the law of rational intercepts or the law of rational indices.

Already in 1669 the Danish crystallographer Steno made a detailed study of ideal and distorted quartz crystals (Figure 9-5). He traced their outlines on paper and found that the corresponding angles of different sections were always the same regardless of the actual sizes and shapes of the sections. Thus, all quartz crystals, however much distorted from the ideal, could result from the same fundamental mode of growth and, accordingly, corresponded to the same inner structure.

Instruments were developed to measure the angles made by the crystal faces. In 1780 the contact goniometer (Figure 9-6a) was already in usage. Later, for more precise measurement of the interfacial angles, the reflecting goniometer was introduced (Figure 9-6b).

Another interesting phenomenon observed early in crystals is their cleavage. It is characteristic that they break along well-defined planes. The French crystallographer Haüy [9-7] noticed that the cleavage rhombs from any calcite crystal always had the same interfacial angles. Thus, he suggested that all calcite crystals could be built of these fundamental cleavage rhombs. This is illustrated in Figure 9-7, which is from Haüy's *Traité de Cristallographie* [9-7]. From the units shown in Figure 9-7, it is possible to build straight edges corresponding to the faces of a cube, as well as inclined edges corresponding to the faces of an octahedron. Edges inclined at other edges may also be built. Let the dimensions of the cleavage unit be a and b (Figure 9-8); then $\tan \Theta_1 = b/a$ and $\tan \Theta_2 = b/2a$, and generally $\tan \Theta = mb/na$, where m and n are rational integers. By extension into the third dimension, we may have a reference face making intercepts a, b, and c on three axes. The intercepts made by any other face must be in the proportion of rational multiples of these intercepts. This is called the law of rational intercepts.

Usually, the crystal faces are described by the reciprocals of the multiples of the standard intercepts, hence the name "the law of rational indices." In Figure 9-8 three lines are adopted as axes which may also be directions of the

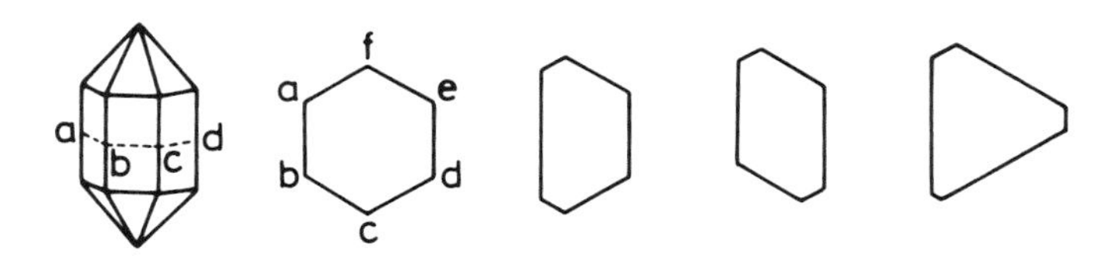

Figure 9-5. Sections of ideal and distorted quartz crystals.

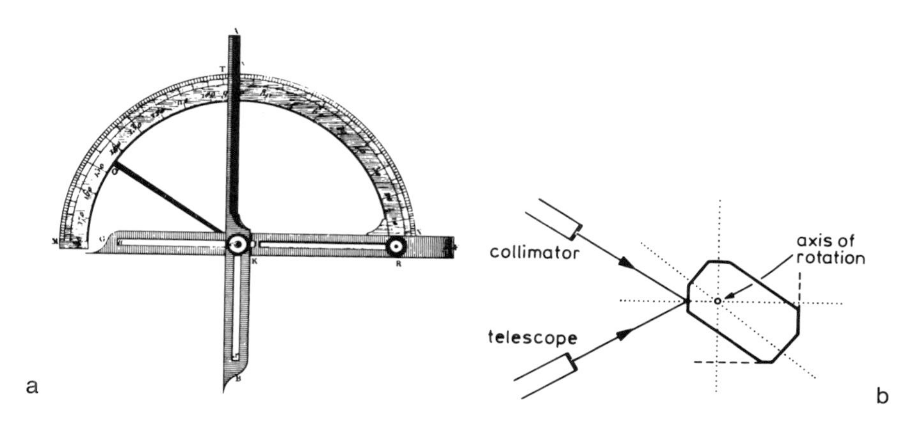

Figure 9-6. (a) Contact goniometer, from Haüy [9-7]. (b) Reflecting goniometer.

crystal edges. A reference face *ABC* makes intercepts a, b, and c on these axes. Another face of the crystal, e.g., *DEC*, can be defined by intercepts a/h, b/k, and c/l. Here h, k, and l are simple rational numbers or zero. They are called Miller indices. The intercept is infinite if a face is parallel to an axis, and h or k or l will be zero. For orthogonal axes the indices of the faces of a cube are (100), (010), and (001). The indices of the face *DEC* in Figure 9-8 are (231).

The simple cleavage model of Haüy indeed revealed a lot about the structure of crystals. However, it was not generally applicable since cleavages do not always lead to cleavage forms which can necessarily fill space by repetition, and, as is known, there is only a limited number of space-filling polyhedra.

The characterization of the regularities in the outer form of crystals led to the recognition of three-dimensional periodicity in their inner structure. This was long before the possibility of determining the atomic arrangements in crystals by diffraction techniques had materialized.

It was 200 years before Dalton and 300 years before X-ray crystallography that Kepler discussed the atomic arrangement in crystals. In his *Strena seu de nive sexangula* [9-8] he presented arrangements of close-packed spheres. These are reproduced in Figure 9-9. Incidentally, close packing of spheres was invoked and illustrated in Dalton's works (Figure 9-10) in relation to gas absorption [9-9]. A close-packed arrangement of cannon balls and a sculpture apparently expressing close packing are shown in Figure 9-11. The fundamental importance of Kepler's idea is that he correlated, for the first time, the external forms of solids with their inner structure. Kepler's search for harmonious proportions is the bridge between his epoch-making discoveries in

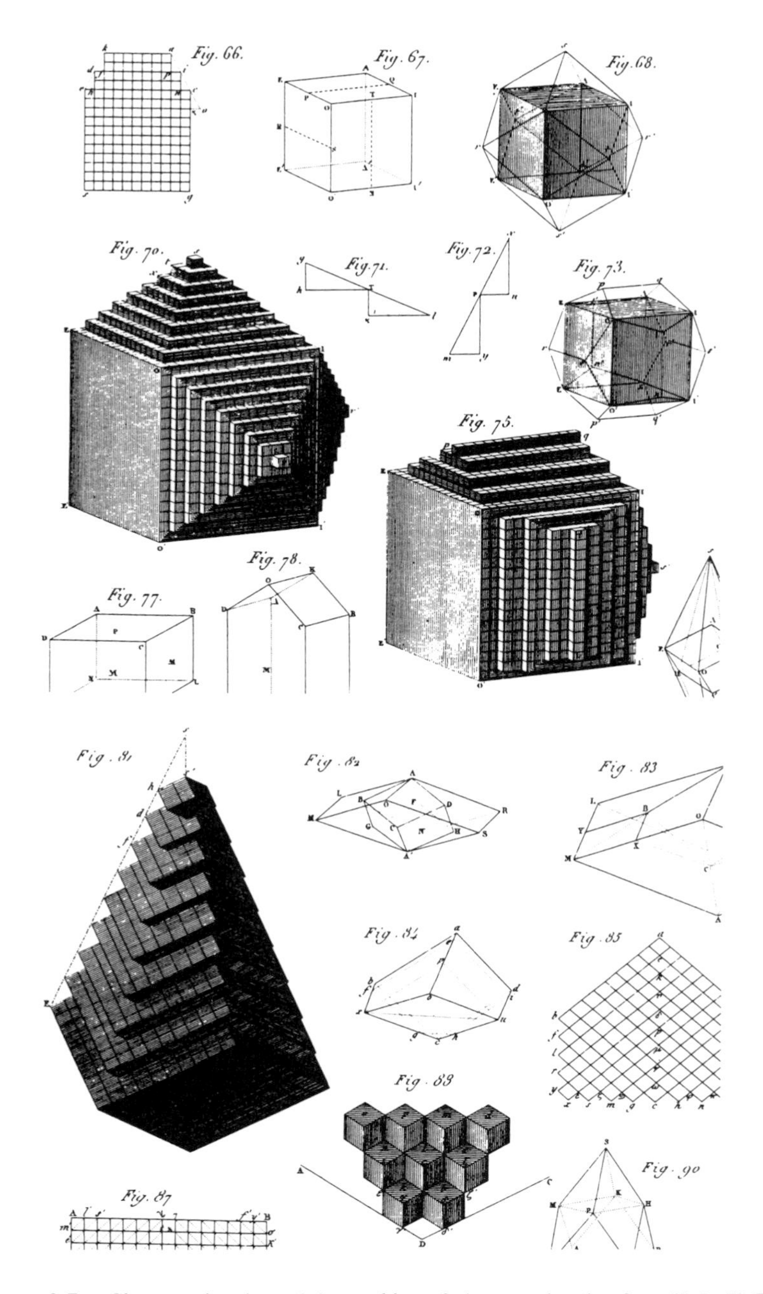

Figure 9-7. Cleavage rhombs and the stacking of cleavage rhombs, from Haüy [9-7].

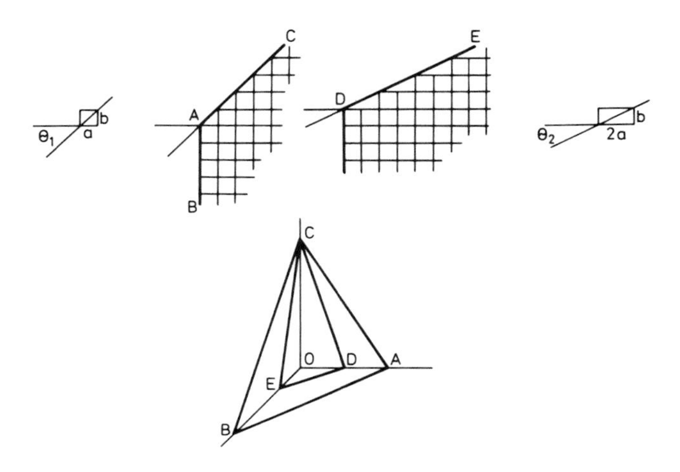

Figure 9-8. Inclined edges from cleavage units and illustration for the law of rational intercepts.

heavenly mechanics and his less widely known but nonetheless seminal ideas in what is called today crystallography. As Schneer has remarked [9-10], the renaissance era provided a stimulating background for the beginnings of the science of crystals.

It is to be noted that even after the discovery of Haüy's model, attention

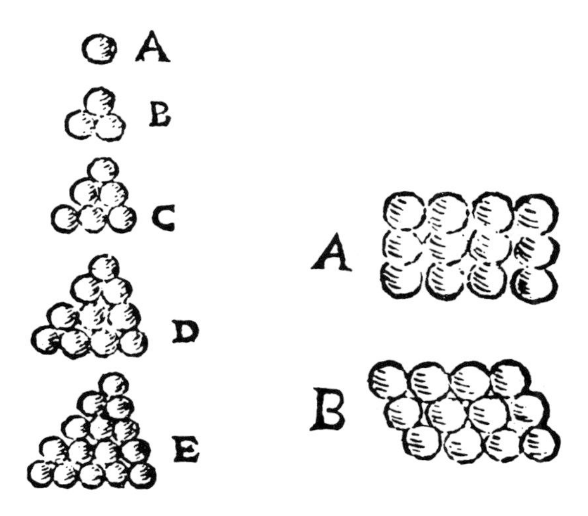

Figure 9-9. Illustration of closely packed spheres by Kepler [9-8].

Figure 9-10. Illustration of closely packed spheres by Dalton [9-9].

was focused on the packing in crystals. The aim was to find those arrangements in space which are consistent with the properties of the crystals.

The most important characteristic of the crystal structure is the three-dimensional periodicity of the atomic arrangement, for which we find an explanation in the dense packing of the participating species.

The symmetry of the form of the crystal is a consequence of its structure. The same high symmetry of the form, however, may be easily achieved for a piece of glass by artificial mechanical intervention. By acquiring the same outer form as is typical for a piece of diamond, the piece of glass will not

a 

Figure 9-11. Examples of close-packed arrangements: (a) Cannon balls, Laconia, New Hampshire; (b) in an open-air sculpture garden near Pécs, Hungary. Photographs by the authors.

acquire all the other properties that the diamond possesses. The difference in value has long been recognized. In the India of the sixth century portrayed by *Kama Sutra* of Vatsayana, one of the arts which a courtesan had to learn was mineralogy (along with chemistry). If she were paid in precious stones, she had to be able to distinguish real crystals from paste [9-11].

It is primarily the structure, and, accordingly, the outer and inner symmetry properties of the crystal, that determines its many outstanding physical properties. The mechanical, electrical, magnetic, and optical properties of crystals are all in close conjunction with their symmetry properties (see, e.g., Ref. [9-12]).

In an actual crystal the atoms are in permanent motion. However, this motion is much more restricted than that in liquids, let along gases. As the nuclei of the atoms are much smaller and heavier than the electron clouds, their motion can be well described by small vibrations about the equilibrium positions. In our discussion of crystal symmetry, as an approximation, the structures will be regarded as completely rigid. However, in modern crystal molecular structure determination, atomic motion must be considered (see, e.g., Ref. [9-13]). Both the techniques of structure determination and the interpretation of the results must include the consequences of the motion of atoms in the crystal. Let the poet crystallographer be cited here [9-14]:

My molecule is sick
And I have caught the illness too.
Two atoms have temperatures
Which are negative,
And two are not resolved at all.
How can I find a cure—
The *R*-factor is enormous
And direct methods fail me?
Perhaps it is not my métier,
To be a structure analyst.

9.2 THE 32 CRYSTAL GROUPS

Although the word crystal in its everyday usage is almost synonymous with symmetry, there are severe restrictions on crystal symmetry. While there are no restrictions in principle on the number of symmetry classes of molecules, this is not so for crystals. All crystals, as regards their form, belong to one or another of only 32 symmetry classes. They are also called the *32 crystal point groups*. Figures 9-12 and 9-13 illustrate them by examples of actual minerals and by stereographic projections with symmetry elements, respectively.

Triclinic and Monoclinic	Orthorhombic	Tetragonal
C_1 1 Sr-tartrate tetrahydrate		C_4 4 Wulfenite
C_i $\bar{1}$ Axinite		C_{4h} 4/m Scheelite
C_2 2 Sucrose	D_2 222 Epsomite	D_4 422 Nickel sulfide
C_s m Hilgardite	C_{2v} mm2 Hemimorphite	C_{4v} 4mm Diabolite
C_{2h} 2/m Augite	D_{2h} mmm Topaz	D_{4h} 4/mmm Cassiterite
		S_4 $\bar{4}$ Cahnite
		D_{2d} $\bar{4}2m$ Chalcopyrite

Figure 9-12. Representation of the 32 crystal point groups by actual minerals, after Ref. [9-15]. (*Continued on next page*)

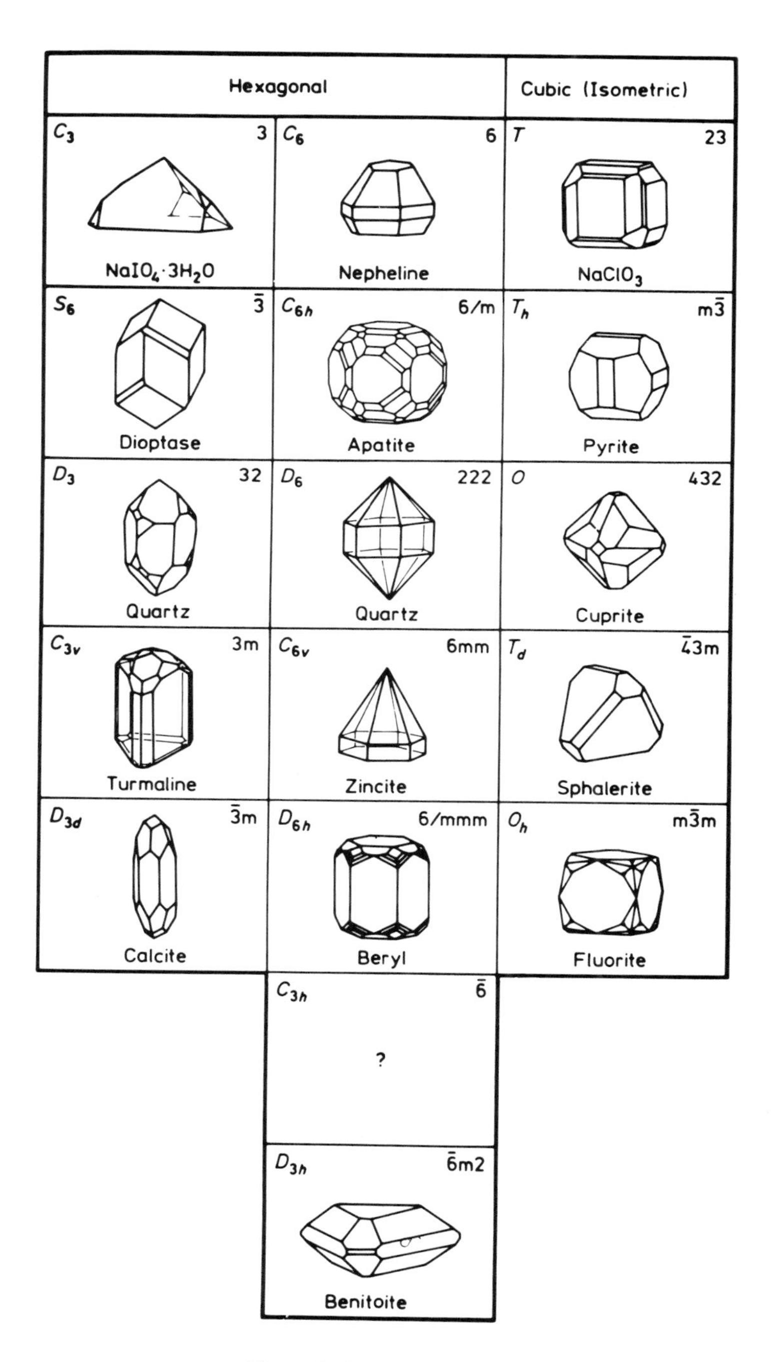

Figure 9-12. *(Continued)*

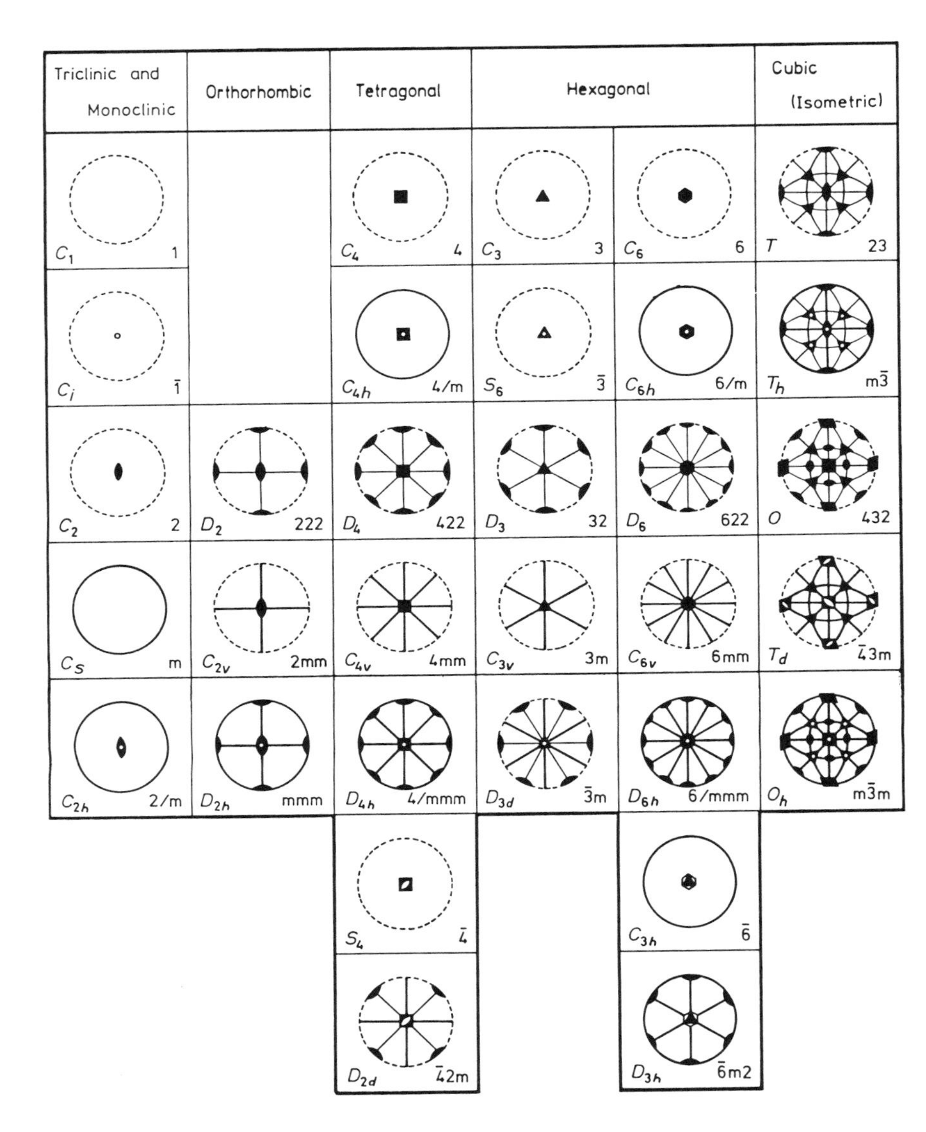

Figure 9-13. Representation of the 32 crystal point groups by stereographic projections.

Stereographic projection starts by representing the crystal through a set of lines perpendicular to its faces. The introduction of this method of representation followed soon after the invention of the reflecting goniometer. Let us place the crystal in the center of a sphere and extend its face normals to meet the surface of the sphere as seen in Figure 9-14a. A set of points will occur on the surface of the sphere representing the faces of the crystal. Join now all the points in the northern hemisphere to the South Pole, and mark the points on the equatorial plane where these connecting lines intersect this plane. This will create a representation of the faces on the upper half of the crystal within a single circle as seen in Figure 9-14b. Performing a similar operation for the

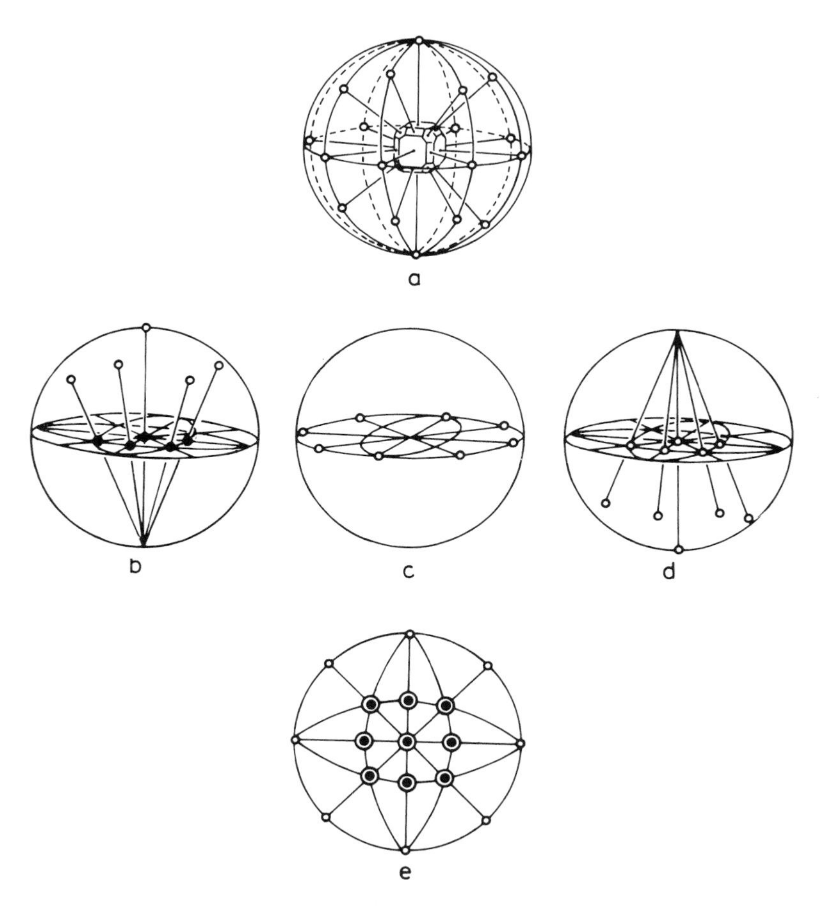

Figure 9-14. The preparation of the stereographic representation. See text for details.

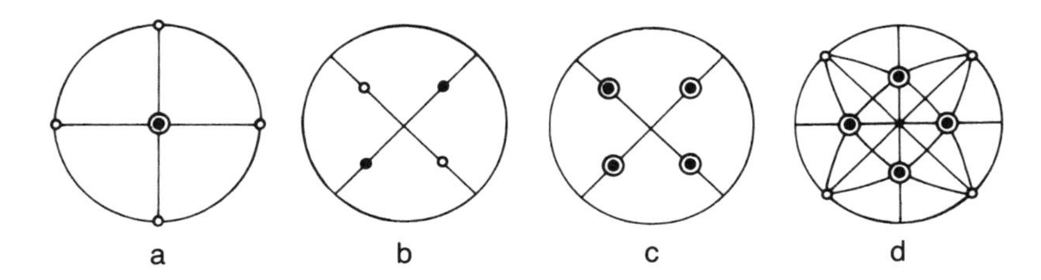

Figure 9-15. Representations of some simple highly symmetrical shapes: (a) Cube; (b) tetrahedron; (c) octahedron; (d) rhombic dodecahedron.

points of the equator (Figure 9-14c) and for the points in the southern hemisphere (Figure 9-14d), we arrive at the representation of the whole crystal within the circle (Figure 9-14e). The points from the northern hemisphere are marked by dots, and those from the southern hemisphere by small circles. Some examples for simple polyhedra are shown in Figure 9-15.

9.3 RESTRICTIONS

To have 32 symmetry classes for the external forms of crystals is a definite restriction, and it is obviously the consequence of inner structure. The translation periodicity limits the symmetry elements that may be present in a crystal. The most striking limitation is the absence of fivefold rotation in the world of crystals. Consider, for example, planar networks of regular polygons (Figure 9-16). Those with threefold, fourfold, and sixfold symmetry cover the

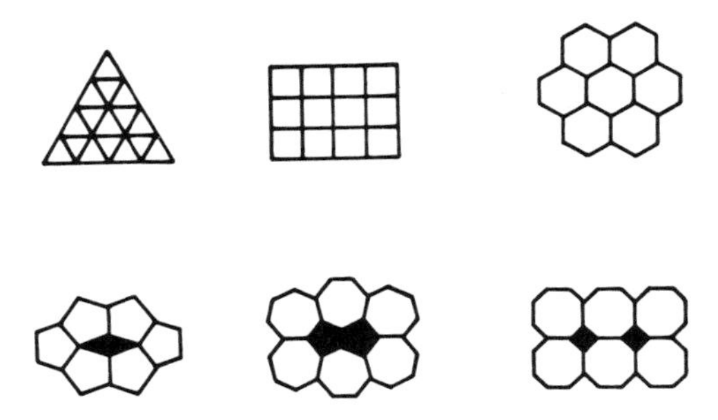

Figure 9-16. Planar networks of regular polygons with up to eightfold symmetry.

available surface without any gaps, while those with fivefold, sevenfold, and eightfold symmetry leave gaps on the surface. Figure 9-17 presents two planar networks of octagons. It is evident that the regular octagons cannot cover the surface without gaps. There are smaller squares among the octagons.

Let us examine now the possible types of symmetry axes in space groups (cf., e.g., Ref. [9-17]). Figure 9-18 shows a lattice row with a period t. An n-fold rotation axis, C_n, is placed on each lattice point. Since n rotations, each by an angle φ, must lead to superposition, it does not matter in which direction the rotations are performed. Two rotations by φ about two axes but in opposite directions are shown in Figure 9-18. The two new lattice points produced this way are labeled p and q. These two new points are equidistant from the original row, and hence the line joining them is parallel to the original lattice row. The length of the parallel line joining p and q must be equal to some integer multiple m of the period t. Were it not, then the line joining the two new lattice points p and q would not be a translation of the lattice, and the resulting array would not be periodic.

Using Figure 9-18, it is possible to determine the possible values that the rotation angle φ can have in the lattice,

$$mt = t + 2t\cos\varphi \qquad m = 0, \pm 1, \pm 2, \pm 3, \ldots$$

where $+m$ or $-m$ is taken depending on the direction of the rotation:

$$\cos\varphi = \frac{m-1}{2}$$

b

Figure 9-17. Octagonal planar networks: (a) Detail of a Tintoretto painting in the Prado Museum, Madrid; photograph by the authors; (b) Hungarian needlework [9-16].

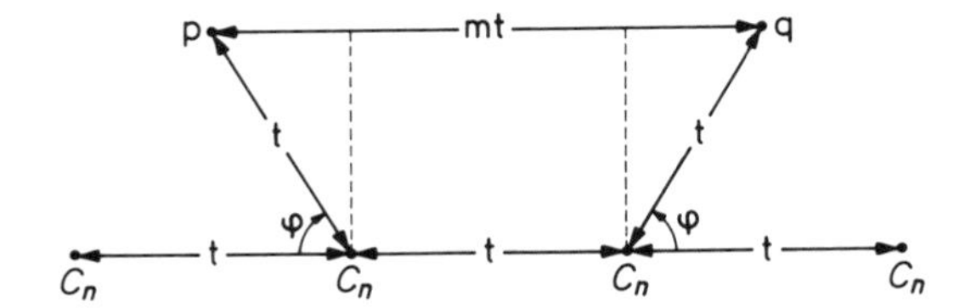

Figure 9-18. Illustration of the determination of the possible throws that rotation axes can have in space groups. After Azaroff [9-17]. Copyright (1960) McGraw-Hill, Inc. Used with permission.

Only the solutions corresponding to the range

$$-1 \leqslant \cos\varphi \leqslant 1$$

need be considered, and these are shown in Table 9-1. Five solutions are possible, and, accordingly, only five kinds of rotation axes are compatible with a lattice. Thus, not only fivefold symmetry is not allowed in crystal structures, but all periods larger than six are impossible. Naturally, this applies to the planar networks as well.

The permissible periods of mirror-rotation axes have the same limitations as those of the proper rotation axes.

Let us examine now the limitations on the screw axes. In a lattice the screw axes must be parallel to a translation direction. After n rotations by an angle φ and n translations by the distance T, that is, after n translations along the screw axis, the total amount of translation distance in the direction of this axis must be equal to some multiple of the lattice translation mt,

$$nT = mt$$

Table 9-1. Allowed Rotation Axes n in a Lattice

Possible values of $m - 1$	$\cos\varphi$	φ (°)	n
-2	-1	180	2
-1	$-\frac{1}{2}$	120	3
0	0	90	4
$+1$	$+\frac{1}{2}$	60	6
$+2$	$+1$	360 or 0	1

where n and m are integers. Rearranging this equation,

$$T = \frac{mt}{n}$$

where m, of course, may be 0, 1, 2, 3, etc., but n may only be 1, 2, 3, 4, or 6. It is then possible to determine the permissible values of the pitch of the screw axes in lattices. They are summarized in Table 9-2, taking also into consideration that $(\frac{3}{2})t = t + (\frac{1}{2})t$, $(\frac{5}{4})t = t + (\frac{1}{4})t$, etc. There are only 11 screw axes that are allowed in a lattice, n_m, according to Table 9-2. The subscript in the notation is the m of the expression $T = (mt)/n$. The proper rotation axes may be considered to be special cases of the screw axes, with $m = 0$ and $m = n$. The 11 screw axes are shown in perspective in Figure 9-19. It is seen there that some pairs are identical except for the direction of the screw motion. Such screw axes are enantiomorphous. The enantiomorphous screw axis pairs are the following:

- 3_1 and 3_2
- 4_1 and 4_3
- 6_1 and 6_5
- 6_2 and 6_4

Finally, the only remaining symmetry element is considered, the glide-symmetry plane. It causes glide reflection as a result of reflection *and*

Table 9-2. Possible Values of the Pitch T of an n-Fold Screw Axis

A. Possible values of T								
$n = 1$	$0t$,	$1t$,	$2t$, . . .					
$n = 2$	$0t$,	$(1/2)t$,	$(2/2)t$,	$(3/2)t$, . . .				
$n = 3$	$0t$,	$(1/3)t$,	$(2/3)t$,	$(3/3)t$,	$(4/3)t$, . . .			
$n = 4$	$0t$,	$(1/4)t$,	$(2/4)t$,	$(3/4)t$,	$(4/4)t$,	$(5/4)t$, . . .		
$n = 6$	$0t$,	$(1/6)t$,	$(2/6)t$,	$(3/6)t$,	$(4/6)t$,	$(5/6)t$,	$(6/6)t$,	$(7/6)t$, . . .
B. Possible values of T (redundancies eliminated)								
$n = 1$								
$n = 2$		$(1/2)t$,						
$n = 3$		$(1/3)t$,	$(2/3)t$					
$n = 4$		$(1/4)t$,	$(2/4)t$,	$(3/4)t$				
$n = 6$		$(1/6)t$,	$(2/6)t$,	$(3/6)t$,	$(4/6)t$,	$(5/6)t$		
C. Notation of screw axes allowed in a lattice								
$n = 2$		2_1						
$n = 3$		3_1	3_2					
$n = 4$		4_1	4_2	4_3				
$n = 6$		6_1	6_2	6_3	6_4	6_5		

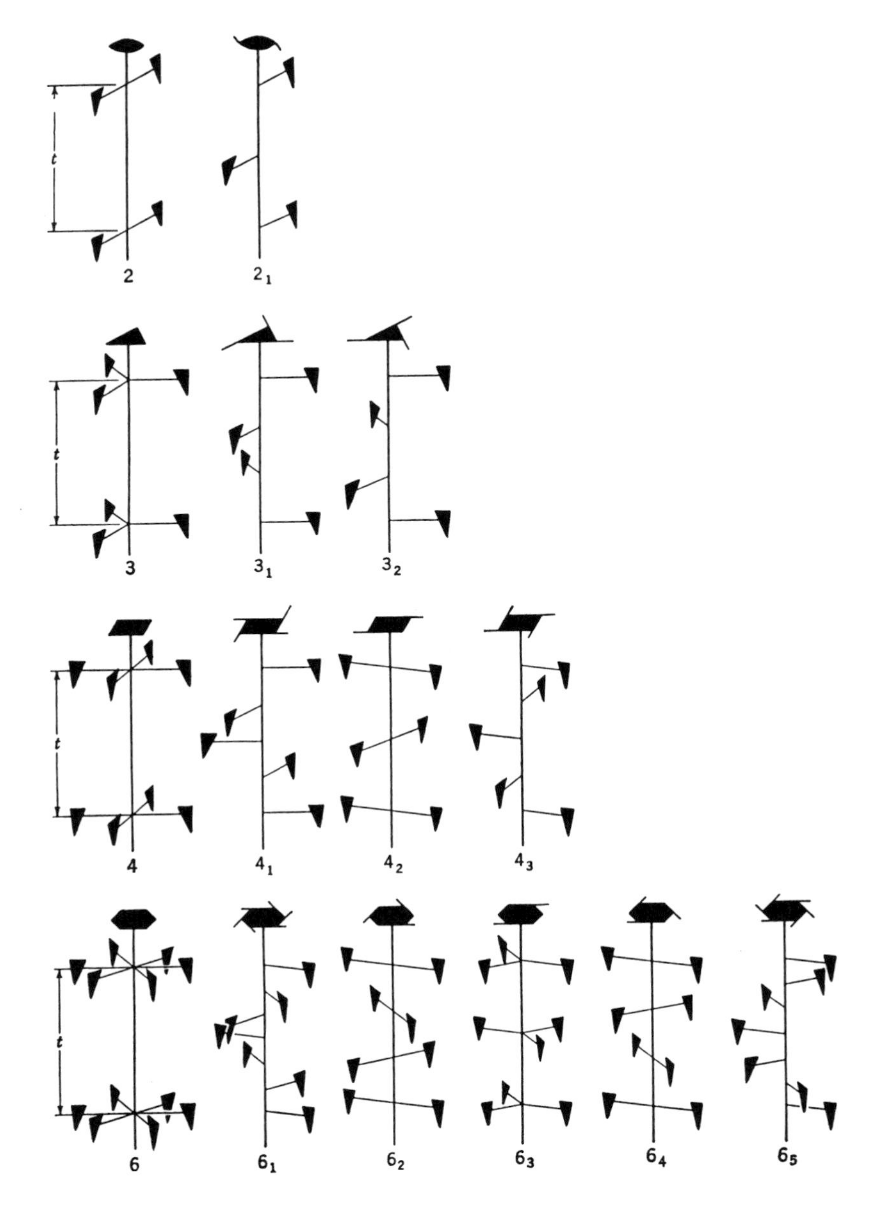

Figure 9-19. The 11 screw axes. The simple twofold, threefold, fourfold, and sixfold axes are also shown for completeness. After Azaroff [9-17]. Copyright (1960) McGraw-Hill, Inc. Used with permission.

Table 9-3. Possible Glide Planes

Glide type	Symbol	Translation component
Axial	a	$a/2$
Axial	b	$b/2$
Axial	c	$c/2$
Diagonal	n	$a/2 + b/2$; $b/2 + c/2$; or $c/2 + a/2$
Diamond[a]	d	$a/4 + b/4$; $b/4 + c/4$; or $c/4 + a/4$

[a]Translation component is one-half of the true translation along the face diagonal of a centered plane lattice.

translation. The translation component T of a glide plane is one-half of the normal translation of the lattice in the direction of the glide. A glide along the a axis is $T = (\frac{1}{2})a$, and this is called an a glide. Similarly, a diagonal glide can have $T = (\frac{1}{2})a + (\frac{1}{2})c$. The different possible glide planes are summarized in Table 9-3.

The fact that the crystal has a lattice framework imposes strict limitations

Table 9-4. Characterization of Crystal Systems

System	Minimal symmetry (diagnostic symmetry elements)	Relations between edges and angles of unit cell	Lattice type	Numbering in Figure 9-20
Triclinic	1 (or $\bar{1}$)	$a \neq b \neq c$ $\alpha \neq \beta \neq \gamma \neq 90°$	P	1
Monoclinic	2 (or $\bar{2}$)	$a \neq b \neq c$ $\alpha = \gamma = 90° \neq \beta$	P C (or A)	2 3
Orthorhombic	222 (or $\overline{222}$)	$a \neq b \neq c$ $\alpha = \beta = \gamma = 90°$	P C (or B or A) I F	4 5 6 7
Trigonal (rhombohedral)	3 ($\bar{3}$)	$a = b = c$ $\alpha = \beta = \gamma \neq 90°$	R	8
Hexagonal	6 ($\bar{6}$)	$a = b \neq c$ $\alpha = \beta = 90°$, $\gamma = 120°$	P	9
Tetragonal	4 (or $\bar{4}$)	$a = b \neq c$ $\alpha = \beta = \gamma = 90°$	P I	10 11
Cubic	Four 3 (or $\bar{3}$)	$a = b = c$ $\alpha = \beta = \gamma = 90°$	P I F	12 13 14

on the symmetry of its outer form. On the other hand, the question arises as to whether it is possible to derive any information about the crystal lattice from the knowledge of the symmetry of its outer form.

The 32 crystal point groups can be classified by symmetry criteria. They are usually grouped according to the highest ranking rotation axis that they contain. The resulting groups are called crystal systems. There are altogether seven of them, and they are listed in Table 9-4. The crystal point groups have to be combined with all possible space lattices in order to produce the space groups.

9.4 THE 230 SPACE GROUPS

There are 14 infinite lattices, called Bravais lattices, in three-dimensional space. They are shown in Figure 9-20. These lattices are the analogs of the five infinite lattices in two-dimensional space (Figure 8-27). The Bravais lattices are presented as systems of points at vertices of parallelepipeds. The corresponding parallelepipeds are capable of filling space without any gaps or overlap. The representation of the lattices by systems of points is especially useful as it makes it possible to join the lattice points in any desired way conforming with the symmetry requirements. In this way, not only the original parallelepipedal forms but any other possible figures may be used as building units for the space lattice.

The 14 Bravais lattices are classified in Table 9-4 as the following types: primitive (P, R), side-centered (C), face-centered (F), and body-centered (I). The numbering of the Bravais lattices in Table 9-4 corresponds to that in Figure 9-20. The lattice parameters are also enumerated in the table. In addition, the distribution of lattice types among the crystal systems is shown.

The actual infinite lattices are obtained by parallel translations of the Bravais lattices as unit cells. Some Bravais cells are also primitive cells, others are not. For example, the body-centered cube is a unit cell but not a primitive cell. The primitive cell in this case is an oblique parallelepiped constructed by using as edges the three directed segments connecting the body center with three nonadjacent vertices of the cube.

The three-dimensional space groups are produced by combining the 32 crystallographic point groups with the Bravais lattices. Since the symmetry elements in a space lattice can have translation components, indeed not only the 32 groups but also the analogous groups, which have screw axes and glide planes, have to be considered. There are altogether 230 three-dimensional space groups! Their complete description can be found in the *International Tables for X-Ray Crystallography* [9-18]. Only a few examples are discussed here.

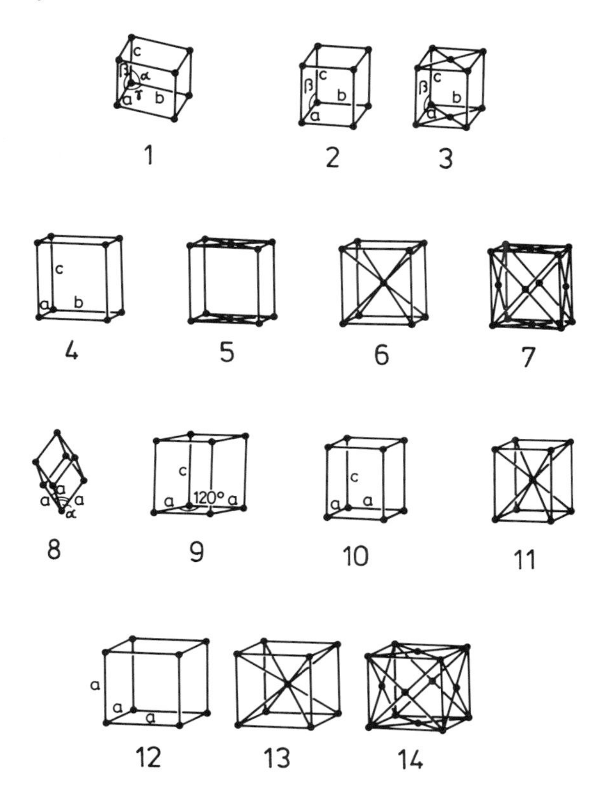

Figure 9-20. The 14 Bravais lattices.

There are only two combinations possible for the triclinic system. They are named $P1$ and $P\bar{1}$. For the monoclinic system three point groups are to be considered and two lattice types. Combining P and I lattices, on one hand, and point group 2 and symmetry 2_1, on the other hand, the four possible combinations are $P2$, $P2_1$, $I2$, and $I2_1$. The latter two, however, are equivalent; only their origins differ.

The description of the symmetry elements of the space groups is similar to that of the point groups [9-19]. The main difference is that the order in which the symmetry elements of the space groups are listed may be of great importance, except for the triclinic system. The order of the symmetry elements expresses their relative orientation in space with respect to the three crystallographic axes. For the monoclinic system, the unique axis may be the c or the b axis. For the $P2$ space group, the complete symbol may be $P112$ or

$P121$, depending on this choice and using the sequence *abc*. The two variations are called first setting and second setting, respectively. The ordering of symbols for the orthorhombic system is especially important. The symmetry elements are usually listed in the order *abc*. The space groups which belong to the crystal class $2mm$ are properly presented as $Pmm2$, c being the unique axis.

In the tetragonal system, the c axis is the fourfold axis. The sequence for listing the symmetry elements is c, a, [110], since the two crystallographic axes orthogonal to c are equivalent. For example, the three-dimensional space group notation $P\bar{4}m2$ has the following meaning: the unique axis in a primitive tetragonal lattice is a $\bar{4}$ axis, the two a axes are parallel to m, and the [110] direction has twofold symmetry. A similar sequence is used for listing the symmetry elements of the hexagonal system, for which the c axis again is the unique axis and the other two are equivalent. P denotes the primitive hexagonal lattice while R denotes the centered hexagonal lattice in which the primitive rhombohedral cell is chosen as the unit cell.

All three crystallographic axes are equivalent in the cubic system. The order of listing the symmetry elements is a, [111], [110]. When the number 3 appears in the second position, it merely serves to distinguish the cubic system from the hexagonal one.

It may be of interest to add some new symmetry to a group or to decrease its symmetry and examine the consequences. If the addition produces a new group, it is called a supergroup of the original group. If eliminating symmetry leads to a new group, it is usually a subgroup of the original one. For example, the point group 1 is obviously a subgroup of all the other 31 groups as it has the lowest possible symmetry. On the other hand, the highest symmetry cubic group can have no supergroups.

It is important to distinguish between the symmetry of the lattice and the symmetry of the actual building elements of the crystal—the atoms, ions, or molecules. In the illustration of Figure 9-21, the lattice positions are occupied by spheres, which have the highest possible symmetry. However, the building elements usually have lower symmetries, especially in molecular crystals. Brock and Lingafelter [9-20] pointed out commonly existing misconceptions about the difference between the crystal and its lattice. A crystal is an array of units (atoms, ions, or molecules) in which a structural motif is repeated in three dimensions. A lattice is an array of points, and every point has the same environment of points in the same orientation. Each crystal has an associated lattice, whose origin and basis vectors can be chosen in various ways. From the above it is clear, for example, that it would be improper to speak about "interpenetrating lattices," while it is correct to talk about interpenetrating arrays of atoms [9-20].

As we have reached in our discussion the system of the 230 three-dimensional space groups, it appears, as it indeed is, a perfect system. It was

Figure 9-21. Artistic expression of atomic arrangement in extended structure. Stainless steel sculpture *Cosmonergy* in downtown Seoul by Professor Kawan-Mo Chung. Photograph by the authors.

established a long time ago, in fact, well before X-ray diffraction could have been applied to the determination of crystal structure. That these 230 three-dimensional space groups were derived in their entirety by Fedorov, Schoenflies, and Barlow, working independently at the end of the 19th century, will always be considered a great scientific feat. No crystal can ever be produced, either in nature or artificially, whose structure would not fall into one or another of these 230 groups.

An interesting statistical test was performed concerning the total number of three-dimensional space groups some time in the mid-sixties [9-21]. It was a uniquely appropriate point in the history of crystallography for such a test: Already a large number of crystal structures had been determined, but examples for all the space groups had not yet been found among actual crystals. The total number of three-dimensional space groups had long before been firmly established. Thus, the test was considered as much to be a check of the applied statistical method as to be a source of crystallographic information.

Although there are 230 space groups, not all of them are in practice distinguishable. So 11 enantiomorphous groups were excluded from the count as were two more groups for other reasons. Thus, the number of space groups to be considered was 217. The 3782 crystal structures that were reviewed showed a wide variation in the frequency of occurrence of the different space groups. One group occurred 355 times, while 33 groups occurred only once each. It was also interesting that only 178 groups out of the total of 217 occurred. Based on the available data for the distribution of the space groups among the determined structures, the findings were extrapolated to an indefinitely large sample. The statistical test led to an extrapolated value of 216. The estimated accuracy of the procedure was 2%. Thus, the estimate agreed with the accepted value of the total number of practically distinguishable space groups of 217.

The statistical analysis was also applied separately to the data on inorganic and organic crystals. In both cases the extrapolated estimate for the total number of three-dimensional space groups was smaller than when all the data had been considered together. The total numbers estimated for the inorganic and organic structures were 209 and 185, respectively. Thus, the conclusion could be made that the inorganic and organic crystals belong to space groups with different population distributions. Statistical analysis of population distributions among the three-dimensional space groups, according to various criteria, has remained an important research tool (see, e.g., Refs. [9-22] and [9-23]).

In the following sections, we will look in some more detail at the symmetry systems of two fundamentally important crystals, rock salt and diamond, following their descriptions by Shubnikov and Koptsik [9-19]. The descriptions that will be given are far from complete. They are intended to give some flavor for the characterization of these two highly symmetrical structures rather than to be a rigorous treatment.

9.4.1 Rock Salt

The unit cell of the rock salt structure and the projection of this structure along the edges of the unit cell onto a horizontal plane are shown in Figure 9-22a and b. The equivalent ions are related by translations $a = b = c$ along the edges of the cube, or $(a + b)/2$, $(a + c)/2$, $(b + c)/2$ along the face diagonals. All this corresponds to the face-centered cubic group (F). The structure coincides with itself not only after these translations, but also after the operations of the point group $m\bar{3}m$ (or in other notation $\tilde{6}/4$). The point-group symmetry elements are shown also in Figure 9-22c. The symmetry elements of this group intersect at the centers of all ions, and thus they become symmetry elements for the whole unit cell and, accordingly, for the whole crystal.

Among the projected symmetry elements in Figure 9-22c, there are some which are derived from the generating elements. This is the case, for example, for vertical glide-reflection planes with elementary translations $a/2$ and $b/2$ (represented by broken lines), translations (dot-dash lines), vertical screw axes 2_1 and 4_2, and symmetry centers (small hollow circles, some of which lie above the plane by $\frac{1}{4}$ of the elementary translation).

Two very simple descriptions of the rock salt crystal structure are also given. According to one, the sodium and chloride ions occupy positions with point-group symmetry $m\bar{3}m$ forming a checkered pattern in the $Fm\bar{3}m$ space group. According to the other description, the structure consists of two cubic sublattices in parallel orientation, one of sodium ions and the other of chloride ions.

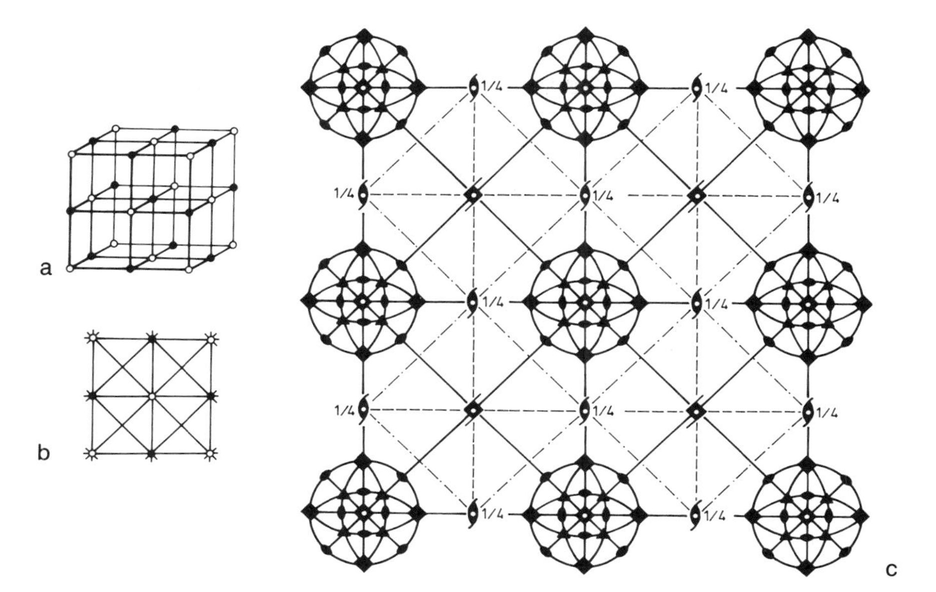

Figure 9-22. The crystal structure of rock salt. (a) A unit cell. (b) Projection of the structure along the edges of the unit cell onto a horizontal plane. (c) Projection of some symmetry elements of the $Fm\bar{3}m$ space group onto the same plane. The vertical screw axes 2_1 and 4_2 are marked by the symbols ◊ and ◘, respectively. After Shubnikov and Koptsik [9-19].

9.4.2 Diamond

Figure 9-23 illustrates the diamond structure. It can be regarded as a set of two face-centered cubic sublattices displaced relative to each other by $\frac{1}{4}$ of the body diagonal of the cube. Each of the two sublattices has the $F\bar{4}3m$ space group, and, in addition, there are some operations transforming one to the other. The complete diamond structure has the space group $Fd\bar{3}m$, where "d" stands for a "diamond" plane.

Among the projected symmetry elements in Figure 9-23c, there are again some which are produced by the generating elements. Special for the diamond structure are the symmetry elements which connect the two subgroups $F\bar{4}3m$. They include vertical left-handed and right-handed screw axes, 4_1 and 4_3, respectively, symmetry centers (small hollow circles, $\frac{1}{8}$ and $\frac{3}{8}$ of the elementary translation c above the plane), vertical "diamond" glide-reflection planes d represented by dot-dash lines with arrows, and similar systems of connecting elements in the horizontal directions.

The subgroup $F\bar{4}3m$ is common to both the rock salt space group $Fm\bar{3}m$ and the diamond space group $Fd\bar{3}m$. The space group $Fd\bar{3}m$ is obtained from

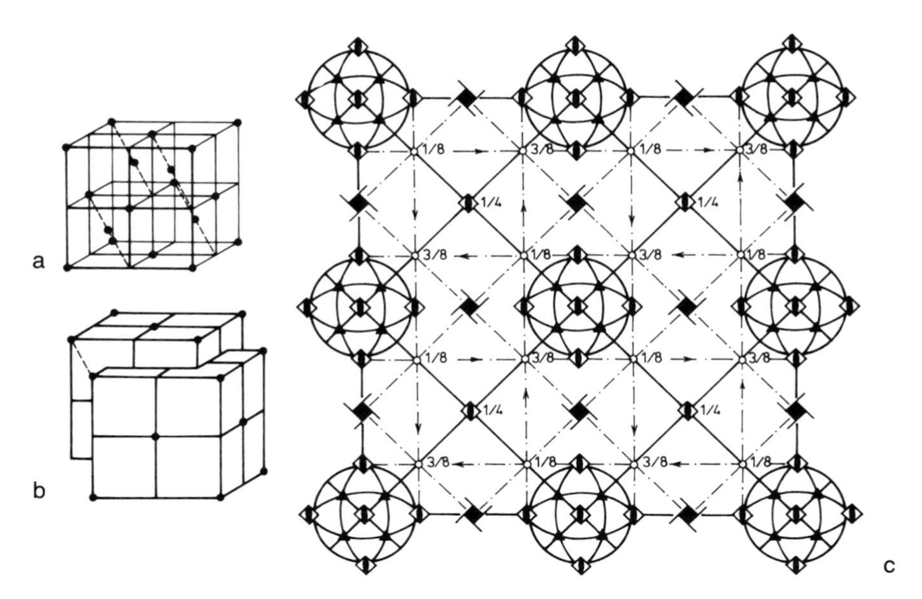

Figure 9-23. The diamond structure. (a) A unit cell; the edges of the cube are the a, b, and c axes. (b) Two face-centered cubic sublattices displaced along the body diagonal of the cube. (c) Projection of some symmetry elements of the $Fd\bar{3}m$ space group onto a horizontal plane. The vertical screw axes 4_1 and 4_3 are marked by the symbols and , respectively. After Shubnikov and Koptsik [9-19].

$Fm\bar{3}m$ by replacing the symmetry planes m by glide-reflection planes d with the latter displaced $\frac{1}{8}$ along the cube edges.

9.5 DENSE PACKING

Dalton [9-24] envisaged the structural difference between water and ice in packing properties. Figure 9-24 reproduces a drawing from his 1808 book *A New System of Chemical Philosophy*. According to Dalton, the "atoms" of ice arrange themselves in a hexagonal scheme, while the "atoms" of water do not. In any case it is remarkable that the principal difference between the water and ice structures is expressed in the packing density. Figure 9-25 originates from a different age [9-25]. It shows the atomic and molecular arrangements in the crystals of 2Zn-insulin from the work of Dorothy Hodgkin and her associates. The molecular structure of insulin is extremely complicated, but the molecular packing, especially the arrangement of the insulin hexamers, reminds us of Dalton's hexagonal ice.

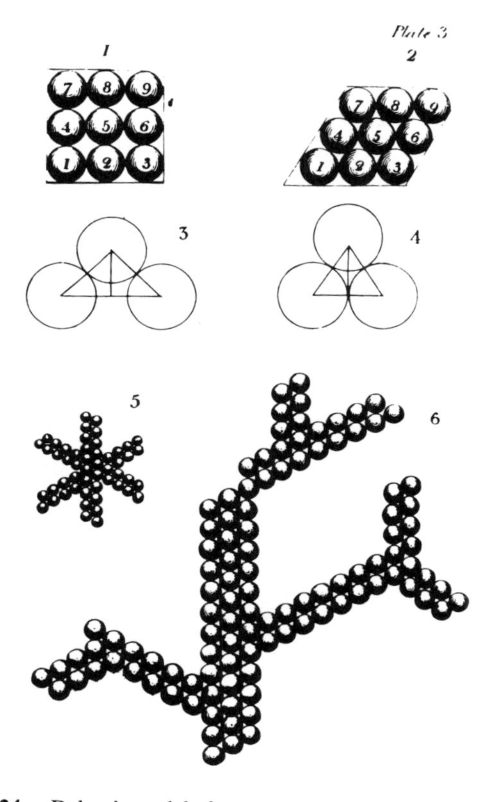

Figure 9-24. Dalton's models for water (1, 3) and ice (2, 4–6) [9-24].

The symmetry of the crystal structure is a direct consequence of dense packing. The densest packing is when each building element makes the maximum number of contacts in the structure. First, the packing of equal spheres in atomic and ionic systems will be discussed. Then molecular packing will be considered. Only characteristic features and examples will be dealt with here since systematic treatises on crystal symmetries are available for consultation [9-17, 9-22, 9-26].

9.5.1 Sphere Packing

The most efficient packing results in the greatest possible density. The density is the fraction of the total space occupied by the packing units. Only those packings are considered in which each sphere is in contact with at least six neighbors. The densities of some packings are given in Table 9-5. There are

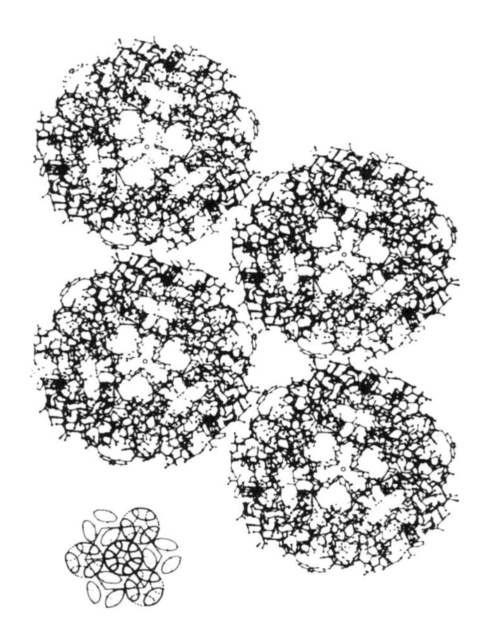

Figure 9-25. Atomic arrangement in the 2Zn-insulin crystal. The smaller projection drawing shows the molecular packing in the insulin hexamers. Courtesy of Professor D. Hodgkin [9-25].

stable arrangements with smaller numbers of neighbors, meaning lower coordination numbers, when directed bonds are present. In our discussion, however, the existence of chemical bonds is not a prerequisite at all.

For three-dimensional six-coordination, the most symmetrical packing is when the spheres are at the points of a simple cubic lattice (Figure 9-26a). Each sphere is in contact with six others situated at the vertices of an octahedron. For

Table 9-5. Densities of Sphere Packing[a]

Coordination number	Name of packing	Density
6	Simple cubic	0.5236
8	Simple hexagonal	0.6046
8	Body-centered cubic	0.6802
10	Body-centered tetragonal	0.6981
12	Closest packing	0.7405

[a]After Wells [9-26].

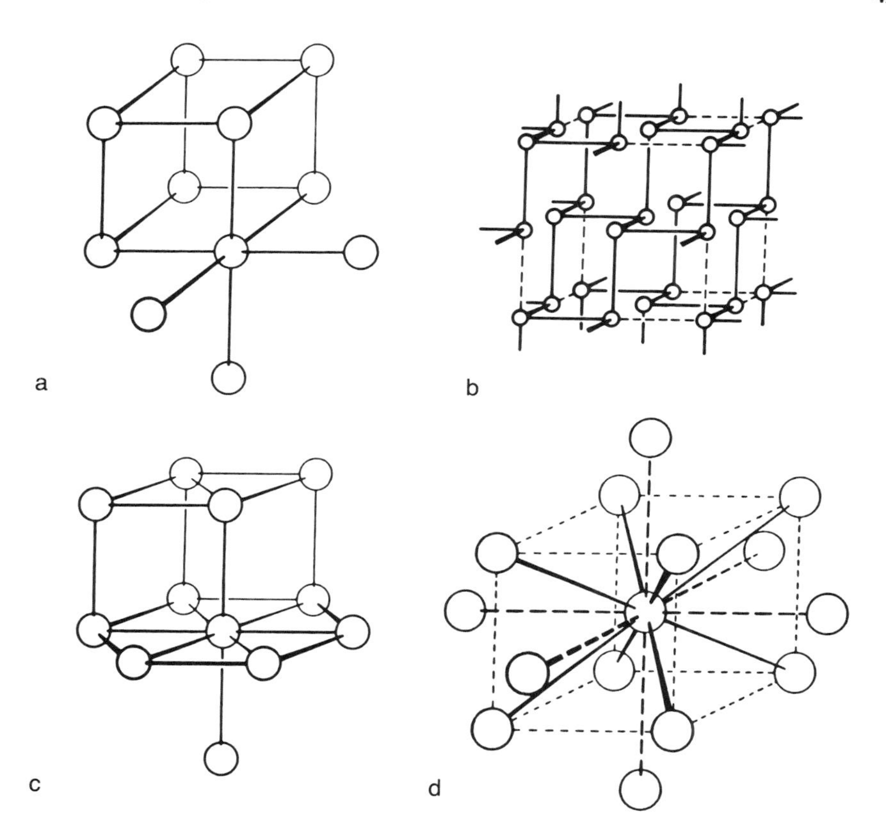

Figure 9-26. Various types of sphere packing: (a) Simple cubic; (b) the somewhat distorted cubic packing of arsenic; (c) simple hexagonal; (d) body-centered cubic. After Wells [9-26]. Reproduced with permission.

the sake of clarity, the atoms are shown separated in the figure. The packing is more realistically represented when the spheres touch each other. Already Kepler (Figure 9-9) and later Dalton (Figures 9-10 and 9-24) also employed such representations.

The structure of crystalline arsenic provides an example of somewhat distorted simple cubic packing. It is illustrated in Figure 9-26b. Each atom has three nearest and three more distant neighbors. The layers formed by the nearest bonded atoms may also be derived from a plane of hexagons. These layers buckle as the bond angle decreases from 120°.

The simple hexagonal sphere packing is shown in Figure 9-26c. The coordination number is eight. This packing is not very important for crystal structures.

Figure 9-26d shows the body-centered packing with eight-coordination. For the central atom, the six next nearest neighbors are at the centers of neighboring unit cells. In terms of polyhedral domains, a truncated octahedron is adopted here. The central atom, in fact, has a coordination number of 14.

It may often be convenient to describe the crystal structure in terms of the domains of the atoms [9-26]. The domain is the polyhedron enclosed by planes drawn midway between the atom and each neighbor, these planes being perpendicular to the lines connecting the atoms. The number of faces of the polyhedral domain is the coordination number of the atom, and the whole structure is a space-filling arrangement of such polyhedra.

The closest packing of equal circles on a plane surface has already been considered. The closest packing of spheres on a plane surface is a similar problem. Again, the densest arrangement is when a sphere is in contact with six others. Layers of spheres may then be superimposed in various ways. The closest packing is when each sphere touches three others in each adjacent layer, the total number of contacts then being 12. Closest packing is thus based on closest packed layers. Figure 9-27 illustrates this. The spheres in one layer are labeled *A*, and a similar layer can be placed above the first so that the centers of the spheres in the upper layer are vertically above the positions *B* (or *C*). The third layer can be placed in two ways. The centers of the spheres may lie above either the *C* or the *A* positions. The two simplest sequences of layers are then *ABABAB* . . . and *ABCABC* . . .. They will have the same density (0.7405).

The packing based on the sequence *ABAB* . . . is called hexagonal closest

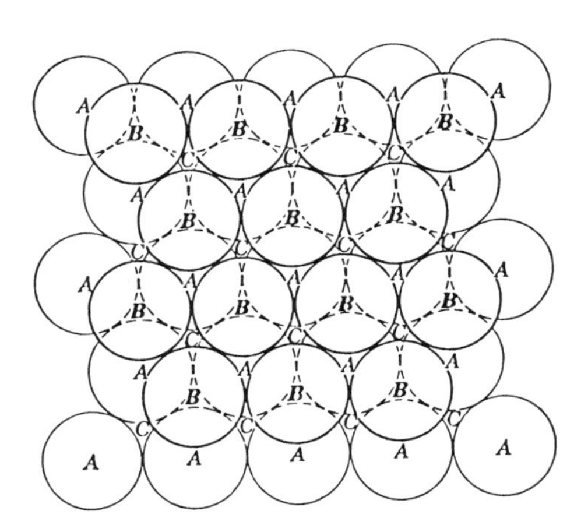

Figure 9-27. Closest packing of *ABC* layers. After Wells [9-26]. Reproduced with permission.

packing and is illustrated by Figure 9-28a. Each sphere has 12 neighbors situated at the vertices of a coordination polyhedron.

The packing based on the sequence *ABCABC* . . . is called cubic closest packing. It is illustrated in Figure 9-28b and is characterized by cubic symmetry.

The closest packing of equal spheres is achieved in an arrangement in which each sphere touches three others in each adjacent layer. The total number of neighbors is then 12. Although the packing in any layer is evidently the densest possible packing, this is not necessarily true of the space-filling arrangements resulting from stacking such layers. Thus, consider the addition of a fourth sphere to the most closely packed triangular arrangement [9-26]. The maximum number of contacts is three in the emerging tetrahedral group. The space-filling arrangement would require each tetrahedron to have faces common with four other tetrahedra. However, regular tetrahedra are not suitable to fill space without gaps or overlaps because the angle of the tetrahedron, 70°32′, is not an exact submultiple of 360°.

Alternatively, continue placing spheres around a central one, all spheres having the same radius. The maximum number that can be placed in contact with the first sphere is 12. However, there is a little more room around the central sphere than just for 12, but not enough for a 13th sphere. Because of the extra room, there are an infinite number of ways of arranging the 12 spheres [9-26].

9.5.2 Icosahedral Packing

The most symmetrical arrangement is to place the 12 spheres at the vertices of a regular icosahedron, which is the only regular polyhedron with 12 vertices. Thus, the icosahedral packing is the most symmetrical. However, it is

Figure 9-28. Close packing of spheres: (a) Hexagonal closest packing; (b) cubic closest packing. After Shubnikov and Koptsik [9-19].

Figure 9-29. Icosahedral polyoma virus, drawn after Ref. [9-27].

not the densest packing. Also, it is not a crystallographic packing. When icosahedra are packed together, they will not form a plane but will gradually curve up and will eventually form a closed system as is illustrated in Figure 9-29 [9-27].

Buckminster Fuller recognized early the importance of icosahedral construction and its great stability in geodesic shapes as well as in viruses [9-28]*:

> This simple formula governing the rate at which balls are agglomerated around other balls or shells in closest packing is an elegant manifest of the reliably incisive transactions, formings, and transformings of Universe. I made that discovery in the late 1930s and published it in 1944. The molecular biologists have confirmed and developed my formula by virtue of which we can predict the number of nodes in the external protein shells of all the viruses, within which shells are housed the DNA–RNA-programmed design controls of all the biological species and of all the individuals within those species. Although the polio virus is quite different from the common cold virus, and both are different from other viruses, all of them employ frequency to the second power times ten plus two in producing those most powerful structural enclosures of all the biological regeneration of life. It is the structural power of these geodesic-sphere shells that makes so lethal those viruses unfriendly to man. They are almost indestructible.

Indeed, the discoverers of virus structures Caspar and Klug [9-29] stated:

*We thank Professor H.-U. Nissen, Zürich, for this quote.

> The solution we have found . . . was, in fact, inspired by the geometrical principles applied by Buckminster Fuller in the construction of geodesic domes . . . The resemblance of the design of geodesic domes . . . to icosahedral viruses had attracted our attention at the time of the poliovirus work . . . Fuller has pioneered in the development of a physically orientated geometry based on the principles of efficient design.

The length of an edge of a regular icosahedron is some 5% greater than the distance from the center to a vertex. Thus, the sphere of the outer shell of 12 makes contact only with the central sphere. Conversely, if each sphere of an icosahedral group of 12, all touching the central sphere, is in contact with its five neighbors, then the central sphere must have a radius some 10% smaller than the radius of the outer spheres. The relative size considerations are important in the structures of free molecules as well if the central atom or group of atoms is surrounded by 12 ligands [9-30].

An interesting case, and a step forward from the isolated molecule toward more extended systems, is when an icosahedron of 12 spheres about a central sphere is surrounded by a second icosahedral shell exactly twice the size of the first [9-31]. This shell will contain 42 spheres and will lie over the first so that spheres will be in contact along the fivefold axes. Further layers can be added in the same fashion. The third layer is shown in Figure 9-30 as an example of icosahedral packing of equal spheres. The layers of spheres succeed each other in cubic close-packing sequence on each triangular face. Each sphere which is

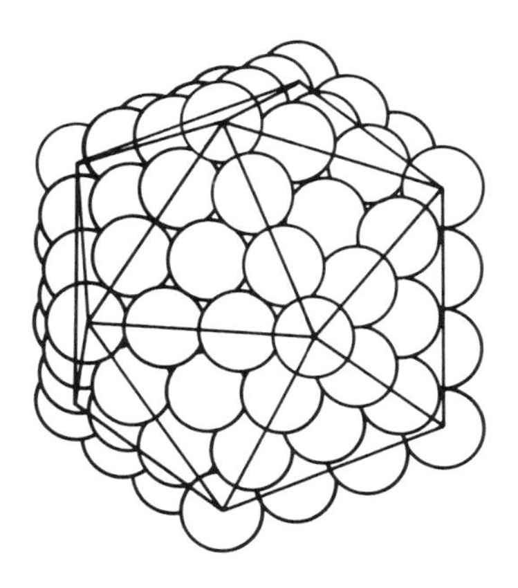

Figure 9-30. Illustration of icosahedral packing, after Mackay [9-31]: Icosahedral packing of spheres showing the third shell.

not on an edge or vertex touches only six neighbors, three above and three below. Each such sphere is separated by a distance of 5% of its radius from its neighbors in the plane of the face of the icosahedron. The whole assembly can be distorted to cubic close packing in the form of a cuboctahedron. This distortion may be envisaged as a reversible process by the kind of transformation discussed earlier. Herbert Hauptman (Figure 9-31a), a mathematician turned crystallographer and chemistry Nobel laureate in 1985, has devoted a lot of attention to close packing of spheres in the icosahedron. Figure 9-31b shows one of his beautiful stained-glass models.

While the most symmetrical arrangement of 12 neighbors, viz., the icosahedral coordination, does not lead to the densest possible packing, other arrangements do. The cuboctahedron and its "twinned" version, alone or in combination, lead to infinite sphere packing with the same high density (0.7405). Both coordination polyhedra are shown in Figure 9-32. The "twinned" polyhedron is obtained by reflecting one-half of a cuboctahedron cut parallel to a triangular face across the plane of section.

9.5.3 Connected Polyhedra

There are, of course, more complex forms of closest packing than those considered so far. Besides, the species to be packed need not be identical.

a

Figure 9-31. (a) Herbert Hauptman (1989). Photograph by the authors. (b) Stained-glass model of icosahedron with densely packed spheres. Photograph courtesy of Dr. Hauptman, Buffalo, New York.

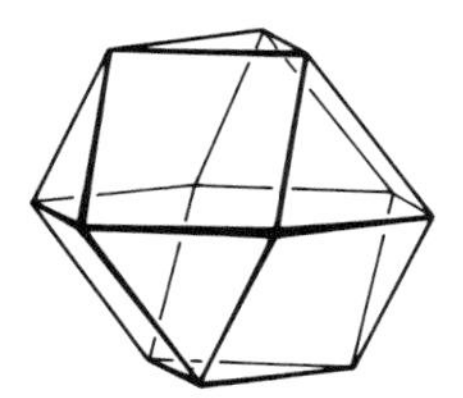
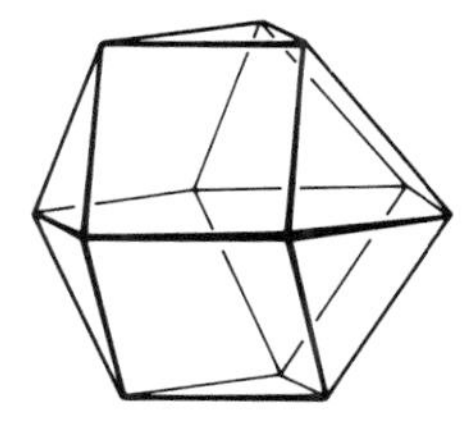

Figure 9-32. Cuboctahedron and "twinned" cuboctahedron.

Thus, close packing of atoms of two kinds could be considered. Close-packed structures with atoms in the interstices are also important. The interstice arrays may have very different arrangements in various structures. A shorthand notation of some configurations has been worked out to facilitate the description of more complicated systems (see, e.g., Ref. [9-26]). Such a notation is illustrated in Figure 9-33. Suppose, for example, that in a compound with composition AX_2, each atom A is bonded to four X atoms and that all four X atoms are equivalent. Each X atom must then be bonded to two A atoms. The lines of the squares in Figure 9-33 do not represent chemical bonds; rather, these squares stand for polyhedral arrangements. Among the AX_n polyhedral groups, the most common are the AX_4 tetrahedra and AX_6 octahedra. They may appear in various orientations in crystal structures. Similar structural features have already been discussed for the polyhedral molecular geometries. Whereas in molecules only two, or at most a few, polyhedra were joined, here we deal with their infinite networks.

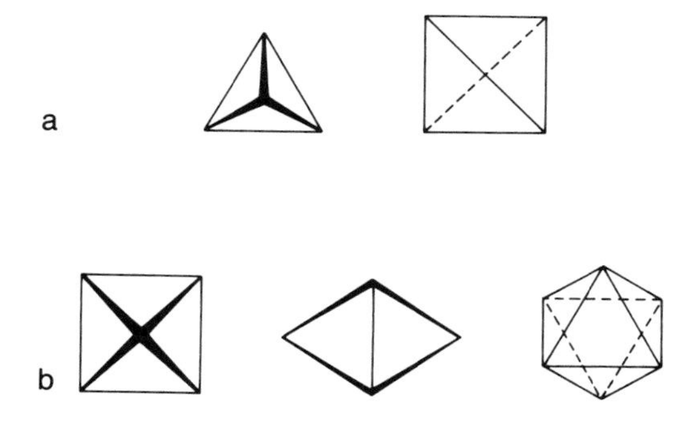

Figure 9-33. Shorthand notations for the tetrahedron (a) and the octahedron (b). After Wells [9-26].

Many crystal structures may be built from the two most important coordination polyhedra, the tetrahedron and octahedron. They may share vertices, edges, or faces. The ways in which the polyhedra are connected introduce certain geometrical limitations with important consequences as to the variations of the interatomic distances and bond angles [9-26].

Examples are shown in Figures 9-34–9-39 for a variety of ways in which tetrahedral and octahedral units may be connected. Tetrahedra share two vertices or/and three vertices in Figure 9-34. For one of these, decorations analogous to its projection are shown in Figure 9-35. Octahedra share adjacent vertices and form a tetramer in two representations in Figure 9-36a and b. Two more examples show infinite chains of octahedra sharing adjacent (Figure 9-36c) and nonadjacent vertices (Figure 9-36d). Octahedra sharing two, four, or six edges are presented in Figure 9-37. An example of octahedra sharing faces and edges is seen in Figure 9-38 together with an analogous pattern from a Formosan basket weaving [9-34]. Finally, a composite structure from tetrahedra and octahedra is shown in Figure 9-39.

The tetrahedra and octahedra are important building blocks of crystal structures. The great variety of structures combining these building blocks, on one hand, and the conspicuous absence of some of the simplest structures, on the other hand, together suggest that the immediate environment of the atoms is

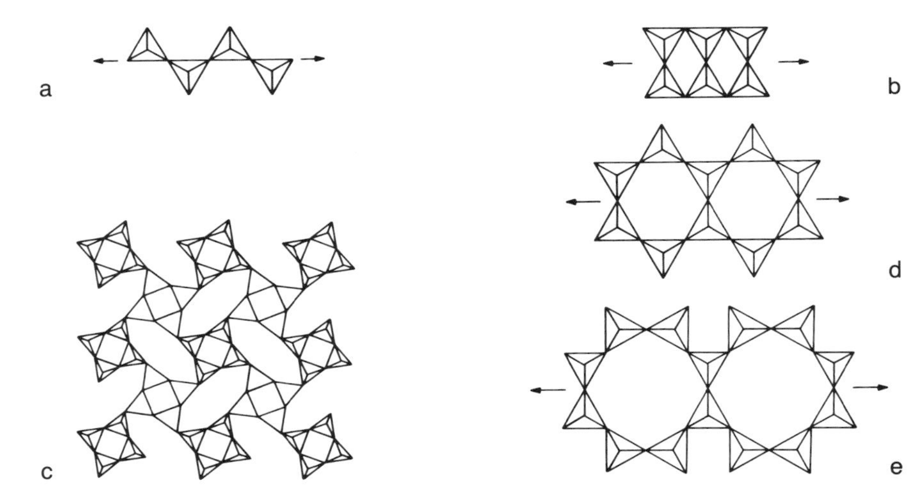

Figure 9-34. Connected tetrahedra. (a) All tetrahedra share two vertices. (b) and (c) All tetrahedra share three vertices. After Ref. [9-32]. (d) and (e) Some tetrahedra share two and others share three vertices.

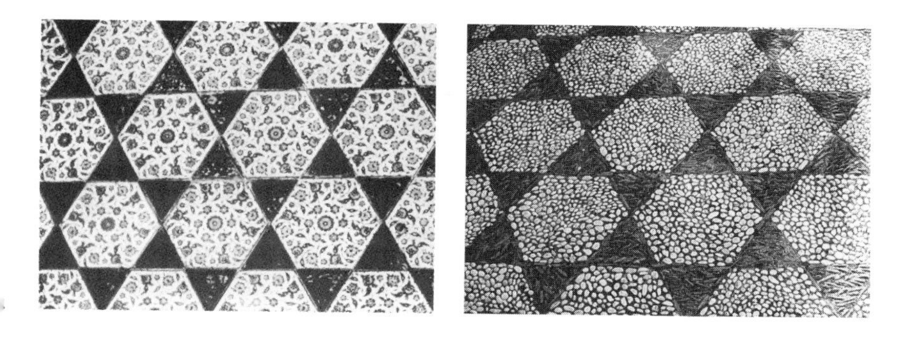

Figure 9-35. Decorations, analogous to the one-dimensional pattern of Figure 9-34d but extending in two dimensions: (a) Islamic decoration, drawn after Ref. [9-33]; (b) pavement pattern in Granada, Spain; photograph by the authors.

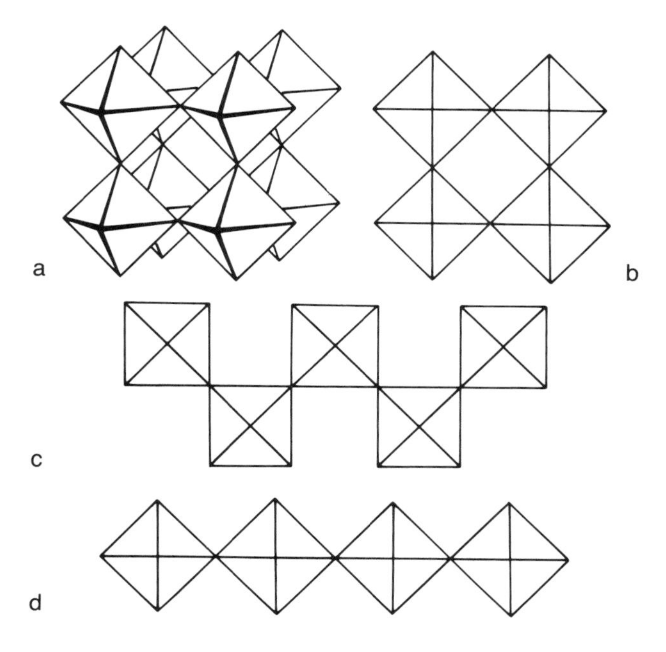

Figure 9-36. Connected octahedra. (a) and (b) Two representations of four octahedra sharing adjacent vertices and forming a tetramer. (c) Infinite chain of octahedra connected at adjacent vertices. (d) Infinite chain of octahedra connected at nonadjacent vertices.

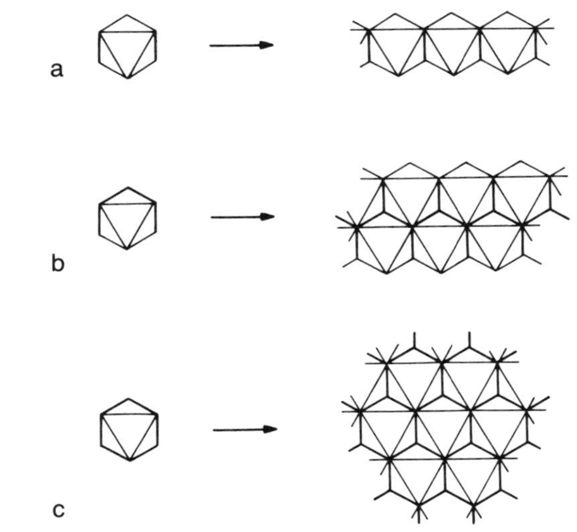

Figure 9-37. Octahedra sharing two (a), four (b), and six edges (c).

not the only factor that determines these structures. Indeed, the relative sizes of the participating atoms and ions are of great importance.

9.5.4 Atomic Sizes

The interatomic distances are primarily determined by the position of the minimum in the potential energy function describing the interactions between the atoms in the crystal. The question is then, what are the sizes of the atoms and ions? Since the electron density for an atom or an ion extends indefinitely, no single size can be rigorously assigned to it. Atoms and ions change relatively little in size when forming a strong chemical bond, and even less for weak bonds. For the present discussion of crystal structures, the atomic and ionic radii should, when added appropriately, yield the interatomic and interionic distances characterizing these structures.

Covalent and metallic bondings suppose a strong overlap of the outermost atomic orbitals, and so the atomic radii will be approximately the radii of the outermost orbitals. The atomic radii [9-35] are empirically obtained from interatomic distances. For example, the C–C distance is 1.54 Å in diamond, the Si–Si distance is 2.34 Å in disilane, and so on. The consistency of this approach is shown by the agreement between the Si–C bond lengths determined experimentally and calculated from the corresponding atomic radii. The interatomic distances appreciably depend on the coordination. With decreasing

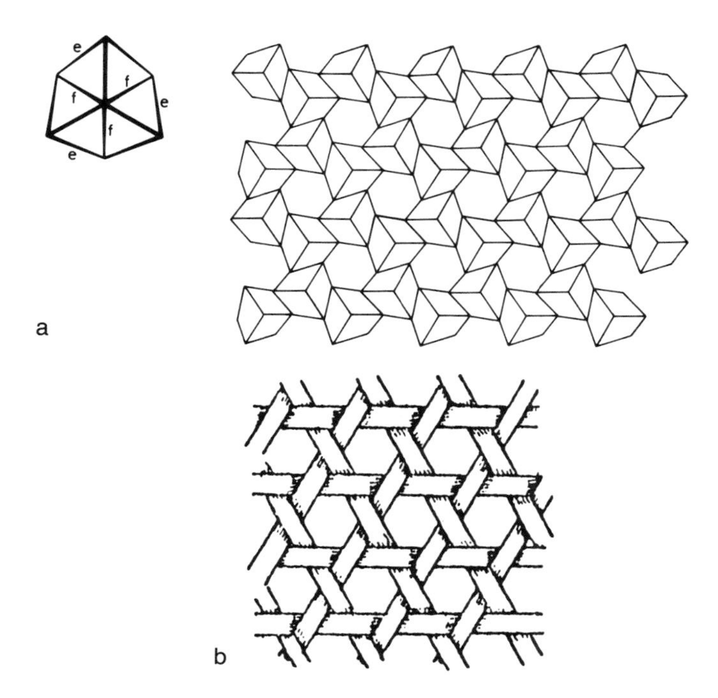

Figure 9-38. Joined octahedra sharing faces and edges: (a) Nb_3S_4 crystal, after Ref. [9-26]; (b) analogy from Formosan basket weaving pattern, after Ref. [9-34].

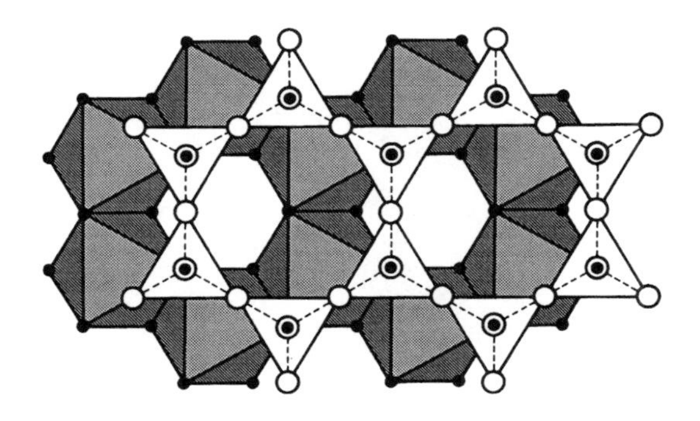

Figure 9-39. Joined tetrahedra and octahedra: A composite structure (kaolin) built from tetrahedra and octahedra, after Ref. [9-26]. Reproduced with permission.

coordination number, the bonds usually get shorter. For coordination numbers of 8, 6, and 4, the bonds get shorter by about 2, 4, and 12%, respectively, as compared with the bonds for a coordination number of 12.

The covalent bond is directional, and multiple covalent bonds are considerably shorter than the corresponding single ones. For carbon as well as for nitrogen, oxygen, or sulfur, the decrease on going from a single bond to a double and a triple bond amounts to about 10 and 20%, respectively.

Establishing the system of ionic radii is an even less unambiguous undertaking than that for atomic radii. The starting point is a system of analogous crystal structures. Such is, for example, the structure of sodium chloride and the analogous series of other alkali halide face-centered crystals. In any case the ionic radii represent relative sizes, and if the alkali and halogen ions are chosen as the starting point, then the ionic radii of all ions represent the relative sizes of the outer electron shells of the ions as compared with those of the alkali and halogen ions.

Consider now the sodium chloride crystal structure shown in Figure 9-40. It is built from sodium ions and chloride ions, and it is kept together by electrostatic forces. The chloride ions are much larger than the sodium ions. As equal numbers of cations and anions build up this structure, the maximum number of neighbors will be the number of the larger chloride ions that can be accommodated around the smaller sodium ion. The opposite would not work: although more sodium ions could surround a chloride ion, the same coordination could not be achieved around the sodium ions. Thus, the coordination number will obviously depend on the relative sizes of the ions. In the simple ionic structures, however, only such coordination numbers may be accomplished that make a highly symmetrical arrangement possible. The relative

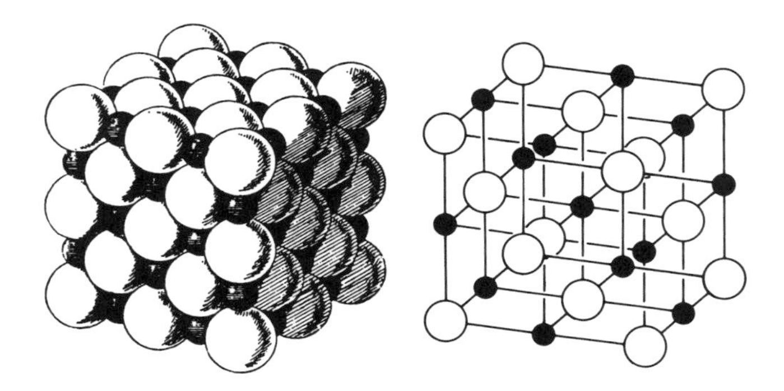

Figure 9-40. The sodium chloride crystal structure in two representations. The space-filling model is from W. Barlow [9-36].

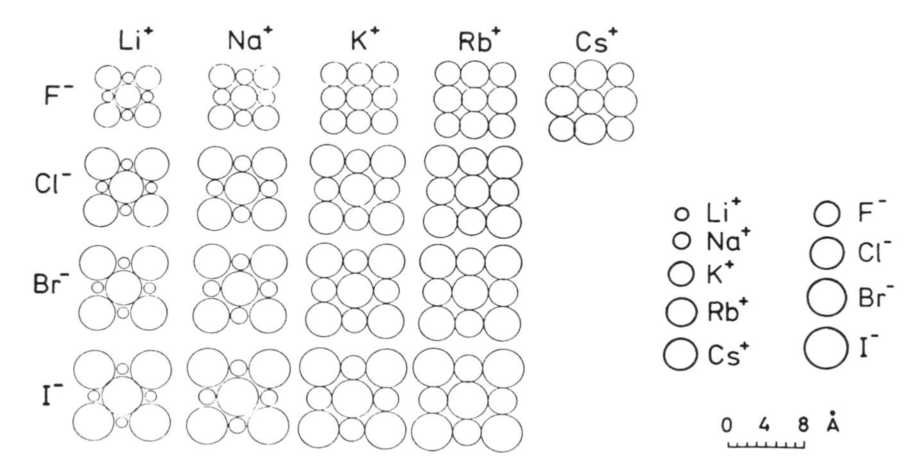

Figure 9-41. The arrangement of ions in cube-face layers of alkali halide crystals with the sodium chloride structure. Adaptation from Pauling [9-35]. Copyright (1960) Cornell University. Used by permission of the publisher, Cornell University Press.

sizes of the sodium and chloride ions allow six chloride ions to surround each sodium ion in six vertices of an octahedron. Figure 9-41 shows the arrangement of ions in cube-face layers of alkali halide crystals with the sodium chloride structure. As the relative size of the metal ion increases with respect to the size of the halogen anion, greater coordination may be possible. Thus, for example, the cesium ion may be surrounded by eight chloride ions in eight vertices of a cube in the cesium chloride crystal as shown in Figure 9-42.

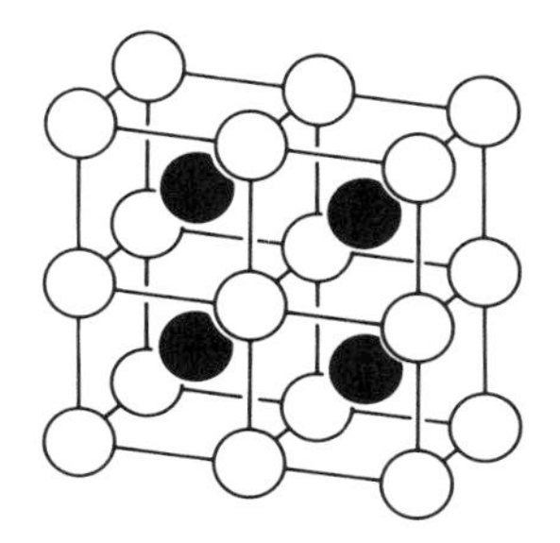

Figure 9-42. Cesium chloride crystal structure.

9.6 MOLECULAR CRYSTALS

A molecular crystal is built from molecules. It is easily distinguished from an ionic/atomic crystal on a purely geometrical basis. At least one of the intramolecular distances of an atom in the molecule is significantly smaller than its distances to the adjacent molecules. Every molecule in the molecular crystal may be assigned a certain well-defined space in the crystal. In terms of interactions, there are the much stronger intramolecular interactions and the much weaker intermolecular interactions. Of course, even among the intramolecular interactions, there is a range of interactions of various energies. Bond stretching, for instance, requires a proportionately higher energy than angular deformation, and the weakest are those interactions that determine the conformational behavior of the molecule [9-37]. On the other hand, there are differences among the intermolecular interactions as well. For example, intermolecular hydrogen bond energies may be equal to or even greater than the conformational energy differences. Thus, there may be some overlap in the energy ranges of the intramolecular and intermolecular interactions.

The majority of molecular crystals are organic compounds. There is usually very little electronic interaction between the molecules in these crystals, although, as will be discussed later, even small interactions may have appreciable structural consequences. The physical properties of the molecular crystals are primarily determined by the packing of the molecules.

9.6.1 Geometrical Model

As structural information for large numbers of molecular crystals has become available, general observations and conclusions have appeared [9-22]. An interesting observation was that there are characteristic shortest distances between the molecules in molecular crystals. The intermolecular distances of a given type of interaction are fairly constant. From this observation a geometrical model was developed for describing molecular crystals. First, the shortest intermolecular distances were found, and then the so-called "intermolecular atomic radii" were postulated. Using these quantities, spatial models of the molecules were built. Fitting together these models, the densest packing could be found empirically. A simple but ingenious device was even constructed for fitting the molecules. A packing example is shown in Figure 9-43. The molecules are packed together in such a way as to minimize the empty space among them. The concave part of one molecule accommodates the convex part of the other molecule. The example is the packing of 1,3,5-triphenylbenzene molecules in their crystal structure. The arrangement of the areas designated to the molecules is analogous to a characteristic decoration pattern, an example of which is also shown in Figure 9-43. The analogy is not quite superficial. The

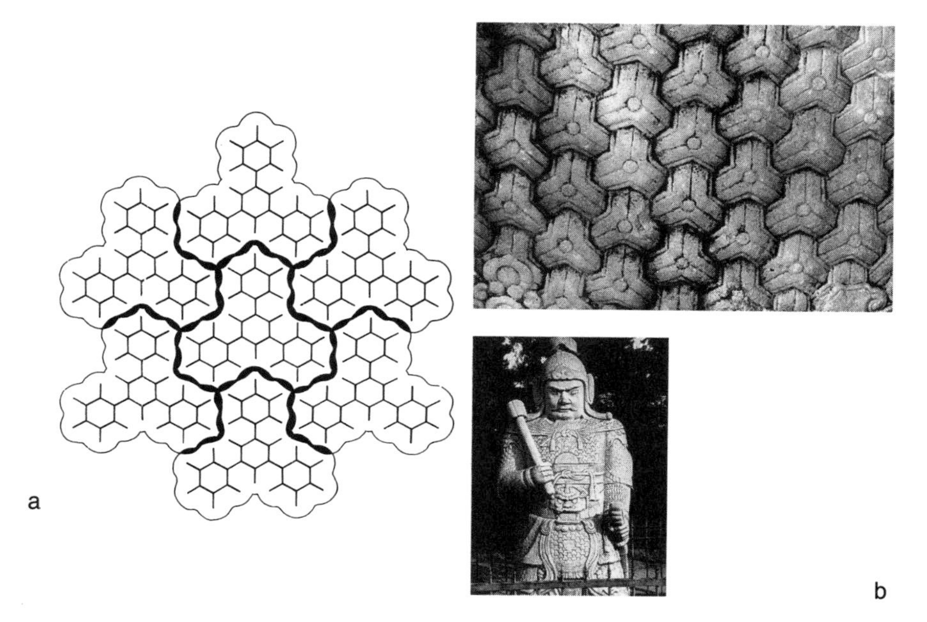

Figure 9-43. (a) Dove-tail packing. Dense packing of 1,3,5-triphenylbenzene molecules, after Ref. [9-22]. (b) Chinese decoration from a sculpture in the sculpture garden of the Ming tombs, near Beijing. Photographs by the authors.

decoration is from the metal-net dress of a warrior. The dress was made of small units to maintain flexibility, the small units were identical for economy, and they covered the whole surface without gaps to ensure protection.

The complementary character of molecular packing is well expressed by the term *dove-tail packing* [9-38]. The arrangement of the molecules in Figure 9-44a can be called *head-to-tail*. On the other hand, the molecules of a similar compound are arranged *head-to-head* as seen in Figure 9-45a. The head-to-head arrangement is less advantageous for packing. This is well seen in Figs. 9-44b and 9-45b, displaying the arrangement of the molecules in the crystal after Wundl and Zellers [9-39, 9-40].

Many of Escher's periodic drawings with interlocking motifs are also excellent illustrations for the dove-tail packing principle. Figure 9-46 reproduces one of them. Note how the toes of the black dogs are the teeth of the white dogs and vice versa in this figure [9-41].

Because of the interlocking character, the packing in organic molecular crystals is usually characterized by large coordination numbers, i.e., by a

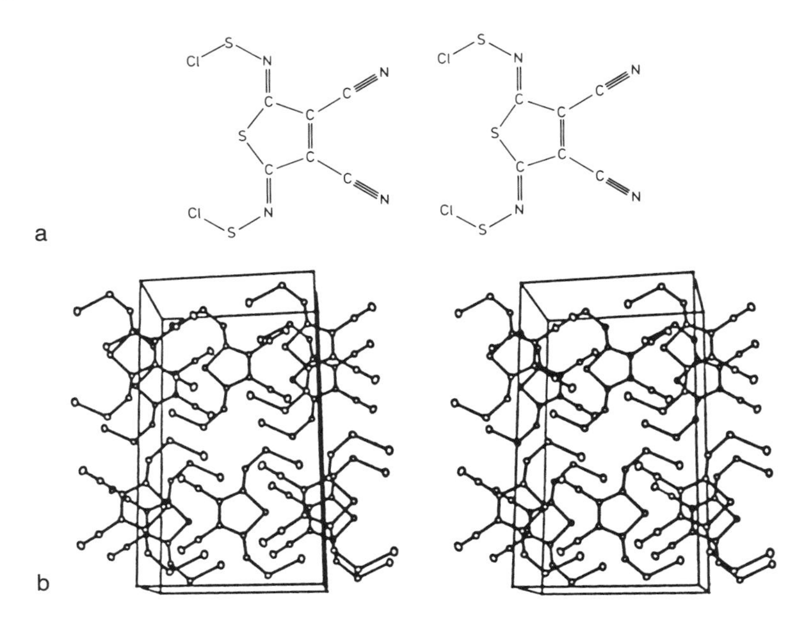

Figure 9-44. Head-to-tail packing (a) and arrangement of molecules in the crystal (b). After Ref. [9-39]. Copyright (1980) American Chemical Society.

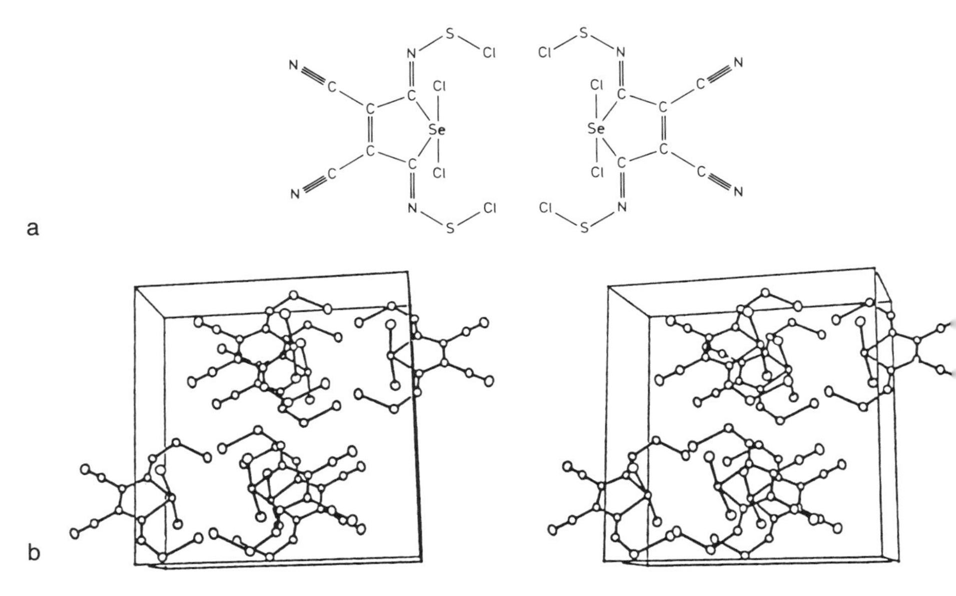

Figure 9-45. Head-to-head packing (a) and arrangement of molecules in the crystal (b). After Ref. [9-40]. Copyright (1980) American Chemical Society.

Figure 9-46. Escher's periodic drawing of dogs, from MacGillavry's book [9-41]. Reproduced with permission from the International Union of Crystallography.

relatively large number of adjacent or touching molecules. Experience shows that the most often occurring coordination number in organic structures is 12, so it is the same as for the densest packing of equal spheres. Coordination numbers of 10 or 14 occur also but less often.

A. I. Kitaigorodskii was a true pioneer in the field of molecular crystals. First of all, he gave real sizes and volumes to the molecules by accounting for the hydrogen atoms, however poorly their positions could be determined. The whole molecule was considered in examining the packing, rather than the heavy-atom skeleton only. Figuratively speaking, and using Kitaigorodskii's own expression [9-42], he "dressed the molecules in a fur-coat of van der Waals spheres." This was in complete agreement with the molecular models introduced already in the early thirties by Stuart and Briegleb [9-43] to represent the space-filling nature of molecular structures. They are illustrated in Figure 9-47 by a plastic model and a palm-tree fruit which looks remarkably similar to the space-filling molecular models.

The geometrical model allowed Kitaigorodskii [9-22, 9-38] to make predictions of the structure of organic crystals in numerous cases, knowing only the cell parameters and, obviously, the size of the molecule itself. In the age of fully automated, computerized diffractometers, this may not seem to be so important, but it has indeed enormous significance for our understanding of the packing principles in molecular crystals.

The packing as established by the geometrical model is what is expected to be the ideal arrangement. Usually, it does not differ from the real packing as determined by X-ray diffraction measurements. When there are differences between the ideal and experimentally determined packings, it is of interest to

Figure 9-47. Illustrations of space-filling molecular models: (a) Plastic model; (b) palm-tree fruit in Hawaii. Photographs by the authors.

examine the reasons for their occurrence. The geometrical model has some simplifying features. One of them is that it considers uniformly the intermolecular atom $\cdots$ atom distances. Another is that it considers interactions only between adjacent atoms.

The development of experimental techniques and the appearance of more sophisticated models have recently pushed the frontiers of molecular crystal chemistry much beyond the original geometrical model. Some of the limitations of this model will be mentioned later. However, its simplicity and the facility of visualization ensure this model a lasting place in the history of molecular crystallography. It has also exceptional didactic value.

The so-called coefficient of molecular packing (k) has proved useful in characterizing molecular packing. It is expressed in the following way:

$$k = \frac{\text{molecular volume}}{\text{crystal volume/molecule}}$$

The molecular volume is calculated from the molecular geometry and the atomic radii. The quantity crystal volume/molecule is determined from the X-ray diffraction experiment. For most crystals k is between 0.65 and 0.77. This is remarkably close to the coefficient of the dense packing of equal spheres. The density of closest packing of equal spheres is 0.7405 [9-26]. If the form of the molecule does not allow the coefficient of molecular packing to be greater than 0.6, then the substance is predicted to transform into a glassy state with decreasing temperature. It has also been observed that morphotropic changes associated with loss of symmetry led to an increase in the packing density. Comparison of analogous molecular crystals shows that sometimes the decrease in crystal symmetry is accompanied by an increase in the density of packing.

Another interesting comparison involves benzene, naphthalene, and anthracene. When their coefficient of packing is greater than 0.68, they are in the solid state. There is a drop in this coefficient to 0.58 when they go into the liquid phase. Then, with increasing temperature, their k decreases gradually down to the point where they start to boil. The fused-ring aromatic hydrocarbons have served as targets of a systematic analysis of packing energies and other packing characteristics [9-44].

Recently, geometrical considerations have gained additional importance due to their role in molecular recognition, which implies "the (molecular) storage and (supramolecular) retrieval of molecular structural information," according to J.-M. Lehn [9-45]. The formation of supramolecular structures necessitates commensurable and compatible geometries of the partners. The molecular structure of the inclusion complex formed by *para-tert*-butylcalix-[4]arene and anisole [9-46] is shown in Figure 9-48. The representation is a combination of a line drawing of the calixarene molecule and a space-filling model of anisole.

The supramolecular formations and the molecular packing in the crystals show close resemblance, and the nature of the interactions involved is very much the same. There is great emphasis on weak interactions in both. According to Lehn [9-47], "beyond molecular chemistry based on the covalent bond lies supramolecular chemistry based on molecular interactions—the associations of two or more chemical entities and the intermolecular bond."

The relevance of supramolecular structures to molecular crystals and molecular packing was eloquently expressed by J. D. Dunitz [9-48]:

> A crystal is, in a sense, the supramolecule *par excellence*—a lump of matter, of macroscopic dimensions, millions of molecules long, held together in a periodic arrangement by just the same kind of non-bonded interactions as those that are responsible for molecular recognition and complexation at all levels. Indeed, crystallization

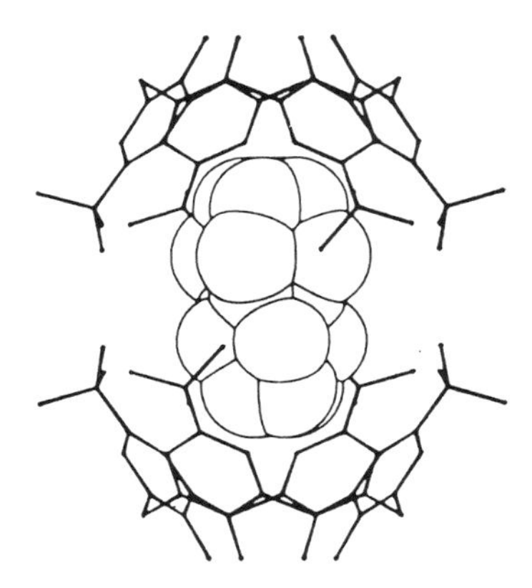

Figure 9-48. Two *para-tert*-butylcalix[4]arene molecules envelope an anisole molecule. After Andreetti *et al.* [9-46]. Reprinted by permission of Kluwer Academic Publishers.

itself is an impressive display of supramolecular self-assembly, involving specific molecular recognition at an amazing level of precision.

9.6.2 Densest Molecular Packing

Kitaigorodskii [9-22, 9-38] examined the relationship between densest packing and crystal symmetry by means of the geometrical model. He determined that real structures will always be among those that have the densest packing. First of all, he established the symmetry of those two-dimensional layers that allow a coordination number of six in the plane at an arbitrary tilt angle of the molecules with respect to the axes of the layer unit cell. In the general case for molecules with *arbitrary* form, there are only two kinds of such layers. One has inversion centers and is associated with a nonorthogonal lattice. The other has a rectangular net, from which the associated lattice is formed by translations, plus a second-order screw axis parallel to a translation. The next task was to select the space groups for which such layers are possible. This is an approach of great interest since the result will answer the question as to why there is a high occurrence of a few space groups among crystals while many of the 230 space groups hardly ever occur.

We present here some of the highlights of Kitaigorodskii's considerations [9-22]. First, the problem of dense packing is examined for the plane groups of symmetry. The distinction between dense-packed, densest packed, and maximum density was introduced for the plane layer of molecules. The plane was called dense-packed when a coordination number of six was achieved for the molecules. The term densest packed meant six-coordination with any orientation of the molecules with respect to the unit cell axes. The term maximum density was used for the packing if six-coordination was possible for any orientation of the molecules with respect to the unit cell axes while the molecules retained their symmetry.

For the plane group *p*1 it is possible to achieve densest packing with any molecular form if the translation periods t_1 and t_2 and the angle between them are chosen appropriately as illustrated in Figure 9-49. The same is true for the plane group *p*2, shown also in Figure 9-49. On the other hand, the plane groups *pm* and *pmm* are not suitable for densest packing. As is seen in Figure 9-50, the molecules are oriented in such a way that their convex parts face the convex parts of other molecules. This arrangement, of course, counteracts dense packing. The plane groups *pg* and *pgg* may be suitable for six-coordination as shown by the example in Figure 9-51a. This layer is not of maximum density, and in a different orientation of the molecules, only four-coordination is achieved, as seen in Figure 9-51b. For the plane groups *cm*, *cmm*, and *pmg*, six-coordination cannot be achieved for a molecule with arbitrary shape. For higher symmetry groups, for example, tetragonal *p*4 or hexagonal *p*6, the axes of the unit cell are equivalent, and the packing of the molecules is not possible without overlaps. This is illustrated for group *p*4 in Figure 9-52.

If the molecule, however, retains a symmetry plane, then it may be packed with six-coordination in at least one of the plane groups, *pm*, *pmg*, or *cm*. The form shown in Figure 9-53 is suitable for such packing in *pmg* and *cm*, though not in *pm*. Thus, depending on the molecular shape, various plane groups may be applicable in different cases.

The criteria for the suitability as well as for the incompatibility of plane groups for achieving molecular six-coordination have been considered and illustrated with examples. The next step is to apply the geometrical model to the examination of the suitability of three-dimensional space groups for densest

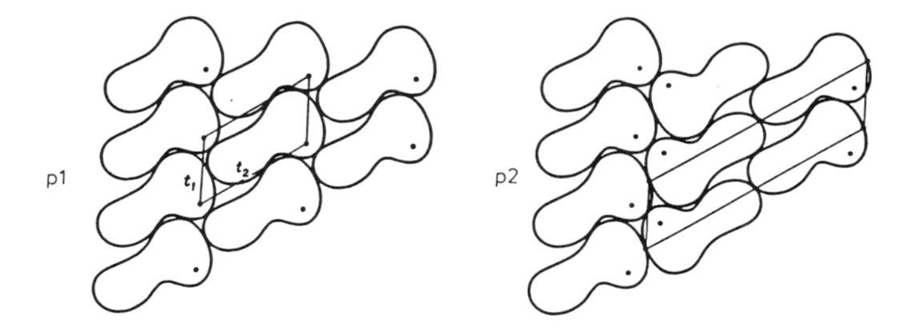

Figure 9-49. Densest packing with space groups *p*1 and *p*2, after Ref. [9-22].

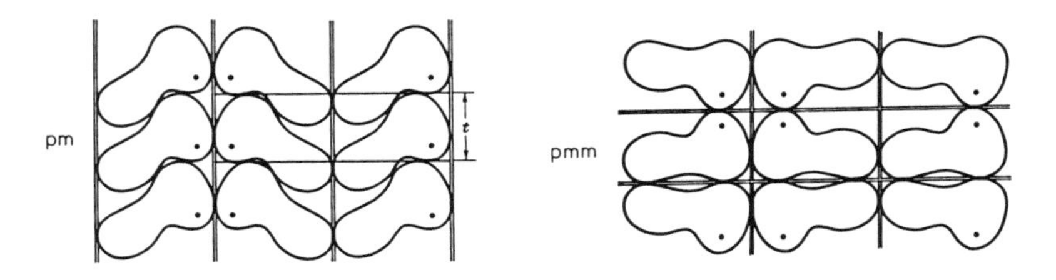

Figure 9-50. The symmetry planes in the space groups *pm* and *pmm* prevent dense packing [9-22].

packing. The task in this case is to select those space groups in which layers can be packed in such a manner as to allow the greatest possible coordination number. Obviously, for instance, mirror planes would not be applicable for repeating the layers.

Low-symmetry crystal classes are typical for organic compounds. Densest packing of the layers may be achieved either by translation at an arbitrary angle formed with the layer plane or by inversion, glide plane, or screw-axis rotation. In rare cases closest packing may also be achieved by twofold rotation.

Kitaigorodskii [9-22] has analyzed all 230 three-dimensional space groups from the point of view of densest packing. Only the following space

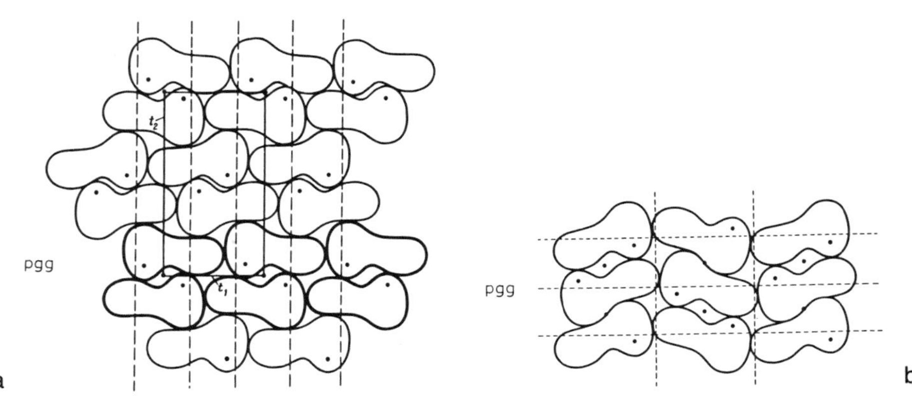

Figure 9-51. Two forms of packing with *pgg* space groups: (a) Densest packing of molecules with arbitrary shape; (b) another orientation of the molecules which reduces the coordination number to four. After Ref. [9-22].

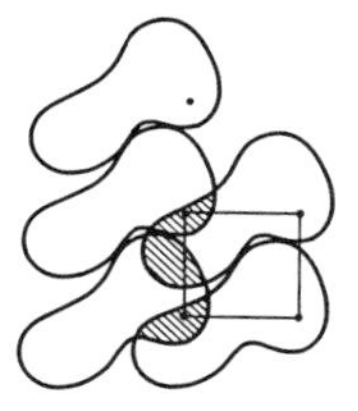

Figure 9-52. Molecules of arbitrary shape cannot be packed in space group *p*4 without overlaps. After Ref. [9-22].

groups were found to be available for the densest packing of molecules of arbitrary form:

$$P\bar{1},\ P2_1,\ P2_1/c,\ Pca,\ Pna,\ P2_12_12_1$$

For molecules with symmetry centers, there are even fewer suitable three-dimensional space groups, namely:

$$P\bar{1},\ P2_1/c,\ C2/c,\ Pbca$$

In these cases all mutual orientations of the molecules are possible without losing the six-coordination.

The space group $P2_1/c$ occupies a strikingly special position among the organic crystals. This space group has the unique feature that it allows the formation of layers of densest packing in all three coordinate planes of the unit cell.

The space groups $P2_1$ and $P2_12_12_1$ are also among those providing densest packing. However, their possibilities are more limited than those of the space

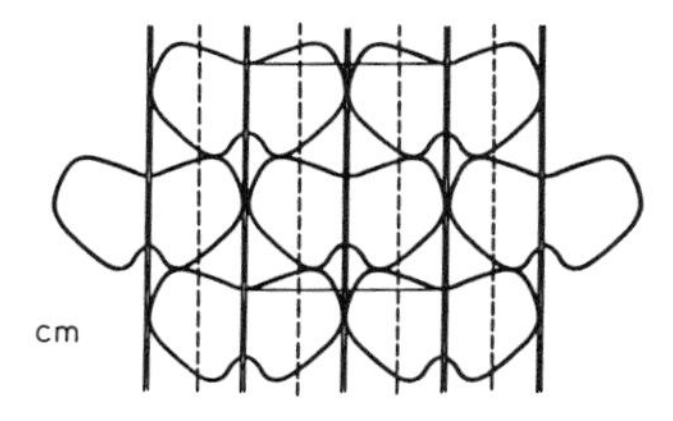

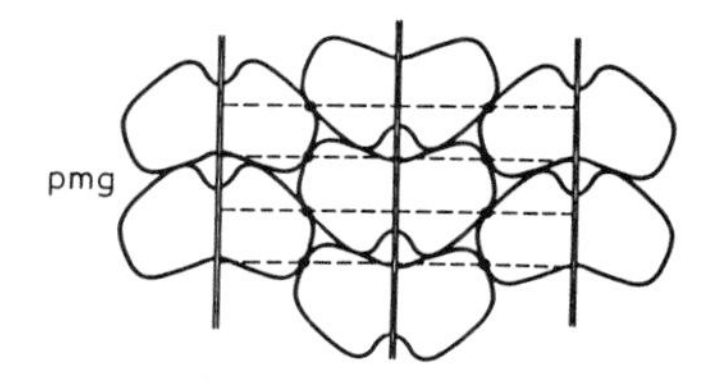

Figure 9-53. Molecules with a symmetry plane achieve six-coordination in the space groups *cm* and *pmg*. After Ref. [9-22].

group $P2_1/c$, and these space groups occur only for molecules that take either left-handed or right-handed forms.

According to statistical examinations performed some time ago, these three space groups are the first three in frequency of occurrence.

An interesting and really fundamental question is the conservation of molecular symmetry in the crystal structure. Densest packing may often be facilitated by partial or complete loss of molecular symmetry in the crystal structure. There are, however, space groups in which some molecular symmetry may "survive" densest packing in building of the crystal. Preserving higher symmetry though usually results in too great a sacrifice of packing density. On the other hand, there may be some energetic advantage of some well-defined symmetrical arrangements. The alternative to the geometrical model for discussing and establishing molecular packing in organic crystals has been energy calculations, based on carefully constructed potential energy functions (see, e.g., Ref. [9-49] and references therein).

9.6.3 Energy Calculations and Crystal Structure Predictions

It is important to be able to determine *a priori* the arrangement of molecules in crystals. The correctness of such predictions is a test of our understanding of how crystals are built. A further benefit is the possibility of calculating even those structures that are not amenable to experimental determination. However, even as part of an experimental study, it is instructive to build good models, which can then be refined. The main advantages of the geometrical model have been seen above. Its main limitations are the following. It cannot account for the structural variations in a series of analogous compounds. It is very restricted in correlating structural features with various other physical properties. Finally, it cannot be used to make detailed predictions for unknown structures. Calculations seeking the spatial arrangement of molecules in the crystal corresponding to the minimum of free energy have become a much used tool. If the system is considered as completely rigid, the molecular packing may be determined by minimizing the potential energy of intermolecular interactions.

Considering the molecules to be rigid, that is, ignoring the vibrational contribution, the energy of the crystal structure is expressed as a function of geometrical parameters including the cell parameters, the coordinates of the centers of gravity of the symmetrically independent molecules, and parameters characterizing the orientation of these molecules. In particular cases, the number of independent parameters can be reduced. On the other hand, consideration of the nonrigidity of the molecules necessitates additional parameters. Minimizing the crystal structure energy leads to structural parameters corresponding to optimal molecular packing. Then it is of great interest to compare these findings with those from experiment.

To determine the deepest minimum on the multidimensional energy surface as a function of many structural parameters is a formidable mathematical task. Usually, simplifications and assumptions are introduced concerning, for example, the space-group symmetry. Accordingly, the conclusions from these theoretical calculations cannot be considered to be entirely *a priori*.

The considerations of the intermolecular interactions can be conveniently reduced to considerations of atom–atom nonbonded interactions. Although these interactions can be treated by nonempirical quantum-mechanical calculations, empirical and semiempirical approaches have also proved successful in dealing with them. In the description of the atom–atom nonbonded interactions, it is supposed that the van der Waals forces originate from a variety of sources.

In addition to the intermolecular interactions, the intramolecular interactions may also be taken into account in a similar way. This rather limited approach may nevertheless be useful for calculating molecular conformation and even molecular symmetry. Deviations from the ideal conformations and symmetries may also be estimated in this way, provided they are due to steric effects.

By summation over the interaction energies of the molecular pairs, the total potential energy of the molecular crystal may be obtained in an atom–atom potential approximation. The result is expected to be approximately the same as the heat of sublimation extrapolated to 0 K provided that no changes take place in the molecular conformation and vibrational interactions during evaporation.

In many of the molecular packing studies, the crystal classes are taken from experimental X-ray diffraction determinations. The optimal packing is then determined for the assumed crystal class. In other cases, the crystal classes have also been established in the optimization calculations.

Depero [9-50] has summarized the various approaches developed historically in molecular packing calculations. Her scheme has the following five subdivisions:

1. Geometrical model: The molecular energy is considered to be in a deep minimum, and, accordingly, all intramolecular changes are ignored.
2. Atom–atom potential method: The packing energy is expressed by summing all intermolecular interactions, the pairwise potentials are determined especially, and, again, the molecular structure is kept unchanged.
3. Force fields: All atoms, not only those participating in intermolecular interactions, are included.
4. Pseudo-potential calculations: The Schrödinger equation is applied, the core electrons are described by empirical potentials, and only the valence electrons are taken into account in the calculations.

5. *Ab initio* calculations: All electrons are considered in the application of the Schrödinger equation; no empirical parameters are involved.

Ideally, it should be possible to predict molecular packing, and thus the crystal structure, from the knowledge of the composition of a compound and the symmetry and geometry of its molecules. It has proved, however, a rather elusive task. Only a few years ago the frustration over the difficulty in predicting crystal structures was expressed by the Editor of *Nature* in the following words [9-51]: "One of the continuing scandals in the physical sciences is that it remains impossible to predict the structure of even the simplest crystalline solids from a knowledge of their chemical composition." There has been considerable progress in this respect, however, mainly due to the utilization of the wealth of information from data banks, and in particular, from the Cambridge Crystallographic Data Centre [9-52].

It has also proved fruitful to use the energy calculations with computer graphic analysis. Plausible crystal-building scenarios have been described which, while not being necessarily unique solutions, seem to point in the right direction in conquering this important frontier of structural science. An example is the construction of organometallic crystals by Braga and Grepioni [9-53], illustrated here with $Ru_3(CO)_{12}$ in Figure 9-54. The construction, which is a simultaneous process in reality, is broken down into three steps in the model. First, a row of molecules is constructed in a head-to-tail arrangement. The second step involves adding rows to form a layer utilizing interlocking interactions. Finally, whole layers are added to form a crystal.

A concerted use of geometry and energy considerations, as demonstrated by the crystal building of $Ru_3(CO)_{12}$, seems most promising. Extending such studies may accomplish a "Kitaigorodskian dream," as they "provide the starting point for the formulation of a generalized force field for intermolecular interactions in organic crystals" [9-49].

There seems to be a remarkable consistency between Buckminster Fuller's evaluation of chemistry ("chemists consider volumes as material domains and not merely as abstractions"; see Chapter 1) and Kitaigorodskii's geometrical model of crystal structures. In one of his last statements, Kitaigorodskii (Figure 9-55) said (when asked about his most important achievements in science): "I've shown that the molecule is a body.* One can take it, one can hit with it—it has mass, volume, form, hardness. I followed the ideas of Democritos . . ." [9-42].

*A similar statement is also attributed to A. I. Kitaigorodskii, "The molecule also has a body; when it's hit, it feels hurt all over." This implied the possibility of structural changes in the molecule upon entering the crystal structure, a symbolic departure from Kitaigorodskii's earlier views about the constancy of molecular geometry regardless of whether in the gas phase or in the crystal.

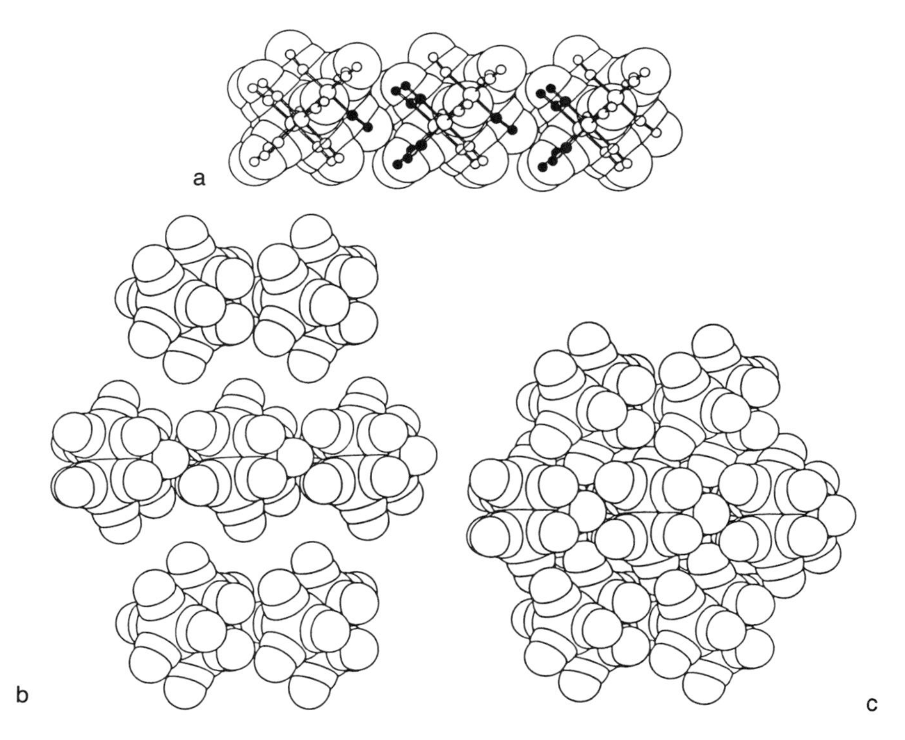

Figure 9-54. Building a crystal of $Ru_3(CO)_{12}$, according to Braga and Grepioni [9-53]: (a) Row of molecules; (b) forming a layer; (c) extending in three dimensions. Copyright (1991) American Chemical Society.

Figure 9-55. (a) A. I. Kitaigorodskii (1914–1985) among his students in the late 1960s in Moscow. (b) Democritos (ca. 460–ca. 370 B.C.) on Greek stamp: "Nothing exists except atoms and empty space; everything else is opinion." (See Mackay [9-54].)

9.6.4 Hypersymmetry

There are some crystal structures in which further symmetries are present in addition to those prescribed by their three-dimensional space group. The phenomenon is called hypersymmetry and has been discussed in detail by Zorky and co-workers [9-55, 9-56]. Thus, hypersymmetry refers to symmetry features not included in the system of the 230 three-dimensional space groups. For example, phenol molecules, connected by hydrogen bonds, form spirals with threefold screw axes as indicated in Figure 9-56. This screw axis does not extend, however, to the whole crystal, and it does not occur in the three-dimensional space group characterizing the phenol crystal.

A typical characteristic of hypersymmetry operations is that they exercise their influence in well-defined discrete domains. These domains do not overlap—they do not even touch each other. The usual hypersymmetry elements lead to point-group properties. This means that no infinite molecular chains could be selected, for example, to which these hypersymmetry operations would apply. They affect, instead, pairs of molecules or very small groups of molecules. Thus, they can really be considered as local point-group operations. These hypersymmetry elements, accordingly, divide the whole crystalline system into numerous small groups of molecules or transform the crystal space into a layered structure.

A prerequisite for hypersymmetry is that there should be chemically identical (having the same structural formula), but symmetrically indepen-

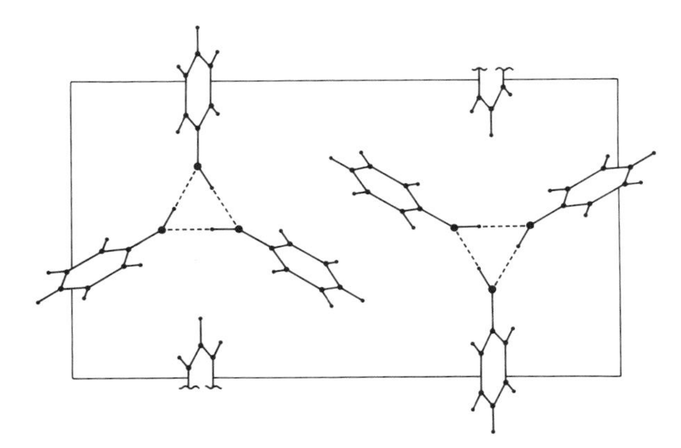

Figure 9-56. The molecules in the phenol crystal are connected by hydrogen bonds and form spirals with a threefold screw axis. This symmetry element is not part of the three-dimensional space group of the phenol crystal. After Ref. [9-55].

dent, molecules in the crystal structure—symmetrically independent, that is, in the sense of the three-dimensional space group to which the crystal belongs. The question then arises as to whether these symmetrically independent but chemically identical molecules will have the same structure or not. Only if they do have the same structure and conformation as well as bond configuration can we talk about the validity of the hypersymmetry operations. Here, preferably, quantitative criteria should be introduced, which is the more difficult since, for example, with increasing accuracy, structures that could be considered identical before may no longer be considered so later when more accurate data become available.

On the other hand, since even a slightly different environment will have some influence on the molecular structure, the hypersymmetry operations will not be absolute. In this, the hypersymmetry operations are somewhat different from the usual symmetry operations. The ultimate goal is to find a generalized formulation of the space-group system that would allow the simultaneous consideration of the usual symmetry as well as the hypersymmetry. When such a generalized formulation of space groups encompassing usual and hypersymmetry operations becomes available, the task of discovering crystals with hypersymmetry will be greatly facilitated.

A special case of hypersymmetry is when the otherwise symmetrically independent molecules in the crystal are related by hypersymmetry operations, making them enantiomorphous pairs.

Hypersymmetry is a rather widely observed, and sometimes ignored, phenomenon which is not restricted to a special class of compounds. It may be supposed, however, that certain types of molecules are more apt to have this kind of additional symmetry in their crystal structures than others.

There are hypersymmetry phenomena in some crystal structures that are characterized by extra symmetry operations applicable to infinite chains of molecules. This kind of hypersymmetry has proved to be more easily detectable and has been reported often in the literature.

Hypersymmetry may be interpreted on the basis of the symmetry of the potential energy functions describing the conditions of the formation of the molecular crystal. The molecules around a certain starting molecule will be related by the symmetry of the potential function itself or the symmetry of certain combinations of the potential energy functions. The occurrence of some screw axes of rotation by hypersymmetry elements has been successfully interpreted in this way. In some instances, energy calculations as well as geometrical reasoning have shown the physical importance of hypersymmetry. For example, hypersymmetry may be related to stronger chemical bonding among molecules. Hypersymmetry may often be described as involving layered structure of a molecular crystal. This, again, may have advantages for geometrical and energy considerations. Thus, the phenomenon of hypersymmetry is another good example of how symmetry properties and other properties are related to each other.

9.6.5 Crystal Field Effects

Elucidating the effects of intermolecular interactions may greatly facilitate our understanding of the structure and energetics of crystals. The geometrical changes of molecular structures cover a wide range in energy. Molecular shape, symmetry, and conformation change more easily upon the molecule becoming part of a crystal than do bond angles and especially bond lengths.

Kitaigorodskii [9-38] suggested four approaches to investigating the effect of the crystalline field on molecular structure: (1) comparison of gaseous (i.e., free) and crystalline molecules; (2) comparison of symmetrically (i.e., crystallographically) independent molecules in the crystal; (3) analysis of the structure of molecules whose symmetry in the crystal is lower than their free molecular symmetry; and (4) comparison of the molecular structure in different polymorphic modifications. It is also possible that the molecule has higher symmetry in the crystal than as a free unit in the gas. Thus, for example, biphenyl has a higher molecular symmetry (a coplanar structure) in the crystal [9-57] than in the vapor [9-58] (where the two benzene rings are rotated by about 45° relative to each other, as shown in Figure 9-57).

9.6.5.1 Structure Differences in Free and Crystalline Molecules

Points 1 and 3 above both refer to the comparison of the structure of free and crystalline molecules. Such comparisons provide, perhaps, the most straightforward information, since the structure of the free molecule is determined exclusively by intramolecular interactions. Any difference that is reliably detected will carry information as to the effects of the crystal field on the molecular structure. However, before discussing more subtle structural differences in molecular crystals as compared with free molecules, it is appropriate to point out some more striking differences between ionic crystals and the corresponding vapor-phase molecules.

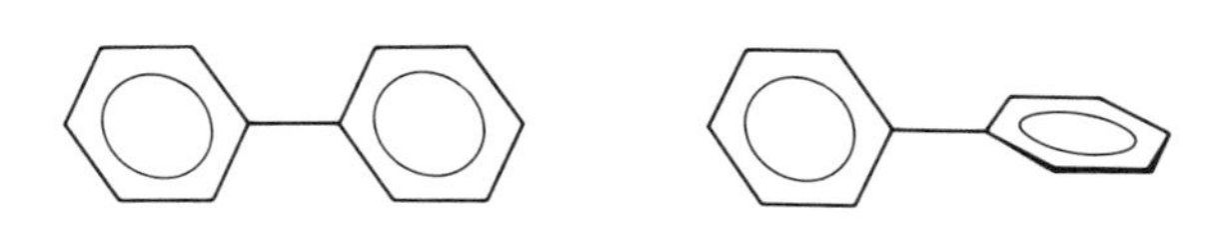

Figure 9-57. The molecular structure of biphenyl is coplanar in the crystal [9-57] while its two benzene rings are rotated by about 45° relative to each other in the vapor [9-58].

Although molecules cannot be identified as the building blocks of ionic crystals, the free molecules of *some* compounds may be considered as if they were taken out of the crystal. A nice example is sodium chloride, whose main vapor components are monomeric and dimeric molecules. They are indicated in the crystal structure in Figure 9-58, as is a tetrameric species. Mass-spectrometric studies of cluster formation determined a great relative abundance of a species with 27 atoms in the cluster. The corresponding $3 \times 3 \times 3$ cube may, again, be considered as a small crystal [9-59].

Another series of simple molecules whose structure may easily be traced back to the crystal structure is shown in Figure 9-59. It is evident, for example, that various MX_2 and MX_3 molecules may take different shapes and symmetries from the same kind of crystal structure. The crystal structure is represented by the octahedral arrangement of six "ligands" around the "central atom."

There seems to be even less structural similarity for many other metal halides when the crystalline systems are compared with the molecules in the vapor phase. Aluminum trichloride, for example, crystallizes in a hexagonal layer structure. Upon melting and then evaporation at relatively low temperatures, dimeric molecules are formed. At higher temperatures, they dissociate into monomers (Figure 9-60) [9-60]. The coordination number decreases from six to four and then to three in this process.

Under closer scrutiny, even the dimeric aluminum trichloride molecules can be derived from the crystal structure. Figure 9-61 shows another representation of crystalline aluminum trichloride which facilitates the identification of the dimeric units. Correlation between the molecular composition of the vapor and the source crystal has been established for some metal halides [9-62].

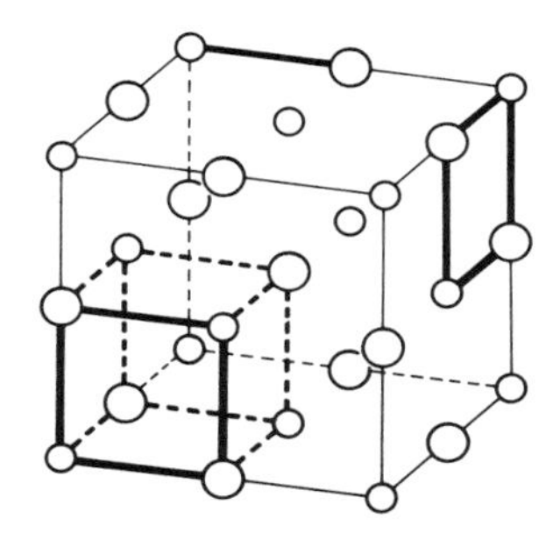

Figure 9-58. Part of sodium chloride crystal structure with NaCl, $(NaCl)_2$, and $(NaCl)_4$ units indicated. The species of $3 \times 3 \times 3$ ions has a high relative abundance in cluster formation.

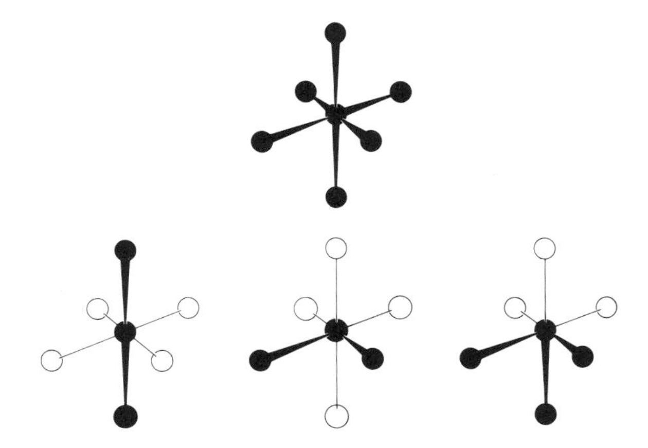

Figure 9-59. Different shapes of MX_2 and MX_3 molecules derived from the crystal structure in which the central atom has an octahedral environment.

Gas/solid differences of a different nature may occur in substances forming molecular crystals. In some cases, for example, the vapor contains more rotational isomers than the crystal. Thus, for example, the vapor [9-63] of ethane-1,2-dithiol consists of *anti* and *gauche* forms with respect to rotation about the central bond while only the *anti* form was found in the crystal [9-64].

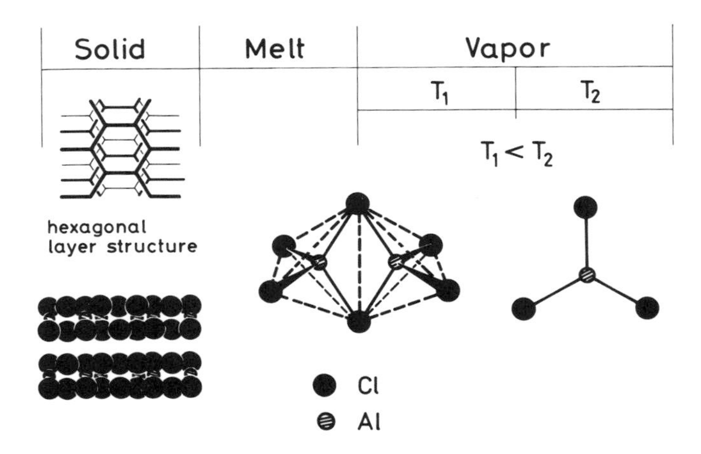

Figure 9-60. Structural changes upon evaporation of aluminum trichloride.

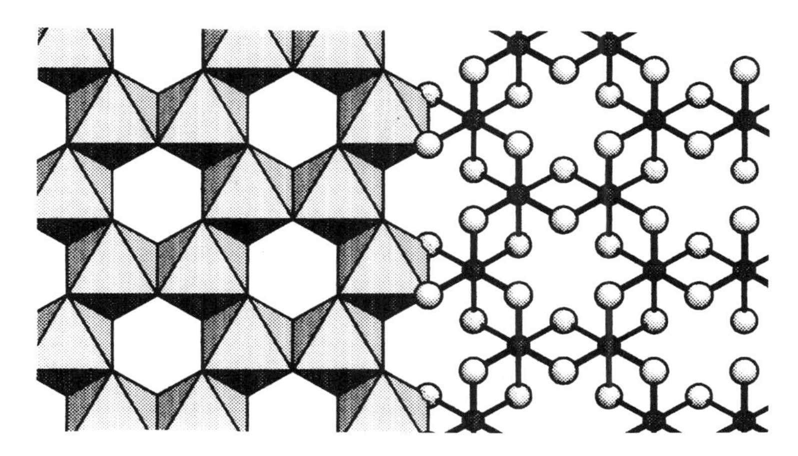

Figure 9-61. The crystal structure of aluminum trichloride, after Ref. [9-61]. The dimeric unit with a four-membered ring is discernible. Copyright (1993) John Wiley & Sons. Used by permission.

The comparison of the structures of free and crystalline molecules is obviously based on the application of various experimental techniques with theoretical calculations playing an increasing role. Thus, it is important to comment upon the inherent differences in the physical meaning of the structural information originating from such different sources [9-65]. The consequences of intramolecular vibrations on the geometry of free molecules have already been mentioned. The effects of molecular vibrations and librational motion in the crystal are not less important. To minimize their effects, it is desirable to examine the crystal molecular structure at the lowest possible temperatures. Also, the corrections for thermal motion are of great importance. Especially when employing older data in comparisons and discussing subtle effects, these problems have to be considered. There is another important source of differences in structural information, which may have no real structural implications. Apparent differences may originate from the difference in the physical meaning of the physical phenomena utilized in the experimental techniques. When all sources of apparent differences have been eliminated, and the molecular structure still differs in the gas and the crystal, the intermolecular interactions in the crystal may indeed be responsible for these differences [9-66, 9-67].

Variations in the ring angular deformations of gaseous and crystalline substituted benzene derivatives have been found to be a sensitive indicator of intermolecular interactions [9-68]. This was especially well indicated when different molecular packings were found to influence the benzene ring deformations to different extents for similar derivatives.

These findings may serve as a stimulus for the structural chemist in search of gas/crystal differences in accurately determined systems and for the theoretical chemist who may build models and perform calculations on them in which both the intramolecular and intermolecular interactions are adequately represented. The gas/crystal structural changes obviously depend on the relative strengths of the intramolecular and intermolecular interactions. More pronounced changes under the influence of the crystal field are expected, for example, in relatively weak coordination linkages than in stronger bonds. Thus, the N–B bond of donor–acceptor complexes is considerably longer in the gas than in the crystal. The difference is about 0.05 Å for $(CH_3)_3N–BCl_3$ (**9-1**)

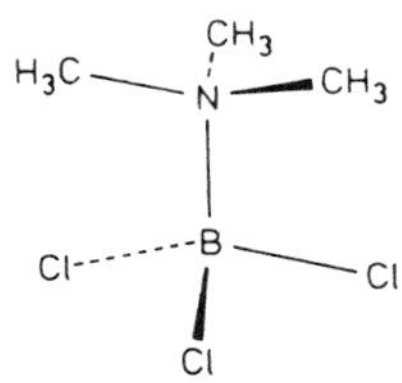

CH_3 H_3C N CH_3 B Cl Cl Cl

9-1

[9-69], and it may be supposed that the intermolecular forces somewhat compress the molecule along the coordination bond in the crystal. An extreme case of an 0.84-Å difference was reported for $HCN–BF_3$ [9-70].

Another case in point is the silatrane structures, where, again, the relatively weak N–Si dative bond is much longer in the gas than in the crystal. The difference is 0.28 Å for 1-fluorosilatrane [9-71], represented here (**9-2**) by the heavy-atom skeleton.

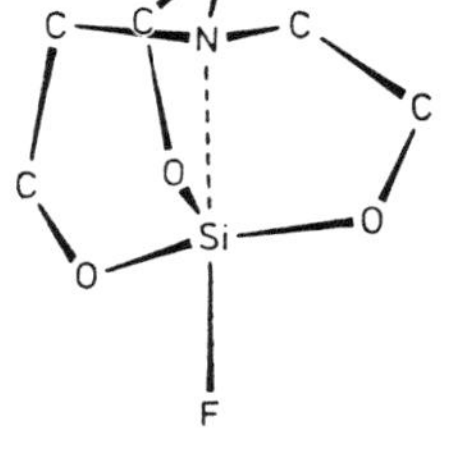

C C C N C C C O O Si O F

9-2

Gas/crystal comparisons are as of yet mainly confined to registering structural differences. The interpretation of these results is at a qualitative initial stage. Further investigation of such differences will enhance our understanding of the intermolecular interactions in crystals.

9.6.5.2 Conformational Polymorphism

The investigation of different rotational isomers of the same compound in different crystal forms (*polymorphs*) is also an efficient tool in elucidating intermolecular interactions. The phenomenon is called conformational polymorphism. The energy differences between the polymorphs of organic crystals are similar to the free energy differences of rotational isomers of many free molecules, viz., a few kilocalories per mole. When the molecules adopt different conformations in the different polymorphs, the change in rotational isomerism is attributed to the influence of the crystal field since the difference in the intermolecular forces is the single variable in the polymorphic systems.

Polymorphism is ubiquitous (see, e.g., Ref. [9-48]), and most compounds can exist in more than one crystalline form. Bernstein and co-workers [9-37] have extensively studied conformational polymorphism of various organic compounds with a variety of techniques in addition to X-ray crystallography. Among the molecules investigated were *N*-(*p*-chlorobenzylidene)-*p*-chloroaniline [**9-3**, X = Cl (**I**)], which exists in at least two forms, and *p*-methyl-*N*-(*p*-methylbenzylidene)aniline [**9-3**, X = CH_3 (**II**)], which exists in at least three forms. For **I**, a

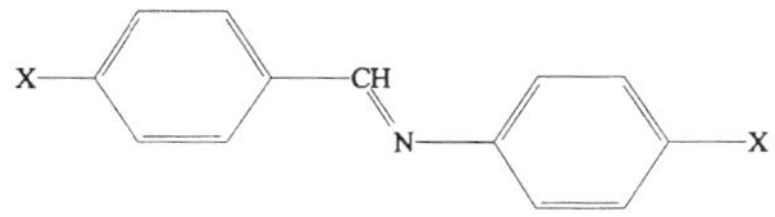

9-3

high-energy planar conformation was shown to occur with a triclinic lattice. A lower energy form with normal exocyclic angles was found in the orthorhombic form. It was an intriguing question as to why molecule **I** would not always pack with its lowest energy conformation.

The X-ray diffraction work has been augmented by lattice energy calculations employing different potential functions. The results did not depend on the choice of the potential function, and they showed that the crystal packing and the (intra)molecular structure together adopt an optimal compromise. The minimized lattice energies were analyzed in terms of partial atomic contributions to the total energy. Even for the trimorphic molecule **II**, the relative energy contributions of various groups were similar in all polymorphs. However, this obviously could only be achieved in some lattices by adopting a conformation different from the most favorable with respect to the structure of the isolated molecule. The investigation of conformational polymorphism proved to be a suitable and promising tool for investigating the nature of those crystal forces influencing molecular conformation and even molecular structure, in a broader sense.

Obviously, possible variations in bond angles and bond lengths have been ignored in the considerations described above. The energy requirements for changing bond angles and bond lengths are certainly higher than those for conformational changes [9-37] and, accordingly, higher than what may be available in polymorphic transitions. However, some relaxation of the bond configuration may take place, especially considering that the (intra)molecular structure is also adopted as a compromise by the bond configurations and the rotational forms.

Bond configuration relaxation during internal rotation has been investigated by quantum-chemical calculations for a series of 1,2-dihaloethanes [9-72]. The bond angle C–C–X may change by as much as 4° during internal rotation according to these calculations. If there is then a mixture of, say, *anti* and *gauche* forms, as is often the case, and the relaxation of the bond configuration is ignored, this may lead to considerable errors in the determination of the *gauche* angle of rotation.

9.7 BEYOND THE PERFECT SYSTEM

The 230 space groups exhaustively characterize all the symmetries possible for infinite lattice structures. So "exhaustively" that according to some views this perfect system is a little too perfect and a little too rigid. These views may well point toward the further development of our ideas on structures and symmetries [9-11, 9-73].

There is an inherent deficiency in crystal symmetry in that crystals are not really infinite. Mackay [9-73] argues that the crystal formation is not the insertion of components into a three-dimensional framework of symmetry elements; on the contrary, the symmetry elements are the consequence. The crystal arises from the local interactions between individual atoms. He furthermore says that a regular structure should mean a structure generated by simple rules, and the list of rules considered to be simple and "permissible" should be extended. These rules would not necessarily form groups. Furthermore, Mackay finds the formalism of the International Tables for X-Ray Crystallography [9-18] to be too restrictive and quotes Bell, the historian of mathematics, on the rigidity of the Euclidean geometry formalism: "The cowboys have a way of trussing up a steer or a pugnacious bronco which fixes the brute so that it can neither move nor think. This is the hog-tie and it is what Euclid did to geometry."

Mackay presented a long list [9-74] covering a whole range of transitions from classical crystallographic concepts to what is termed the modern science of structure at the atomic level. This list is reproduced in Table 9-6. Notice some resonance of several of Mackay's ideas with other directions in modern chemistry, where the nonclassical, the nonstoichiometrical, the nonstable, the nonregular, the nonusual, and the nonexpected are gaining importance. For crystallography it seems to be a long way yet to perform all the suggested

Table 9-6. Mackay's List of Transitions from the Classical Concepts of Crystallography to the Modern Concepts of a Science of Structure[a]

Classical concepts	Modern concepts
Absolute identity of components	Substitution and nonstoichiometry
Absolute identity of the environment of each unit	Quasi-identity and quasiequivalence
Operations of infinite range	Local elements of symmetry of finite range
"Euclidean" space elements (plane sheets, straight lines)	Curved space elements. Membranes, micelles, helices. Higher structures by curvature of lower structures
Unique dominant minimum in free energy configuration space	One of many quasiequivalent states; metastability recording arbitrary information (pathway); progressive segregation and specialization of information structure
Infinite number of units. Crystals	Finite numbers of units. Clusters; "crystalloids"
Assembly by incremental growth (one unit at a time)	Assembly by intervention of other components ("crystalase" enzyme). Information-controlled assembly. Hierarchical assembly
Single level of organization (with large span of level)	Hierarchy of levels of organization. Small span of each level
Repetition according to symmetry operations	Repetition according to program. Cellular automata
Crystallographic symmetry operations	General symmetry operations (equal "program statements")
Assembly by a single pathway in configuration space	Assembly by branched lines in configuration space. Bifurcations guided by "information," i.e., low-energy events of the hierarchy below.

[a]Ref. [9-74].

transitions, but the initial breakthroughs are fascinating and promising. Impressive progress has been reported in the studies of liquids, amorphous materials, and metallic alloys as regards the description of their structural regularities (cf. Ref. [9-75] and see also Section 9.8).

Liquid structures, for example cannot be characterized by any of the 230 three-dimensional space groups, and yet it is unacceptable to consider them as possessing no symmetry whatsoever. Bernal noted [9-76] that the major structural distinction between liquids and crystalline solids is the absence of long-range order in the former. A generalized description should also characterize liquid structures and colloids, as well as the structures of amorphous substances. It should also account for the greater variations in their physical properties as compared with those of the crystalline solids. Bernal's ideas [9-77] have greatly encouraged further studies in this field, which may be called generalized crystallography. Referring to Bernal's geometrical theory of liquids, Belov [9-78] noted in Bernal's obituary: "His last enthusiasm was for the laws of lawlessness."

The paradoxical incompleteness and inadequacy of perfect symmetry, compared with less-then-perfect symmetry, are well expressed in a short poem entitled "Gift to a Jade" by the English poet Anna Wickham [9-79]:

> For love he offered me his perfect world.
> This world was so constricted and so small
> It had no loveliness at all,
> And I flung back the little silly ball.
> At that cold moralist I hotly hurled
> His perfect, pure, symmetrical, small world.

The structures intermediate between the perfect order of crystals and the complete disorder of gases are not merely rare exceptions. On the contrary, they are often found in substances which are very common in our environment or are widely used in various technologies. These include plastics, textiles, and rubber. Glass is an especially fascinating material whose amorphous atomic network was discussed by Zachariasen [9-80] over six decades ago in a contribution which is considered still valid [9-81]. Figure 9-62 shows Zachariasen's two-dimensional representations of crystalline and amorphous structures of compounds of the same composition, A_2O_3. Guinier [9-82] envisaged a continuous passage from the exact scheme of neighboring atoms in a crystal to the very flexible arrangement in an amorphous body. The term paracrystal was coined for domains with approximate long-distance order in the range of a few tens to a few hundreds of atomic diameters. Figure 9-63 is a schematic representation of a paracrystal lattice with one atom per unit cell. The blackened areas indicate the regions where an atom is likely to be found around the atom fixed at the origin. At greater distances, the neighboring sites first overlap, and then merge, and thus eventually the long-range order vanishes completely.

One of the most fascinating examples of nonperiodic regular arrangements is described in Mackay's paper [9-73] *De nive quinquangula*—on the pentagonal snowflake. A regular but "noncrystalline" structure is built from regular pentagons in a plane. It starts with a regular pentagon of given size (zeroth-order pentagon). Six of these pentagons are combined to make a larger one (first-order pentagon). As is seen in Figure 9-64, the resulting triangular gaps are covered by pieces from cutting up a seventh zeroth-order pentagon. This indeed yields five triangles plus yet another regular pentagon of the order -1. This construction is then repeated on an ever increasing scale as indicated in Figure 9-64. The hierarchical packing of pentagons builds up like a pentagonal snowflake, shown in Figure 9-65 from a computer drawing. Attempts of pentagonal tiling of the plane were already quoted in Chapter 1 and will be referred to again in Section 9.8.

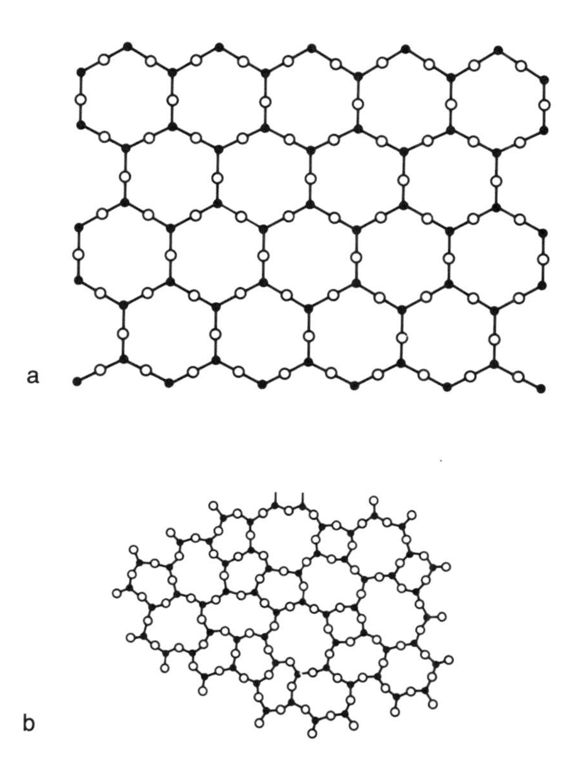

Figure 9-62. Zachariasen's representation of the atomic arrangement in the crystal (a) and glass (b) of compounds of the same composition, A_2O_3. After Ref. [9-80].

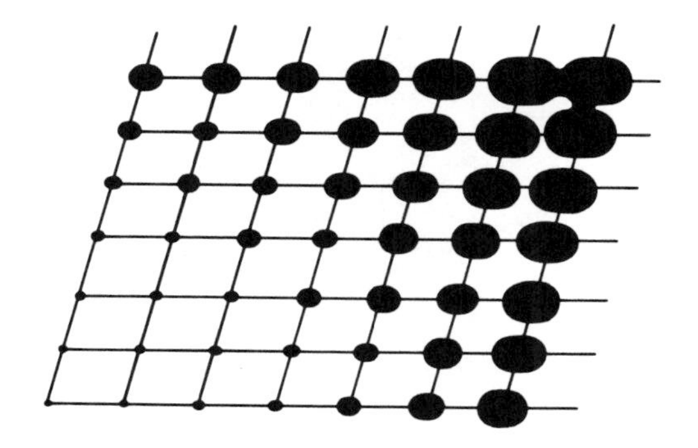

Figure 9-63. Paracrystal lattice with one atom per unit cell. After Guinier [9-82]. Used with permission.

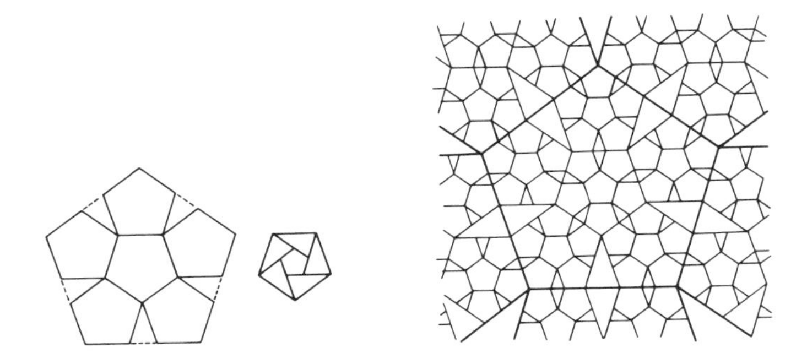

Figure 9-64. Tiling with regular pentagons, after Mackay [9-73].

Figure 9-65. *Pentagonal Snowflake*. Computer drawing, courtesy of Robert H. Mackay, London, 1982.

Mackay [9-73, 9-74] called attention to yet another limitation of the 230-space-group system. It covers only those helices that are compatible with the three-dimensional lattices. All other helices that are finite in one or two dimensions are excluded. Some important virus structures with icosahedral symmetry are among them. Also, there are very small particles of gold that do not have the usual face-centered cubic lattice of gold. They are actually icosahedral shells. The most stable configurations contain 55 or 147 atoms of gold. However, icosahedral symmetry is not treated in the International Tables, and crystals are only defined for infinite repetition.

Crystals are really advantageous for the determination of the structure of molecules. A crystal provides an amplification which multiplies the scattering of the X rays from a single molecule by the number of molecules in the array, perhaps by 10^{15}. It also minimizes the damage to individual molecules by the viewing radiation. The spots are emphasized in the diffraction pattern, and the background is neglected. The damaged molecules transfer their scattering contribution to the background as do those which are not repeated with regular lattice periodicity. However, defects and irregularities may be important and may well be lost in present-day sophisticated structural analyses.

It is perhaps worth pointing out that every crystal is in fact defective, even if its only defect is that it has surfaces. However, if a crystal is only a ten-unit-cell cube, about half of the unit cells lie in the surface and thus have environments very different from those of the other half. The physical observation is that very small aggregates need not be crystalline, although they may nevertheless be perfectly structured. Mackay's proposal is to apply the name *crystalloid* to them. His proposed definitions are as follows [9-73]:

- *Crystal*: The unit cell, consisting of one or more atoms, or other identical components, is repeated a large number of times by three noncoplanar translations. Corresponding atoms in each unit cell have almost identical surroundings. The fraction of atoms near the surface is small, and the effects of the surface can be neglected.
- *Crystallite*: A small crystal where the only defect is the existence of the external surface. The lattice may be deemed to be distorted, but it is not dislocated. Crystallites may further be associated into a mosaic block.
- *Crystalloid*: A configuration of atoms, or other identical components, finite in one or more dimensions, in a true free energy minimum, where the units are not related to each other by three lattice operations.

The above ideas are being further developed mainly by translating them into more quantitative descriptions that are being applied to various structural problems. They can also be compared with similarly new definitions mentioned in Section 9.8. These new attempts of taxonomy by no means belittle the great importance of the 230 three-dimensional space groups and their wide applicability. What is really expected is that they will eventually help in the systematization and characterization of the less easily handled systems with varying degrees of regularity in their structures.

The appearance of *quasicrystals* on the scene of materials has given a great thrust to these developments.

9.8 QUASICRYSTALS

The term "quasicrystal" was coined by Paul Steinhardt, who studied the structure of metallic glasses by theoretical means and modeling [9-83] (for an overview, see Ref. [9-84]). What he meant by this term was to have something expressing the connection between crystals on the one hand and *quasiperiodic* long-range translational order, on the other. Here, long-range translational order means that the position of a unit cell far away in the lattice is determined by the position of a given unit cell. In a crystal structure there is only one unit cell, whereas in a quasiperiodic structure there is more than just one. The repetition of the unit cell is regular in the crystal whereas it is not regular, nor is it random, in the quasiperiodic structure. In two-dimensional space, this is accomplished by the Penrose tiling [9-85], which was originally created more as recreational mathematics than as the extraordinarily important scientific tool that it has eventually become. A Penrose tiling is shown in Chapter 1 in Figure 1-11. Some attempts of pentagonal tiling over the centuries have already been mentioned in Chapter 1. There is a detailed and systematic discussion of pentagonal tilings in Grünbaum and Shephard's book [9-86].

The discovery of the Penrose tilings was a breakthrough in that pentagonal symmetry occurred in a pattern otherwise described by space-group symmetry. Curiously, the Penrose tiling was first communicated not by its inventor but by Martin Gardner in the January 1977 issue of *Scientific American* [9-87]. Mathematical physicist Roger Penrose himself subsequently published a paper in a university periodical, which was then reprinted in a mathematical magazine [9-85]. The title of the communication was rather telling, *Pentaplexity*, with a more somber subtitle, *A Class of Non-Periodic Tilings of the Plane*.

Mackay [9-88] made the connection with crystallography. He designed a pattern of circles based on a quasilattice to model a possible atomic structure. An optical transformation then created a simulated diffraction pattern exhibiting local tenfold symmetry (see Figure 1-12 in Chapter 1). In this way, Mackay virtually predicted the existence of what were later to be known as quasicrystals and issued a warning that such structures may be encountered but may stay unrecognized if unexpected [9-89]!

The moment of discovery came in April 1982 when Dan Shechtman was doing some electron diffraction experiments on alloys, produced by very rapid cooling of molten metals. In the experiments with molten aluminum with added magnesium, cooled rapidly, he observed an electron diffraction pattern with tenfold symmetry (see Figure 1-13 in Chapter 1). It was as great a surprise as it can be imagined to have been for any well-trained crystallographer. Shechtman's surprise was recorded with three question marks in his lab diary, "10-fold???" [9-90].

Fortunately, Mackay's fear that quasicrystals may be encountered but may stay unrecognized did not materialize because, although Shechtman was not familiar with the Penrose tiling and its potential implications for three-dimensional structures, he had what Louis Pasteur called a *prepared mind* for new things.* He did not let himself be discouraged by the seemingly well-founded disbelief of many though he did not attempt to publish his observations until he and his colleagues found a model that could be considered a possible origin of the experimental observation [9-84]. Ilan Blech constructed a three-dimensional model of icosahedra filling space almost at random and added restrictions to the model stipulating that the adjacent icosahedra touch each other at edges, or, in a later version, at vertices. Blech's model produced a simulated diffraction pattern that was consistent with Shechtman's observations.

The first report about Shechtman's seminal experiment did not appear until two and a half years after the experiment. The delay was caused by the authors' cautiousness and by some journal editors' skepticism. The paper [9-91] was titled modestly "Metallic Phase with Long-Range Orientational Order and No Translational Symmetry." It starts with the following sentence: "We report herein the existence of a metallic solid which diffracts electrons like a single crystal but has point group symmetry $m\overline{35}$ (icosahedral) which is inconsistent with lattice translations." The three-page report was followed by an avalanche of papers [9-92, 9-93], conferences, schools, special journal issues (see, e.g., Ref. [9-94]), and monographs (see, e.g., Refs. [9-95]–[9-98]), of which only a sample is mentioned here. Figure 9-66 shows some beautiful representatives of quasicrystals [9-99–9-101], and an artistic expression of what could be a quasicrystal [9-102].

Independent of Mackay's predictions and Shechtman's experiments, there was another line of research by Steinhardt and Levine, leading to a model encompassing all the features of *shechtmanite* (the original quasiperiodic alloy was eventually named so) and other materials that are symmetric and icosa-

*"Dans les champ de l'observation, l'hasard ne favorise que les esprits préparés." (In the field of observation, chance only favors those minds which have been prepared.) *Encyclopaedia Britannica* 1911, 11th edition, Vol. 20, quoted here after Mackay [9-54].

Figure 9-66. Quasicrystals: (a) Flowerlike icosahedral quasicrystals in a quenched Al/Mn sample, from Csanády *et al*. [9-99]; (b) pentagonal dodecahedron in quasicrystalline Al/Cu/Ru sample obtained by slow cooling from melt; photograph courtesy of Professor H.-U. Nissen, Zürich (cf. Ref. [9-100]). (*Continued on next page*)

hedral and nonperiodic [9-83]. It was perfect timing that as soon as Steinhardt and Levine built up their model and produced its simulated diffraction pattern, they could see its proof from a real experiment.

Steinhardt [9-103], like Mackay [9-74] before (see Section 9.7), felt the need for redefinition of materials categories. These categories now included the newly discovered quasicrystals. Steinhardt has succinctly characterized crystals, glassy materials, and quasicrystals as follows:

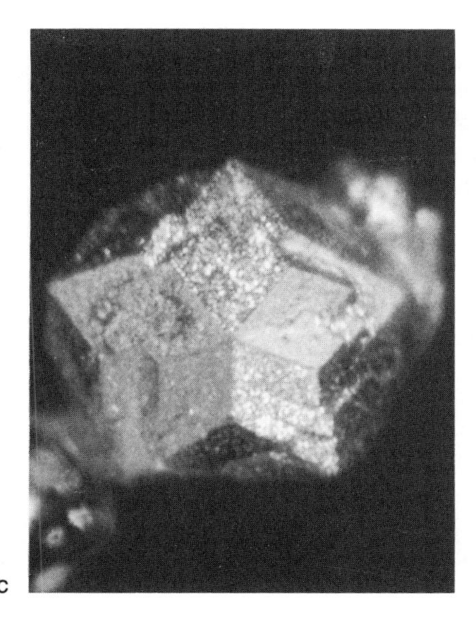

Figure 9-66. (*Continued*) (c) Triacontahedral Al/Li/Cu quasicrystal; photograph courtesy of Professor F. Dénoyer, Orsay (cf. Ref. [9-101]). (d) Sculpture resembling a quasicrystal, by Peter Hächler, Switzerland. Photograph by the authors [9-102].

- *Crystal*: "highly ordered, with its atoms arranged in clusters which repeat periodically, at equal intervals, throughout the solid."
- *Glassy material*: "highly disordered, with atoms arranged in a dense but random array."
- *Quasicrystal*: "highly ordered atomic structure, yet the clusters repeat in an extraordinarily complex nonperiodic pattern."

The appearance of quasicrystals has caused a minirevolution in crystallography. The lack of periodicity was a major obstacle in applying the traditional terms and approaches to this domain of materials. This was an interesting development also from the point of view of Mackay's suggestions for generalized crystallography (cf. Section 9.7). He truly anticipated the breakdown of the perfect traditional system, which he felt was a little too perfect.

It has been suggested that quasicrystals be treated as three-dimensional sections of materials that are periodic in more than three dimensions. On the other hand, a new and more general formulation of crystallography has also been proposed [9-104] which would stay within the realm of three-dimensionality and would not have the concept of periodicity at the focus of its foundation. David Mermin [9-104] compared abandoning the traditional classification scheme of crystallography, based on periodicity, to abandoning the Ptolemaic view in astronomy and likened changing to a new foundation to astronomy's adopting the Copernican view—hence the title of Mermin's communication, "Copernican Crystallography." The suggestion was to build the new foundation on the three-dimensional concept of point-group operations that would have the concept of indistinguishable densities at its focus, rather than identical densities, to correspond to the character of quasisymmetries, describing, among others, the quasicrystals.

Concluding, we quote again Mackay [9-105], who stated:

> Amorphous materials may be shapeless, but they are not without order. Order, like beauty, is in the eyes of the beholder. If you look only with X-ray diffraction eyes, then all you see is translational order, to wit crystals. . . . there is a wider range of structures, between those of crystals and those of gases, . . . Other structures need not be failed crystals but are *sui generis*.

In the words of the poet crystallographer [9-14],

> We cruise through the hydrosphere
> Our world is of water, like the sea,
> But the molecules more sparsely spread,
> Not independent, not touching
> But somewhere in between,
> Clustering, crystallizing, dispersing
> In the delicate balance of radiation
> And the adiabatic lapse rate.

As crystallography is becoming more general, transforming itself into the science of structures, so may we anticipate a broadening application of the symmetry concept in the description and understanding of all possible structures.

9.9 FORGET ME NOT

Lest the classical crystallographer feels our discussion too esoteric, let us return now to those exquisite shapes that we think of when the word *crystal* is mentioned. The words of the 19th century English writer John Ruskin (after Azaroff [9-17]) and drawings of C. Bunn [9-106] are cited here (Figure 9-67).

> And remember, the poor little crystals have to live their lives, and mind their own affairs, in the midst of all this, as best they may.

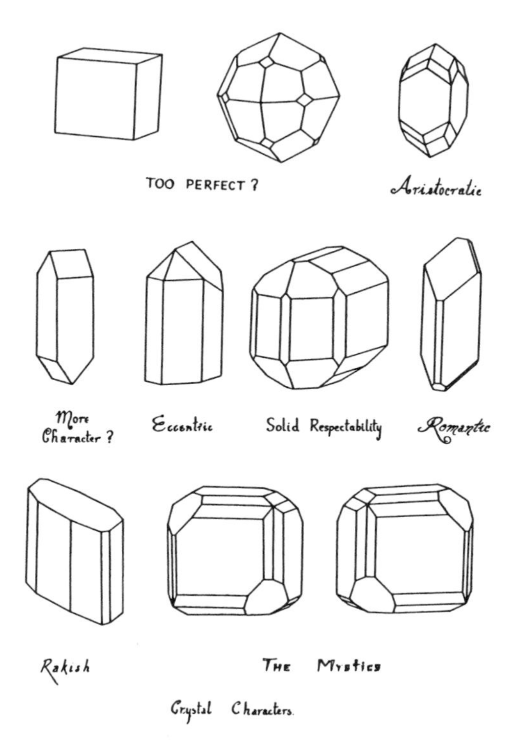

Figure 9-67. *Crystal Characters* from C. Bunn's book [9-106]. Reproduced with permission.

They are wonderfully like humane creatures—forget all that is going on if they don't see it, however dreadful; and never think what is to happen tomorrow. They are spiteful or loving, and indolent or painstaking, with no thought whatever of the lava or the flood which may break over them any day; and evaporate them into air-bubbles, or wash them into a solution of salts. And you may look at them, once understanding the surrounding conditions of their fate, with an endless interest. You will see crowds of unfortunate little crystals, who have been forced to constitute themselves in a hurry, their dissolving element being fiercely scorched away; you will see them doing their best, bright and numberless, but tiny. Then you will find indulged crystals, who had had centuries to form themselves in, and have changed their mind and ways continually; and have been tired, and taken heart again; and have been sick, and got well again; and thought they would try a different diet, and then thought better of it; and made but a poor use of their advantages, after all.

And sometimes you may see hypocritical crystals taking the shape of others, though they are nothing like in their minds; and vampire crystals eating out the hearts of others; and hermitcrab crystals living on the shells of others; and parasite crystals living on the means of others; and courtier crystals glittering in the attendance upon others; and all these, besides the two great companies of war and peace, who ally themselves, resolutely to attack, or resolutely to defend. And for the close, you see the broad shadow and deadly force of inevitable fate, above all this: you see the multitudes of crystals whose time has come; not a set time, as with us, but yet a time, sooner or later, when they all must give up their crystal ghost—when the strength by which they grew, and the breath given them to breathe, pass away from them; and they fail, and are consumed, and vanish away; and another generation is brought to life, framed out of their ashes.

REFERENCES

[9-1] K. Čapek, *Anglické Listy*, Československý Spisovatel, Prague (1970).
[9-2] P. C. Ritterbush, *Nature (London)* **301**, 197 (1983).
[9-3] A. L. Mackay, *Nature (London)* **301**, 652 (1983).
[9-4] E. Schröder, *Dürer, Kunst und Geometrie*, Akademie Verlag, Berlin (1980).
[9-5] B. Ernst, *The Magic Mirror of M. C. Escher*, Ballantine Books, New York (1976).
[9-6] J. Lima-de-Faria (ed.), *Historical Atlas of Crystallography*, Kluwer, Dordrecht (1990).
[9-7] R. J. Haüy, *Traité de Cristallographie*, Bachelier et Huzard, Paris (1822). [Reprinted by Culture et Civilisation, Brussels (1968).]
[9-8] J. Kepler, *Strena seu de nive sexangula*, Godefridum Tampach, Frankfurt am Main (1611). [English translation, *The Six-Cornered Snowflake*, Clarendon Press, Oxford (1966).]
[9-9] J. Dalton, *Manchester Memoirs* **6** (1805).
[9-10] C. J. Schneer, *Am. Sci.* **71**, 254 (1983).
[9-11] A. L. Mackay, *Izv. Jugosl. Cent. Kristallogr.* **10**, 15 (1975).
[9-12] L. A. Shuvalov, A. A. Urosovskaya, I. S. Zheludev, A. V. Zaleskii, S. A. Semiletov, B. N. Grechushnikov, I. G. Chistyakov, and S A. Pikin, *Sovremennaya Kristallografiya*, Vol. 4, *Fizicheskie Svoistva Kristallov*, Nauka, Moscow (1981).
[9-13] K. N. Trueblood, in *Accurate Molecular Structures. Their Determination and Importance* (A. Domenicano and I. Hargittai, eds.), p. 199, Oxford University Press, Oxford (1992); C. M. Gramaccioli, *ibid.*, p. 220.
[9-14] A. L. Mackay, *The Floating World of Science*, Poems, The RAM Press, London (1980).
[9-15] M. J. Buerger, *Elementary Crystallography, An Introduction to the Fundamental Geometrical Features of Crystals*, 4th printing, John Wiley & Sons, New York (1967); E. S. Dana, *A Textbook of Mineralogy*, 4th ed. (revised and enlarged by W. E. Ford), John Wiley & Sons, New York (1932); P. M. Zorky, *Arkhitektura Kristallov*, Nauka, Moscow (1968).

[9-16] Gy. Lengyel, *Kézimunkák*, Kossuth, Budapest (1978).
[9-17] L. V. Azaroff, *Introduction to Solids*, McGraw-Hill (1960).
[9-18] *International Tables for X-Ray Crystallography*, 3rd ed., Kynoch Press, Birmingham, England (1969).
[9-19] A. V. Shubnikov and V. A. Koptsik, *Symmetry in Science and Art*, Plenum Press, New York (1974). [Russian original: *Simmetriya v nauke i iskusstve*, Nauka, Moscow (1972).]
[9-20] C. P. Brock and E. C. Lingafelter, *J. Chem. Educ.* **57**, 552 (1980).
[9-21] A. L. Mackay, *Acta Crystallogr.* **22**, 329 (1967).
[9-22] A. I. Kitaigorodsky, *Molecular Crystals and Molecules*, Academic Press, New York (1973). [Russian original: A. I. Kitaigorodskii, *Molekulyarnie Kristalli*, Nauka, Moscow (1971).]
[9-23] A. D. Mighell, V. L. Himes, and J. R. Rodgers, *Acta Crystallogr., Sect. A* **39**, 737 (1983); J. Donohue, *Acta Crystallogr., Sect. A* **41**, 203 (1985); R. Srinivasan, *Acta Crystallogr., Sect. A* **47**, 452 (1991); C. P. Brock and J. D. Dunitz, *Acta Crystallogr., Sect. A* **47**, 854 (1991); A. J. C. Wilson, *ACH—Models in Chemistry* **130**, 183 (1993); C. P. Brock and J. D. Dunitz, *Mol. Cryst. Liq. Cryst.* **242**, 61 (1994).
[9-24] J. Dalton, *A New System of Chemical Philosophy*, Manchester, England (1808), p. 128, plate III.
[9-25] D. Hodgkin, *Kristallografiya* **26**, 1029 (1981).
[9-26] A. F. Wells, *Structural Inorganic Chemistry*, 5th ed., Clarendon Press, Oxford (1984).
[9-27] K. W. Adolph, D. L. D. Caspar, C. J. Hollingshed, E. E. Lattman, W. C. Phillips, and W. T. Murakami, *Science* **203**, 1117 (1979).
[9-28] R. B. Fuller, *Synergetics: Explorations in the Geometry of Thinking*, Macmillan, New York (1975), p. 37.
[9-29] D. L. D. Caspar and A. Klug, *Cold Spring Harbor Symp. Quant. Biol.* **27**, 1 (1962).
[9-30] R. E. Benfield and B. F. G. Johnson, *J. Chem. Soc., Dalton Trans.* **1980**, 1743.
[9-31] A. L. Mackay, *Acta Crystallogr.* **15**, 916 (1962).
[9-32] B. C. Chakoumakos, R. J. Hill, and G. V. Gibbs, *Am. Mineral.* **66**, 1237 (1981).
[9-33] I. El-Said and A. Parman, *Geometric Concepts in Islamic Art*, World of Islam Festival Publ. Co., London (1976).
[9-34] Chen Chi-Lu, *Material Culture of the Formosan Aborigines*, The Taiwan Museum, Taipei (1968).
[9-35] L. Pauling, *The Nature of the Chemical Bond*, 3rd ed., Cornell University Press, Ithaca, New York (1973).
[9-36] W. Barlow, *Z. Kristallogr.* **29**, 433 (1898).
[9-37] J. Bernstein, in *Accurate Molecular Structures. Their Determination and Importance* (A. Domenicano and I. Hargittai, eds.), p. 469, Oxford University Press, Oxford (1992).
[9-38] A. I. Kitaigorodsky, in *Advances in Structure Research by Diffraction Methods*, Vol. 3 (R. Brill and R. Mason, eds.), p. 173, Pergamon Press, Oxford, and Friedr. Vieweg and Sohn, Braunschweig (1970).
[9-39] F. Wundl and E. T. Zellers, *J. Am. Chem. Soc.* **102**, 4283 (1980).
[9-40] F. Wundl and E. T. Zellers, *J. Am. Chem. Soc.* **102**, 5430 (1980).
[9-41] C. H. MacGillavry, *Symmetry Aspects of M. C. Escher's Periodic Drawings*, Bohn, Scheltema & Holkema, Utrecht (1976).
[9-42] P. M. Zorky, *ACH—Models in Chemistry* **130**, 173 (1993).
[9-43] H. A. Stuart, *Phys. Z.* **35**, 990 (1934); H. A. Stuart, *Z. Phys. Chem.* **27**, 350 (1934); G. Briegleb, *Fortschr. Chem. Forsch.* **1**, 642 (1950).
[9-44] A. Gavezzotti and G. R. Desiraju, *Acta Crystallogr., Sect. B* **44**, 427 (1988).

[9-45] J.-M. Lehn, *Science* **260**, 1762 (1993).
[9-46] G. D. Andreetti, A. Pochini, and R. Ungaro, *J. Chem. Soc., Perkin Trans. 2* **1983**, 1773; G. D. Andreetti and F. Ugazzoli, in *Calixarenes: A Versatile Class of Macrocyclic Compounds* (J. Vicens and V. Böhmer, eds.), Kluwer, Dordrecht (1991).
[9-47] J.-M. Lehn, *Science* **227**, 849 (1985).
[9-48] J. D. Dunitz, in *Host–Guest Molecular Interactions: From Chemistry to Biology* (D. J. Chadwick and K. Widdows, eds.), p. 92, John Wiley & Sons, Chichester, England (1991).
[9-49] K. Mirsky, *ACH—Models in Chemistry* **130**, 197 (1993); A. Gavezzotti and G. Filippini, *ACH—Models in Chemistry* **130**, 205 (1993).
[9-50] L. E. Depero, in *Advances in Molecular Structure Research*, Vol. 1 (M. Hargittai and I. Hargittai, eds.), JAI Press, Greenwich, Connecticut (1995).
[9-51] J. Maddox, *Nature (London)* **335**, 201 (1988).
[9-52] C. P. Brock, Collected Abstracts, XVI Congress and General Assembly, International Union of Crystallography, Beijing, 1993, Paper MS-06.0.1 (p. 4).
[9-53] D. Braga and F. Grepioni, *Organometallics* **10**, 1254 (1991).
[9-54] A. L. Mackay, *A Dictionary of Scientific Quotations*, Adam Hilger, Bristol, England (1992).
[9-55] P. M. Zorky and V. A. Koptsik, in *Sovremennie Problemi Fizicheskoi Khimii* (Ya. I. Gerasimov and P. A. Akishin, eds.), Izdatel'stvo Moskovskogo Universiteta, Moscow (1979).
[9-56] P. M. Zorky and E. E. Dashevskaya, *ACH—Models in Chemistry* **130**, 247 (1993).
[9-57] G. P. Charbonneau and Y. Delugeard, *Acta Crystallogr., Sect. B* **33**, 1586 (1977); H. Cailleau, J. L. Bauduor, and C. M. E. Zeyen, *Acta Crystallogr., Sect. B* **35**, 426 (1979); H. Takeuchi, S. Suzuki, A. J. Dianoux, and G. Allen, *Chem. Phys.* **55**, 153 (1981).
[9-58] A. Almenningen, O. Bastiansen, L. Fernholt, B. N. Cyvin, S. J. Cyvin, and S. Samdal, *J. Mol. Struct.* **128**, 59 (1985).
[9-59] T. P. Martin, *Phys. Rev.* **95**, 167 (1983).
[9-60] M. Hargittai, *Kém. Közlem.* **50**, 371 (1978).
[9-61] U. Müller, *Inorganic Structural Chemistry*, John Wiley & Sons, Chichester, England (1993).
[9-62] M. Hargittai and G. Jancsó, *Z. Naturforsch., A* **48**, 1000 (1993).
[9-63] G. Schultz and I. Hargittai, *Acta Chim. Hung.* **75**, 381 (1973).
[9-64] M. Hayashi, Y. Shiro, T. Oshima, and H. Murata, *Bull. Chem. Soc. Jpn.* **38**, 1734 (1975).
[9-65] A. Domenicano and I. Hargittai (eds.), *Accurate Molecular Structures. Their Determination and Importance*, Oxford University Press, Oxford (1992).
[9-66] I. Hargittai and M. Hargittai, in *Molecular Structure and Energetics*, Vol. 2 (J. F. Liebman and A. Greenberg, eds.), Chapter 1, VCH Publishers, Deerfield Beach, Florida (1986).
[9-67] M. Hargittai and I. Hargittai, *Phys. Chem. Miner.* **14**, 413 (1987).
[9-68] A. Domenicano and I. Hargittai, *ACH—Models in Chemistry* **130**, 347 (1993).
[9-69] M. Hargittai and I. Hargittai, *J. Mol. Struct.* **39**, 79 (1977).
[9-70] W. A. Burns and K. R. Leopold, *J. Am. Chem. Soc.* **115**, 11622 (1993).
[9-71] G. Forgács, M. Kolonits, and I. Hargittai, *Struct. Chem.* **1**, 245 (1990).
[9-72] P. Scharfenberg and I. Hargittai, *J. Mol. Struct.* **112**, 65 (1984).
[9-73] A. L. Mackay, *Phys. Bull.* **1976**, 495.
[9-74] A. L. Mackay, *Kristallografiya* **26**, 910 (1981).
[9-75] I. Hargittai and W. J. Orville Thomas (eds.), *Diffraction Studies on Non-Crystalline Substances*, Elsevier, Amsterdam (1981).

[9-76] J. D. Bernal, *Acta Phys. Acad. Sci. Hung.* **8**, 269 (1958).
[9-77] J. D. Bernal and C. H. Carlisle, *Kristallografiya* **13**, 927 (1969).
[9-78] N. V. Belov, *Kristallografiya* **17**, 208 (1972).
[9-79] A. Wickham, *Selected Poems*, Chatto and Windus, London (1971).
[9-80] W. H. Zachariasen, *J. Am. Chem. Soc.* **54**, 3841 (1932).
[9-81] A. R. Cooper, *J. Non-Cryst. Solids* **49**, 1 (1982).
[9-82] A. Guinier, in *Diffraction Studies on Non-Crystalline Substances* (I. Hargittai and W. J. Orville Thomas, eds.), p. 411, Elsevier, Amsterdam (1981).
[9-83] D. Levine and P. J. Steinhardt, *Phys. Rev. Lett.* **53**, 2477 (1984).
[9-84] M. La Brecque, *Mosaic* **18**, 1 (1987/88).
[9-85] R. Penrose, Eureka No. 39; *Math. Intell.* **2**, 32 (1979/80). [Reprinted also in *Per. Mineral.* **59**, 69 (1990).]
[9-86] B. Grünbaum and G. C. Shephard, *Tilings and Patterns*, W. H. Freeman & Co., New York (1987).
[9-87] M. Gardner, *Sci. Am.* **236**, 110 (1977).
[9-88] A. L. Mackay, *Physica A* **114**, 609 (1982).
[9-89] A. L. Mackay, Two Lectures on Fivefold Symmetry at the Hungarian Academy of Sciences, Budapest, September, 1982.
[9-90] I. Hargittai, *Per. Mineral.* **61**, 9 (1992).
[9-91] D. Shechtman, I. Blech, D. Gratias, and J. W. Cahn, *Phys. Rev. Lett.* **53**, 1951 (1984).
[9-92] A. L. Mackay, *Int. J. Rapid Solidification* **2**, S1 (1987).
[9-93] M. Ronchetti, *Per. Mineral.* **59**, 219 (1990).
[9-94] L. Loreto and M. Ronchetti (eds.), *Topics on Contemporary Crystallography and Quasicrystals*, *Per. Mineral.* **59**, 1–3, Special Issue (1991).
[9-95] I. Hargittai (ed.), *Quasicrystals, Networks, and Molecules of Fivefold Symmetry*, VCH, New York (1990).
[9-96] M. V. Jaric (ed.), *Aperiodicity and Order*, Vol. 1, *Introduction to Quasicrystals*, Academic Press, San Diego (1988).
[9-97] M. V. Jaric (ed.), *Aperiodicity and Order*, Vol. 2, *Introduction to the Mathematics of Quasicrystals*, Academic Press, San Diego (1989).
[9-98] M. V. Jaric and A. Gratias (eds.), *Aperiodicity and Order*, Vol. 3, *Extended Icosahedral Structures*, Academic Press, San Diego (1989).
[9-99] A. Csanády, K. Papp, M. Dobosy, and M. Bauer, *Symmetry* **1**, 75 (1990).
[9-100] I. Hargittai (ed.), *Fivefold Symmetry*, World Scientific, Singapore (1992), p. xiv.
[9-101] F. Dénoyer, in *Quasicrystals, Networks, and Molecules of Fivefold Symmetry* (I. Hargittai, ed.), VCH, New York (1990).
[9-102] I. Hargittai, *Math. Intell.* **14**, 58 (1992).
[9-103] P. J. Steinhardt, *Endeavour, New Ser.* **14**(3), 112 (1990).
[9-104] N. D. Mermin, *Phys. Rev. Lett.* **68**, 1172 (1992).
[9-105] A. L. Mackay, *J. Non-Cryst. Solids* **97** & **98**, 55 (1987).
[9-106] C. Bunn, *Crystals: Their Role in Nature and Science*, Academic Press, New York (1964).

Index